Paramecium

Edited by H.-D. Görtz

Contributors
A. Adoutte · R. D. Allen · J. Beisson · J. D. Berger
A. Burgess-Cassler · J. Cohen · D. Cummings · A. K. Fok
M. Freiburg · M. Fujishima · H.-D. Görtz · R. Hinrichsen
A. Kitamura · S. Klumpp · C. Kung · W. G. Landis
H. Machemer · K. Mikami · D. Nyberg · R. R. Preston
R. L. Quackenbush · R. Ramanathan · Y. Saimi
A. Sainsard-Chanet · H. J. Schmidt · J. E. Schultz
Y. Takagi · M. Takahashi · Y. Tsukii · J. Van Houten

Foreword by John R. Preer, Jr.

With 125 Figures

Springer-Verlag
Berlin Heidelberg New York
London Paris Tokyo

Prof. Dr. HANS-DIETER GÖRTZ
Zoologisches Institut der Universität Münster
Schloßplatz 5
4400 Münster, FRG

ISBN-13: 978-3-642-73088-7 e-ISBN-13: 978-3-642-73086-3
DOI: 10.1007/978-3-642-73086-3

Library of Congress Cataloging-in-Publication Data. Paramecium. Bibliography: p. Includes index. 1. Paramecium. I. Görtz, H.-D. (Hans-Dieter), 1945– . QL368.H87P37 1988 593.1'72 87-32239.

Softcover reprint of the hardcover 1st edition 1988

Typesetting, printing and bookbinding:
Graphischer Betrieb Konrad Triltsch, Würzburg
2131/3130-543210

Foreword

Why a Book on Paramecium? Biologists usually concentrate their efforts on a single problem and a single organism. There is a difficulty with this practice, however, for as work on a problem proceeds it often becomes more advantageous to study the problem in another organism. Some biologists avoid the difficulty by moving from one organism to the other as the problem demands. However, this tactic also has a disadvantage, for a thorough knowledge of the life cycle and thorough mastery of ways to handle a given organism in the laboratory are obviously of great importance to the researcher, and one can never know several organisms as well as one can know a single one.

Another way of doing research is to pick the organism, learn all one can about it from all points of view, and then assess the significance of the findings. Tracy Sonneborn practiced research in very much this way. He would have found virtually every chapter in this volume about *Paramecium* a fascinating summary of one of his areas of research. Indeed, the beginnings of most of the topics in this book are founded on his studies. With every new fact he learned about *Paramecium*, he carefully assessed the significance of his findings, not only for research on protozoa, but for biology in general. His work, and in a way this book too, are indicative of the success of his strategy.

Is this strategy still useful today? Probably not in the way that Sonneborn practised it, for increases in the depth of our knowledge about these many topics and greater sophistication of the techniques being used for analysis require more skills than any one group is likely to possess. However, although ciliate researchers may not wish to do research in all these areas, there is still much virtue in knowing about the findings of others on *Paramecium*. The persistence of regional and international conferences devoted to the biology of single organisms attests to this fact. Hence, there are many workers in relatively narrow fields with broad interests who will want to read this book from cover to cover in order to learn more about *Paramecium*. For others it will be regarded as a book for specialists, or at least those who wish to broaden their understanding of some special area.

In this preface I will consider briefly some of the advantages of *Paramecium* to the researcher, then cite a few of its disadvantages and some steps being taken to minimize them. I will also try to summarize briefly the relationship between a few of the phenomena that overlap different chapters in this book. In addition I hope to provide a little background and assess the general significance of the work on *Paramecium* in some of the areas of research. Although most of the areas will be touched upon, the account is not intended to be even, but instead reflects the author's major interests. In some cases, new findings made since the chapters were written cast a new light on various problems; whenever possible, this material is also included. References to much of the older literature are not documented in this preface and the reader is referred to the appropriate chapters in this book, or to one of the more recent reviews: Tracy M. Sonneborn (1974) *Paramecium aurelia,* in R. C. King (ed.) *Handbook of Genetics,* vol. 2, Plenum Press, New York, pp. 469–594; Ralph Wichterman (1985), *The Biology of Paramecium,* 2nd ed., Plenum Press, New York; Joseph G. Gall, 1986, *The Molecular Biology of Ciliated Protozoa,* Academic Press, New York.

Paramecium Has Many Advantages. To cell biologists and molecular biologists, *Paramecium* takes its place as another example of a eukaryotic cell. However, of all the eukaryotic cells being investigated, *Paramecium* has numerous features that warrant a very special niche for it. Many of these features are shared with other protists, especially the ciliates, but many are unique. The size of *Paramecium* makes it peculiarly suited for the use of electrodes for physiological work, and for the transfer of fluids and organelles by microinjection techniques. Of all living forms, only *Paramecium* (along with the other ciliates) has a cell cortex so rich in structure and complex in development, micronuclei, and macronuclei with different structures and functions and with amazingly complex patterns of formation, mating type substances of such variety and specificity, life cycle stages of immaturity, maturity and senility so well defined in single-celled organisms, primitive behavioral mechanisms so suited for analysis, cilia and also mitochondria in such abundance, and such a rich variety of endosymbionts. To aid the geneticist there is also autogamy, cytogamy, macronuclear regeneration, delayed separation, and cytoplasmic exchange. Achieving homozygosity of all loci at a single stroke with autogamy or cytogamy is a tremendous aid in studies of heredity. The same is true of the generation of heterokaryons by macronuclear regeneration. Moreover, all of these processes can be controlled by the investigator.

Finally, *Paramecium* is not only a cell, it is also an organism. Its complex anatomical cellular form cannot merely be pinched in two at cell division to generate two daughters like the parent; a complex process of development must occur. It is not bathed in lymph or hemolymph like the cells of most multicellular organisms; it must find and process its food. It has a mouth, a primitive digestive system, and an anal pore. It must respond to its environment in an appropriate way or it will go without food or be destroyed. Its entire set of organismic functions must be integrated. Moreover, it is a unit in ecology, in population biology, and in evolution.

Some Disadvantages and How They Have Been Minimized. In many cases, techniques have been devised to minimize the inherent disadvantages of *Paramecium* as an experimental organism. Although paramecia can be maintained by anyone with relative ease, constant and reliable culture conditions that produce maximum growth rate, high population density, and normal physiological behavior require some effort. Different batches of dessicated lettuce or cereal leaves differ in their suitability for growth of paramecia in bacterized medium, irrespective of whether they are made in the laboratory or purchased commercially. A common strategy to minimize this problem is to obtain a large batch of a dessicated medium and when it has been tested and found suitable, then to store it for use over a long period of time. Moreover, advances in the composition of bacterized culture media are still being made; note the new medium recently described by Enright and Hennessey (1987). Living bacterial cultures which must be maintained in order to serve as a food source may change their characteristics over the years; the prevention of contamination of bacterial cultures that are maintained for long periods of time requires constant vigilance, but storage of stock cultures of bacteria in the presence of glycerol at $-70\,^{\circ}$C is a great aid.

The presence of living bacteria in cultures may be a serious disadvantage in some sorts of work, such as studies on drug resistance or enzymatic composition. However, axenic methods are available. Moreover, the axenic media often yield very high population densities. Nevertheless, they too have their drawbacks. Media are tedious to prepare. Although washing paramecia free of contaminating bacteria is relatively easy, establishment of bacteria-free cultures often requires a difficult period of adjustment by the ciliates themselves. Some strains (e.g., some strains of "killers") fail to grow in most of the available media. Moreover, physiological functions are rather different in the axenic media from what they are in their normal habitat of bacterized cultures. In axenic media, life cycles are not well characterized, growth rates are reduced and mating is achieved with great difficulty, if at all. However, advances in the use of axenic media have also been made in recent years (Allen and Nerad 1978; Schönefeld et al. 1986).

Since cells cannot be cultured on agar, methods for the isolation of many cells, such as plating techniques, cannot be carried out using standard bacterial methods. However, the use of microtiter wells and special transfer apparatus can aid with this problem.

Maintenance of stocks is troublesome, for cells die when dessicated, and freezing is only obtained under very exacting conditions and with the use of complex equipment. However, alternate methods of stock maintenance that are not so time-consuming have been developed. Moreover, most stocks are maintained satisfactorily in national stock culture collections (e.g., the American Type Culture Collection, 12301 Parklawn Drive, Rockville, Md. 20852, USA) and are thus available when necessary.

Although autogamy, which is found in some species, is a very useful process to the experimentalist, one pays a price for it. It makes the study of heterozygotes rather difficult. Moreover, the phenomenon of aging that is encountered in all species means that macronuclear caryonides are not immortal, and variants which are restricted to the macronucleus and not found in the micronuclei must all be lost eventually. Since lines of other ciliates that fail to age have often been found in nature, one wonders if similar lines may not eventually be found or even constructed in the laboratory for *Paramecium*.

Mating Types. In the 200 or so years after its discovery by Leeuwenhoek in the 1600's, our knowledge of *Paramecium* increased very slowly. However, by the early 1900's, the main features of the life cycle and the processes of conjugation had been worked out. The early students of the genetics of *Paramecium*, Jennings, Jollos, and others, found the going difficult because they were unable to control conjugation. Sonneborn's great discovery of mating types in 1937, however, ushered in the modern work. Moreover, his finding in *P. aurelia* of mating type I that mated only with II, mating type III that mated only with IV, all the way up to XIII that mated with XIV, not only made controlled matings possible, but also defined a new set of taxonomic entities. These entities, 14 in number in the *P. aurelia* complex, one for each pair of mating types, were first called varieties, then syngens, and finally were elevated to the rank of species. The discovery of mating types had a profound effect on protozoology, for mating types were soon recognized, not only in other species of paramecia, but in many other protozoa and other single-celled organisms. Indeed, the use of the term "syngen" still persists for many of them.

Although molecular biology has been applied to the genetic problems of mating types in other organisms, molecular studies have not proved feasible in *Paramecium*, where mating types were discovered. This failure stems from the fact that the mating type substances in *Paramecium*, and in *Tetrahymena* as well, appear to be insoluble; they are bound within the membranes of cilia and hence are very difficult to identify. New techniques to identify genes without prior isolation of their products may soon remedy this situation.

Caryonidal Inheritance. Caryonidal inheritance is a remarkable phenomenon peculiar to ciliates. Since several different traits referred to in different chapters in this book are inherited caryonidally, it is useful to consider the various cases together. A caryonide consists of all the individuals derived from a single macronucleus that has been newly formed at autogamy or conjugation. The production of macronuclei is a complex process, involving both amplification and major rearrangements of the DNA. At one stage of both autogamy and conjugation a diploid nucleus is formed by the fusion of two haploid nuclei. This diploid nucleus now undergoes mitosis to produce four micronuclei, which one would expect to be identical. However, two of these nuclei produce macronuclei, while the other two remain as micronuclei. Moreover, in caryonidal inheritance, even though the two macronuclei are derived by mitosis from a single diploid nucleus, the two nuclei may turn out to be different in their genetic properties. The essential feature of caryonidal inheritance is that each macronucleus at its formation may be determined to produce alternative phenotypes. Thus in caryonidal inheritance of mating type in species 1 of the *P. aurelia* complex, the two macronuclear anlagen developing in a single cell may both be determined for mating type I, both for mating type II, or one for mating type I and the other for mating type II. Once determined, however, each macronucleus generally remains constant throughout subsequent generations. Alleles that restrict the number of alternative phenotypes to one, i.e., alleles that effectively eliminate caryonidal inheritance, are often encountered. Caryonidal inheritance has been found to occur for mating type, serotype expression, trichocyst discharge, temperature resistance, and macronuclear DNA sequences. It has been found in *Tetrahymena*, *P. multimacronucleatum*, and in the *P. aurelia* complex.

Caryonidal inheritance was first discovered by Sonneborn for mating types I and II of species 1 of the *P. aurelia complex*. It was later found that mating type was inherited caryonidally in all but species 13, where it follows simple Mendelian rules. Sonneborn found that in species 1, 3, 5, 9, and 11, the two macronuclear anlagen formed in a single cell after conjugation or autogamy are determined for mating type independently of each other, and also independently of the cytoplasm in which they lie, whereas in species 2, 4, 6, 7, 8, 10, 12, and 14, the cytoplasm in which the anlagen develop exerts an influence on the new type, so there is a strong tendency for the two anlagen developing in one cell to be alike. Although this cytoplasmic influence tends to obscure the correlation with caryonide, Nanney (1954) has shown that it can still be seen. Sonneborn called the first group of species the Type A species and the second group the Type B species.

The ability to produce serotype A in the *P. aurelia* complex, a trait showing caryonidal inheritance, was found by Epstein and Forney (1984) to be based on alternate processing of DNA during the development of the new macronuclei at autogamy and conjugation. Variations in the sequences of macronuclear DNA found in different vegetative lines in *Tetrahymena* have the same basis (White and Allen 1985). It has also been suggested by Orias (1981) that caryonidally inherited mating types in *Tetrahymena* are due to alternate paths of DNA processing in the production of macronuclei.

The Cytoplasmic State in Caryonidal Inheritance. In the group B species of the *P. aurelia* complex (but not the group A species) the cytoplasm, or the "cytoplasmic state" exerts a strong effect on the determination of the developing anlagen. In ingenious experiments involving both cytoplasmic exchange and macronuclear regeneration, Sonneborn showed that the cytoplasmic state is itself determined by the old macronucleus of the cells in which the anlagen are developing. Thus the cytoplasmic effect is not due to self-reproducing cytoplasmic determinants, but to a curious nuclear "feedback" acting through the cytoplasm.

Examples. Three more recent cases of caryonidal inheritance clearly exhibit the system of mating type inheritance found in all the Type B species of the *P. aurelia* complex. The first, studied by Sonneborn and Schneller (1979), is a trichocyst variant found in *P. tetraurelia* in which normal trichocyst discharge is prevented. The second case is the mutant d48 studied by Epstein and Forney (1984), which is unable to produce serotype A in *P. tetraurelia.* In the d48 mutation the cytoplasm controls the processing of macronuclear DNA so that the A gene, which is always present in the micronucleus, is completely eliminated from the macronuclear genome when it develops from the micronuclear DNA during conjugation or autogamy. Finally, Doerder and Berkowitz (1987) have shown that in *T. thermophila* the ability to express serotype H exhibits B type caryonidal inheritance.

Dissociation of the Cytoplasmic State and the Phenotype. The first evidence for uncoupling of the phenotype and the cytoplasmic state was discovered by Nanney in the 1950's when he found that some cells of mating type VIII in *P. tetraurelia* appeared to have VII-determining cytoplasm; they always gave rise to VII progeny at autogamy. The most remarkable example, however, was provided by Taub (1963) in his studies on mating type inheritance in species 7 of the *P. aurelia* group. In some stocks, mating type was inherited exactly like mating types VII and VIII in the group B species of *P. tetraurelia.* However, Taub found an allele that produced a radically different pattern of inheritance. This allele, mt^{XIII}, had two distinct effects, one on the mating type itself, the other on the cytoplasmic state. Cells homozygous for the allele were always mating type XIII. Surprisingly, however, it was possible to show with appropriate crosses that the cytoplasm of such cells was always XIV-determining.

Simple Restrictive Alleles. Several restrictive alleles that affect only the phenotype and not the cytoplasmic state are known in the Group B species that show caryonidal inheritance with a cytoplasmic effect. These include Taub's *n* that restricts caryonides to mating type XIII. Byrne (1973) has also isolated a series of mutants, $mt\mathrm{A}^{\mathrm{O}}$, $mt\mathrm{B}^{\mathrm{O}}$, etc. that act similarly to restrict the type to VII. Mutant strains d8, d29 and d12 (Epstein and Forney 1984; Epstein and Forney, pers. commun.) contain alleles which prevent the expression of serotype A in *P. tetraurelia;* in mutant strain d12 (but not in d8 and d29) the gene coding for the serotype A protein has been completely deleted. The fact that the cytoplasmic state is not modified in these mutants (as it is in the d48 mutant referred to above) further emphasizes the independence of the trait and the cytoplasmic state.

Alleles Affecting the Maintenance of Cytoplasmic States. Several alleles other than Taub's mt^{XIII}, described above, have also been found to affect the cytoplasmic state. Brygoo (1977) in studying species 4 of *P. aurelia* found not only cytoplasmic state O (which determines the odd mating type, VII) and its complementary cytoplasmic state E (determining the even mating type, VIII), but also a cytoplasmic state O* (which determines a rather unstable mating type VII). He found that the allele $mt\mathrm{D}^{51}$ is dominant over $mt\mathrm{D}^{32}$ in restricting the cytoplasmic state to O or E, and in excluding the cytoplasmic state O*.

Finally, Doerder's induced mutations in *Tetrahymena* (see above), which will not be described here, appear to restrict the cytoplasmic determining state so that serotype H cannot be expressed in *Tetrahymena.*

Determination of the Micronucleus. The cytoplasmic state can influence not only the development of macronuclear anlagen, but even micronuclei can be partially predetermined. In crosses of amicronucleates to normal cells, Sonneborn (1954) and Brygoo et al. (1980) have shown that the state of the cytoplasm in which micronuclei are derived may cause the micronuclei to influence the determination of the macronuclei to which they give rise. The nature of this effect is completely unknown.

Environmental Effects. Soon after the discovery of caryonidal inheritance, Sonneborn showed that temperature affects the frequency with which mating types I and II appear. Later, it was shown that temperature also affects the frequency with which mating types VII and VIII appear in species 4 of the

P. aurelia complex (Nanney 1954, 1957). In the case of trichocysts discharge, it was shown by Sonneborn and Schneller (1979) that starvation favors the non-discharge state over normal discharge.

Protoplasmic Transfer by Microinjection. The cytoplasmic state has been modified in several cases by the transfer of cytoplasm or macronucleoplasm. Koizumi et al. (1986) have investigated the effect of transfers on the determination of mating types VII and VIII in *P. tetraurelia.* Change of type was effected by transfer of VIII cytoplasm into the cytoplasm of VII recipients, but similar transfer of karyoplasm was ineffective. On the other hand, VII cytoplasm was ineffective when injected into VIII cytoplasm, while VII karyoplasm was effective. These differences appear to be related to the specific times during and near the time of conjugation at which the effective substances are produced, as well as the precise time at which the recipient nuclei are receptive. Harumoto (1986) was able to transform the d48 cytoplasmic state into wild type by the injection of wild-type karyoplasm into the cytoplasm of the d48 serotype mutation in *P. tetraurelia.* Transfer of wild-type cytoplasm was only weakly effective.

Speculations on the Nature of the "Cytoplasmic State". No complete and adequate hypothesis has been proposed for the molecular basis of the cytoplasmic state. It is clear that in the case of the d48 mutation the alternative traits that are selected in caryonidal inheritance are due to modifications in processing of micronuclear chromosomes as they are transformed into macronuclear chromosomes, and the same could be true for all cases of caryonidal inheritance. An explanation for the cytoplasmic state, however, is more difficult. Perhaps the alternative cytoplasmic states themselves also represent the outcome of alternative processing of micronuclear chromosomes. If so, it is reasonable to assume that processing of chromosomes during formation of the new macronuclei at conjugation and autogamy is under the control of the old macronucleus. The cytoplasmic state was shown by Sonneborn to be only a feed-back loop that mimics cytoplasmic inheritance. To account for the cytoplasmic state, one need only assume that the genes in the old macronucleus that control processing of chromosomes determining the particular trait in question (say serotype or mating type) also control their own processing. Although this rather speculative hypothesis provides a rational formal explanation for the cytoplasmic state, it does not explain the apparent high degree of specificity observed for the different traits.

Genome Structure. Recent work, just now beginning to appear in print, on macronuclear DNA of stocks of the *P. aurelia* complex using orthogonal field electrophoresis and related techniques, indicate that the mean size of macronuclear DNA, although variable, is much smaller than expected for the size of intact micronuclear chromosomes (Baroin et al. 1987; Godiska et al. 1987; J. D. Forney, pers. commun.). Telomeres are found (Baroin et al. 1987; Forney and Blackburn 1987) to be an apparently random mixture of repeated C_4A_2 and C_3A_3 sequences, rather than the C_4A_2 repeats found in *Tetrahymena* or the C_4A_4 repeats found in the hypotrichs. In both hypotrichs and in *Tetrahymena,* these or similar sequences do not assume their terminal positions on the macronuclear chromosomes until formation of the macronuclei. The same appears to be true for *Paramecium,* for alternate processing of macronuclear chromosome ends results in different lengths and DNA contents of macronuclear chromosomes known to be derived from the same micronuclear chromosome (Baroin et al. 1987; Forney and Blackburn 1987; Godiska et al. 1987). These observations suggest that in macronuclear formation extensive remodeling of the genome occurs in *Paramecium,* just as it does in *Tetrahymena.* Chromosome breakage, internal deletions and the formation of new telomeres all occur.

One of the oldest questions in the genetics of *Paramecium* is the problem of how genic balance is maintained in the macronucleus in the absence of any apparent mitosis. Four specific models are representative of the numerous theories that have been considered (see Preer 1968; Preer and Preer 1979). According to model No. 1, hidden mitoses actually occur in the macronucleus. Although this model was eliminated in the minds of most observers by observations with the light microscope, it was completely eliminated from consideration by the advent of the electron microscope, which revealed that microtubules are not present in the required numbers and positions.

According to model No. 2 there are so many copies of each chromosome in the macronucleus that random segregation does not have time to build up serious imbalances before death caused by aging occurs. Although mathematical treatment has not ruled out this model for *Paramecium* (Preer 1976), such an explanation has been clearly ruled out for *Tetrahymena* (Preer and Preer 1979). Since the structure of the macronuclei in *Tetrahymena* and *Paramecium* is so similar, investigators have generally looked elsewhere for models.

The most popular theory in the past has been model No. 3, the subunit theory. It was first suggested by Sonneborn when he found that in macronuclear regeneration each one of the macronuclear fragments formed at conjugation or autogamy in the *P. aurelia* complex can form a new macronucleus. According to the subunit theory, macronuclear chromosomes are organized into individual haploid or diploid subunits that are physically connected and remained joined at segregation. The theory has enjoyed its most extensive development in *Tetrahymena,* where extensive mathematical analysis showed that all genetic data on macronuclear segregation of heterozygous genes is consistent with a model of haploid subunits. The cytological or biochemical evidence for such subunits, however, has not been forthcoming. Moreover, the finding that ciliates show extensive reorganization of their chromosomes when macronuclei are formed has not increased acceptance of the model.

According to model No. 4 (Preer and Preer 1979; Brunk 1986), each macronuclear chromosome acts as an independent replicon, regulating its own copy number much as a mixture of plasmids control their own copy number in bacteria. Although it was shown that the genetic data on macronuclear segregation in ciliates is consistent with this model, there has been no molecular evidence as yet that provides a clear test for the model. The recent work on the macronuclei of hypotrichs (Helftenbein 1985) that shows that the copy number for different macronuclear chromosomes within the same genome can differ, is easily accommodated by theory No. 4, while providing a serious difficulty for the subunit hypothesis.

Membranes. H. S. Jenning's book entitled *Behavior of the Lower Organisms,* published in 1931, created considerable excitement among students of animal behavior and psychologists. It set the tone for the work to come. About 15 years ago, Ching Kung initiated the modern work on behavioral genetics in *Paramecium.* It soon merged with electrophysiological studies and work on the biochemistry of membranes. The result is that *Paramecium* has become an important model for general studies on excitable membranes and membrane channels; the results of these studies are summarized in several chapters in this book.

The membranes of vacuoles, many of which are recycled at exceedingly high rates, have also received considerable attention and these studies are also documented here.

Cortical Structures and Organelles. The ciliate cortex provides one of the most beautiful examples of cellular development in all of biology. Although the molecular mechanisms are completely unknown, some of the major principles began to be understood with the work of Tartar (1961) on Stentor. They were extended in the genetic studies of Sonneborn on double animals in *Paramecium,* and by the work of Beisson and Sonneborn (1965) on inverted kinetics. The problems exhibited here go far beyond the scope of molecular biology today and deal with the next highest level of organization. They will surely some day become the successors to the current studies on DNA, RNA, and the formation of proteins. The problem of how a complex asymmetrical structure with its hierarchy of structural units can divide to produce two identical daughter cells is an extremely difficult problem and much of the research today has still, of necessity, not entirely left the descriptive stage. Current studies focus on the chemical constitution and function of the component structures (including cilia and mitochondria), on a genetic dissection of their development, and finally on attempts to view the whole phenomenon and discern important principles and relations. In this book examples of all these approaches are seen.

Aging and the Life Cycle. One of the phenomena recognized by the early workers on *Paramecium* in the 1800's was the long-term changes that occur throughout the different stages of the life cycle. These changes were further elucidated by the studies of Sonneborn and Schneller after the discovery of autogamy. The changes consist of reduced viability and eventual death in the absence of autogamy or conjugation (aging), sexual immaturity and immaturity, and number of fissions between autogamies. Jennings (1929) pointed out that since these changes are very long-lasting, they must involve changes in hereditary mechanisms. Unfortunately, aging appears to be as resistant to experimental analysis in ciliates as it is in the cells of higher eukaryotes. It may be that aging in ciliates is tied up with the peculiar structure of the macronucleus, but even if it is, the phenomenon may be of great interest in understanding the nature of aging in other cells. It appears that the technique of microinjection may well provide a powerful tool for studying the life cycle, both as an assay tool for active molecules [cf. the isolation of immaturin by Haga and Hiwatashi (1981)] and in providing strong evidence that the seat of the aging is indeed in the macronucleus (Aufderheide 1987).

Serotypes. The study of the inheritance of antigenic variation in the surface proteins of *Paramecium* began with the studies of Jollos on dauermodifications [see Preer (1968) for a review of dauermodifications in protozoa], which he thought were long-lasting changes in phenotype that were intermediate between environmental modifications and mutations. The concept, although based on numerous examples in many organisms, was not very useful and has disappeared from use. Other cases of antigenic variation in surface proteins, such as those found in the trypanosomes and in bacteria such as *Salmonella,* are all associated with a parasitic way of life. They are thought to be an aid to escaping the immune system of their hosts. However, it is unknown whether the different antigenic variants of *Paramecium* enjoy a selective advantage over each other under different environmental conditions. The large, abundant, stable immobilization proteins, with their easily isolated mRNA's and genes, are particularly suitable for molecular work, and considerable strides have been made in recent years in their molecular analysis. The finding (Forney et al. 1983) that serotype switching is not due to chromosomal rearrangements as it is in trypanosomes, may not indicate a very great difference between *Paramecium* and the trypanosomes, for switches in surface antigens of trypanosomes that are due not to rearrangements in DNA, but to activation and inactivation without rearrangement are also known (Bernards 1985). Elucidation of the mechanism of serotype switching would be an important contribution to our knowledge of the control of gene action in eukaryotes.

One point that might be emphasized is the usefulness of mutations in the study of the molecular biology of the serotype system. Mendelian deletions of the well-characterized genes that specify serotype specificity provide a simple and unambiguous means of identifying the genes, especially in the presence of pseudogenes and cross-hybridizing sequences of DNA. Moreover, as already pointed out, they are also proving useful in analyzing caryonidal inheritance and cytoplasmic states. A concerted attack on the problem of caryonidal inheritance that includes a search for mutants affecting cytoplasmic states in the serotype system has never been carried out yet. The advanced state of our knowledge of the genetics and molecular biology of serotypes makes them a particularly appropriate system for such studies.

Finally, it is interesting to note the beginning of the molecular genetics of serotypes in *Tetrahymena* with the recent discovery of a cDNA clone for the gene for serotype H in *Tetrahymena* (R. Hallberg and F. P. Doerder, pers. commun.).

Endosymbionts. The study of the intracellular bacterium that was at first considered a cytoplasmic genetic factor, kappa, has provided a major impetus in the study of endosymbiosis. It is significant that the majority of paramecia of *P. biaurelia* freshly collected from nature contain endosymbionts. Moreover, work by Landis (see Chap. 24, this Vol.) suggests that paramecia that bear the endosymbiont kappa have a selective advantage over those that are free of kappa. The story of the virus-like forms and plasmids that infect kappa, and their involvement in the production of "R bodies", toxicity, and specific resistance to toxicity are especially intriguing. The interactions and the detail with which they are known is virtually unparalleled in cases of endosymbiosis in other organisms. The variety of forms living within the cytoplasm, and in some cases in the nucleus of *Paramecium*, is stunning. A major accomplishment in recent years is the finding by Quackenbush (see Chap. 23, this Vol.) that the gene for R bodies is located on the plasmid found within kappa and that it can be subcloned into *Escherichia coli* and expressed to yield R body-producing *E. coli*. Perhaps experiments of the same type will eventually shed light on the old and difficult problems of the nature and action of the toxins produced by kappa and its relatives, and also the means by which the endosymbiont makes its hosts resistant to the specific toxin produced by the endosymbiont infecting the host.

Ecology and Speciation. The few studies on ecology of *Paramecium* have been interesting, and it is hoped that more will be not only of theoretical interest, but even have practical consequences in determination of the quality of water in our streams, ponds, and lakes. Studies on the base sequences of rDNA of ciliates (Sogin and Elwood 1986) promise a better determination of the phylogenetic position of the genus *Paramecium*.

Transformation. Recently, it has been found (Godiska et al. 1987) that microinjection of a plasmid containing the A serotype gene into the macronucleus of the deletion mutant d12 lacking the A gene restores the ability of the paramecia to produce the serotype A surface protein. Expression is normal, and the transformant DNA is present in high copy number and persists until the transformed macronucleus is replaced by a new one at autogamy. Since the transformation frequency is very high (40%) the technique should be of great importance in analyzing both gene expression and the replication of macronuclear DNA in *Paramecium*.

JOHN R. PREER, JR.

References

Allen SL, Nerad TA (1978) Method for the simultaneous establishment of many axenic cultures of *Paramecium*. J Protozool 25:134–139

Aufderheide KJ (1987) Clonal aging in *Paramecium tetraurelia*. II. Evidence of functional changes in the macronucleus with age. Mech Ageing Develop 37:265–279

Baroin A, Prat A, Caron F (1987) Telomeric site position heterogenity in macronuclear DNA of *Paramecium primaurelia*. Nucl Acids Res 15:1717–1728

Beisson J, Sonneborn TM (1965) Cytoplasmic inheritance of the organization of the cell cortex in *Paramecium aurelia*. Proc Natl Acad Sci USA 53:275–282

Bernards A (1985) Antigenic variation of trypanosomes. Biochim Biophys Acta 824:1–15

Brunk CF (1986) Genome reorganization in *Tetrahymena*. Intern Rev Cytol 99:49–83

Brygoo Y (1977) Genetic analysis of mating type differentiation in *Paramecium tetraurelia*. Genetics 87:633–654

Brygoo Y, Sonneborn TM, Keller AM, Dippell RV, Schneller MV (1980) Genetic analysis of mating type differentiation in *Paramecium tetraurelia*. II. Role of the micronuclei in mating-type determination. Genetics 94:951–959

Byrne BC (1973) Mutational analysis of mating type inheritance in syngen 4 of *Paramecium aurelia.* Genetics 74:63–80

Doerder FP, Berkowitz MS (1987) Nucleo-cytoplasmic interaction during macronuclear differentiation in ciliate protists: Genetic basis for cytolasmic control of SerH expression during macronuclear development in *Tetrahymena thermophila.* Genetics 117:13–23

Enright WJ, Hennessey TM (1987) Growth of *Paramecium tetraurelia* in bacterized, monaxenic cultures. J Protozool 34:137–142

Epstein LN, Forney JD (1984) Mendelian and non-Mendelian mutations affecting surface antigen expression in *Paramecium tetraurelia.* Mol Cell Biol 4:1583–1590

Forney JD, Blackburn EH (1987) Developmentally controlled telomere addition in wild type and mutant paramecia. Mol Cell Biol (in press)

Forney JD, Epstein LN, Preer LB, Rudman BM, Widmayer DJ, Klein WH, Preer JR, Jr (1983) Structure and expression of genes for surface proteins in *Paramecium.* Mol Cell Biol 3:466–474

Godiska R, Aufderheide KJ, Gilley D, Hendrie P, Fitzwater T, Preer LB, Polisky B, Preer JR, Jr (1987) Transformation of *Paramecium* by microinjection of a cloned serotype gene. Proc Natl Acad Sci, USA 84:7590–7594

Haga N, Hiwatashi K (1981) A protein called immaturin controlling sexual maturity in *Paramecium.* Nature 289:177–179

Harumoto T (1986) Induced change in a non-Mendelian determinant by transplantation of macronucleoplasm in *Paramecium tetraurelia.* Mol Cell Biol 6:3498–3501

Helftenbein E (1985) Nucleotide sequence of a macronuclear DNA molecule coding for alpha-tubulin from the ciliate *Stylonychia lemnae.* Special codon usage: TAA is not a translation termination codon. Nucl Acids Res 13:415–433

Jennings HS (1929) Genetics of the protozoa. Biblio Genetica 5:105–330

Koizumi S (1986) Analysis of mating type determination by transplantation of O macronuclear karyoplasm in *Paramecium tetraurelia.* Devel Genet 7:187–195

Nanney DL (1954) Mating type determination in *Paramecium aurelia.* A study in cellular heredity. In: Sex in Microorganisms. American Association for the Advancement of Science, Washington, DC, 266–283

Nanney DL (1957) Mating type inheritance at conjugation in variety 4 of *Paramecium aurelia.* J Protozool 4:89–95

Orias E (1981) Probable somatic DNA rearrangements in mating type determination in *Tetrahymena thermophila:* A review and a model. Devel Genet 2:185–202

Preer JR, Jr (1968) Genetics of the Protozoa. In: Chen, TT (ed) Research in Protozoology, pp 129–278

Preer JR, Jr (1976) Quantitative predictions of random segregation models of the ciliate macronucleus. Genet Res 27:227–238

Preer Jr, Jr, Preer LB (1979) The size of macronuclear DNA and its relation to models for maintaining genic balance. J Protozool 26:14–18

Schönefeld U, Alfermann AW, Schultz JE (1986) Economic mass cultivation of *Paramecium tetraurelia* on a 200-liter scale. J Protozool 33:222–225

Sogin ML, Elwood HJ (1986) Primary structure of the *Paramecium tetraurelia* small-subunit rRNA coding region: phylogenetic relationships within the Ciliophora. J Mol Evol 23:53–60

Sonneborn TM (1954) Patterns of nucleocytoplasmic integration in *Paramecium.* Caryologia 6 (Suppl):307–325

Sonneborn TM, Schneller MV (1979) A genetic system for alternative stable characteristics in genomically identical homozygous clones. Dev Genet 1:21–46

Tartar V (1961) "The Biology of *Stento*". (Pergamon Press, New York)

Taub SR (1963) The genetic control of mating type differentiation in *Paramecium.* Genetics 48:815–834

White TC, Allen SL (1985) Macronuclear persistence of sequences normally eliminated during development in *Tetrahymena thermophila.* Develop Genet 6:113–132

Contents

Chapter 1 Introduction

Hans-Dieter Görtz[1]

Paramecium has been known for more than a century and since its discovery outstanding scientists investigated this little animal to learn about its biology. Soon, however, *Paramecium* was also used as a model for the study of basic biological phenomena, many of which are still enigmatic today, and there are numerous reasons why *Paramecium* is the object of choice for the study of very heterogeneous problems. The size of this protozoon favors light microscopic as well as electron microscopic investigations, microsurgical and electrophysiological experiments. The latter are especially tempting, because *Paramecium* is a motile organism with simple but pronounced behavioral reactions. This big ciliate is frequently the host for other microorganisms and therefore suitable for the study of different types of endocytobiosis. *Paramecium* can be grown in mass cultures even in axenic media, thus facilitating studies of its molecular biology and biochemistry (see Chapter 16 by Schultz and Klumpp), and such biochemical studies are often supported by the availability of mutants. In fact, the genetics of Mendelian and cytoplasmic inheritance in *Paramecium* has a long history, and still new results are obtained. The genetic approaches were encouraged by the increasing knowledge on the sexuality and cell cycle of *Paramecium.*

Today, it is almost impossible for an individual scientist to present, value, and discuss the enormous amount of data and observations in the heterogeneous fields of research on *Paramecium.* Therefore, we attempted to do it in cooperation: specialists have reviewed the work done in their own fields. Each chapter is intelligible by itself, although several chapters are thematically related to each other, and the reading of some, e.g., "Cytology" (Chapter 2) by R.D. Allen, will certainly facilitate the understanding of specific phenomena described in others.

The discussion on evolutionary history and relationships of the various species of *Paramecium* is still not settled. The breeding systems, sexuality, life cycle, and genetics are topics of the articles by D. Nyberg (Chapter 3), Y. Tsukii (Chapter 4), M. Fujishima (Chapter 5), and J.D. Berger (Chapter 7). Closely related to their articles is Chapter 8 by K. Mikami concerned with the nuclear dimorphism and the function and genesis of the nuclei, whereas Chapter 6 by A. Kitamura deals with mating type substances. The phenomenon of

1 Zoologisches Institut der Universität, Schloßplatz 5, D-4400 Münster, FRG

Paramecium. Ed. by H.-D. Görtz

aging has long been observed in *Paramecium,* offering experimental access with the capability of distinguishing between germinal and somatic aging (see Chapter 9 by Y. Takagi). Genome organization and molecular genetics of *Paramecium* is reviewed by M. Freiburg (Chapter 10) and, in the context of the widely studied immobilization antigens, by H.J. Schmidt (Chapter 11). In this field the discovery of deviations from the universal genetic code in *Paramecium* was one of the highlights within the last few years. Based on the background of Mendelian genetics and research of nuclear genomes, the mitochondrial function and genetics have been investigated. The genetics, respiratory chain, and genome organization of mitochondria in *Paramecium* are reviewed by A. Sainsard-Chanet and D. Cummings in Chapter 12.

The analysis of motility, excitability, and behavior, of their causalities and genetics is of special interest to cell biologists, geneticists, molecular biologists, and physiologists. The various aspects of this broad field, where *Paramecium* is thoroughly and successfully investigated with different methods, is treated in six chapters, written by H. Machemer (Chapters 13 and 14), R. Ramanathan et al. (Chapter 15), J.E. Schultz and S. Klumpp (Chapter 16), M. Takahashi (Chapter 17), and J. Van Houten and R.R. Preston (Chapter 18). The understanding of results obtained with certain methods as presented in individual chapters is aided by the other chapters, and perhaps the value of *Paramecium* for scientific work will become most obvious in this part of the book.

A cell-biological phenomenon of actual interest is the lysosomal cycle with endo- and exocytosis and the details of digestion and membranous flow. Again these phenomena have been thoroughly investigated in *Paramecium* and are treated in Chapter 19 by A. Fok and R.D. Allen. Somewhat related is Chapter 20 by A. Adoutte, who reviews the literature and current work on trichocysts, their biogenesis and exocytosis. Genetic, microscopic, biochemical, and other methods have been employed to study the genesis, function, and biological meaning of this organelle. Various components of the cytoskeleton which are responsible for cellular structure and intracellular movements have been investigated and described in detail in the *Paramecium* cell. Understanding the cytoskeleton, in fact, can help to understand various functions of *Paramecium* and eukaryotic cells in general; the present knowledge and actual questions are reviewed in Chapter 21 by J. Cohen and J. Beisson.

For many decades *Paramecium* has been used to study cytoplasmic inheritance and endocytobiosis with bacterial, algal, and other microorganisms (Chapter 22 by Görtz), and the "killer" phenomenon – sensitive paramecia are killed by bacterial endosymbionts released by the killer paramecia – is certainly one of the best-studied examples of nonchromosomal inheritance in animal cells. This phenomenon with recent advances in the understanding of its molecular genetics is treated by R.L. Quackenbush (Chapter 23). Endocytobionts may well be of advantage for the host paramecia, but the ecological meaning of this and the ecology of *Paramecium,* in general, are more and more investigated; this volume ends with the chapter "Ecology" by W. Landis to emphasize the importance. To close the circle, the discussion of ecological strategies in this chapter leads back to Chapter 3, "The species concept and breeding system", by D. Nyberg.

With regard to the tremendous progress in recent years the authors have emphasized the actual state in their fields on the background of the basic knowledge necessary for understanding. To obtain sufficient clearness and to stay within the scope of the book, we had to confine the individual chapters by omitting many older observations, side aspects, and certain details. For the same reasons we did not try to achieve completeness. However, it is hoped that this book will stimulate even more biologists to work with *Paramecium.* We have tried to show that the qualities of this interesting protozoon favor the study of general biological phenomena with different methods, and thus contribute to a comprehensive view of an eukaryotic organism.

After finishing his chapter our colleague Dr. A. Kitamura deceased. His scientific work is a substantial contribution to our knowledge on the biology of *Paramecium,* and we are grateful to have his chapter in this book.

I thank all authors for their kind cooperation and, especially the authors from Münster, for many helpful discussions. Furthermore, I am grateful to Dr. K. Hiwatashi for valuable suggestions and help. He and his students – many of them authors of this book – have contributed substantially to our knowledge on *Paramecium.* I am indebted to the staff of the Springer-Verlag, especially Dr. D. Czeschlik, for their expert assistance and generous support. My thanks are due most of all to my teacher, Dr. K. Heckmann, for advice and stimulating discussions and to my wife Monika for her help and encouragement.

Chapter 2 Cytology

RICHARD D. ALLEN[1]

CONTENTS

1 Introduction

Paramecium is a single-celled eukaryotic organism. Being an organism, *Paramecium* carries on all the normal functions required to sustain and propagate life, including growth, metabolism, catabolism, and reproduction. Being a cell, and more specifically an aerobic cell, *Paramecium* contains the typical structures and organelles of aerobic nonphotosynthetic eukaryotic cells. These in-

1 Pacific Biomedical Research Center and Department of Microbiology, University of Hawaii, Honolulu, HI 96822, USA

Paramecium. Ed. by H.-D. Görtz

clude an uninterrupted cell membrane, a cytoplasm containing mitochondria, ribosomes, endoplasmic reticulum, a Golgi apparatus, an endocytic system of coated pits, and shuttle vesicles, a phagosome-lysosome system and food storage bodies, and a nuclear system containing nucleoli, chromatin, and nuclear envelope. In addition, *Paramecium* possesses the elements necessary for movement, both of the cell as a whole and of the organelles within the cell. These elements are the microtubules and microfilaments variously organized into multiple cilium-basal body complexes and into bundles, bands, meshworks, and ribbons within the cytoplasm. *Paramecium* also contains more specialized structures found in only a few eukaryotic cells. These are the contractile vacuole complexes, the dual nuclear system, the alveolar system, and the specialized extrusive organelles called trichocysts, as well as the highly structured oral apparatus and the cytoproct, the site of defecation.

The cytology of *Paramecium,* both at the light and electron microscope levels has been reviewed before (e.g., Wichterman 1953, 1986; Jurand and Selman 1969; Ehret and McArdle 1974; Vivier 1974). This chapter will emphasize recent work (since 1974) on the ultrastructure of *Paramecium* as interpreted by a cell biologist. When possible, the functions of the various structures will be summarized. Problems needing further investigation will be given throughout the chapter.

Due to an error both on my part and on that of the supplier of *Paramecium* the cultures of *Paramecium* used by my lab since 1971 have been incorrectly labeled *P. caudatum* when, in fact, they were *P. multimicronucleatum.* The only observable difference between these two species at the ultrastructural level is in the micronuclear number and structure. For this review I have used the correct species name in the text when citing our work.

2 The Pellicle

2.1 Plasma Membrane, Alveolus, Epiplasm

The overall elongated shape, which varies from slipper to footlike in different *Paramecium* species, is maintained by a semi-rigid pellicle that is capable of elastic deformations. This pellicle, from outside to inside, consists of the plasma membrane, the alveolar system (a mosaic of flattened membrane-bounded sacs), and the epiplasm which is an amorphous layer lining the inner alveolar membranes (Fig. 1). These three closely apposed layers are molded into ridges which form a pattern of hexagons and parallelograms over the cell surface. Cilia, singly or in pairs, emerge from the depressions bordered by these polygon-shaped ridges. Internally short bundles of microfilaments, designated striated bands (Allen 1971) due to a subtle striated appearance, link the sides of the ridges. These bands insert into the epiplasm on one slope of the ridge, pass under the ridge, and insert into the epiplasm on the opposite slope (Fig. 4). Bands fan out around each depression and seem to stabilize the surface ridges. These striated bands are distinct in form and in space from another system of microfilaments which G. Gelei (1937) observed with the light microscope and

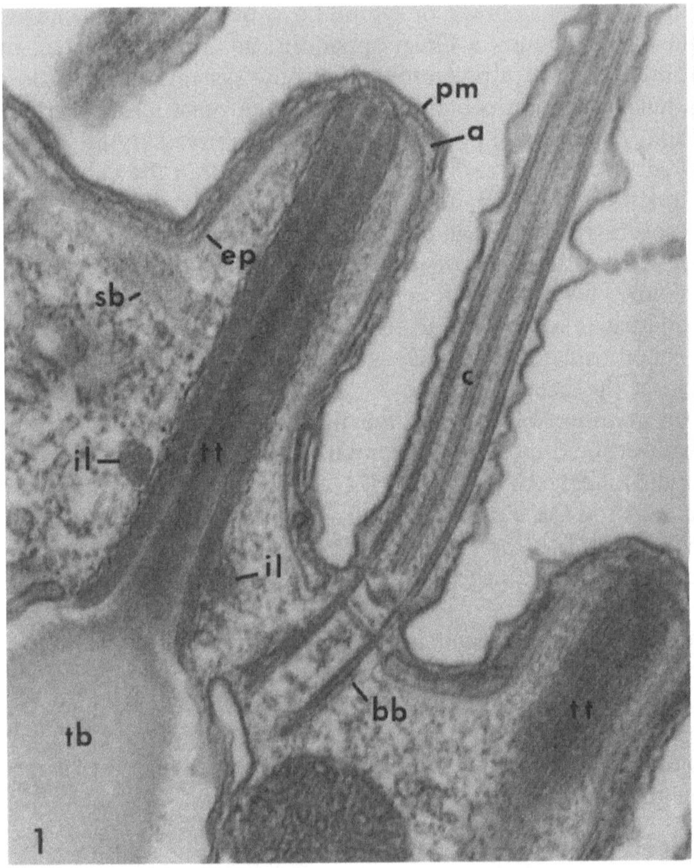
pm
a
ep
sb
c
il
tt
il
bb
tb
tt
1

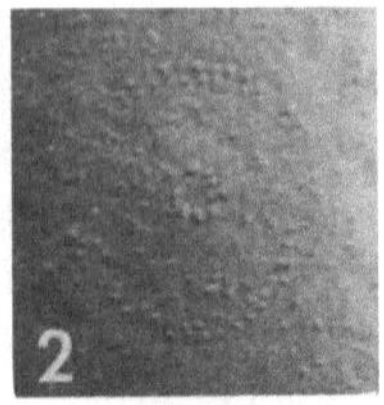
2

3

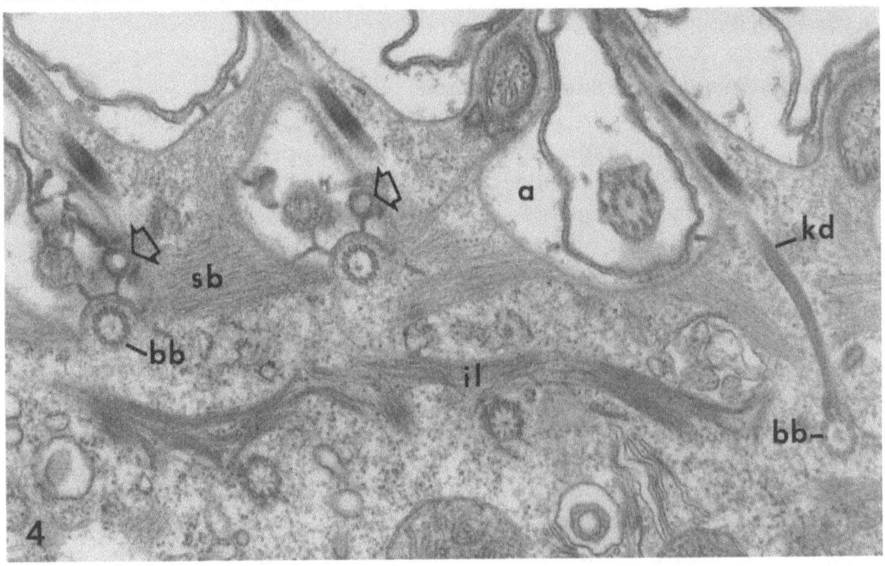
a
kd
sb
bb
il
bb
4

called the infraciliary lattice. This lattice of membrane-unassociated, branching, microfilamentous bundles occupies a plane at the level of the basal bodies (Fig. 4), a few tenths of a micrometer internal to the striated bands. Under periods of extreme mechanical and chemical stress this lattice forms a cage preventing the cytoplasmic organelles from spilling out into membrane-bounded blisters that appear at the cell surface (Sibley and Hanson 1974).

The 9-nm plasma membrane forms an unbroken boundary over the cell. This membrane is coated on its external side by a 15-nm-thick glycocalyx which stains with ruthenium red (Wyroba and Przelecka 1973; Watanabe 1981). As seen in freeze-fracture replicas this membrane contains specializations visualized as arrays of intramembrane particles (IMPs). Double rings of particles surrounding a rosette of 8 to 11 particles are found in the plasma membrane at the docking site of each docked trichocyst (Fig. 2) (Janisch 1972; Plattner et al. 1973; Satir 1974). The rings persist in the absence of a docked trichocyst but assume the shape of parentheses (Beisson et al. 1976). Bands of IMPs are found in the plasma membrane at the margins of both the cytopharynx and cytoproct (Allen 1978) where unusual stress is placed on the cell membrane during vacuole formation and vacuole defecation. Specialized plaques of particles (Allen 1978) arranged in a few short rows, which are separated by a distance of some 25 nm, are found on the ventral surface anterior to the vestibulum (Fig. 3). In several cases specializations of IMPs probably reflect a junction between the plasma membrane and the trichocyst membrane (Plattner et al. 1973). The role of the plaques, though a chemosensory function has been suggested (Allen 1978), has not been confirmed.

The alveolar system is composed of a mosaic of flattened membrane-bounded cisternae and is applied close to the inner surface of the plasma membrane. The individual units of this mosaic are joined laterally to other units by membrane-bounded pores which cross the space (the septum) between the units (Allen 1971). This makes the lumen of the alveolar sacs as well as their membranes continuous. In electron micrographs the volume of this luminal space varies from cell to cell, being at times flattened (Fig. 1) and at other times greatly distended (Fig. 4). What produces this variation has not been determined, nor has a particular appearance been correlated with a given environmental condition. Possibilities are that the space acts as an insulating layer protecting the cell from local changes in the environment (Pitelka 1965; Wyroba 1981) or that this space somehow functions to regulate motile activity of the cell (Allen and

◄

Fig. 1. Portion of pellicle of *P. multimicronucleatum.* Plasma membrane (*pm*), alveolus (*a*), epiplasm (*ep*), trichocyst body (*tb*), trichocyst tips (*tt*), basal body (*bb*), cilium (*c*), striated band (*sb*) and infraciliary lattice (*il*) are evident. ×47 500

Fig. 2. Freeze-fracture PF face of plasma membrane at trichocyst docking site. ×71 000

Fig. 3. Plaques of intramembrane particles on PF face of plasma membrane are found only on anterior ventral surface. ×47 500

Fig. 4. Cross-sections of basal bodies (*bb*) with associated kinetodesmal fibers (*kd*) and parasomal sacs (*arrows*). Alveoli (*a*), striated bands (*sb*), and infraciliary lattice (*il*). ×28 500

Eckert 1969). In nature such an alveolar system appears to be limited to ciliates, sporozoans, and some dinoflagellates.

The alveolar system is absent in the cytopharyngeal region and along a narrow line at the tip of the ridge of the closed cytoproct; otherwise, it covers the whole cell. Basal bodies (kinetosomes), trichocysts, parasomal sacs (coated pits), and contractile vacuole pores do pass through this zone and membrane-bounded openings form in the alveolar system as these structures develop and take up their positions in the cortex.

The epiplasm forms an amorphous or filamentous layer on the cytosolic side of the inner alveolar membranes (Figs. 1 and 4). As with the alveolar system, gaps appear in the epiplasm at the cytopharynx, cytoproct, contractile vacuole pores, parasomal sacs, and trichocyst docking sites. However, the epiplasm is continuous under the septa of the alveolar system and is in intimate contact with the terminal plate that crosses the distal lumen of the basal bodies. Unlike the alveolar system, the epiplasm is discontinuous along the crests of the pellicular ridges (arrow, Fig. 30) (Allen 1971).

The plasma membrane invaginates into the cytoplasm in each surface depression at a point to the right of each cilium or pair of cilia (Pitelka 1965; Hufnagel 1969). These invaginations are called the parasomal sacs (Fig. 4), which are, in fact, coated pits complete with a clathrinlike coat on their cytosolic surfaces (Fig. 24). These sacs, which are normally about 0.3 μm long by 75 to 120 nm in diameter, at times enlarge and pinch off. This has been demonstrated by the incorporation of markers such as horseradish peroxidase (Allen and Fok 1980) and cationic ferritin (Westcot et al. 1985) into the coated vesicles that form from the parasomal sacs (Allen and Fok 1980). The type of material that in nature finds access into the cell through these sacs is unknown. If receptor-mediated endocytosis occurs in *Paramecium* as in higher organisms, these would seem to be the logical sites for such a mechanism of selective molecular concentration and uptake.

2.2 Vestibulum, Contractile Vacuole Pores, Cytoproct

In *Paramecium* the opening to the oral region is located in a midventral position about equidistant from the cell's anterior and posterior ends. A funnel-shaped, laterally flattened indentation, called the vestibulum, bearing a complete pellicle and the regular somatic ciliature but without trichocysts (Wichterman 1986) is the first chamber to be encountered in moving from outside to inside the oral cavity (Fig. 17). At the internal margin of the vestibulum an opening known as the buccal overture leads into a tapering cavity known as the buccal cavity. This cavity will be described in Section 4.1.

Other specialized pellicular regions of *Paramecium* are the two contractile vacuole pores which lie on the dorsal surface and the cytoproct on the ventral surface. The contractile vacuole pores are circular in profile and are about 1.2 μm in depth (Fig. 27). The pores are found along the same row of basal bodies, i.e., the same kinety, in the middorsal region, about one-third the cell's length from the anterior and posterior end. The wall of the pore is lined with helically wound microtubules closely applied to the inner surface of the plasma membrane (McKanna 1973). The bottom of the pore is limited by a single membrane.

The cytoproct is located midway between the vestibulum and the cell's posterior end on the ventral surface. When closed it is composed of a narrow undulating ridge (Allen 1978). The internal slopes of this ridge are covered with extensions of the alveolar system but not the top of the ridge where only a single membrane limits the cell (Allen and Wolf 1974). Short fibers lying perpendicular to the membranes border the slopes of the ridge and distinguish the cytoproct from all other ridges in the pellicle. Single microtubules also extend from the apex of the ridge into the cytoplasm.

3 Cytoplasmic Organelles

Internally, the cytoplasm (Fig. 5) contains a large number of organelles and vesicles as well as ribosomes, food storage reserves, and sometimes symbionts. The cytoplasm contains the basic machinery for synthesis of proteins and for building and attaching complex carbohydrates to these proteins. Thus, it contains the endoplasmic reticulum, ribosomes, and the Golgi apparatus. The cytoplasm also contains sites for catabolism both of endogenous and exogenous material. Phagosomes (also called food or digestive vacuoles) and lysosomes fall into this category. Autophagosomes and, possibly, crystals are a transient part of this catabolic machinery. Mitochondria provide sites for respiration and ATP production. Peroxisomes carry out roles in oxidizing reactions and in detoxifying hydrogen peroxide and, if they are like those in *Tetrahymena,* gluconeogenesis. The cytoplasm also contains the contractile vacuole system that is involved in osmoregulation. Microtubules and microfilaments appear and disappear as needed by the cell to guide vesicular and organellar movements and to mediate the motive force for these movements and for membrane remodeling and flow.

3.1 Mitochondria

A single *Paramecium* contains several thousand mitochondria which though distributed throughout the cell are more abundant in the cortex. These organelles are composed of two membranes, an outer membrane which under normal conditions has few if any outpocketings and an inner membrane which sends off a number of tubular processes into the mitochondrial matrix (Figs. 6–8). Though not lamellar, these tubules are equivalent to mitochondrial cristae of higher organisms. The number of tubules and length of the tubules seem to vary from cell to cell, from species to species, as well as under different culture conditions. In *P. multimicronucleatum* the tubules are at times so abundant and so long that they assume a pattern of parallel waves extending from one end of the mitochondrion to the opposite end (Fig. 6). In *P. caudatum* several tubules may wind around each other or around a straight central tubule (Figs. 7 and 8) as shown also in the hypotrich ciliate *Tachycoma pellionella*

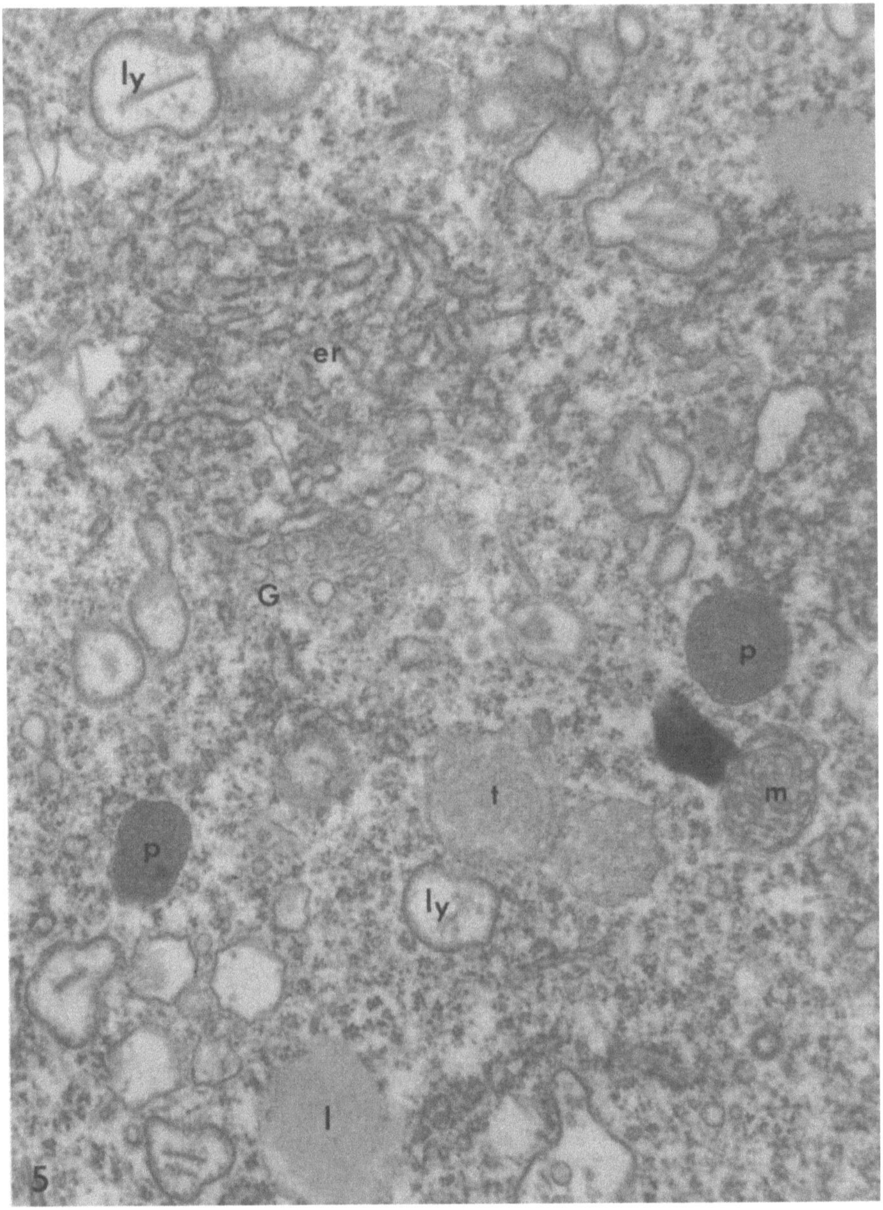

Fig. 5. Cytoplasm of day-2 cell after starving and refeeding. Rough endoplasmic reticulum (*er*) is associated with Golgi stacks (*G*); lysosomes (*ly*), peroxisomes (*p*), mitochondrion (*m*), lipid droplets (*l*), and primordial trichocyst vesicles (*t*) are present. ×38 000

(Wilbert and Heller 1971). Internally, these tubules contain an endogenous peroxidatic activity (Fig. 6) which forms an electron-opaque reaction product when the cells are incubated in diaminobenzidine and H_2O_2. Mitochondria also contain their own ribosomes and DNA.

Mitochondria vary in their appearance depending on their state of activity. Some have an expanded electron-transparent matrix and limited space between the inner and outer membranes. Others have a dense matrix with or without expanded tubules. According to current concepts, mitochondria with expanded, transparent matrices have stored energy while those with a dense matrix and swollen cristae are making ATP or have had their supply of ATP depleted.

Images of what appear to be dividing mitochondria can frequently be seen. Mitochondria increase in length during the first quarter of the cell cycle without other alterations in their fine structure (Perasso and Beisson 1978). Upon doubling in length an equatorial constriction forms which pinches the mitochondrion into two.

Mitochondria in cells taken from stationary phase or clonally aging cultures may contain topographical contortions which make it appear that they contain vesicles when they, in fact, have cup-shaped depressions filled with cytosolic material (e.g., Sundararaman and Cummings 1976a). Mitochondria in varying stages of disruption are a common inclusion of autophagosomes (Fig. 14). Whether these are worn out organelles or are being eliminated from a cell which finds itself in a state of unfavorable nutrition is not known. The latter is possibly more accurate since autophagosomes containing mitochondria are rarely seen in log-phase cells.

3.2 Peroxisomes

Peroxisomes, also known as microbodies, were first identified in *Paramecium* by Stelly et al. (1975) using the diaminobenzidine and hydrogen peroxide incubation which under high alkaline pH produces an electron-opaque reaction product. Like mitochondria, they are scattered throughout the cytoplasm. These organelles are bounded by a single membrane and contain a homogeneous, relatively electron-opaque matrix (Fig. 9). Tubules are found in their granular matrix. No crystalline nucleoids have been reported in peroxisomes of *Paramecium*. In size, peroxisomes are smaller than mitochondria. They are circular to oblong in profile and range from 0.3 to 1 μm in diameter. Though *Paramecium*, when well fed, contains lipid storage granules, no special association of the peroxisomes with these lipid granules has been observed as is frequently observed between glyoxysomes and lipid bodies in cells of germinating seeds. The development of peroxisomes in *Paramecium* has not been studied.

3.3 The Endoplasmic Reticulum and Ribosomes

As in other eukaryotic cells an elaborate system of membranes is present in *Paramecium* which can be distinguished by its association with ribosomes

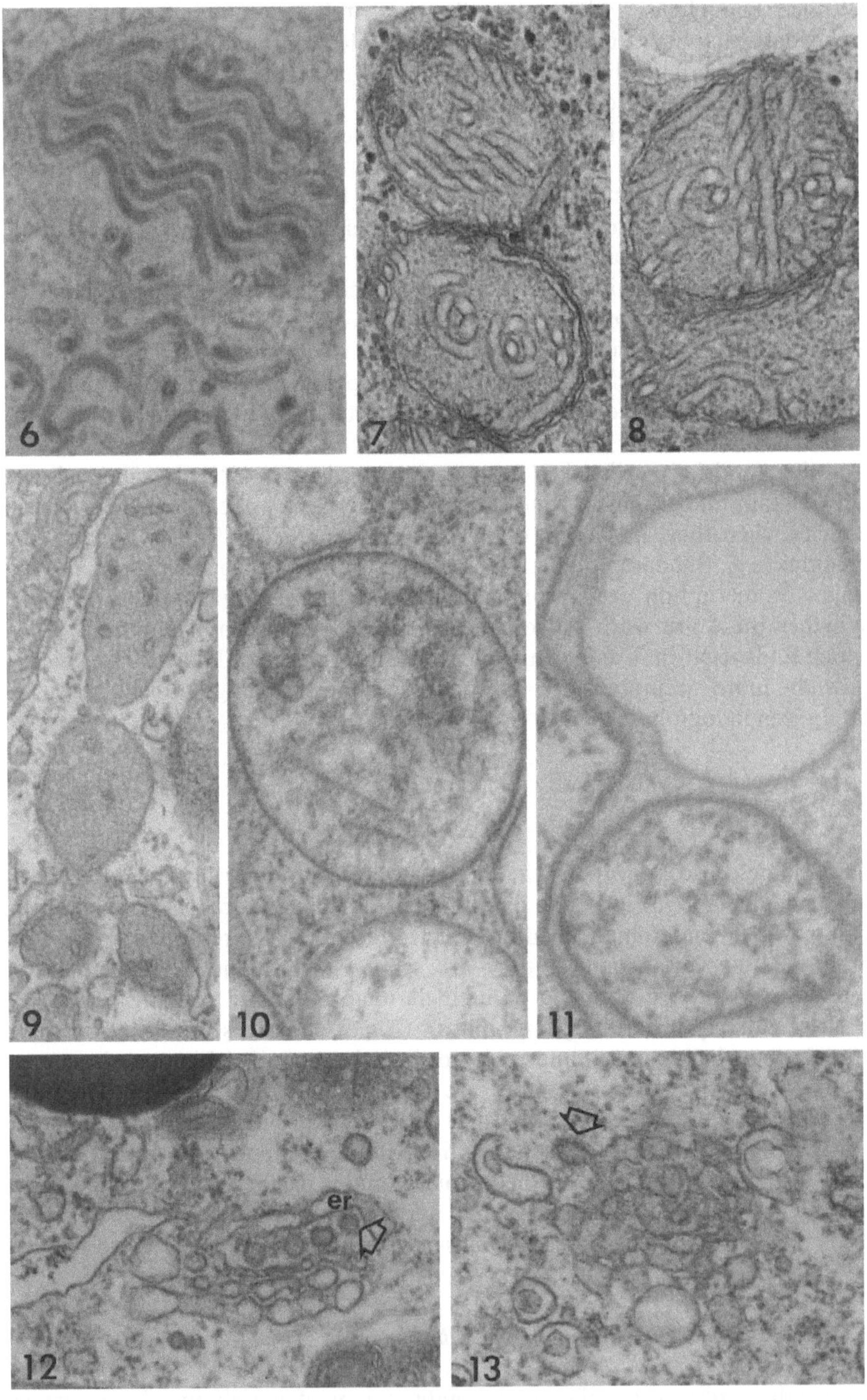
6
7
8
9
10
11
er
12
13

(Fig. 5). This rough endoplasmic reticulum (ER) is especially abundant in the cortical region around the trichocysts. In profile the ER is circular or tubular, large flattened sheets are rare in *Paramecium.* However, large foci of tubular rough ER are seen in 1-day-old cultures following a period of starvation and refeeding. In such cells large electron-opaque patches composed almost solely of rough ER and bounded by Golgi stacks indicate a very active protein synthesis. By day 2 of the exponential phase of growth of *P. multimicronucleatum* the patches are diminished (Fig. 5). Only the usual scattered ER profiles are seen from day 3 to day 5 and throughout stationary phase (Fok and Allen 1981).

Smooth surfaced vesicles can be found in all cells but whether these constitute a smooth ER is questionable. Occasionally, a close association of the rough ER can be seen with the phagolysosome membrane or with other organelles. In these cases the membrane next to the vacuole or organelle is free of ribosomes (smooth), while the opposite membrane of the cisterna is studded with ribosomes (Allen and Fok 1984a). Likewise, the ER membrane bordering the *cis* face of a Golgi stack is smooth (Esteve 1972). Smooth membrane also appears around organelles during autophagosome formation (Fok and Allen 1981), but while this membrane may initially be derived from the ER, the autophagosome membrane may be partially of acidosomal or lysosomal origin. Presumably, smooth ER can be formed but no large masses of smooth membranes accumulate in the cell apart from the contractile vacuoles, Golgi stacks, trichocysts, and products of parasomal sac activity (none of which appear to be an intimate part of the ER system) which would be active in sterol synthesis. The absence of smooth ER for producing sterols as in mammals is not surprising since ciliates such as *Tetrahymena,* and probably *Paramecium* as well, do not make cholesterol or other sterols and must be supplied a sterol in their diet. As in other cells the rough ER probably synthesizes itself by producing membrane proteins on its ribosomes and containing within its structure the necessary enzymes to produce its phospholipids and for core glycosylation of glycoproteins. The associations between the rough ER and other organelles and the myriad vesicles occurring in *Paramecium* have not been worked out, although by analo-

◄

Figs. 6–8. Various mitochondrial tubular configurations. All ×47500

Fig. 6. The mitochondrial tubules contain endogenous peroxidase and sometimes assume a wavy pattern in *P. multimicronucleatum*

Figs. 7, 8. Mitochondria of *P. caudatum* in which the tubules are wrapped into a helix (Fig. 7) or form helices around a straight tubule (Fig. 8)

Fig. 9. Peroxisomes in a late stationary phase cell. ×47500

Fig. 10. Lysosome in an early stationary phase cell. ×47500

Fig. 11. Two acidosomes next to a nascent vacuole membrane demonstrating the different morphologies these organelles display. ×47500

Fig. 12. Cross-section of Golgi stack in a cell from a starved culture. Transition vesicles (*arrow*) next to the ER (*er*) and three fenestrated cisternae make up the stack. ×47500

Fig. 13. Planar view through the Golgi stack. A coated bleb (*arrow*) is continuous with the highly fenestrated cisternae. ×47500

gy with other organisms, it can be assumed that most membranes of the cell have their origin in the rough ER. One exception to this is the inner membrane of mitochondria.

3.4 Golgi Apparatus

The Golgi apparatus was not easily recognized when the first ultrastructural studies of ciliates such as *Paramecium* were made. However, when *Paramecium* is taken from stationary-phase or death-phase cultures the cytoplasm becomes relatively electron transparent and the Golgi units are easily identified.

The Golgi apparatus of *Paramecium* is decentralized and occurs, probably where needed, as hundreds of small stacks of three or so highly fenestrated cisternae lying adjacent to a region of ribosome-free ER (Figs. 12 and 13) (Pitelka 1965; Esteve 1972). Though scattered throughout the cytoplasm the stacks are particularly abundant next to the ER that is adjacent to the trichocysts. In day 1 and 2 cells of a culture age study, Golgi stacks were found around the periphery of extensive fields of rough ER (Fig. 5) (Fok and Allen 1981). The *cis* face of the Golgi stack consists of small transition vesicles that bud from a region of smooth ER and because of their "coats" tend to be slightly more electron-opaque than the cisternae. The three cisternae progressively become more swollen moving away from the *cis* face. The edges of the cisternae balloon out into small bulbous extensions that at times are coated with bristles (Fig. 13). Small vesicles similar to these bulbs are prevalent in the cytoplasm near the Golgi stacks. Most of these vesicles have smooth surfaces when free in the cytosol. Some may be shuttle vesicles on their way from one cisterna to another, while others may be primary lysosomes or vesicles containing secretory material. Frequently these vesicles have indentations causing them to be cup-shaped or donut-shaped when seen in profile (Fig. 15). Such vesicles may be on their way to becoming flattened vesicles. With the new techniques now being used to separate the various cisternae in Golgi stacks and different populations of vesicles from the mixture in the cytoplasm it may soon become possible to track the course of these vesicles.

No special association of the ER with the *trans* face of the Golgi stacks has been observed in *Paramecium* similar to the trans-Golgi network or GERL of higher organisms. Also, unlike some other organisms, acid phosphatase has been identified in all parts of the Golgi stacks from the *cis* face transition vesicles (Fok et al. 1984) to the *trans* face cisterna (Esteve 1970). Small vesicles budding from the Golgi stacks and lying in the vicinity of Golgi stacks contain acid phosphatase (Fok et al. 1984). Cytochemical studies identifying other enzymes have yet to be carried out on the Golgi apparatus of *Paramecium.*

3.5 Lysosomes

In *Paramecium,* lysosomes, broadly defined as vesicles containing acid phosphatase, come in a wide variety of sizes and shapes. Such acid phosphatase-

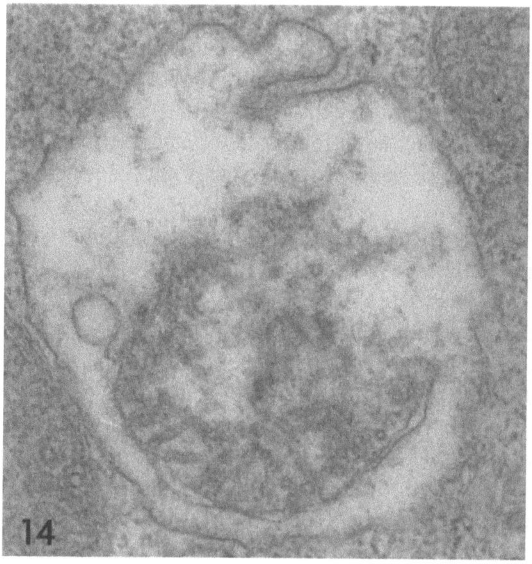

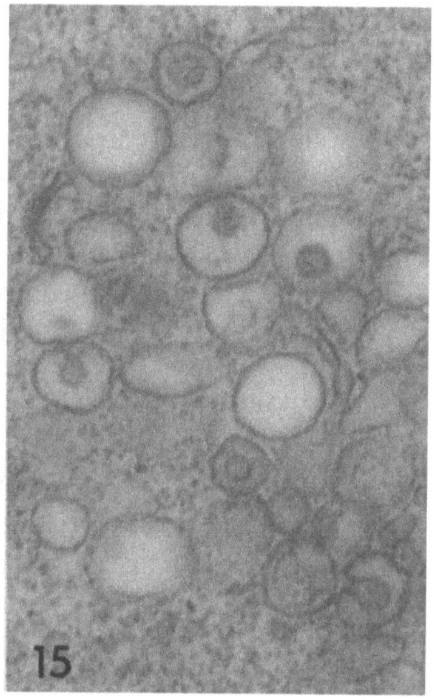

Fig. 14. Autophagosome containing partially digested mitochondrion. ×47 500

Fig. 15. Field of small vesicles like those that accumulate near Golgi stacks. Such vesicles have a thick membrane, a luminal glycocalyx, and membrane invaginations and may be lysosomal precursors. ×47 500

positive bodies include the very small vesicles scattered in the cytoplasm in the vicinity of the Golgi stacks (Fok et al. 1984). These fit the definition of primary lysosomes which contain hydrolases but no substrate on which to act. In addition to these small vesicles, medium-sized vesicles, 0.4 µm long by 0.2 µm wide, which bind to phagosomes that are condensed and acidified (see Sect. 4.1) contain clumps of acid phosphatase reaction product (Fok et al. 1984). These are secondary lysosomes (Figs. 5 and 10). These lysosomes fuse with phagosomes to form phagolysosomes which are large acid phosphatase-positive vacuoles (Allen and Staehelin 1981; Allen and Fok 1984a). Similar, but smaller versions of the secondary lysosomes seem to be retrieved from older vacuoles (Allen and Fok 1984c).

Lysosomes of all types are characterized by a 9-nm membrane which carries a thick fibrous glycocalyx on its luminal surface. An electron-transparent zone can often be found between the glycocalyx and the lysosomal matrix material (Fok and Allen 1979).

A unique feature of the secondary lysosomes of *Paramecium*, so far not reported in other ciliates, is the inclusion of paracrystalline sheets composed of 9 × 32 nm subunits organized into tilted and overlapping rows, the rows have a width of 12.5 nm (Fok and Allen 1979). The molecular nature of this paracrystalline material is unknown. However, to distinguish an early phagolysosome one has only to look at the vacuole's internal content for the presence of this paracrystalline material which signifies that secondary lysosomes have fused with the phagosome. This paracrystalline material disappears from the older defecation-competent vacuoles. Whether primary lysosomes contain molecules that aggregate into paracrystalline material is not known.

The size of lysosomes varies with culture age. Very large lysosomes appear in stationary phase cells. These lysosomes may have several paracrystalline sheets as well as a fairly dense but homogeneous matrix. Upon refeeding these bodies may vesiculate into smaller secondary lysosomes since they disappear in refed cells.

The matrix of the lysosomes varies as it does in the lysosomes of other cells. Lysosomes of some cells contain a compact matrix of fibrous material (Esteve 1970) which, when released into a phagosome, retains its identity for a period of time. Lysosomes frequently enclose small vesicles which are also released into the phagosome upon fusion. As the small compact lysosomes enlarge, their content becomes more dilute and more electron-transparent and paracrystalline sheets become more prominent. On the other hand, different electron-microscope techniques alter the appearance of lysosomes. The low concentrations of glutaraldehyde currently used as a fixative for electron microscopy do not usually preserve the compact nature of the matrix material.

3.6 Acidosomes

A new population of vesicles has recently been discovered in *Paramecium* (Allen and Fok 1983a–c). These are large vesicles, about 0.75 μm in diameter, which have a thick, limiting membrane and an uneven profile (Fig. 11). They bear a very thin luminal glycocalyx which sometimes appears beaded. These vesicles are either entirely electron-transparent or they contain some "wispy" material. The origin of these vesicles has not been determined, however, some can become heavily labeled with HRP within 3 s (Allen and Fok, unpublished observations) of exposing the cell to HRP. This means they were in direct contact with the external medium themselves or had fused with HRP-labeled vesicles derived from the cell membrane in the 3-s interval.

Acidosomes are most abundant in the cell's posterior half, particularly near the cytopharynx. They bind to the forming vacuole's cytosolic surface where they maintain a distance of 20 to 40 nm from the vacuolar membrane. Acidosomes move to the posterior end of the cell bound to the young vacuole. At this point they all fuse with the vacuole simultaneously where they remain attached by short necks for some time (~ 1 min).

The function of the acidosomes is to supply protons as well as the mechanism for further acidification to the young vacuole. New vacuoles are incapable of becoming acid if their fusion with acidosomes is blocked with cytochalasin B (Allen and Fok 1983c), a drug which prevents polymerization of actin. The acidosomes are themselves acid as judged from their ability to accumulate the weak bases acridine orange and neutral red (Allen and Fok 1983c).

Acidosome membranes appear more fragile than many other membranes, frequently breaks or a complete disintegration of those parts of the acidosome membrane which are not bound to the vacuole membrane are seen following conventional fixation. In freeze fracture the EF face of the acidosome membrane is almost free of IMPs, while the PF face is studded with many IMPs. Microfilaments are often found adjacent to the acidosome membrane both when the acidosomes are free in the cytoplasm and, even more so, after they have fused with the young vacuole.

3.7 Trichocysts

The principal secretory organelles of *Paramecium* are the trichocysts. These organelles have received a revival of interest as models of exocytosis (see Chapter 20 by Adoutte, this Vol.). Trichocysts occupy a relatively large volume of the cortex and must require a considerable expenditure of energy for their development, deployment, and extrusion, yet their function remains one of the most long-standing mysteries of this organism (Wichterman 1986).

Trichocysts, when mature, are docked tip first within the pellicular ridges at the plasma membrane (Fig. 1). Trichocysts alternate with single or paired basal bodies along each of the kineties. Since there are 90 to 100 kineties in *P. multimicronucleatum* and in this cell kineties are frequently over 200 µm long and trichocyst docking sites are spaced at intervals of about 1 µm, this large *Paramecium* could potentially contain well over 6000 to 8000 trichocysts (Hausmann and Allen 1976). These 8000 trichocysts would occupy a volume of roughly 25×10^3 μm^3 or about 10% of the total cell volume.

Trichocysts are composed of a body and tip. The body is carrot-shaped, 3–4 µm long and 0.5–1 µm wide (Hausmann et al. 1972); the tip is a pointed shaft surrounded by a sheath. Its complex fine structure was detailed by Bannister in 1972. A trichocyst membrane encloses both the trichocyst body and tip and comes into close apposition with the margins of the alveolus and the plasma membrane. A rosette of 8 to 11 IMPs, each 15 nm in diameter, is present in the PF face of the plasma membrane at all occupied trichocyst docking sites. Double rings of IMPs form a 300 nm diameter circle around this rosette where the plasma membrane covers the margins of the alveolus (Fig. 2). A band of IMPs is also found in the PF face of the trichocyst membrane where this membrane contacts the adjacent alveolar membranes (Allen and Hausmann 1976; Beisson et al. 1976). The epiplasm ends just below this zone permitting a membrane-to-membrane approach. The distal half of the membrane around the trichocyst tip is thickened with additional material, the collar, applied to its cytosolic face. Triggered by a wide variety of stimuli (chemical, mechanical, or electrical), the trichocyst membrane will fuse with the plasma membrane and the trichocyst body will undergo a profound transformation from a compact body to a long needle-shaped shaft eight times the length of the resting body. The unchanged tip remains at the distal end of this extended shaft.

As soon as a trichocyst has cleared the cell's body, the plasma membrane and trichocyst membrane reseal so that the trichocyst membrane is released back into the cell (Hausmann and Allen 1976; Plattner et al. 1985). Markers such as horseradish peroxidase, if present in the surrounding medium, will be trapped within the retrieved trichocyst membrane sac. These membranes eventually vesiculate, often remaining near the Golgi stacks (Allen and Fok

1984 b) with which they were associated before trichocyst extrusion. Most of the horseradish peroxidase and presumably at least part of the trichocyst membrane find their way to lysosomes and phagolysosomes where the membrane may become part of the lysosomal membrane system. Direct recycling of trichocyst membrane back to developing trichocysts has not been observed (Allen and Fok 1984 b).

During trichocyst development fields of vesicles of various diameters containing an amorphous material can be found in the cell's cytoplasm. The smallest of the vesicles (Fig. 5) are near Golgi stacks. Fusion of these smaller vesicles produces the large spherical 2 to 3 µm diameter vesicles in which an electron-opaque paracrystalline core begins to grow (for references, see Hausmann 1978). The core grows into an elongated shaft in the middle of the primordial mass. Eventually, a more opaque tip forms on one end of this shaft and the primordial material disappears as it becomes incorporated into the core. The surrounding membrane is now tightly pressed to the core. Such young trichocysts exhibit a distinct crystalline pattern with 7-nm periodicity (Selman and Jurand 1970). They maintain this appearance until some time after they are docked. They then lose their paracrystalline pattern and become electron-transparent. Plattner et al. (1982) believe that trichocysts are directed to unoccupied docking sites along individual microtubules which extend from basal bodies into the cytoplasm.

4 Cytoplasmic Organelle Systems

4.1 Digestive System

4.1.1 Buccal Cavity, Cytopharynx, Phagosomes, Phagolysosomes, and Cytoproct

The buccal overture (Fig. 17) is the opening leading from the vestibulum into the buccal cavity. The buccal cavity (Fig. 16) is set at right angles to the vestibulum and is widest anteriorly, narrowing posteriorly (Figs. 17–19). In the buccal cavity the pellicle is complete but the ciliary pattern no longer resembles that of the somatic cortical region. Rather, the cilia and basal bodies are arranged into more tightly packed fields or membranelles. There are three such membranelles. The first membranelle, consisting of four rows of basal bodies and cilia, called the ventral peniculus, curves gently from the anterior left margin of the buccal overture dorsally and along the left side of the buccal cavity and then ventrally to end near the posterior margin of the vestibulum (Fig. 16 a, b). The dorsal peniculus is also composed of four rows of packed cilia and closely borders the dorsal, right side of the ventral peniculus. In *P. multimicronucleatum,* the two ventral rows of cilia of the dorsal peniculus end just beyond the posterior margin of the vestibulum, but the two dorsal rows continue on to near the posterior end of the buccal cavity (Allen 1974). A third set of four rows of cilia, the quadrulus, passes over the anterior and dorsal surfaces of the buccal cavity bending to the left across the dorsal surface as it

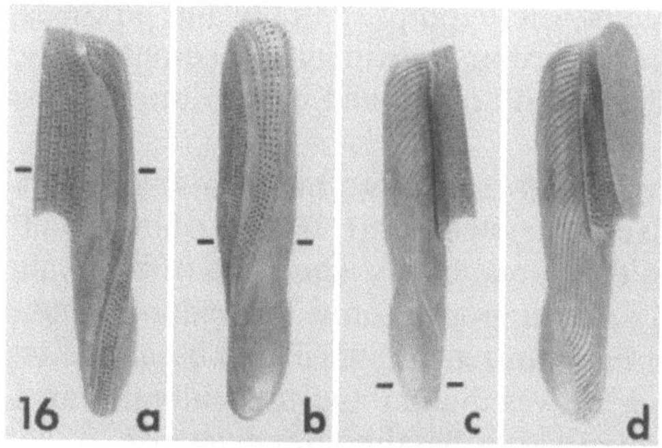

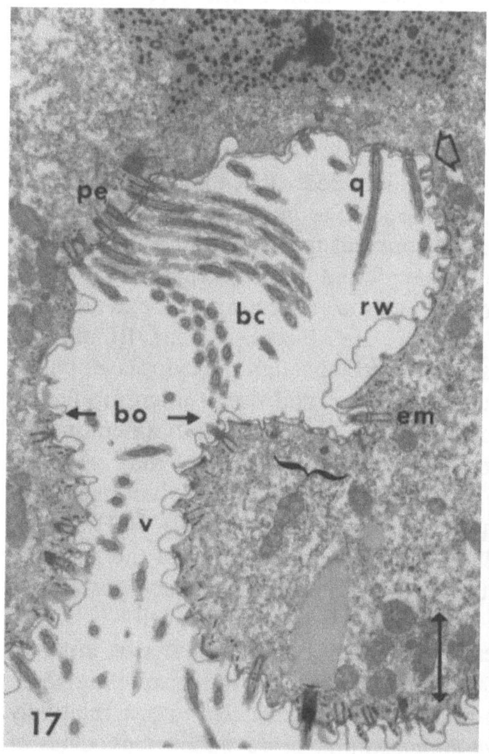

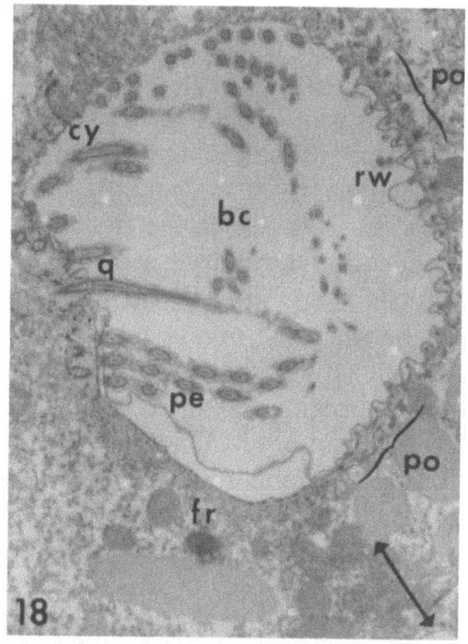

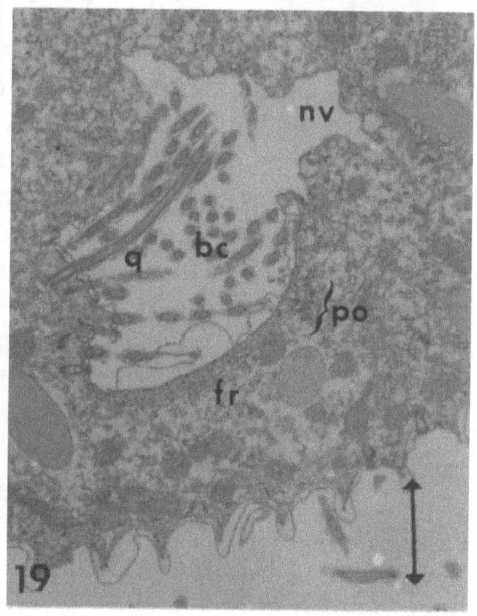

Fig. 16 a – d. A model of the oral region showing its left side (**a**), dorsal surface (**b**), right side (**c**), and tilted to show the right ventral aspect (**d**) (Allen 1974)

Figs. 17 – 19. Cross-sections of the buccal cavity of the same cell at levels indicated at lines in Fig. 16 a – c respectively. The vestibulum (*v*), buccal cavity (*bc*), buccal overture (*bo*), endoral membranelle (*em*), penniculi (*pe*), quadrulus (*q*), ribbed wall (*rw*), cytopharynx (*cy*), filamentous reticulum (*fr*), nascent vacuole membrane (*nv*), and postoral fibers (*po*) are labeled. An open *arrow* in Fig. 17 points to the single membrane-limited extension of the cytopharyngeal membrane. *Bracket* in Fig. 17 indicates site of anarchic field. Each dorsal-ventral axis is indicated by *double-headed arrows*. All ×6800

passes the posterior end of the vestibulum where it dips ventrally and proceeds along the left wall (Fig. 16a, b). These four rows are widely spaced anteriorly but tighten ranks posteriorly where they wind around and end on the narrow posterior wall of the buccal cavity.

The right wall of the buccal cavity is corrugated into many grooves, each subtended by a segment of the alveolar system, this is the ribbed wall (Figs. 16–18). Along the right side of the buccal cavity where the ribbed wall borders the vestibulum a row of cilia set in a groove forms the endoral membranelle (Figs. 16 and 17). A region of unorganized and unciliated basal bodies lies between the organized ciliature of the vestibulum and the endoral membranelle (Figs. 16 and 17). This anarchic field of basal bodies gives rise to the new oral region.

The ventral left floor of the buccal cavity posterior to the vestibulum bears no ridges or cilia and is covered only by a pellicle consisting of a plasma membrane, a usually inflated alveolus and an epiplasm (Figs. 18 and 19). A net of fine filaments with regularly spaced electron-opaque nodes covers the cytosolic surface of this part of the buccal cavity and extends as a thin layer over the cytosolic face of the ribbed wall where it appears to be firmly anchored to the bases of the grooves of the ribs, this may prevent the ribs from flattening. Elements of this filamentous reticulum are also found at the proximal ends of the basal bodies of the membranelles. The filamentous reticulum thus extends over the same area of the buccal cavity as does the alveolar system. Bundles of microtubules, the postoral fibers, pass posteriorly through the cytoplasm paralleling the grooves of the ribbed wall, picking up additional microtubules as they go. Microtubules on the side of a bundle bordering the filamentous reticulum are bound to this reticulum by electron-opaque links and may originate from the reticulum in a manner similar to the origin of the cytopharyngeal ribbons which arise out of this same filamentous reticulum (Allen 1974). However, in *P. trichium (putrinum)* these bundles arise from basal bodies of the anarchic field (Patterson 1981). A large group of these bundles continues posteriorly from the end of the buccal cavity toward the posterior end of the cell.

The elaborate orientation of these ciliary membranelles and ribs must be designed for the efficient movement of food particles into the developing vacuole and for concentrating a large number of these particles into a relatively small volume. Protozoa such as *Acanthamoeba* and professional phagocytes of mammals take up bacteria by a progressive binding of the cell membrane to the bacterium resulting in the engulfment of one bacterium at a time (Steinman et al. 1983). These cells have solved the problem of trapping, concentrating, and packaging food derived from a dilute source differently from the large filter feeding ciliates. Though much time and effort was spent in the earlier part of this century on attempts to understand the forces and currents involved in feeding in *Paramecium* (for references, see Mast 1947), this process still remains incompletely understood.

A smooth, elongated, teardrop-shaped region covering the dorsal surface over the posterior part of the buccal cavity (Fig. 16b) is limited only by a single membrane. The plane of the opening defined by the termination of alveoli and epiplasm and the beginning of the single membrane-limited cytopharynx is termed the cytostome. The pointed end of the "teardrop" is anterior and extends between the quadrulus and the ribbed wall as a narrow single membrane-limited gap (arrow, Fig. 17) extending forward to the buccal overture (Allen 1974). From a plane near the end of the vestibulum the "teardrop" begins to widen and expand over the dorsal posterior surface of the buccal cavity. This is the developing (nascent) food vacuole or phagosome. An elaborate system of microtubular ribbons, the cytopharyngeal ribbons, and bands composed of two microtubules each are arrayed over a 1-μm-wide strip along the left margin of this expanded single membrane-limited region from about the middle of the long axis of the buccal cavity to its posterior end. This forms the left side of the cytopharynx. The right side of the cytopharynx does not bear highly organized microtubular ribbons but has a number of loosely organized pairs of microtubules originating from the ends of the ribs of the ribbed wall that pass posteriorly paralleling the long axis of the right margin of the cytopharynx.

New vacuole membrane is only added to the growing vacuole between the ends of the cytopharyngeal ribbons on the left side of the cytopharynx (Fig. 18). This membrane is obtained from a pool of discoidal vesicles which accumulate between the ends of the cytopharyngeal ribbons and under a microfilamentous bundle called the cytostomal cord (Allen 1974). Here, they fuse with the growing phagosome (Fig. 20).

Phagosomes are released from the cytopharynx in association with acidosomes. They move rapidly to the posterior end of the cell along the postoral fibers. Acidosomes fuse and membrane is retrieved from the young vacuoles as the vacuole condenses and becomes acidic (Allen and Fok 1983a–c). The vacuole then travels forward, moved by cyclosis, as it acquires a

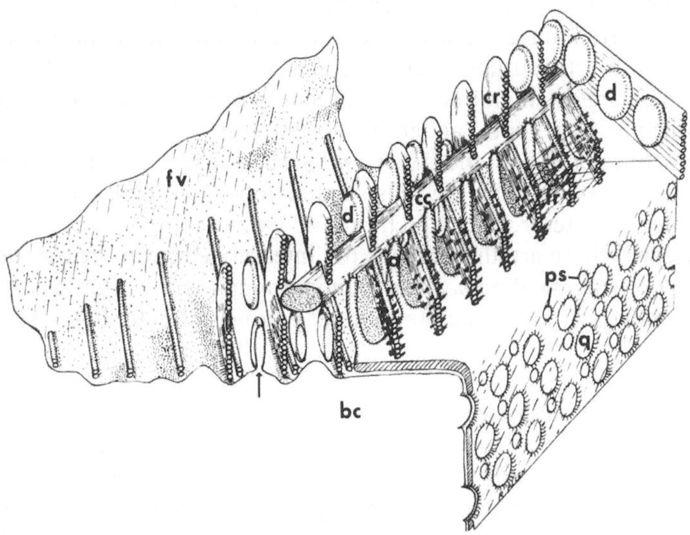

Fig. 20. Drawing of section of left cytopharyngeal region. Discoidal vesicles (*d*) align along the cytopharyngeal ribbons (*cr*) and accumulate under the cytostomal cord (*cc*) next to the membrane of the cytopharynx. Here, they fuse (*arrow*) and enter the growing vacuole (*gv*) membrane. Buccal cavity (*bc*), alveolus (*a*), parasomal sac (*ps*), quadrulus (*q*) (Allen 1974)

layer of lysosomes. The lysosomes fuse with the condensed vacuole and digestion commences as the pH of the vacuole rises (Fok et al. 1982; Allen and Fok 1984a). Both lysosomal membrane and acid phosphatase are retrieved following digestion (Allen and Fok 1984c) as the vacuoles become defecation-competent (Fok et al. 1982). The vacuole is finally guided to the cytoproct along microtubules where fusion of the vacuole membrane with the plasma membrane results in defecation (Allen and Wolf 1974). The phagosomal-phagolysosomal system will be discussed in greater detail in Chapter 19 (this Vol.) by Fok and Allen.

4.1.2 Storage Bodies and Crystals

Paramecium from time to time contains at least three types of storage materials (glycogen, lipid, and crystals) whose presence or absence depend on the nutritional condition of the cell. Glycogen (Fig. 21) is stored as small electron-opaque beta particles scattered throughout the cytoplasm (Esteve 1969). This product is frequently dissolved during the fixation and dehydration steps used for electron microscopy. In well-fed and axenically grown cells lipid (Fig. 23) is accumulated and stored as nonmembrane-bounded droplets of about 1 µm in diameter (Fok and Allen 1979, 1981). When a large number of these collect in one place they will appear as very large fields of lipid when stained for light microscopy. These lipid droplets disappear or become depleted in stationary and death phase cells.

Another regular feature of stationary phase cells are membrane-bound crystals (Figs. 22 and 23) which accumulate in the anterior and particularly the posterior end of cells. These crystals disappear when axenically grown cells are transferred to fresh growth medium and reappear in mid-log-phase cells.

A fourth structure accumulates in stationary and death phase cultures. These are small globules which seem to be surrounded by a membrane and frequently have both an electron-transparent zone and an intensely electron-opaque zone (Fok and Allen 1981). These show some resemblance to lipofuscin granules of mammalian cells. Presumably, they represent an accumulation of an intermediate product which the cell can no longer process completely. It is not clear if these are the same as the intensely electron-opaque bodies observed in clonally aging *Paramecium* (Sundararaman and Cummings 1976a) whose contents appear more like the residual products of autophagocytosis.

4.2 Endocytic System

Although receptor-mediated endocytosis has not been studied in *Paramecium* at the molecular level, morphologically a mechanism does exist which is similar to that present in mammalian cells. The essential components for this type of activity include receptors at the cell surface, coated pits into which the ligand-bound receptors will move, and coated vesicles which transport the receptors to an intermediate vesicle system, the endosomes. The endosomes provide an

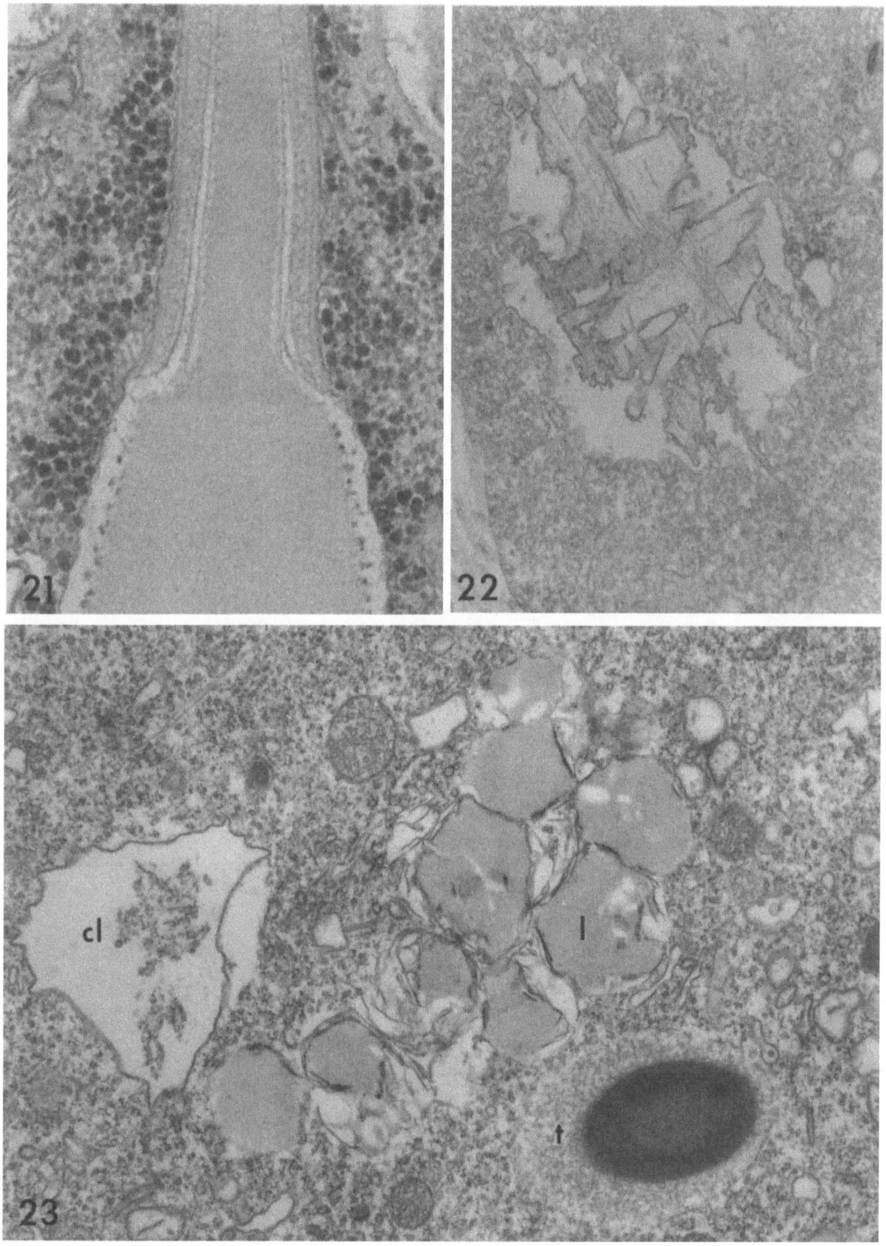

Fig. 21. Glycogen stored around a trichocyst in *P. caudatum*. ×57 000

Fig. 22. A crystal typical of those in the anterior and posterior poles of the cell. ×47 500

Fig. 23. A field of lipids (*l*) next to a crystal (*cl*) in a cell from day 2 after refeeding a starved culture. An immature trichocyst (*t*) is also present. ×19 000

acidic compartment which releases the ligands from their receptors. The unbound receptors are then retrieved from the endosome membrane by the removal of small vesicles which are returned to the cell membrane. Somehow the ligands are passed on to the trans-reticular system of the Golgi or to lysosomes for processing.

In *Paramecium* some part of the 15-nm-thick glycocalyx, consisting of complex carbohydrates attached to glycoproteins on the cell surface, may function as the receptors. The parasomal sacs (Fig. 24) are coated pits, as they are in *Tetrahymena* (Nilsson and Van Deurs 1983), in which the ligand-bound receptors could accumulate. The parasomal sacs pinch off to form coated vesicles (Fig. 24) (Allen and Fok 1980; Westcot et al. 1985). These vesicles quickly lose their coats and fuse with other intermediate vesicles which are either flattened cisternae or, more often, ring-shaped, i.e., profiles of two membrane circles, one within the other (Figs. 25 and 26). These intermediate vesicles acquire ferritin or horseradish peroxidase activity soon after label is internalized at the parasomal sacs (Westcot et al. 1985). Small coated blebs are frequently observed extending from these rings (Fig. 26). These may form vesicles for recycling the putative, unbound receptors back to the plasma membrane. Where would such

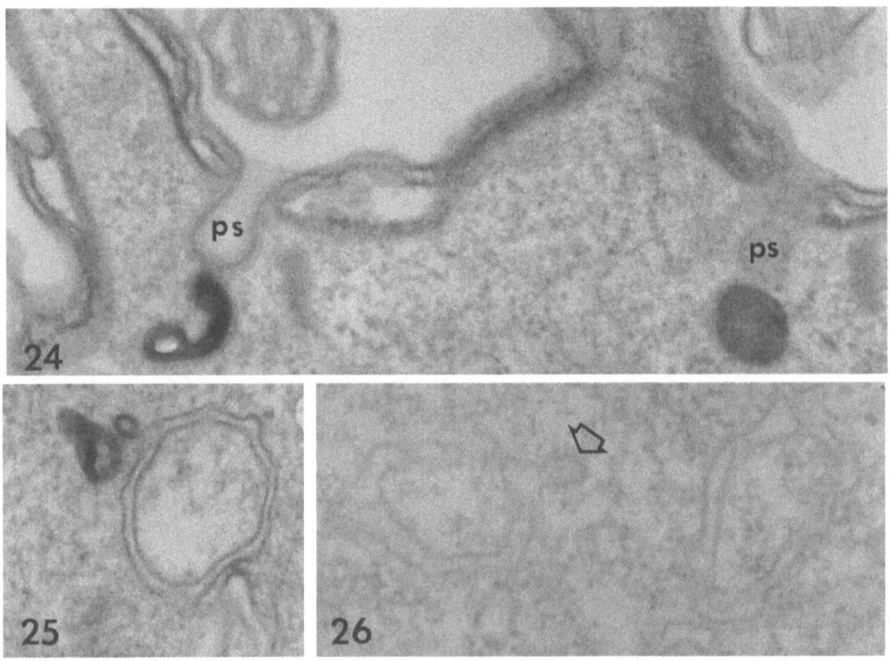

Fig. 24. Coated pits or parasomal sacs (*ps*) are sites of endocytosis at the plasma membrane. The vesicles that form from the pits during an HRP pulse become labeled. Though faint both pits and vesicles can be seen to bear a clathrinlike coat. ×47 500

Figs. 25, 26. Coated vesicles fuse with cisternae which have ringlike profiles. Coated pits can be found on these cisternae (*arrow*). Such cisternae may be receptosomes involved in recycling membrane and uncoupling ligands from receptors. Both ×47 500

recycling vesicles fuse with the plasma membrane? Since the only sites at which the plasma membrane penetrates through the alveolar mosaic are at the parasomal sacs it would appear that vesicles bearing recycled receptors would need to fuse again with the parasomal sacs. Yet the parasomal sacs appear always to be coated. Whether vesicles can fuse with coated membranes is not clear. It is a fact that the membranes of the cell surface double in area in each division cycle by some as yet unobserved route. Thus, recycled membrane may be inserted into the membrane in a manner similar to or simultaneously with newly synthesized membrane. Fundamental problems to be addressed include the question of the existence of receptor-mediated endocytosis in *Paramecium.* Also, what types of molecules would need to enter the cell by such a selective process? Some work along these lines has been done in *Tetrahymena* (Csaba 1985) but not in *Paramecium.*

4.3 Contractile Vacuole System

Like the digestive vacuole system the contractile vacuole system is composed of pools of distinct membranes which are associated sequentially in space. However, unlike the digestive vacuole system whose members are more dissociated and in a constant state of flux, the membranes of the contractile vacuole system remain essentially fixed in position where some undergo configurational changes.

Most species of *Paramecium* have two contractile vacuoles, one in each half of the cell, which open to the surface, each through a cylindrical pore (Fig. 27). The vacuole itself when filled is roughly spherical and is bound to the pore by 6 to 12 ribbons each composed of 20 to 30 microtubules. The microtubules have their origin on the cytosolic side of the pore from which they curve in a left-handed helix around the pore (McKanna 1973) and pass over the distal hemisphere of the vacuole where each ribbon leaves the vacuole proper to pass over an ampulla and finally out along a narrow and elongated collecting canal (Hausmann and Allen 1977).

The collecting canal, ampulla, and the short connecting canal between the ampulla and vacuole form three parts of the same tube, but their diameters swell or contract to differing extents as fluid is stored in them or released into the vacuole. The microtubular ribbon begins to fork as it passes over the outer surface of the ampulla (Hausmann and Allen 1977) and by the time it reaches the collecting canal the ribbon has divided into several groups of two to six microtubules, some of which extend along both the upper and lower surfaces of this canal (Fig. 28). The collecting canals end at the distal extremities of the microtubules. The lengths of the canals seem to be dependent on the lengths of the microtubules.

During diastole, the contractile vacuole is filled with fluid (Fig. 27) and its surface is composed of a single membrane. No noticeable coating such as a glycocalyx or microfilamentous layer has been seen either on its luminal or cytosolic surface. During the filling cycle, the ampullae open into the vacuole through the narrow necks called the connecting canals each at the point where a

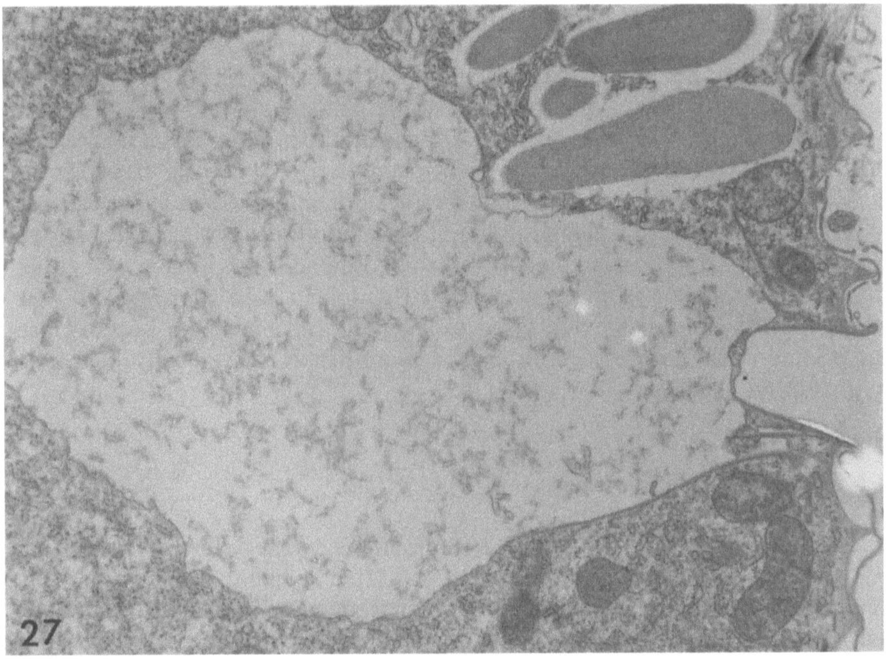

Fig. 27. Contractile vacuole in diastole limited by a single membrane and affixed to the pore by microtubular ribbons. ×9500

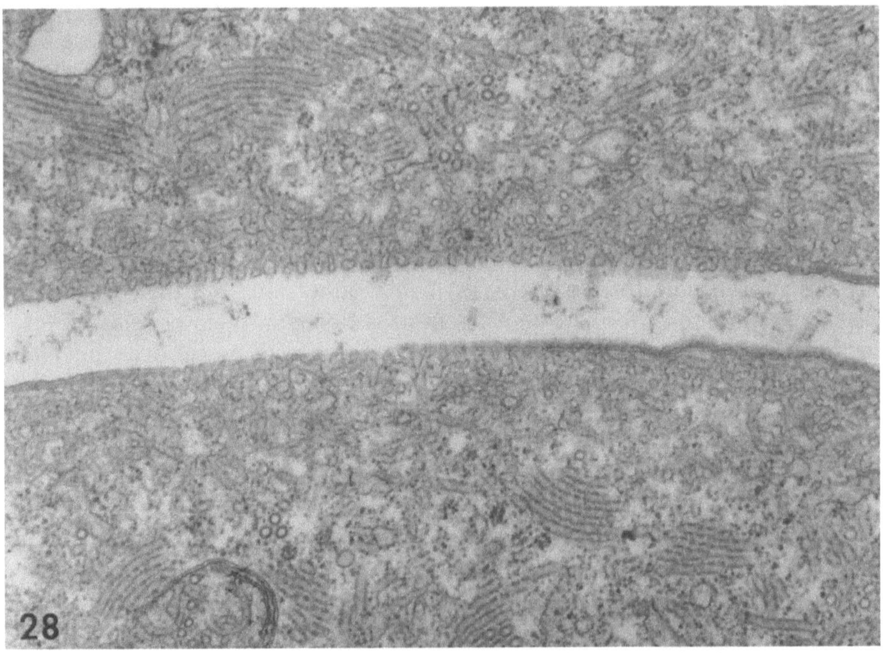

Fig. 28. Collecting canal with rows of connections to the spongiome in regions between forked microtubular ribbons. Decorated tubules form bundles peripheral to the spongiome. ×28 500

ribbon of microtubules separates from the vacuole. The close association of the vacuole membrane with several ribbons of microtubules allows the membrane of the vacuole to extend up along the ribbons beyond the proximal end of the pore; this assures a close approach of the vacuole membrane and the plasma membrane that forms the diaphragm across the inner orifice of the pore. These two membranes are particularly close around the pore's inner edge (Fig. 27).

During vacuole diastole, the ampullae, at first reduced in diameter, begin to enlarge. Their membrane is derived from a reticulum of tubules, the spongiome, which also surrounds the microtubule reenforced tube of the collecting canal. The spongiome is composed of a three-dimensional network of 40–45 nm diameter tubules (McKanna 1976) which have no obvious association with ribosomes or microfilaments (Fig. 28). This network of tubules empties into the collecting canal by way of regularly spaced openings arranged in linear fashion along the collecting canal between the bundles of the forked microtubular ribbons (Hausmann and Allen 1977). On the opposite side, the spongiome is associated with bundles of larger, straight, 50-nm-diameter tubules. These are called decorated tubules because they bear pairs of helically wound pegs on their surfaces which can be seen in thin sections as well as freeze-fracture replicas of the fractured tubules (Hausmann and Allen 1977). Decorated tubules (Fig. 28) empty at one end into ducts which are said to open into the spongiome network (McKanna 1976). At their opposite ends they end blindly. Decorated tubules do not occur around the ampullae or connecting canals.

Diastole begins when the fluid is first delivered from the collecting canal and ampullae to the contractile vacuole which partially fills the vacuole. A slower filling process follows until the vacuole rounds up (Patterson 1977). Thus, the contractile vacuole is at its greatest diameter, while the ampullae and collecting canals are at their smallest diameter.

At the beginning of systole the vacuole membrane fuses with the plasma membrane, most likely at the periphery of the diaphragm of the somewhat widened pore (McKanna 1973; Patterson 1977). The fluid is expelled and the vacuole seems to disappear when viewed in the light microscope (Fig. 29). In electron micrographs it is evident that one wall of the vacuole does not simply collapse against the opposite wall, rather the membrane undergoes a configurational change from a smooth topography to a highly tubularized network which is collapsed against the ribbons of microtubules radiating from the pore (Allen and Fok, unpublished observations). The pore is quickly closed by a resealing of the plasma membrane over the inner surface. The space formerly occupied by the vacuole is now filled with cytoplasm containing many ribosomes and small cytoplasmic elements but few larger cytoplasmic organelles (Fig. 29).

As the ampullae (Fig. 29) swell during vacuole systole they use up the membrane of the spongiomelike network that occupies the space next to the microtubular ribbons in this region. The collecting canals also enlarge, but not to the same extent that all of the spongiome membrane is utilized.

There seems to be no specialized structure such as a sphincterlike band of microfilaments at the junction between the contractile vacuole and ampullae to account for the inability of the fluid to pass from the ampullae into the con-

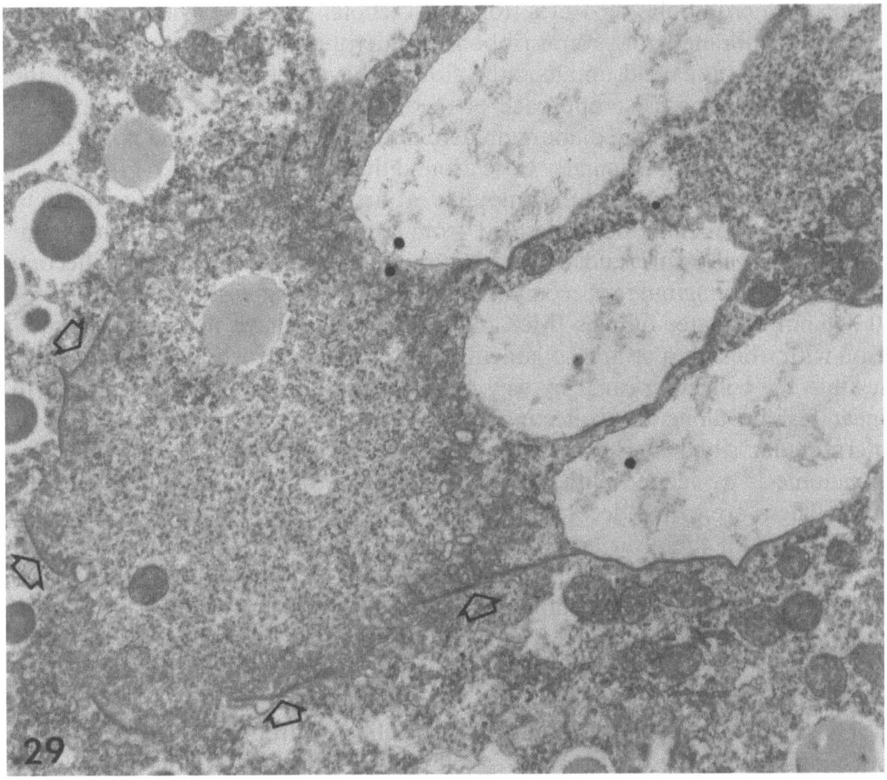

Fig. 29. Contractile vacuole in systole surrounded by and linked to 11 engorged ampullae (only parts of four ampullae are seen here). The contractile vacuole membrane forms a network of tubules that remains attached to the microtubular ribbons (*arrows*). ×11400

tractile vacuole tubules while the ampullae fill. A discontinuity between the tubularized membrane of the vacuole and the membrane of the ampullae must occur when the contractile vacuole membrane is transformed into a tubular network. The forces involved in the movement of fluid within the contractile vacuole system seem to be located within the molecular architecture of the membrane system itself since no external microfilamentous system has ever been observed either in thin sections or by immunofluorescence microscopy (Kersken et al. 1986).

The "backbone" of the contractile vacuole system is the array of 6 to 12 ribbons of microtubules which provide the rigid framework for the attachment of the tubular membranous networks. One *Paramecium* species, *P. trichium,* does not have the elongated collecting canal typical of the other *Paramecium* species (Patterson 1980). Presumably, this cell does not have a well-developed microtubular cytoskeleton. Microtubules may also be involved in the indentation of the pellicle prior to systole (Patterson 1977) in a similar way to that proposed for the spreading of the cytoproct lips prior to defecation (Allen and Wolf 1974).

Many questions remain about the actual sequestration and movement of fluid and the actual composition of the fluid in the contractile vacuole complex. It is assumed that the fluid initially passes through the membranes of the decorated tubules (McKanna 1976). The fluid then passes into the spongiome network and on into the collecting canals, the ampullae, and finally into the contractile vacuole from which it is excreted through the pores. The time seems ripe for a more complete study of this process, as we have recently raised a monoclonal antibody to the decorated tubules which should permit the isolation and study of these intriguing membranes.

4.4 Basal Bodies and Cilia

Probably the feature of living *Paramecium* which is the first to attract attention is the movement of the beating cilia covering the entire cell (see Tamm 1972). These cilia are arranged in rows on the cell's cortex. The rows of cilia over the dorsal surface of the cell follow the long axis of the cell, while those on the sides and ventral surface more or less curve around the oral region and end along the anterior and posterior sutures.

Internally each cilium is an outgrowth of a basal body (Pitelka 1965). Basal bodies (Figs. 1 and 30), which are homologous to centrioles of higher animal cells, are composed of a series of nine triplet fibers arranged in a cylinder (Fig. 4). Each fiber is composed of one complete microtubule, the A subfiber, and two overlapping microtubules (B and C subfibers) which share 3 of their 13 protofilaments with each microtubule they join. The proximal end of the basal body, which is the primordial end, contains a cartwheel, i.e., a central cylinder with nine radiating filaments when observed in cross-sections. Accessory structures attach to the outer surface of the triplets at this end of the basal body. A long, striated fiber, the kinetodesmal fiber (Fig. 4), passes anteriorly from the basal body's right anterior quadrant and enters a bundle of overlapping kinetodesmal fibers derived from other basal bodies of the same kinety. This bundle lies along the right side of the kinety. A small ribbon of postciliary microtubules arises from the right posterior quadrant and passes up toward the epiplasm where it ends. A transverse ribbon of microtubules arises from the anterior left side of the basal body and crosses to the left where it ends under the epiplasm. These structures anchor the cilium during ciliary beating. If two basal bodies are present, the posterior basal body will have all the attached accessory structures and the anterior basal body will have its own developing transverse ribbon of microtubules.

The internal lumen of the basal body contains electron-opaque, amorphous clumps which disappear upon either pronase or RNase treatment (Dippell 1976). At its distal end the basal body is connected to the pellicle. A diaphragm of material, continuous with the epiplasm, forms a terminal plate (Figs. 1 and 30) across the lumen of the basal body at the level where the triplet fibers become the doublet fibers of the cilium. Dute and Kung (1978) have described the basal body-cilium transition region in detail. Distal to the terminal plate is an axosomal plate and axosome into which only one of the two central tubules of the ciliary axoneme is embedded.

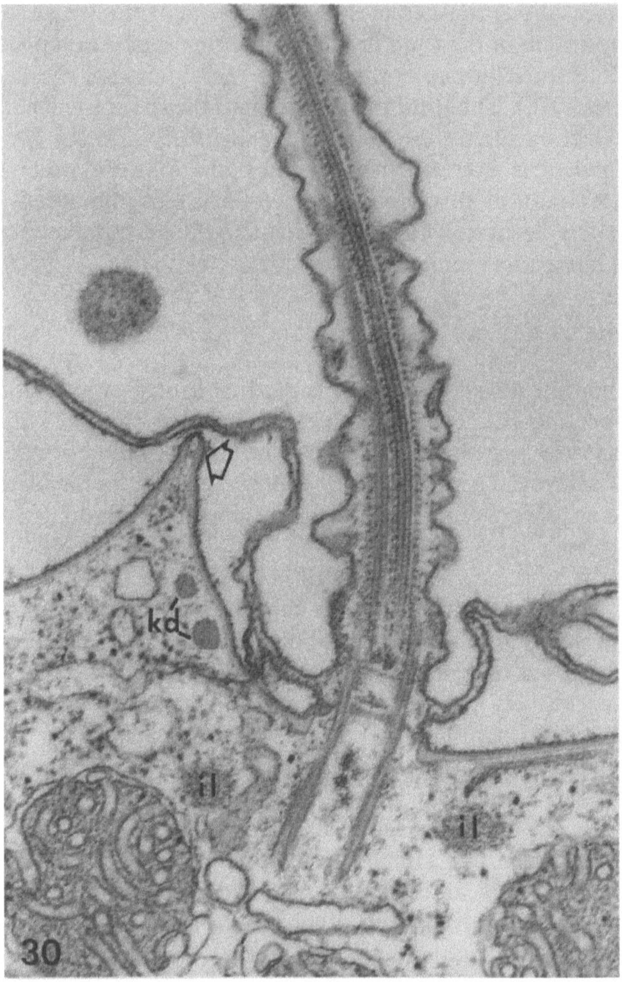

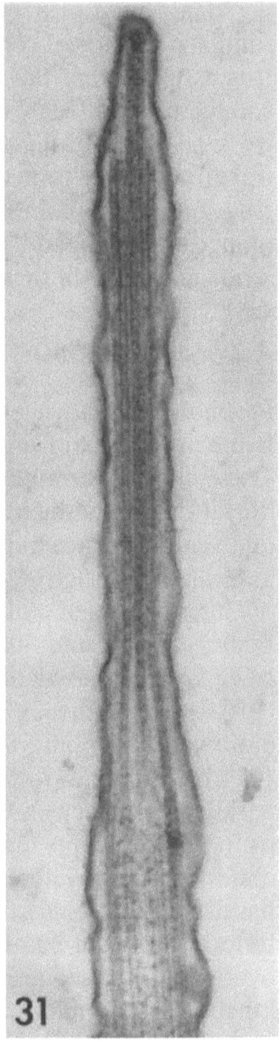

Fig. 30. Cilium and basal body. Rows of arms along subfiber A of the peripheral doublets are evident in this grazing longitudinal section. The epiplasm is absent from the crest of the pellicular ridge (*arrow*). Kinetodesmal fibers (*kd*), infraciliary lattice (*il*). ×57 000

Fig. 31. Tip of cilium illustrating how the fibers of the axoneme terminate. ×57 000

Freeze-fracture replicas of this transition zone reveal nine plaques of IMPs on the PF face of the ciliary membrane in a region just distal to the axosomal plate (Plattner 1975), while two rings of IMPs form a ciliary "necklace" in the membrane at the level of the axosomal plate. The location of the plaques, each plaque is composed of three vertical rows of five IMPs in each row, corresponds to the location of anionic sites on the outer surface of the ciliary membrane (Dute and Kung 1978).

The cilium is composed of nine doublet microtubules arranged in a cylinder with two singlet microtubules in the center of the cylinder. An array of spokes and arms (Figs. 1 and 30) tie these microtubules together. The ciliary axoneme is enclosed by a ciliary membrane which is an extension of the plasma membrane.

At the tip of the cilium, the B subfiber of each of the nine doublets ends before the A subfiber (Fig. 31). The two central singlets end last and are surmounted at their tips by a cap (described in *Tetrahymena* by Sale and Satir 1977, Dentler 1980). Thus, cross-sections of the distal ends of cilia differ. Some contain a mixture of double and single peripheral microtubules enclosing the two central singlets. Others are of nine or less peripheral singlets bordering the two central singlets, while the most distal cross-sections will have a membrane around only two singlet microtubules (see Satir 1968).

As the peripheral doublets slide along each other (Mogami and Takahashi 1983) the cilium bends. The motile force is associated with the ATPase activity in the two rows of arms of the A subfiber (Fig. 30). Chemical bonds progressively form and break between the arms and a series of linearly arranged binding sites on the B subfiber of the adjacent doublet. A sliding movement is generated.

Cilia are capable of moving in a three-dimensional pattern and also of reversing the power strokes. Further details on some of these types of activities and their control mechanisms can be found in Chapters 13 and 14 (this Vol.) by Machemer.

Basal bodies develop next to the anterior proximal surface of preexisting basal bodies (Dippell 1968). The proximal end of the basal body is formed first and its axis is oriented perpendicular to the mature basal body. This is then followed by an elongation of the peripheral triplets and tilting of the body toward the cell surface. In *Paramecium* several nascent basal bodies can be found in a row anterior to a mature basal body in the cell's equatorial region. The basal bodies will move apart along the kinety and accessory structures will form and insert into the preexisting cortical patterns as they take up their final positions. If this pattern is experimentally reversed, the nascent basal bodies within the reversed cortical regions will develop accessory structures and be deployed according to the reversed pattern (Beisson and Sonneborn 1965).

4.5 Cytomatrix

Microtubules occur in *Paramecium* both as permanent and temporary structures. As permanent structures they appear as the triplet fibers of basal bodies, the 9 + 2 axonemes of cilia, and the ribbons or bundles of microtubules linked together by bridges and linkers now called microtubule accessory proteins. Microtubular ribbons and bundles include the postciliary and transverse ribbons arising from the proximal ends of basal bodies, the cytopharyngeal ribbons, and postoral bundles of microtubules of the oral region, bundles of microtubules arising from the basal bodies bordering the cytoproct and the ribbons of microtubules making up the backbone of the contractile vacuole complex. Most of these ribbons at some point bind to membranes. Individual microtubules are found radiating from the cortex during trichocyst docking, from the tip of the cytoproct ridge, and scattered in the cytoplasm where they sometimes make contact with organelles such as mitochondria, lysosomes, and peroxisomes. Single microtubules may be more labile and be polymerized only at the time they are needed. The mitotic and meiotic microtubule spindles in the dividing micronucleus and the microtubules in the dividing macronucleus are temporary as are the cytospindle bands of microtubules in the pellicular crests of recently divided cells (Ehret et al. 1964; Cohen et al. 1982). The consensus of most students of microtubules is that they are involved in establishing assymmetric cell shape and in producing frameworks along which various movements occur rather than in motive force production. All microtubules in *Paramecium* seem to fit one or the other of these functions.

Microfilaments are present in *Paramecium* either as permanently organized bands, notably the striated bands and infraciliary lattice in the cortex and the filamentous reticulum and cytostomal cord in the oral region, or as temporary meshworks. Some of the permanent systems contain actin, while others do not (Kersken et al. 1986). The microfilaments that surround early phagosomes and the spent phagolysosomal membrane at open cytoprocts are of the temporary meshwork variety and contain actin (Cohen et al. 1984a, b; Kersken et al. 1986). Actin seems to be involved in phagosome release from the cytopharynx, the fusion of acidosomes with phagosomes and membrane replacement of early phagosomes and the retrieval of membrane from the open cytoproct (Cohen et al. 1984a, b; Allen and Fok 1983c, 1985). Actin is also involved in movement of the phagosome along postoral microtubules (Cohen et al. 1984b) and in cyclosis (Wasik and Sikora 1980; Plattner et al. 1985), although a role of actin in cyclosis is still controversial (Cohen et al. 1984b).

The presence of a microtrabecular system, i.e., a proteinaceous cytoplasmic matrix in the cytosol as reported for mammalian cells (Wolosewick and Porter 1979) has not been reported in *Paramecium*. This remains to be investigated. Chapter 21 (this Vol.) by Cohen and Beisson covers the cytoskeleton in greater detail.

5 Nucleus

5.1 Macronucleus

Paramecium, like other ciliates, has two types of nuclei. A larger macronucleus is present which contains many copies of the genome and also many nucleoli. This nucleus produces both ribosomal RNA and messenger RNA for protein synthesis. The macronucleus occupies a stationary position in the middle of the cell slightly more anterior than posterior. It usually comes into fairly close association with the anterior dorsal surface of the buccal cavity. Although ribbons of microtubules have been reported to link these two bodies, no compelling evidence that microtubules serve a macronuclear anchoring function has been given. Why the macronucleus is immobile remains an open question.

The macronuclear envelope, composed of two membranes, contains closely packed pores which, like other nuclear pores, are not simple openings but are bridged by filamentous material radiating into the pores from a ring of eight subunits around each pore margin. Many pores have another subunit in their center which may be ribonucleoprotein (RNP) material on its way out of the nucleus. The pore is the site of RNP translocation into the cytoplasm (e.g., Clawson et al. 1985).

Internally the macronucleus is filled with thousands of small 0.1 to 0.2 μm bodies (Fig. 32) which contain DNA. Larger 0.5 to 1 μm electron-opaque bodies scattered between the DNA bodies, the nucleoli, are composed mainly of RNP surrounding a clump of DNA. In stationary phase and death phase cells these nucleoli appear to be hollow or disassemble and sheets of particles peel off the surface (Fok and Allen 1981). These particles are probably ribosomal RNP in an advanced stage of processing.

Paramecium cannot survive long without a macronucleus. During division the macronucleus undergoes a sequence of positional and shape changes. It first moves to the dorsal surface and elongates into an oval shape as a marginal band of nucleoplasmic microtubules is formed. It then changes into an elongated sausage shape as the marginal band disassembles and intranuclear microtubules assemble (Tucker et al. 1980). Nucleoli sometimes attach to the microtubular framework. Though discrete chromosomes are not observed, a fairly equitable segregation of chromatin material to the daughter macronuclei results (Tucker et al. 1980). The macronuclear envelope remains intact throughout division pinching into two nuclei just before or at the time of cytokinesis.

In aging cell lines of *P. aurelia* where autogamy is prevented, the nucleoli frequently appear to aggregate into large ring-shaped or coiled loops (Sundararaman and Cummings 1976b). The chromatin bodies also appear to be smaller and there appears to be fewer per unit area. Threads of granules similar to those seen in stationary phase cells peel from the nucleoli and accumulate in the macronucleus. These may be ribosomal RNP which cannot be transported to the cytoplasm.

5.2 Micronucleus

The number of micronuclei per cell varies from species to species in *Paramecium* and within individual cells of some species such as *P. multimicronucleatum* and the *P. aurelia* complex. *P. caudatum, P. bursaria,* and *P. trichium* have micronuclei which are relatively large and have threads of electron-opaque chromatin spread throughout the micronucleus. The other *Paramecium* species have a vesicular type of micronucleus. In this type the chromatin occurs as a central vesicular mass and a wide electron-transparent zone separates this mass from the nuclear envelope (Fig. 33). Nucleoli are not found in micronuclei but the nuclear envelope is perforated with numerous pores like the macronucleus.

The two types of nuclei usually retain a particular spatial relationship to each other during interphase, e.g., the micronuclei can often be found in indentations of the macronucleus just anterior to the anterior end of the buccal cavity in *P. multimicronucleatum* and *P. caudatum.* Inaba and Kudo (1972) found microtubules connecting the micro- and macronuclei in this zone of con-

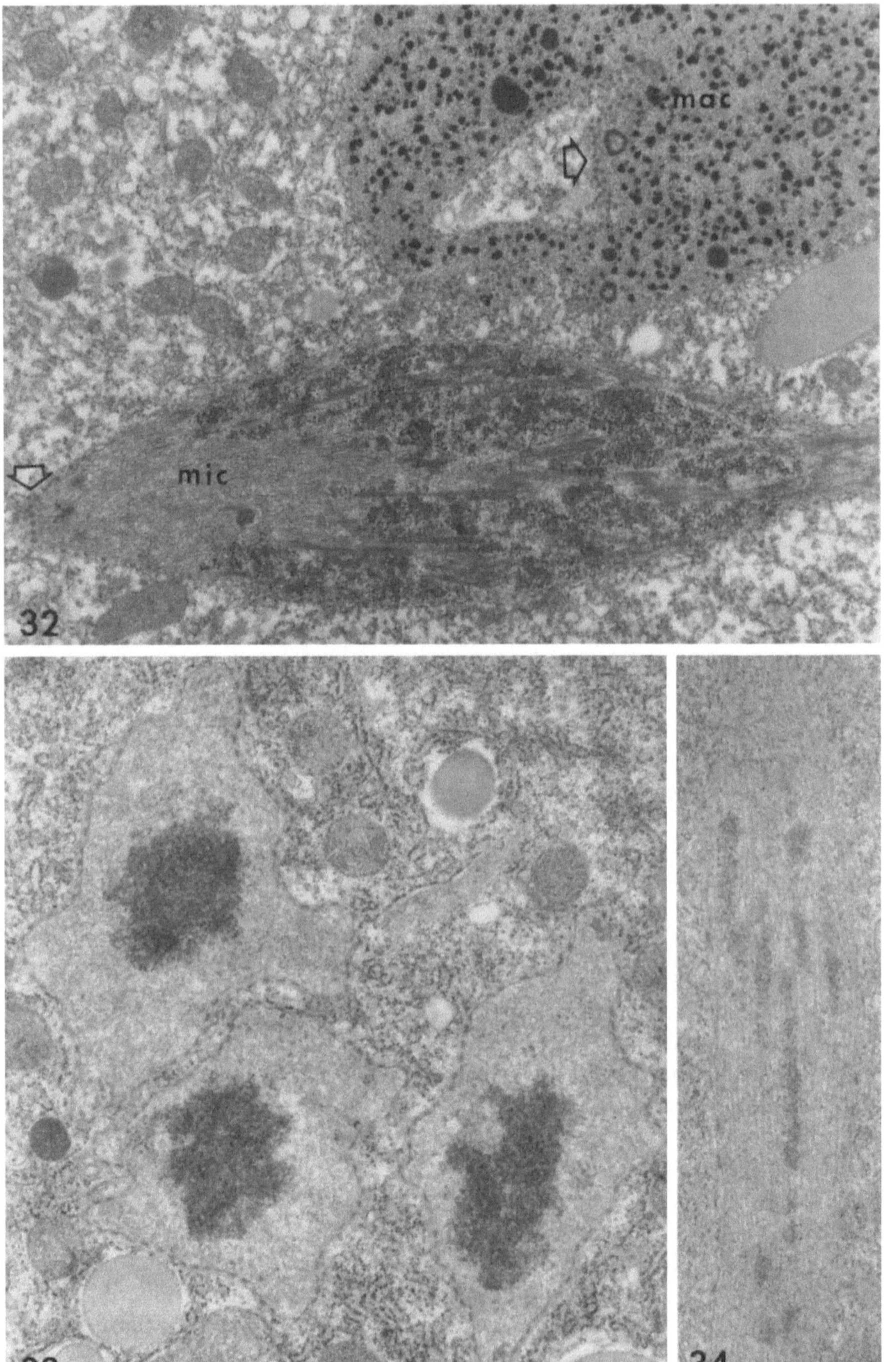
mac
mic
32
33
34

tact in *P. multimicronucleatum.* I have occasionally seen intimate contact of the micro- and macronuclear envelopes in this same species. The two nuclei usually retain a particular spatial relationship to each other during interphase, e.g., the micronuclei can often be found next to the macronucleus just anterior to the anterior end of the buccal cavity in *P. multimicronucleatum* and *P. caudatum.*

Micronuclei are much smaller than macronuclei and are diploid. Their major role appears to be carried out during and immediately following conjugation at which time the micronucleus undergoes meiosis, pronuclei are exchanged, and new macronuclei arise from mitotic products of the synkaryon, the product of fusion of pronuclei. Before division microtubules line the inner surface of the nuclear envelope of the micronuclei of *P. aurelia* (Jurand and Selman 1970; Stevenson and Lloyd 1971). At prophase the condensed chromatin core of this species fragments and the peripheral microtubules disappear. At metaphase individual chromosomes (over 100 in *P. aurelia*) are lined up on the equatorial plate where they are attached to bundles of microtubules (for *P. caudatum* see Fig. 32). A tenfold increase in micronuclear length occurs during anaphase as the continuous microtubules lengthen and daughter chromosomes (Fig. 34) are separated. This forms a 30-µm-long separation spindle in *P. aurelia.* Daughter nuclei pinch from this separation spindle in telophase and the spindle subsequently disappears. Except for the absence of a separation spindle, the two divisions of meiosis are similar to mitosis. Synaptonemal complexlike structures joining homologous chromosomes have been reported in the first meiotic division (Stevenson 1972). More detailed accounts of these and other aspects of the nuclei of *Paramecium* may be found in several chapters of this book.

6 Symbionts

It is not uncommon for *Paramecium* to harbor other living organisms in a symbiotic or mutualistic association. The most notable of these associations is that between *P. bursaria* and *Chlorella* which is beneficial to both organisms. Living *Chlorella* are enclosed by a perialgal vacuole membrane which is a modification of the host's digestive vacuole membrane system into which *Chlorella* are first trapped. This membrane has been investigated by Meier et al. (1980, 1984) in freeze-fracture replicas and has been found to differ from typical digestive vacuoles. Such a membrane modification may account for the sym-

Fig. 32. Macronucleus (*mac*) next to dividing micronucleus (*mic*). Chromatin bodies and the larger nucleoli are evident in the macronucleus. The micronucleus is filled with bundles of microtubules and chromatin. Both nuclei have pores in their nuclear envelopes (*arrows*). ×19000

Fig. 33. Three micronuclei of the vesicular type in *P. multimicronucleatum.* A fourth was present in the same cell. Nuclear pores are evident in all three micronuclei. ×19000

Fig. 34. One end of a highly elongated micronucleus in division. Electron-opaque chromosomes are evident in this late anaphase nucleus. ×19000

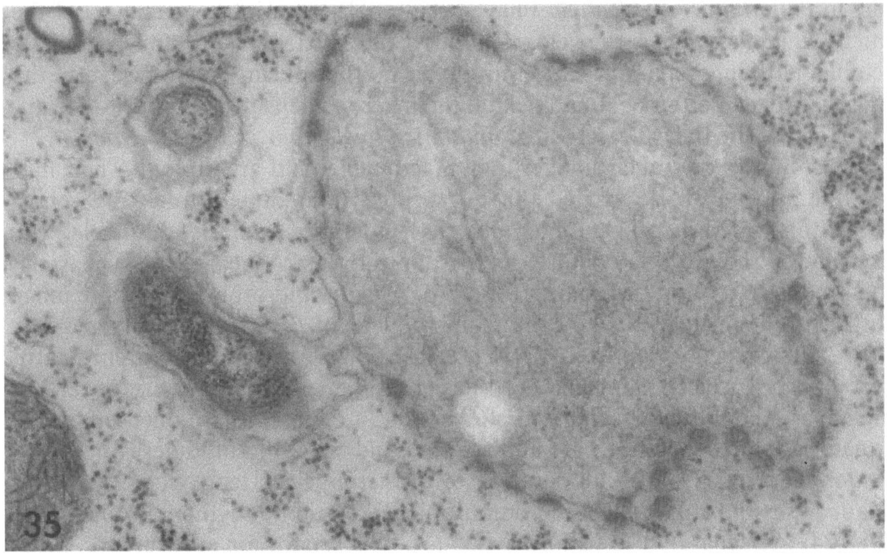

Fig. 35. Symbiotic bacterium located in the perinuclear space of a micronucleus of *P. multimicronucleatum*. The outer membrane of the nuclear envelope encloses the bacterium. ×47 500

biont's ability to bypass the cell's normal defense mechanisms allowing it to either prevent the acidification of the newly formed phagosome or to prevent the subsequent stage of phagosomal-lysosomal fusion and digestion. In the latter case the symbiont would first need to be able to survive the very acid environment of the phagosome since it can live for only a short time at pH 4.5 (Reisser 1975). Perialgal vacuoles do not contain acid phosphatase so they have not fused with lysosomes (Karakashian and Karakashian 1973) nor do they fuse with thorotrast-containing secondary lysosomes (Karakashian and Rudzinska 1981). The presence of the symbiont in the vacuole must somehow alter the vacuole membrane so this membrane can no longer fuse with lysosomes. In *Tetrahymena* particles which have a highly anionic surface are capable of preventing the fusion of lysosomes with phagosomes (Weidner and Sibley 1985).

Many bacterial and fungal symbionts are known to inhabit specific locations in *Paramecium*. Some bacteria inhabit the macronucleus, some are specific for the micronucleus, one is found within the intermembrane (perinuclear) space of nuclear envelopes (Fig. 35) and others live freely in the cytoplasm or inside vacuoles. Viruses of symbiotic bacteria living within *Paramecium* are also well known. More on these symbionts will be found in Chapter 22 (this Vol.) by Görtz.

Acknowledgments. Dr. Agnes K. Fok has contributed to this chapter both in research performed and in many discussions. Our current research is supported by the National Science Foundation, USA, through grants PCM 8402881 and DCB 8502212.

References

Allen RD (1971) Fine structure of membranous and microfibrillar systems in the cortex of *Paramecium caudatum.* J Cell Biol 49:1 – 20

Allen RD (1974) Food vacuole membrane growth with microtubule-associated membrane transport in *Paramecium.* J Cell Biol 63:904 – 922

Allen RD (1978) Particle arrays in the surface membrane of *Paramecium:* junctional and possible sensory sites. J Ultrastruct Res 63:64 – 78

Allen RD, Eckert R (1969) A morphological system in ciliates comparable to the sarcoplasmic reticulum-transverse tubular system in striated muscles. J Cell Biol 43:4a – 5a

Allen RD, Fok AK (1980) Membrane recycling and endocytosis in *Paramecium* confirmed by horseradish peroxidase pulse-chase studies. J Cell Sci 45:131 – 145

Allen RD, Fok AK (1983a) Phagosome fusion vesicles of *Paramecium.* I. Thin-section morphology. Eur J Cell Biol 29:150 – 158

Allen RD, Fok AK (1983b) Phagosome fusion vesicles of *Paramecium.* II. Freeze-fracture evidence for membrane replacement. Eur J Cell Biol 29:159 – 165

Allen RD, Fok AK (1983c) Nonlysosomal vesicles (acidosomes) are involved in phagosome acidification in *Paramecium.* J Cell Biol 97:566 – 570

Allen RD, Fok AK (1984a) Stages of digestive vacuoles in *Paramecium:* membrane surface differences and location. Eur J Cell Biol 35:149 – 155

Allen RD, Fok AK (1984b) Membrane behavior of exocytic vesicles. III. Flow of horseradish peroxidase labeled trichocyst membrane remnants in *Paramecium.* Eur J Cell Biol 35:27 – 34

Allen RD, Fok AK (1984c) Retrieval of lysosomal membrane and acid phosphatase from phagolysosomes of *Paramecium caudatum.* J Cell Biol 99:1955 – 1959

Allen RD, Fok AK (1985) Modulation of the digestive lysosomal system in *Paramecium caudatum.* III. Morphological effects of cytochalasin B. Eur J Cell Biol 37:35 – 43

Allen RD, Hausmann K (1976) Membrane behavior of exocytic vesicles. I. The ultrastructure of *Paramecium* trichocysts in freeze-fracture preparations. J Ultrastruct Res 54:224 – 234

Allen RD, Staehelin LA (1981) Digestive system membranes: freeze-fracture evidence for differentiation and flow in *Paramecium.* J Cell Biol 89:9 – 20

Allen RD, Wolf RW (1974) The cytoproct of *Paramecium caudatum:* structure and function, microtubules, and fate of food vacuole membranes. J Cell Sci 14:611 – 631

Bannister LH (1972) The structure of trichocysts in *Paramecium caudatum.* J Cell Sci 11:899 – 929

Beisson J, Sonneborn TM (1965) Cytoplasmic inheritance of the organization of the cell cortex in *Paramecium aurelia.* Proc Natl Acad Sci USA 53:275 – 282

Beisson J, Lefort-Tran M, Pouphile M, Rossignol M, Satir B (1976) Genetic analysis of membrane differentiation in *Paramecium.* Freeze-fracture study of the trichocyst cycle in wild-type and mutant strains. J Cell Biol 69:126 – 143

Clawson GA, Feldherr CM, Smuckler EA (1985) Nucleocytoplasmic RNA transport. Mol Cell Biochem 67:87 – 100

Cohen J, Adoutte A, Grandcamp S, Houdebine LM, Beisson J (1982) Immunocytochemical study of microtubular structures throughout the cell cycle of *Paramecium.* Biol Cell 44:35 – 44

Cohen J, Garreau de Loubresse N, Beisson J (1984a) Actin mobilization during phagocytosis in *Paramecium.* J Submicrosc Cytol 16:103 – 104

Cohen J, Garreau de Loubresse N, Beisson J (1984b) Actin microfilaments in *Paramecium:* localization and role in intracellular movements. Cell Motil 4:443 – 468

Csaba G (1985) The unicellular *Tetrahymena* as a model cell for receptor research. Int Rev Cytol 95:327 – 377

Dentler WL (1980) Structures linking the tips of ciliary and flagellar microtubules to the membrane. J Cell Sci 42:207 – 220

Dippell RV (1968) The development of basal bodies in *Paramecium.* Proc Natl Acad Sci USA 61:461 – 468

Dippell RV (1976) Effects of nuclease and protease digestion on the ultrastructure of *Paramecium* basal bodies. J Cell Biol 69:622 – 637

Dute R, Kung C (1978) Ultrastructure of the proximal region of somatic cilia in *Paramecium tetraurelia*. J Cell Biol 78:451 – 464
Ehret CF, McArdle EW (1974) The structure of *Paramecium* as viewed from its constituent levels of organization. In: Van Wagtendonk WJ (ed) *Paramecium*. A current survey. Elsevier, Amsterdam London New York, pp 263 – 338
Ehret CF, Alblinger J, Savage N (1964) Development and ultrastructural studies of cell organelles. Argonne Nat Lab Biol Med Div Annu Rep 6971:62 – 70
Esteve J-C (1969) Observations sur l'ultrastructure et le mètabolisme du glycogène de *Paramecium caudatum*. Arch Protistenk 111:195 – 203
Esteve J-C (1970) Distribution of acid phosphatase in *Paramecium caudatum*: its relations with the process of digestion. J Protozool 17:24 – 35
Esteve J-C (1972) L'appareil de Golgi des Cilies. Ultrastructure, particulièrement chez *Paramecium*. J Protozool 19:609 – 618
Fok AK, Allen RD (1979) Axenic *Paramecium caudatum*. I. Mass culture and structure. J Protozool 26:463 – 470
Fok AK, Allen RD (1981) Axenic *Paramecium caudatum*. II. Changes in fine structure with culture age. Eur J Cell Biol 25:182 – 192
Fok AK, Lee Y, Allen RD (1982) The correlation of digestive vacuole pH and size with the digestive cycle in *Paramecium caudatum*. J Protozool 29:409 – 414
Fok AK, Muraoka JH, Allen RD (1984) Acid phosphatase in the digestive vacuoles and lysosomes of *Paramecium caudatum*: a timed study. J Protozool 31:216 – 220
Gelei G von (1973) Ein neues Fibrillensystem in Ectoplasma von Paramecium. Arch Protistenk 89:133 – 162
Hausmann K (1978) Extrusive organelles in protists. Int Rev Cytol 52:197 – 276
Hausmann K, Allen RD (1976) Membrane behavior of exocytic vesicles. II. Fate of the trichocyst membranes in *Paramecium* after induced trichocyst discharge. J Cell Biol 69:313 – 326
Hausmann K, Allen RD (1977) Membranes and microtubules of the excretory apparatus of Paramecium caudatum. Cytobiologie 15:303 – 320
Hausmann K, Stockem W, Wohlfarth-Bottermann K-E (1972) Cytologische Studien an Trichocysten. II. Die Feinstruktur ruhender und gehemmter Spindeltrichocysten von Paramecium caudatum. Cytobiologie 5:228 – 246
Hufnagel LA (1969) Cortical ultrastructure of *Paramecium aurelia*. Studies on isolated pellicles. J Cell Biol 40:779 – 801
Inaba F, Kudo N (1972) Electron microscopy of the nuclear events during binary fission in *Paramecium multimicronucleatum*. J Protozool 19:57 – 63
Janisch R (1972) Pellicle of *Paramecium caudatum* as revealed by freeze etching. J Protozool 19:470 – 472
Jurand A, Selman GG (1969) The anatomy of *Paramecium aurelia*. Macmillan, London; St. Martin's press, New York
Jurand A, Selman GG (1970) Ultrastructure of the nuclei and intranuclear microtubules of *Paramecium aurelia*. J Gen Microbiol 60:357 – 364
Karakashian MW, Karakashian SJ (1973) Intracellular digestion and symbiosis in *Paramecium bursaria*. Exp Cell Res 81:111 – 119
Karakashian SJ, Rudzinska MA (1981) Inhibition of lysosomal fusion with symbiont-containing vacuoles in *Paramecium bursaria*. Exp Cell Res 131:387 – 393
Kersken H, Vilmart-Seuwen J, Momayezi M, Plattner H (1986) Filamentous actin in *Paramecium* cells: mapping by phalloidin affinity labeling in vivo and in vitro. J Histochem Cytochem 34:443 – 454
Mast SO (1947) The food vacuole in *Paramecium*. Biol Bull 92:31 – 72
McKanna JA (1973) Fine structure of the contractile vacuole pore in *Paramecium*. J Protozool 20:631 – 638
McKanna JA (1976) Fine structure of fluid segregation organelles of *Paramecium* contractile vacuoles. J Ultrastruct Res 54:1 – 10
Meier R, Reisser W, Wiessner W, Lefort-Tran M (1980) Freeze-fracture evidence of differences between membranes of perialgal and digestive vacuoles in *Paramecium bursaria*. Z Naturforsch 35c:1107 – 1110

Meier R, Lefort-Tran M, Pouphile M, Reisser W, Wiessner W (1984) Comparative freeze-fracture study of perialgal and digestive vacuoles in *Paramecium bursaria.* J Cell Sci 71:121–140
Mogami Y, Takahashi K (1983) Calcium and microtubule sliding in ciliary axonemes isolated from *Paramecium caudatum.* J Cell Sci 61:107–121
Nilsson JR, Van Deurs B (1983) Coated pits and pinocytosis in *Tetrahymena.* J Cell Sci 63:209–222
Patterson DJ (1977) On the behaviour of contractile vacuoles and associated structures of *Paramecium caudatum* (Ehrbg). Protistologica 13:205–212
Patterson DJ (1980) Contractile vacuoles and associated structures: their organization and function. Biol Rev 55:1–46
Patterson DJ (1981) On the origin of the postoral microtubules of *Paramecium putrinum* (Hymenostomatida, Ciliophora). Protistologica 17:525–531
Perasso R, Beisson J (1978) Temporal pattern of mitochondrial multiplication during the cell cycle of *Paramecium.* Biol Cell 32:275–290
Pitelka DR (1965) New observations on cortical ultrastructure in *Paramecium.* J Microsc 4:373–394
Plattner H (1975) Ciliary granule plaques: membrane-intercalated particle aggregates associated with Ca^{2+}-binding sites in *Paramecium.* J Cell Sci 18:257–269
Plattner H, Miller F, Bachmann L (1973) Membrane specializations in the form of regular membrane-to-membrane attachment sites in *Paramecium.* A correlated freeze-etching and ultrathin-sectioning analysis. J Cell Sci 13:687–719
Plattner H, Westphal C, Tiggemann R (1982) Cytoskeleton-secretory vesicle interactions during the docking of secretory vesicles at the cell membrane in *Paramecium tetraurelia* cells. J Cell Biol 92:368–377
Plattner H, Pape R, Haacke B, Olbricht K, Westphal C, Kersken H (1985) Synchronous exocytosis in *Paramecium* cells. VI. Ultrastructural analysis of membrane resealing and retrieval. J Cell Sci 77:1–17
Reisser W (1975) On the taxonomy of an auxotrophic *Chlorella* sp. isolated from *Paramecium bursaria.* Arch Microbiol 104:293–296
Sale WS, Satir P (1977) The termination of the central microtubules from the cilia of *Tetrahymena pyriformis.* Cell Biol Int Rep 1:45–49
Satir B (1974) Membrane events during the secretory process. Symp Soc Exp Biol 28:399–418
Satir P (1968) Studies on cilia. III. Further studies on the cilium tip and a "sliding filament" model of ciliary motility. J Cell Biol 39:77–94
Selman GG, Jurand A (1970) Trichocyst development during the fission cycle of *Paramecium.* J Gen Microbiol 60:365–372
Sibley JT, Hanson ED (1974) Identity and function of a subcortical cytoskeleton in *Paramecium.* Arch Protistenk 116:221–235
Steinman RM, Mellman IS, Muller WA, Cohn ZA (1983) Endocytosis and the recycling of plasma membrane. J Cell Biol 96:1–27
Stelly N, Balmefrezol M, Adoutte A (1975) Diaminobenzidine reactivity of mitochondria and peroxisomes in *Tetrahymena* and in wild-type and cytochrome oxidase-deficient *Paramecium.* J Histochem Cytochem 23:686–696
Stevenson I (1972) Ultrastructure of nuclear division in *Paramecium aurelia.* III. Meiosis in the micronucleus during conjugation. Aust J Biol Sci 25:775–799
Stevenson I, Lloyd FP (1971) Ultrastructure of nuclear division in *Paramecium aurelia.* I. Mitosis in the micronucleus. Aust J Biol Sci 24:963–975
Sundararaman V, Cummings DJ (1976a) Morphological changes in aging cell lines of *Paramecium aurelia.* I. Alterations in the cytoplasm. Mech Age Dev 5:139–154
Sundararaman V, Cummings DJ (1976b) Morphological changes in aging cell lines of *Paramecium aurelia.* II. Macronuclear alterations. Mech Age Dev 5:325–338
Tamm SL (1972) Ciliary motion in *Paramecium.* A scanning electron microscope study. J Cell Biol 55:250–255
Tucker JB, Beisson J, Roche DLJ, Cohen J (1980) Microtubules and control of macronuclear "amitosis" in *Paramecium.* J Cell Sci 44:135–151

Vivier E (1974) Morphology, taxonomy and general biology of the genus *Paramecium*. In: Van Wagtendonk WJ (ed) *Paramecium*. A current survey. Elsevier, Amsterdam London New York, pp 1–89
Wasik A, Sikora J (1980) Effects of cytochalasin B and colchicine on cytoplasmic streaming in *Paramecium bursaria*. Acta Protozool 19:103–110
Watanabe T (1981) Electron microscopy of cell surfaces of *Paramecium caudatum* stained with ruthenium red. Tissue Cell 13:1–7
Weidner E, Sibley LD (1985) Phagocytized intracellular microsporidian blocks phagosome acidification and phagosome-lysosome fusion. J Protozool 32:311–317
Westcot DM, Fok AK, Allen RD (1985) Flow of membrane markers from the coated pits in *Paramecium caudatum*. J Cell Biol 101:52a
Wichterman R (1953) The biology of *Paramecium*. Blakiston & McGraw-Hill, New York
Wichterman R (1986) The biology of *Paramecium*, 2nd edn. Plenum, New York London
Wilbert N, Heller G (1971) Der Feinbau der Mitochondrien von *Tachysoma pellionella* (Muller-Stein, 1859) (Ciliata, Hypotricha). Z Naturforsch 26b:1389
Wolosewick JJ, Porter KR (1979) Microtrabecular lattice of the cytoplasmic ground substance. Artifact or reality. J Cell Biol 82:114–139
Wyroba E (1981) Alveolar system of *Paramecium*. I. Trapping of polycationic dye as a result of membrane impairment. Acta Histochem 69:132–148
Wyroba E, Przelecka A (1973) Studies on the surface coat of *Paramecium aurelia*. I. Ruthenium red staining and enzyme treatment. Z Zellforsch 143:343–353

Chapter 3 The Species Concept and Breeding Systems

DENNIS NYBERG[1]

CONTENTS

1 Background

1.1 An Overview of Taxonomic Development

People group and organize entities or observations as a way of understanding the world. We also believe that the structures we generate doing this fundamental taxonomic activity approach the "true" organization of the natural world. Initially few characters, often of a single class, are available for organizing individuals into groups. As new characters, and especially those of a different nature or class, become available, the effectiveness of the original classification is tested and evaluated. An ideal taxonomy results in the same classifi-

1 Department of Biological Sciences, University of Illinois at Chicago, Chicago, IL 60680, USA

Paramecium. Ed. by H.-D. Görtz

cation regardless of the nature of the characters used. In biology the species level was more effective than higher levels of classification in forming "natural" groups, i.e., those in which different classes of characters resulted in the same or similar groupings. In this century the greater effectiveness of the species level of classification could be rationalized through the gene pool shared by members of this evolutionary unit. Along with this new evolutionary definition of species came the discovery that morphologically similar *Drosophila* could sometimes be divided into more than one genetic species. These became known as sibling species.

During this same era important discoveries were made which would revolutionize *Paramecium* taxonomy. First, the ability to culture and maintain pure lines provided access to new classes of nonmorphological characters. Second, Sonneborn's (1937) discovery of mating types revealed that all stocks did not share a common gene pool; there were at least three groups. After the groups were defined it was apparent that stocks within groups typically shared physiological properties (many of which concerned mating) which also distinguished the groups (Sonneborn 1939). Sonneborn (1939) recognized that these groups were separate species but maintained they should not be named because it was too burdensome to identify them.

1.2 The Species of Paramecium à la Sonneborn 1957

Sonneborn (1957) described 16 varieties in the *P. aurelia-P. multimicronucleatum* complex. The major thesis of his monumental analysis was that the 16 varieties, named syngens in that paper, were good species with no gene flow between them. Information that accumulated in the 20 years since the 1937 discovery of mating types proved that at least some varieties could be discriminated on the basis of size, geographic distribution, temperature of mating reactions, time of day of mating reactions, immaturity periods, form of sex during senility, the presence of and sensitivity to various killers, intervarietal mating reactions, and mode of mating type determination.

While some varieties exhibited no mating reactions to other varieties, the varieties which did usually did not form true heterotypic conjugants, but varieties 1 and 7 formed a true F1. In this case neither backcrosses nor F2-by-autogamy yielded viable progeny so the varieties were inferred to be genetically isolated.

The character Sonneborn (1957) chose to establish groupings among the species was the mode of mating type determination or inheritance. All species of the *P. aurelia* complex have two mating types, odd and even. Almost all genotypes have the potential to express both mating types. In group A species each developing macronucleus (caryonide) differentiates independently to express one or the other type soon after fertilization. Thus, among the four caryonidal clones which develop from each pair of conjugants one has a binomial distribution of odd and even clones. This pattern is also referred to as caryonidal inheritance. In group B species mating types are determined to differentiate in the new macronucleus to the same state as that of the old macronucleus. This results in the cytoplasmic pattern of inheritance with the typical result of two sister caryonidal clones with the odd type and two of the even type from each synclone or pair. At the time of Sonneborn's (1957) review the odd-numbered varieties (of the first eleven) all had the group A pattern and the even-numbered varieties had the group B pattern. The third pattern, known as group C, results from genotypic determination of mating type; thus, because all caryonidal clones of a pair have the same genotype, they have the same mating type. The split between the group A species and group B species was also associated with serotype inheritance and the absence or presence of endosymbionts, and was thus believed to be ancient.

1.3 The Breeding Systems of Paramecium à la Sonneborn 1957

In the process of domestication of the varieties of *P. aurelia* considerable information concerning the life cycle and breeding system or genetic system accumulated. Sonneborn's (1957) synthesis of this information found correlation among characters which allowed each species to take a place along an inbreeding-outbreeding continuum. The contrasts that were emphasized were: mating type inheritance, A or B in inbreeders, C in outbreeders; number of mating types, two in inbreeders, many in outbreeders; immaturity period, short in inbreeders, long in outbreeders; maturity period, short in inbreeders, long in outbreeders; old age sexual "escapes", autogamy in inbreeders, none or selfing in outbreeders; and fission rate, high in inbreeders, low in outbreeders. When the *P. aurelia* complex is compared to the *P. bursaria* complex, these contrasts work perfectly, but in other species the correlations are less strong. When species were being compared, one considered the suite of characters, but the single most important character in Sonneborn's (1957) evaluation was the immaturity interval, because it represented a measure of distance between the place of birth and the place at which sexual progeny were produced. Greater distances were assumed to represent greater "strangeness" or genetic differentiation between the two members of a pair.

The most extreme inbreeders were *P. decaurelia* and *P. dodecaurelia.* They had no immaturity period and were rare and presumably found in only limited localities. Within the *P. aurelia-multimicronucleatum* complex the most extreme outbreeders were *P. multimicronucleatum* syngen 2 (varieties 15 and 16) and *P. jenningsi* (variety 12). The rest of the *P. aurelia* complex species fell between these extremes, although *P. tetraurelia* and *P. octaurelia* were also considered extreme inbreeders. All species with autogamy were primarily inbreeders. The varieties of *P. caudatum,* except possibly syngen 3, were all considered inbreeders. The varieties of *P. bursaria* were all considered outbreeders.

1.4 Evolutionary Strategies and Breeding Systems

Agreeing with Darlington (1958) and Stebbins (1950), Sonneborn (1957) believed that only outbreeders had long-term evolutionary potential. Inbreeders were considered to be specialized, which gave them a short-term advantage over outbreeders. Inbreeding sexual species and asexual diploids were associated with stability, and outbreeding sexuals and haploid asexuals were associated with variety. Sonneborn (1957, p. 297) made some predictions about characteristics to be found with more inbreeding, *"the number of closely related syngens per species increases; the range of distribution of a syngen, especially in latitude, becomes more and more restricted, the number of local populations per syngen becomes smaller and smaller; the genetic divergence of each local population becomes greater and greater."*

In this chapter I will update our knowledge of the relationships among stocks and among genetic species of the genus *Paramecium.* I will evaluate the theoretical background of the breeding system concept, as expressed by Nyberg (1974), Nanney (1980), and Dini (1984), and evaluate the empirical usefulness of breeding system information in understanding the ecology and evolution of these species.

2 Updating the Taxonomy and Breeding Systems of *Paramecium*

2.1 Relationships Between Species Complexes

New techniques have provided new characters through which relationships may be evaluated. Relationships among and within species complexes can be updated with sequences of proteins or DNA, studies of isozymes, transfer of parts from one cell to another by microinjection, and techniques of inducing mating between species.

The genus *Paramecium* has traditionally been divided into two groups (Vivier 1974), the cigar- or slipper-shaped "aurelia" group including *P. aurelia, P. caudatum, P. jenningsi,* and *P. multimicronucleatum,* and the shorter, wider "bursaria" group. This morphological distinction has been supported by recent studies. Miyake (1968) found that the chemical agents inducing conjugation in *P. caudatum, P. multimicronucleatum, P. aurelia,* and *P. jenningsi* were not effective in inducing mating in *P. bursaria, P. calkinsi,* or *P. polycarum.* Furthermore, chemically induced cells of *P. multimicronucleatum* would form pairs with both *P. caudatum* and *P. aurelia* complex species. These pairs did not, presumably, result in genetic exchange as the progeny resembled their parent and were assumed to result from self-fertilization (Miyake 1968).

Using injections of cytoplasm from immature cells into mature cells of different species, Miwa (1979a) has shown that immaturin from *P. caudatum* syngen 3 induces immaturity in other syngens of *P. caudatum* and in *P. multimicronucleatum* syngen 2, but not in *P. bursaria.* He also found (Miwa 1979b) that immature cells of *P. primaurelia* possessed a cytoplasmic factor capable of inducing immaturity in *P. tetraurelia, P. caudatum* syngen 3, and *P. multimicronucleatum* syngen 2, but not in *P. bursaria.*

Haga et al. (1983) found that proteins in the cytoplasm of CNR mutants of *P. caudatum* syngen 3 could complement all three pawn loci in *P. tetraurelia* and vice versa. Cytoplasmic transfers have generally indicated a high degree of similarity among the slipper-shaped *Paramecium*, but Beale and Tait (1981) remarked that transfers of mitochondria from *P. primaurelia* to *P. caudatum* were unsuccessful.

Within the "aurelia" group Allen et al. (1983c) found that *P. jenningsi* stocks had all five esterases and phosphatases found in *P. aurelia* complex species, and therefore all these species were closely related, but five syngens of *P. multimicronucleatum* and one stock of *P. caudatum* shared only three or four of the five *P. aurelia* complex patterns.

2.2 *Paramecium aurelia* Complex

2.2.1 *Mitochondrial Transfers and Sequences*

Beale and Knowles (1976) investigated relationships among species of the complex by evaluating transfers of mitochondria. Mitochondria from *P. primaurelia* and *P. pentaurelia* could be transferred between those two species and also to *P. septaurelia*, but mitochondria from *P. septaurelia* failed to develop in *P. primaurelia* or *P. pentaurelia.* They concluded that the incompatibility was mainly under control of the nuclear genome. Cummings and collaborators have greatly extended the study of relationships with their studies of mitochondrial DNA of *P. primaurelia, P. tetraurelia, P. pentaurelia,* and *P. septaurelia.* They found a high degree of homology between the mitochondrial DNA of *P. primaurelia, P. pentaurelia,* and *P. septaurelia,* but *P. tetraurelia* showed less than 50% homology to *P. primaurelia* (Cummings et al. 1980). All four species had a high level of homology of the rRNA genes. Cummings (1980) quantified the divergence estimates based on four 6-bp restriction enzymes and one 4-bp enzyme. *P. tetraurelia* had an estimated mitochondrial DNA sequence divergence from the other three species of between 6.8 and 10.3% (both extremes with *P. pentaurelia* depending on the number of bases recognized by the enzyme). Mitochondrial DNA from *P. primaurelia, P. pentaurelia,* and *P. septaurelia* yielded divergence estimates between 3.7 to 5.4% among the pairwise comparisons. There was no intraspecific variability in *P. primaurelia* or *P. pentaurelia,* but two types of mitochondrial DNA (1.4% divergent) were found among four stocks of *P. tetraurelia* (Cummings 1980). Subsequently, both the small and large subunit rRNA of mitochondria in *P. primaurelia* and *P. tetraurelia* have been sequenced. The small subunit gene was 94% conserved between the two species (Seilhamer et al. 1984a) and the large subunit mitochondrial genes of *P. primaurelia* and *P. tetraurelia* are 96% identical in base sequence (Seilhamer et al. 1984b). Sogin and Elwood (1986) have sequenced the small subunit rRNA encoded by the nucleus of *P. tetraurelia,* but no other *Paramecium* sequences have yet been published.

2.2.2 *Isozymes*

The only comparative studies that have utilized the entire diversity of the *P. aurelia* complex are those involving isozymes. The comparisons of species are facilitated by the infrequency of variation within species, an observation made in the earliest studies of isozymes (Tait 1970; Allen et al. 1971) and extensively documented in recent studies. *P. primaurelia, P. tetraurelia,* and *P. octaurelia* have a frequency of isozyme variation in esterases and phosphatases much lower than that of other organisms (Allen et al. 1983a). *P. biaurelia* is hypervariable for two esterases and most of these variants are unlike the patterns seen in other species. In the other species most of the variants of one species are like a common form or variant in another species (Allen et al. 1982). Stocks from all over the world have been studied and there is no evidence for geographic differentiation in these four species except that in all species a higher frequency of variants occurs among Central and South American stocks (Allen et al. 1983a).

The lack of variation within species facilitates comparisons and groupings of species. Early studies of esterases (Allen et al. 1973) questioned Sonneborn's groups A, B, and C. Recently, Allen et al. (1983b) have summarized the distribution of enzyme subtypes for all 14 species for 11 enzymes. From these groupings they calculated pairwise genetic distances and then constructed a dendrogram estimating the minimum tree length for all 14 species. Four clusters emerged from the dendrogram. The deepest division was between a group including *P. tetraurelia, P. octaurelia, P. decaurelia,* and *P. dodecaurelia* and the other three groups. *P. biaurelia* and *P. triaurelia* form a group. *P. primaurelia, P. pentaurelia, P. sexaurelia,* and *P. novaurelia* also form a group, and *P. tredecaurelia, P. quadecaurelia, P. septaurelia,* and *P. undecaurelia* form the fourth group (Allen et al. 1983b). Except for the *P. tetraurelia* group, all of which are group B species, the other isozyme groups each have at least one group A species and one group B species.

2.2.3 Tolerance Limits

Nyberg (1974) reported median tolerance limits to six salts, low pH, and high temperature of stocks of *P. primaurelia, P. biaurelia, P. tetraurelia,* and one stock of *P. decaurelia.* Among the pairwise comparisons of species with multiple stocks, the following differences were significant ($\alpha = 0.05$): *P. primaurelia* had a greater tolerance to calcium chloride and high temperature than both *P. biaurelia* and *P. tetraurelia.* The stock of *P. decaurelia* had the lowest tolerance of any *P. aurelia* stock to high temperature, calcium, and sodium. That stock had a relatively high tolerance to potassium.

Large differences in tolerance among stocks within a species were discovered in *P. tetraurelia* and *P. primaurelia* for copper (Nyberg 1974). A subsequent survey of lab stocks of *P. tetraurelia* and genetic analysis (Nyberg 1975) demonstrated that two of the wild stocks were highly resistant to copper and that the resistance was attributable to a single allele. Using freshly collected stocks of *P. primaurelia* and *P. triaurelia* and assaying their tolerance to heavy metals, Nyberg and Bishop (1983) found significant variation among stocks of *P. primaurelia* in tolerance to high temperature, copper, cobalt, mercury, zinc, nickel, and cadmium. Among *P. triaurelia* stocks there was significant variation of tolerances to copper, mercury, chromium, cobalt, and cadmium. Between the two species, *P. primaurelia* was significantly more resistant to high temperature and cadmium and more sensitive to nickel than *P. triaurelia*. These results were interpreted as strong evidence for differentiation of local populations. A subsequent genetic study has, however, resulted in a substantial reduction of the estimated genotypic variability among stocks. Nyberg and Bogar (1986) found that large differences in tolerance arose within stocks at the time of autogamy. This among-clones-within-stocks variation was so large that evidence for allelic differences among stocks remained only for mercury and copper. For the other metals the apparent local differentiation probably resulted from interclonal variation within stocks (Nyberg and Bishop 1983). It now seems that genotypic differences in copper tolerance are frequent, but genetic differentiation with respect to other environmental stresses is rare.

2.2.4 Morphology and Other Characters

Sonneborn (1957) reported size differences among the varieties and Hairston and Kellerman (1965) described a consistent morphological difference between *P. biaurelia* and *P. triaurelia*. Powelson et al. (1975) systematically studied morphological differences among stocks of *P. primaurelia, P. biaurelia, P. tetraurelia*, and *P. sexaurelia*. While no single measurement could separate the stocks into four species, a multivariate analysis easily separated these four species. The multivariate technique could not, however, separate the closely related species *P. primaurelia* and *P. pentaurelia* (Gates and Berger 1976).

Sonneborn (1957) mentioned growth rate as a difference between species, but subsequent studies have focused on differences within species. Siegel (1958) reported large differences (1.45 to 3.68 fissions per day) among stocks of *P. primaurelia*. Nyberg (1982) also reported highly significant differences in growth rate (3.75 to 4.58 fpd) among five *P. primaurelia* stocks. Both authors report that F1 progeny from two slow parents grow significantly faster than either parent. Since each wild stock of the *P. aurelia* complex is a homozygous exautogamous genotype derived from the genotype of the wild cell rather than being an actual wild genotype, the growth rate differences may not reflect population differentiation.

Phenotypic and genetic differentiation among wild stocks has also been examined with respect to trichocyst discharge. Nyberg (1978) reported that 8 of 28 *P. tetraurelia* stocks were non-discharge. Genetic analysis revealed that six different loci were involved. The only two stocks which were genetically identical had been collected at the same time and place. Thus, here was another "phenotypic" character which was consistent with considerable genetic differentiation among stocks of *P. tetraurelia*. All 26 *P. primaurelia* stocks and 10 *P. triaurelia* stocks studied by Nyberg and Bishop (1983) were trichocyst discharge (unpublished data).

2.2.5 Distribution

Hairston (1958; 1967) did extensive sampling of the Huron River in Ann Arbor, Michigan. He found the expected species, *P. primaurelia, P. biaurelia, P. triaurelia,* and *P. pentaurelia.* He also found a single stock of *P. tetraurelia* which represented a modest range extension, but no other species. Sampling in Europe discovered two species, *P. triaurelia* and *P. pentaurelia* (Przyboś 1968) previously restricted to North America. *P. pentaurelia* has also been collected in Australia (Sonneborn 1975). The range of *P. tetraurelia* was also extended to Europe (Przyboś 1980). There has been extensive sampling in Poland. Only four species have been found in many collections. *P. biaurelia* is most frequently collected, followed by *P. novaurelia, P. primaurelia,* and *P. tetraurelia,* the latter being rare and restricted to southern Poland (Komala and Przyboś 1984). It now seems likely that each of the first eight species, except *P. septaurelia,* has a cosmopolitan distribution. *P. novaurelia* is puzzlingly restricted to Europe but is abundant there. Pringle and Beale (1960) studied a population of this species in Scotland. Heterozygotes were present but less frequent than expected by random mating. The only new species added to the *P. aurelia* complex is *P. sonneborni* (Aufderheide et al. 1983), a large cell with a long life cycle collected in Texas, USA.

2.2.6 Interpretation and Conclusions

Sonneborn (1957) argued that the biological species of the *P. aurelia* complex were good species with no gene flow between them. Haggard (1974) reexamined the possibility of gene flow between *P. tetraurelia* and *P. octaurelia* and confirmed the original conclusion. *P. primaurelia, P. biaurelia, P. triaurelia, P. tetraurelia, P. sexaurelia, P. octaurelia,* and *P. novaurelia* are all abundant species with broad distributions that can be distinguished in a number of ways. The conditionality of mating reactions according to environmental conditions found in *Tetrahymena* by Orias (1959) and the separation of viability and mating reactions in *P. caudatum* (Sect. 2.3) have provided reasons to consider the possibility that some of the current species might move to a subspecific status of a common species, but the isozyme and mitochondrial DNA evidence suggests that this is very unlikely.

Sonneborn (1957) used the pattern of mating type determination as the major criteria for subdividing the aurelia complex. These patterns are known as caryonidal, cytoplasmic, or genic; A, B, and C, respectively. Each pattern has a high (1 – 10%) rate of exceptions to the rule. *P. primaurelia, P. triaurelia, P. pentaurelia, P. novaurelia, P. undecaurelia,* and *P. quadecaurelia* belong to the A group, while *P. tredecaurelia* is the sole member of the C group. *P. septaurelia* having moved from a tentative A to B between 1957 and 1975 [see Butzel (1974) for a description of this change]. There are some interspecific mating reactions and these were not completely concordant with the A, B division. Specifically, *P. primaurelia* and *P. triaurelia* (A group members) responded to *P. septaurelia* and *P. octaurelia* (B group members) and *P. pentaurelia* also responds to *P. octaurelia.* The four groups proposed by Allen et al. (1983b) include three groups with both A and B species. A wide variety of criteria indicate that *P. primaurelia* and *P. pentaurelia* are very closely related and all available criteria indicate that *P. tetraurelia, P. octaurelia, P. decaurelia,* and *P. dodecaurelia* are a natural group. Beyond these areas of agreement the most glaring discrepancies concern *P. septaurelia.* The mitochondrial DNA evidence indicates it is closely related to *P. primaurelia* but it is not grouped with this species in the isozyme analysis. Another "problem" with the isozyme analysis is the grouping together of *P. biaurelia,* the species with the most endosymbionts, with *P. triaurelia,* a species with no endosymbionts and which is most sensitive to their killing action. Eventually, degrees of relationship will be estimated by sequence similarity and the isozyme analysis is a step in that direction.

Sonneborn's (1957) prediction that many more species of the inbreeding *P. aurelia* complex would be discovered has not proven true. Only one new species has been described (Aufderheide et al. 1983). Instead, species previously believed to have regional distributions are now known to be cosmopolitan. It is not unusual to find more than one species in the same collecting jar. The few physiological differences between species cannot be related to microhabitats and "specializations" of the inbreeding species have not clearly emerged from tolerance or biochemical studies (see also Chap. 24 [this Vol.] by Landis).

Sonneborn's (1957) prediction of considerable genetic differentiation among stocks in inbreeding species was based on low viability of F2-by-autogamy progeny from interstock crosses and Dippell's (1954) discovery of different chromosome numbers in different stocks. It was also based on the repeated collections of stocks restricted to the odd mating type from a stream, from which he inferred there was little dispersal of individuals. Differences among stocks in copper tolerance and genetic variation concerning trichocyst discharge given in recent studies are consistent with relatively large amounts of local differentiation. The failure of isozymes to reveal local differentiation is not. Crosses done in both the phenotype and isozyme studies have, however, corroborated the original observations of some crosses with low F2 viability. I have not, however, ever encountered inviability in the original cross (F1) between *P. aurelia* complex stocks and this is the cross from which incompatibility is inferred in outbreeders. Furthermore, Nyberg (1982) argued that viability may be less important than growth rate in an increasing population. Nevertheless, the preponderance of evidence is that there is probably not a great deal of gene flow in these inbreeding species. The isozymic uniformity probably has a primary cause other than gene flow.

2.3 *Paramecium caudatum* Complex

Gilman (1941) first described the mating relationships of *P. caudatum.* At that time he described four varieties and a probable fifth. Gilman continued to find new varieties and at the time of Sonneborn's (1957) review 16 syngens were recognized. Each syngen had two mating types. The stocks could not be classified into nonoverlapping groups, syngens 2, 8, 9, 10, 14, 15, and 16 were known as the "variety 2 complex" because of asymmetric mating relationships among various stocks. All syngens, except possibly syngen 3, were considered to be inbreeders by Sonneborn (1957). The lack of marker genes, lack of autogamy, and frequent occurrence of macronuclear regeneration after conjugation make it very difficult to evaluate postzygotic genetic isolation, and I infer that the classification scheme was based solely on mating reactivity.

During Gilman's studies Hiwatashi (1949) found four syngens in Japan which later proved to correspond to Gilman's numbers 1, 3, 12, and 13. The factors mentioned above plus the formation of homotypic as well as heterotypic pairs and the discovery that the dominant mating type allele usually permits the expression of both mating types (Hiwatashi 1968) make *P. caudatum* less tractable genetically than *P. tetraurelia. P. caudatum* syngen 3 has a long immaturity period of 50 fissions (Takagi 1970). It now seems probable that Sonneborn's (1957) reports of a lack of an immaturity period in the other syngens was a consequence of macronuclear regeneration.

A variety of technical advances has made it feasible to investigate the relationships among the *P. caudatum* syngens, and Tsukii and co-workers have found some startling results. Among 44 stocks representing all four syngens only a single variant stock (from syngen 3) was found in the LDH isozymes (Agatsuma and Tsukii 1980). Chemically inducing mating and using behavioral marker genes in syngen 3 to select intersyngenic pairs, Tsukii and Hiwatashi (1983) found that crosses between syngens had the same viability (53–93%) as crosses within syngens. From that observation they suggested that *P. caudatum* might be a single biological species. But later they found that a high proportion of the intersyngenic hybrids, which of course were viable and growing, were trisomic or monosomic, which suggested more genetic differentiation than was initially apparent (Tsukii and Hiwatashi 1985; see also Chap. 4 by Tsukii, this Vol.). But while the Japanese investigators moved back from their radical taxonomic proposal, Khadem and Gibson (1985) found that syngens 1, 3, 12, and 13 could not be distinguished from each other on the basis of isozyme variation of six enzymes. Khadem and Gibson (1985) also reported matings between syngens 1, 3, 12, and 13 and between the known stocks and some unknowns. Other unknowns did not mate with the known stocks. These could be divided into two groups based on their mating reactions. Each of the new groups was also homogeneous isozymically, but differentiated from each other and from group I (the knowns). Thus, Khadem and Gibson (1985) claimed there are at least three species of a *P. caudatum* complex but that the four syngens of Hiwatashi are members of the same species. Recent events have thus made a situation which seemed clear and unremarkable 5 years ago into a murky area which will require new collections and extensive study before the nature of relationships in *P. caudatum* is known.

2.4 Other Species of *Paramecium*

P. bursaria was the archetypical outbreeder in Sonneborn's (1957) analysis of breeding systems. Siegel demonstrated that the synclonal pattern of mating type inheritance in syngen 1 was determined genotypically by a pair of alleles at two independent loci and determined the immaturity period to be 50 to 100 fissions, 2 to 4 weeks under maximum growth rate conditions (Siegel 1963). Bomford (1966) reviewed and updated the taxonomic situation. Of the six syngens, numbers 1 and 3 have four mating types and the others all have eight. Syngens 2, 4, and 5 are a group linked by intersyngenic mating reactions of some combinations of mating types, but all exconjugants from intersyngenic matings died without dividing. Bomford (1966) collected three syngens in Scotland. Each syngen was common. None of these three syngens had been collected in the well-sampled United States and apparently the New World and Old World have different species. Landis (1982), who did not differentiate the syngens, found *P. bursaria* to have a more uniform spatial and temporal distribution than *P. aurelia* complex individuals.

Paramecium multimicronucleatum syngen 2 was described as the most outbreeding member of the *P. aurelia* complex in the body of Sonneborn's (1957) paper, but reseparated in the Addenda. This species has a long life cycle with an immaturity interval of 40–50 fissions (Kroll and Barnett 1968; Takagi 1970). Allen et al. (1983c) compared the esterase and acid phosphatase isozymes of five syngens of *P. multimicronucleatum*. Stocks of syngens 1 and 5 could not be distinguished from one another. They were distinct from syngens 2, 3, and 4, which were also distinct from one another. They suggested that four species of the *P. multimicronucleatum* complex should be recognized but they did not name them. Syngens 2, 3, and 4 of *P. multimicronucleatum* had a significantly higher frequency of isozyme variants than stocks of the *P. aurelia* complex and they show some evidence of geographic differentiation (Allen et al. 1983d).

Ammermann (1966) reported on the breeding system of *P. trichium* and *P. woodruffi*. He found only two mating types in *P. woodruffi*, but *P. trichium*, also in the *P. bursaria* group, has five. Neither species had a genic system of mating type determination.

2.5 Summary and Interpretation

Relationships among stocks can now be evaluated using techniques, microinjection, isozymes, and sequencing, not available 30 years ago. The new information has corroborated the distinctness of the genetic species of the *P. aurelia* complex as they were erected based on mating relationships. The more recent studies have not corroborated the importance of the distinction between groups A and B and new clusterings have been proposed. Recent information suggests that the mating reactions result in an overclassification of *P. caudatum*. Eventually, sequencing should be able to provide a solid evolutionary tree of relationships, but this goal is far off. There are problems with a taxonomy based on viability and the suggested roles of breeding systems have not held up well.

The taxonomy of "genetic species" has been based on mating reactions and viability of progeny from matings. These criteria rest on a solid foundation of evolutionary theory, but the observed results can be conditional on the environment or other subtle features. Miyake (1968) found that fairly simple chemical changes could induce mating in many stocks, and the induced stocks had lowered specificity. Orias (1959) found that a change of environmental conditions permitted mating between two *Tetrahymena* syngens. Tsukii and Hiwatashi (1983) were able to create environmental conditions permitting intersyngenic matings, and Khadem and Gibson (1985) reported no mating barriers between syngens, suggesting a change of relationships under different environmental conditions. These facts allow the question, "What happens in nature?" to emerge as a reasonable question. It is unfortunately very difficult to answer.

The results of viability studies can also be difficult to interpret. In species in which stocks are not homozygous the traditional criterion of genetic isolation between stocks which do mate is "F 1" inviability. If some progeny of a cross between two stocks survive, the cross is compatible. In *P. aurelia* this criterion would join *P. primaurelia* and *P. septaurelia.* Their distinctness is inferred from the failure of the F 2-by-autogamy generation. Another complication is a situation in which stocks which were members of the same species would seem to be incompatible by the viability criterion. Görtz and Fujishima (1983) found that stocks infected by a micronuclear symbiont, *Holospora elegans,* mated normally but could not produce viable progeny. Thus, the traditional criterion defining genetic species is not without problems.

A major thesis of Sonneborn's (1957) review was that the characters relating to the breeding system were correlated and could be rationalized as adaptations to inbreeding or outbreeding. The movement of *P. caudatum* from inbreeding status to outbreeding status based on immaturity and maturity intervals generates a new (besides *P. multimicronucleatum* syngen 2) exception to the association of multiple mating types with outbreeding. It does, however, reinforce the association of long life cycles with large body size and low (maximum) fission rates among the well-known members of the genus *Paramecium.*

Sonneborn's (1957) expectations about the distribution of species of the genus have not been supported. Many of the inbreeding *P. aurelia* group species have proved to have cosmopolitan distributions. The syngens of *P. bursaria* apparently do not. Only one new species, *P. sonneborni* (Aufderheide et al. 1983), has joined the "aurelia" group, while in the outbreeding *Tetrahymena pyriformis* complex the number of species has increased from 9 to 17 (Simon et al. 1985).

3 Problems and Future Directions

3.1 The Species Problem

The question of the distinctness of the genetic species of the *P. aurelia* complex is solved. All 14 species are molecularly distinct and typically have very low levels of isozyme variation within the species. The average genetic distance is greater than that of sibling species of *Drosophila* (Allen et al. 1983 b). At a minimum there are eight species with substantial natural populations of wide distribution. Typically, multiple samples from the same area yield two to five species in polymorphic frequencies. Stocks of co-occurring species may be differentiated in ways other than mating reactivity, but so far the physiological differences between species have not been strongly connected with their ecology. Although we cannot offer very satisfactory explanations of why there are so many species, we can be sure that the "genetic species" are evolutionarily distinct and are "good" species. While Sonneborn's original conclusions about the *P. aurelia* complex have been supported, studies of *P. caudatum* have found that mating specificities are not necessarily associated with molecular distinctness and inviability. The hypotrichous ciliates also have more complex interrelations of morphology, mating reactivity, and molecular distance (Machelon and Demar 1984; Genérmont et al. 1985; Schlegel 1985).

The problems of degree of relationship within and between species seem very approachable in this age of sequencing. Sequences of homologous DNA or protein provide measures of evolutionary distance with a minimum of ambiguity. They seem to be definitive in providing a pattern of descent in ways that studies of morphology, tolerance, and isozymes can never be. These later types of studies retain importance, however, because they seem to provide the best information to approach the ecological "problem" of why there are so many species within a complex and the ecological relationships of the species complexes. They are also important because they measure variability at different levels of organization. They allow one to ask about an association between, for instance, variability among stocks in tolerance and variability in isozymes. Only by having both sets of data do we see how inappropriate it can be to talk about "variable species" and "stable species". Species whose stocks are more variable for one set of characters are not necessarily more variable for all sets.

3.2 Breeding Systems

Breeding systems among plants and animals exhibit a great deal of diversity. Interest in these systems has focused on evolution (Stebbins 1950; Darlington 1958) rather than population genetics. The terms used, mutational variety, recombinational variety, stability, strangers, and close relatives, were never defined in terms of measurable quantities. Part of the reason that these ideas were never meshed with population genetics can be attributed to the focus of population genetics on heterozygosity and temporal structure; the ideas of strangers and distance imply geographic structure. Population genetics expresses answers on a per generation basis, while the breeding system reasoning is often rooted in real time. Attempts to build genetic models of breeding system ideas exposes considerable gaps in our knowledge of the population biology and ecology of *Paramecium*. Other problems seem apparent at a more qualitative level.

While Sonneborn's (1957) use of the immaturity interval as a quantitative measure of position on an inbreeding-outbreeding scale was admirable in providing an empirical measure of the breeding system, it seems theoretically weak. The ecological association of distance with time is reasonable, but given two individuals who mated at one spot and 1 year later are at another spot (the same distance because of the same time), it is not clear whether their intervening sexual episodes are very important. The rate of gene flow would seem to be dependent on rates of dispersal of individuals. It is not clear whether one can construct a population model in which the interval between matings plays a large role.

Traditionally, more mating types is called an outbreeding characteristic because more equally frequent mating types increase one's probability of being "compatible" with a stranger. It also increases one's probability of being compatible with a member of one's "home pond" population. If one uses heterozygosity to measure the distance of an outcross, then by analogy with the fact that a separate sexes population is trivially more heterozygous than a randomly mating monoecious population, the number of mating types will affect heterozygosity only at loci very close to the mating type locus. Furthermore, increased compatibility with other cells encountered will reduce the time between matings and from that perspective should be considered an inbreeding characteristic. Even the traditional division, self-incompatible versus self-compatible, does not always give one a simple association with heterozygosity, as autogamy in *Euplotes crassus* retains parental heterozygosity (Dini 1984).

The development of a theoretical treatment of breeding systems in population genetics terms is hampered by the lack of information on the population dynamics and timing of sex in natural populations. Populations of *P. aurelia* complex species seem to go through dramatic increases in density or blooms (Hairston 1967; Landis 1981). If sex is occurring only after blooms, and the individuals in a bloom are descended from one or a few individuals, then most matings may be among "close relatives" regardless of breeding system. On the other hand, it is possible that sex is occurring only at some particular season. Maybe individuals of most species typically mate once a year so that the length of the immaturity period is not very important. Alternatively, it may be that a small proportion of the population is mating at all times, like humans, so that sexual births are occurring continuously. Although the consequences of these different models have not been worked out mathematically, it would seem that they could substantially affect the regulation of variety.

3.3 Conclusion

Sonneborn's recognition that stocks of the "species" *P. aurelia* could be divided into many "genetic species" was an insight that continues to resolve new and old problems concerning eukaryotic microbes. His thesis, expounded in 1957, that the term "species" should be reserved for taxonomists and the evolutionary biologist should use the term "syngen" for genetically studied species was not adopted by the biological community. I believe a major reason was that it was embarrassing to reveal how few species had been genetically studied, but there were other reasons as well (Hairston 1958). While the species of the *P. aurelia* complex were defined by their mating reactions and viability upon crossing, they did prove to be distinct in many ways. The "genetic species" have now been given species names (Sonneborn 1975). The assessment of distinctness of stocks of the *P. aurelia* complex is considerably enhanced by the fact that each stock is completely homozygous, only heterotypic pairs are formed, and after conjugation the exconjugant clones are genotypically identical. If parental stocks are heterozygous, then the different synclones (pairs) of a cross have different genotypes and a number of "F 1" progeny needs to be studied individually to assess viability. Whatever the reasons, the situation in *P. caudatum* has never been as clear as that in the *P. aurelia* complex, and currently it seems that syngens 1, 3, 12, and 13 are at least closely related and may exchange genes in nature. Crosses between syngens and progeny of F 2's and backcrosses result in genotypes with dual specificities and others which are mating-typeless (Tsukii and Hiwatashi 1983). These cells may help explain why natural collections of *P. caudatum* have sometimes proved difficult to group (Sonneborn 1957). It is paradoxical that outbreeding species like those of the *P. caudatum* complex may be mating-typeless, while one does not ever seem to find stocks of the *P. aurelia* complex, all of which have autogamy, which do not mate. Combined with the extensive studies of marine *Euplotes* (Luporini and Dini 1977; Machelon and Demar 1984) further studies of the taxonomy and relationships within and among *Paramecium* and other ciliate species complexes is likely to generate controversy and new models of speciation.

Sonneborn (1957) interpreted many of the taxonomic problems in terms of inbreeding and outbreeding. Outbreeding should result in a few clear-cut species, while inbreeding melts into the taxonomic morass of asexuality. Neither the distribution nor number of species supports this prediction. The contrast of breeding systems has been used by many authors (including myself) to rationalize diverse observations comparing different species. In many cases ideas about the breeding system can rationalize some of the variation, but the characterization of a breeding system rests on vague ideas of "strangeness" and "stability". These ideas may ultimately prove to be important evolutionary concepts which can be empirically measured. Until the breeding system can be defined in an operational way or breeding system theory is tied to traditional population genetics, however, breeding system reasoning will retain too much individual peculiarity to be an explanatory force worthy of widespread attention.

References

Agatsuma T, Tsukii Y (1980) Genetic control of lactate dehydrogenase isozymes in *Paramecium caudatum.* Biochem Genet 18:77–85

Allen SL, Byrne BC, Cronkite DL (1971) Intersyngenic variations in the esterases of bacterized *Paramecium aurelia.* Biochem Genet 5:135–150

Allen SL, Farrow SW, Golembiewski PA (1973) Esterase variations between the 14 syngens of *Paramecium aurelia* under axenic growth. Genetics 73:561–573

Allen SL, Lau ET, Nerad TA, Rushford CL (1982) Esterase variants in four species of the *Paramecium aurelia* complex. J Protozool 29:604–611

Allen SL, Nerad TA, Rushford CL (1983 a) Intraspecies variability in the esterases and acid phosphatases of four species of the *Paramecium aurelia* complex. J Protozool 30:131–143

Allen SL, Adams J, Rushford CL (1983 b) Interspecies relationships in the *Paramecium aurelia* complex: acid phosphatase variation. J Protozool 30:143–147

Allen SL, Nerad TA, Rushford CL (1983c) Comparison of the esterases and acid phosphatases in *Paramecium multimicronucleatum*, syngens 1–5, *P. jenningsi, P. caudatum*, and the *P. aurelia* complex. J Protozool 30:148–154

Allen SL, Rushford CL, Nerad TA, Lau ET (1983d) Intraspecies variability in the esterases and acid phosphatases of *Paramecium jenningsi* and *Paramecium multimicronucleatum:* assignment of unidentified *Paramecia;* comparison with the *P. aurelia* complex. J Protozool 30:155–163

Ammermann D (1966) Das Paarungssystem der Ciliaten *Paramecium woodruffi* and *Paramecium trichium.* Arch Protistenk 109:139–146

Aufderheide KJ, Daggett P-M, Nerad TA (1983) *Paramecium sonneborni* n. sp., a new member of the *Paramecium aurelia* species complex. J Protozool 30:128–131

Beale GH, Knowles JKC (1976) Interspecies transfer of mitochondria in *Paramecium aurelia.* Mol Gen Genet 143:197–201

Beale GH, Tait A (1981) Mitochondrial genetics of *Paramecium aurelia.* Int Rev Cytol 71:19–40

Bomford R (1966) The syngens of *Paramecium bursaria:* new mating types and intersyngenic mating reactions. J Protozool 13:497–501

Butzel HM, Jr (1974) Mating type determination and development in *Paramecium aurelia.* In: Van Wagtendonk WJ (ed) *Paramecium,* a current survey. Elsevier, Amsterdam, pp 91–130

Cummings DJ (1980) Evolutionary divergence of mitochondrial DNA from *Paramecium aurelia.* Mol Gen Genet 180:77–84

Cummings DJ, Maki RA, Conlon PJ, Laping J (1980) Anatomy of mitochondrial DNA from *Paramecium aurelia.* Mol Gen Genet 178:499–510

Darlington CD (1958) Evolution of genetic systems, 2nd edn. Oliver & Boyd, Edinburgh, 265 pp

Dini F (1984) On the evolutionary significance of autogamy in the marine *Euplotes.* Am Nat 123:151–162

Dippell RV (1954) A preliminary report on the chromosomal constitution of certain variety 4 races of *Paramecium aurelia.* Caryologia 6 (Suppl): 1109–1111

Gates MA, Berger J (1976) Morphometric inseparability of *Paramecium primaurelia* and *P. pentaurelia.* Trans Am Microsc Soc 95:507–514

Genérmont J, Machelon V, Demar C (1985) The "vannus" group of the genus *Euplotes.* Sibling species and related forms; evolutionary significance and taxonomic implications. Att Soc Tosc Sci Nat Mem (Ser B) 92:53–65

Gilman LC (1941) Mating types in diverse races of *Paramecium caudatum.* Biol Bull 80:384–402

Görtz H-D, Fujishima M (1983) Conjugation and meiosis of *Paramecium caudatum* infected with the micronucleus-specific bacterium *Holospora elegans.* Eur J Cell Biol 32:86–91

Haga N, Saimi Y, Takahashi M, Kung C (1983) Intra- and interspecific complementation of membrane-inexcitable mutants of *Paramecium.* J Cell Biol 97:378–382

Haggard B (1974) Interspecies crosses in *Paramecium aurelia* (syngen 4 by syngen 8). J Protozool 21:152–159

Hairston NG (1958) Observations on the ecology of *Paramecium,* with comments on the species problem. Evolution 12:440–450

Hairston NG (1967) Studies on the limitation of a natural population of *Paramecium aurelia.* Ecology 48:904–910

Hairston NG, Kellerman SL (1965) Competition between varieties 2 and 3 of *Paramecium aurelia:* the influence of temperature in a food limited system. Ecology 46:134–139

Hiwatashi K (1949) Studies on the conjugation of *Paramecium caudatum.* I. Mating types and groups in the races obtained in Japan. Sci Rep Tohoku Univ (4th Ser) 18:137–140

Hiwatashi K (1968) Determination and inheritance of mating type in *Paramecium caudatum.* Genetics 58:373–386

Khadem N, Gibson I (1985) Enzyme variation in *Paramecium caudatum.* J Protozool 32:622–626

Komala Z, Przyboś E (1984) Distribution of the *Paramecium aurelia* species complex in the Carpathian Chain of Poland. Zool Scr 13:161–163

Kroll RJ, Barnett A (1968) The effect of different fission rates on the onset of maturity in *Paramecium multimicronucleatum.* J Protozool 15 (Suppl): 10
Landis WG (1981) The ecology, role of the killer trait, and interactions of five species of the *Paramecium aurelia* complex inhabiting the littoral zone. Can J Zool 59:1734–1743
Landis WG (1982) The spatial and temporal distribution of *Paramecium bursaria* in the littoral zone. J Protozool 29:159–161
Luporini P, Dini F (1977) The breeding system and genetic relationship between autogamous and non-autogamous sympatric populations of *Euplotes crassus.* Monitore Zool Ital 11:119–154
Machelon V, Demar C (1984) Electrophoretic variations among the genus *Euplotes* (Ciliata, Hypotrichida): comparative data for the sibling species complex *Euplotes vannus* and survey of infrageneric variability. J Protozool 31:74–82
Miwa I (1979a) Specificity of the immaturity substances of *Paramecium.* J Cell Sci 36:253–260
Miwa I (1979b) Immaturity substances in *Paramecium primaurelia* and their specificity. J Cell Sci 38:193–199
Miyake A (1968) Induction of conjugation by chemical agents in *Paramecium.* J Exp Zool 167:359–380
Nanney DL (1980) Experimental ciliatology. John Wiley & Sons, New York, 304 pp
Nyberg D (1974) Breeding systems and resistance to environmental stress in ciliates. Evolution 28:367–380
Nyberg D (1975) Genetic analysis of copper resistance in *Paramecium aurelia* syngen 4. Genetics 80:463–473
Nyberg D (1978) Genetic analysis of trichocyst discharge of the wild stocks of *Paramecium tetraurelia.* J Protozool 25:107–112
Nyberg D (1982) Sex, recombination, and reproductive fitness: an experimental study using *Paramecium.* Am Nat 120:198–217
Nyberg D, Bishop P (1983) High levels of phenotypic variability of metal and temperature tolerance in *Paramecium.* Evolution 37:341–357
Nyberg D, Bogar AE (1986) Genotypic and subgenotypic variation in heavy metal tolerance in *Paramecium.* Am Nat 127:615–628
Orias E (1959) Mating interaction between varieties 6 and 8 of *Tetrahymena pyriformis.* J Protozool 6 (Suppl): 19
Powelson EE, Gates MA, Berger J (1975) A biometrical analysis of 22 stocks of four syngens of *Paramecium aurelia.* Can J Zool 53:19–32
Pringle CR, Beale GH (1960) Antigenic polymorphism in a wild population of *Paramecium aurelia.* Genet Res 1:62–68
Przyboś E (1968) The occurrence of syngens of *Paramecium aurelia* in Rumania. Folia Biol 16:131–136
Przyboś E (1980) Distribution of species of the *Paramecium aurelia* complex in Spain. Folia Biol 28:405–412
Schlegel M (1985) Comparative study of allozyme variation in eight species of hypotrichous ciliates. Z Zool Syst Evolutionsforsch 23:171–183
Seilhamer JJ, Olsen GJ, Cummings DJ (1984a) *Paramecium* mitochondrial genes. I. Small subunit rRNA gene sequence and microevolution. J Biol Chem 259:5167–5172
Seilhamer JJ, Gutell RR, Cummings DJ (1984b) *Paramecium* mitochondrial genes. II. Large subunit rRNA gene sequence and microevolution. J Biol Chem 259:5173–5181
Siegel RW (1958) Hybrid vigor, heterosis, and evolution in *Paramecium aurelia.* Evolution 12:402–416
Siegel RW (1963) New results on the genetics of mating types in *Paramecium bursaria.* Genet Res 4:132–142
Simon EM, Meyer EB, Preparata RM (1985) New wild *Tetrahymena* from southeast Asia, China, and North America, including *T. malaccensis, T. asiatica, T. nanneyi, T. caudata,* and *T. silvana* n. spp. J Protozool 32:183–189
Sogin ML, Elwood HJ (1986) Primary structure of the *Paramecium tetraurelia* small-subunit rRNA coding region: phylogenetic relationships within the ciliophora. J Mol Evol 23:53–60

Sonneborn TM (1937) Sex, sex inheritance and sex determination in *Paramecium aurelia.* Proc Natl Acad Sci USA 23:378–385
Sonneborn TM (1939) *Paramecium aurelia:* mating types and groups; lethal interactions; determination and inheritance. Am Nat 73:390–413
Sonneborn TM (1957) Breeding systems, reproductive methods, and species problems in protozoa. In: Mayr E (ed) The species problem. AAAS, Washington, DC, pp 155–324
Sonneborn TM (1975) The *Paramecium aurelia* complex of fourteen sibling species. Trans Am Microsc Soc 94:155–178
Stebbins GL, Jr (1950) Variation and evolution in plants. Columbia Univ Press, New York, 643 pp
Tait A (1970) Enzyme variation between syngens in *Paramecium aurelia.* Biochem Genet 4:461–470
Takagi Y (1970) Expression of the mating-type trait in the clonal life history after conjugation in *Paramecium multimicronucleatum* and *Paramecium caudatum.* Jpn J Genet 45:11–20
Tsukii Y, Hiwatashi K (1983) Genes controlling mating type specificity in *Paramecium caudatum:* three loci revealed by intersyngenic crosses. Genetics 104:41–62
Tsukii Y, Hiwatashi K (1985) Meiotic nondisjunction and aneuploids in intersyngenic hybrids of *Paramecium caudatum.* Genetics 111:779–794
Vivier E (1974) Morphology, taxonomy and general biology of the genus *Paramecium.* In: Van Wagtendonk WJ (ed) *Paramecium,* a current survey. Elsevier, Amsterdam, pp 1–89

Chapter 4 Mating-Type Inheritance

YUUJI TSUKII[1]

CONTENTS

1 Introduction

Discovery of mating type in *Paramecium* (Sonneborn 1937) was the first demonstration of sex in unicellular animals. Initial genetic studies showed peculiar modes of mating-type inheritance, which were subsequently traced to macronuclear differentiations in which genes controlling the potentiality of the expression of mating type were repressed or derepressed (Sonneborn 1947, 1957; Beale 1954; Nanney 1954). Such nuclear differentiations have been extensively investigated as a model system of cell differentiation in metazoa. Genetic control of mating-type substances, complementary macromolecules on ciliary surfaces which confer mating-type specificities on the cells, has also been extensively studied. Recently, genetic analyses using intersyngen (interspecies) hybrids revealed genes controlling the specificity of the mating-type substances (Tsukii and Hiwatashi 1983).

Mating-type genetics has already been reviewed by Preer (1968), Sonneborn (1974), Butzel (1974), Nanney (1977), Hiwatashi and Kitamura (1985), and others. Here, I note those main features, and then focus on recent discoveries of

1 Laboratory of Biology, Hosei University, Chiyoda-ku, Tokyo 102, Japan

Paramecium. Ed. by H.-D. Görtz

genes controlling the specificity of mating type and the polymorphism of mating type, both of which will be discussed in relation to the molecular structure of mating-type substances.

2 Genetic Basis of Mating-Type Inheritance

Each morphological species of *Paramecium* is known to consist of a number of subgroups, designated syngens, which are reproductively isolated by the specificity of the mating type and other breeding relations (Sonneborn 1938). In all *Paramecium* species so far examined, except *P. bursaria* and *P. trichium*, each syngen has two complementary mating types, which were originally designated by a series of pairs of numbers, one odd and one even number. Currently, they are designated O types (for odd numbered) and E types (for even numbered) because of homologies between different syngens (Sonneborn 1975). In *P. bursaria* and in *P. trichium*, each syngen contains four, eight, or more complementary mating types. Cells of one mating type can mate with those of any other mating type of the same syngen. Genetic and physiological studies suggest that these multiple mating-type systems are produced by a duplication of the two mating-type system (Nanney 1980).

In the species with two mating types, it is generally accepted that a pair of alleles with simple dominance controls the expression of mating type. The recessive allele, mt^o restricts homozygotes to type O while the dominant allele, mt^+ permits the expression of type E (Sonneborn 1974). In the species with multiple mating types, the situation is essentially the same except that two or more pairs of alleles with similar functions are involved in the determination of mating types (Siegel and Larison 1960; Cohen and Siegel 1963). For example, four mating types of *P. bursaria* syngen 1 are determined by combinations of two pairs of alleles at independent loci, which are assumed to control two pairs of complementary mating-type substances, A-a and B-b. The four mating types, I, II, III, and IV, can be explained by combinations of these mating-type substances as AB, aB, ab, and Ab, respectively (Metz 1954).

3 Macronuclear Differentiations for the Expression of Mating Type

3.1 Determination of Mating Type in Sexual Reproduction

In sexual reproduction (conjugation or autogamy), the macronuclei developing in a pair of exconjugant cells (or a single exautogamous cell) are genotypically identical. Therefore, all caryonides (clones each descended from a single cell which contains one of the macronuclei developed from a single fertilized nucleus) have the same genotype for the *mt* locus. However, in some species, the expression of the dominant allele, mt^+ is determined by cytoplasmic factors, environmental factors, or actions of other loci. The patterns of mating-type inheritance are thus grouped into three different systems in different species.

3.1.1 Synclonal System

The first, simplest system occurs in *P. tredecaurelia* (Sonneborn 1974), *P. caudatum* (Hiwatashi 1958, 1964, 1968), and *P. bursaria* (Siegel and Larison 1960). In this system, caryonides do not differentiate for mating type. Mating type is determined only by the genotypes at the *mt* locus, and all cells of a synclone (a set of caryonides derived from one conjugating pair) are of the same mating type. This mode of mating-type inheritance is thus referred to as a synclonal system.

In the next two systems, almost all stocks are homozygous for the dominant allele, mt^{+} and caryonides are randomly or cytoplasmically determined to express either type E or type O.

3.1.2 Caryonidal System

The second, caryonidal system is found in *P. primaurelia, P. triaurelia, P. pentaurelia, P. novaurelia, P. undecaurelia, P. quadecaurelia* (Sonneborn 1974, 1977), and acycler strains of *P. multimicronucleatum* syngen 2 (Barnett 1966). In this system, the new macronuclei developing in a single fertilized cell (exconjugant) are independently and randomly determined for their mating types. Therefore, caryonides become either E or O. The ratio of E to O differs in different stocks and varies with temperature. At least one gene affecting the ratio has been found (Butzel 1955). As mentioned above, in this system, most stocks are homozygous for the dominant allele, mt^{+} which controls the potentiality of the expression of type E. In *P. primaurelia* (species 1 of the *P. aurelia* complex), however, certain wild stocks homozygous for a recessive allele, mt^{o} are incapable of expression of type E (Butzel 1955).

3.1.3 Clonal System

The third, clonal system is found in *P. biaurelia, P. tetraurelia, P. sexaurelia, P. septaurelia, P. octaurelia, P. decaurelia,* and *P. dodecaurelia* (Sonneborn 1974, 1977). In this system, each exconjugant gives rise to a progeny clone of the same mating type as that of the cytoplasmic parent. Mating types are determined by cytoplasmic factors which themselves are under nuclear control (Sonneborn 1954). Several O-restricted mutations are known in *P. tetraurelia* and *P. septaurelia* (Taub 1963, Byrne 1973), and an E-restricted mutation is also known in *P. tetraurelia* (Brygoo and Keller 1981). These suggest that not only mt^{+} but also other genes are involved in the determination of mating type. With the exception of one mutation, none of the the O-restricted mutations affect cytoplasmic factors and maintain equally an O- or E-determining cytoplasm according to their cytoplasmic origin. This indicates that macronuclei are stably differentiated not only for mating type, but also for production of the corresponding cytoplasmic factors.

The presence of the cytoplasmic factors was initially shown by a change of mating type which occurred after the exchange of cytoplasm between conjugants (Sonneborn 1947; Butzel 1973). Direct evidence of cytoplasmic E-determining factors has been obtained by transplanting E cytoplasm into O cells just before development of the new macronucleus, which induced transformation of O to E in the progeny clones. On the other hand, transplantation of O cytoplasm into E cells failed to induce a change from E to O (Koizumi 1971). However, when the O macronucleus was transplanted into the cytoplasm of E cells, E exautogamous clones were transformed to O. These results suggest that in O cells, O-determining factors are present only in the macronucleus. Such O-determining "karyoplasmic" factors may be produced or activated at conjugation and then decrease in activity in subsequent fissions after the development of a new macronucleus (Koizumi and Kobayashi 1984).

It has also been suggested that gametic nuclei are predetermined for mating type before nuclear exchange (Sonneborn 1954; Brygoo 1977; Brygoo et al. 1980). In crosses between amicronucleate clones and normal micronucleate clones, O amicronucleate cells became E when they received a micronucleus from a E partner. This can be explained by the predetermination of the gametic nuclei in E cells. However, when the micronuclei from vegetative cells were transplanted into amicronucleate cells of different mating type, exautogamous clones expressed the same mating type as the recipients. This indicates that micronuclei in the asexual cell cycle are undetermined for mating type (Mikami and Koizumi 1982).

3.2 Modification of the Expression of Mating Type During Asexual Reproduction

After sexual reproduction, *Paramecium* cells enter the asexual cell cycle and express their mating types. However, actual expression of mating type is affected not only by the clonal life stages but also by cultural stages or diurnal changes. Furthermore, though cells usually reproduce true to mating types which were determined at sexual reproduction, activity of the dominant gene, mt^+ is, in some cases, repressed or derepressed by various intra- or extracellular factors, and thus mating-type switching occurs.

3.2.1 Control of Mating Activity

Mating activity usually appears when two or three conditions are fulfilled. First, mating-type expression occurs only when cells are sexually mature. In most cases, *Paramecium* cells do not express their mating types during an immaturity period which precedes the maturity period. The length of the immaturity period varies in different species (syngens) and is measured in terms of the number of cell divisions rather than in absolute time (Sonneborn 1957; Miwa and Hiwatashi 1970). Immaturity is totally lacking in *P. tetraurelia* and *P. octaurelia.* Second, even in the maturity period, cells express their mating activity only when moderately starved. Almost nothing is known about how starvation induces mating activity. Furthermore, strains of *P. bursaria, P. biaurelia, P. triaurelia* and *P. undecaurelia* show a circadian rhythm of mating activity (Cohen 1965; Karakashian 1968; Sonneborn 1975).

3.2.2 Control of the Activity of the mt^+ Gene

In advance of cellular aging, mating-type expression may become unstable so that a reversible change from E to O or from O to E occurs within a culture. Such mating-type instability, i.e., instability of the activity of the mt^+ gene, is common in stocks of E type in *P. caudatum* (Hiwatashi 1969; Myohara and Hiwatashi 1975). The change from E to O occurs without segregating clones of different mating types. Time of onset of the instability is affected by slow growth, actinomycin D, ribonuclease, and puromycin (Hiwatashi 1969). A modifier gene is required for the mating-type change in *P. caudatum* syngen 3 (Hiwatashi and Myohara 1976). Mating-type instability was also found in *P. pentaurelia* and *P. septaurelia* (Taub 1966a, b; Bleyman 1967).

A unique circadian rhythm of mating-type reversal is known in syngen 2 of *P. multimicronucleatum* (Sonneborn and Barnett 1958). Certain strains of this species (cyclers) show a daily cycle of changes of the two mating types. These changes of mating-type expression can be controlled by the setting of the biological clock, i.e., activity of an mt^+ is switched under the control of a biological clock. A dominant allele *C* is necessary for the appearance of the circadian rhythm (Barnett 1966).

4 Genetic Control of Mating-Type Substances

4.1 Butzel's Hypothesis; the Precursor-Product Hypothesis

In *P. primaurelia*, Butzel (1953, 1955, 1974) proposed an hypothesis that the O mating-type substance is a precursor of the E mating-type substance, and the mt^+ gene controls the conversion of O substance to E substance. In a homozygote for the recessive allele, mt^o O substance could not be converted to E, and all caryonides would be type O. His hypothesis was later recognized as capable, with minor modifications, of accounting for all three systems of mating-type inheritance (Sonneborn 1974). The mt^+ gene in the synclonal system could exist in only one state capable of derepression. In the other two systems, the activity of the mt^+ gene can be repressed or derepressed by chance (caryonidal) or by the cytoplasmic factors (clonal). The O substance would be constitutively produced in any cell, while the conversion of O to E can be induced to function under the action of mt^+.

This precursor-product hypothesis explains how genes control the difference of complementary mating types, however, there is no direct evidence for proving the hypothetical action of mt^+. As already mentioned, mt^+ is only known to control the expression of E type. We do not know whether mt^+ directly controls the specificity of E type or regulates the expression of E type. On the other hand, it is unlikely that the recessive allele mt^o at the *mt* locus controls the specificity of type O, because cells homozygous for the dominant allele mt^+ can express both E and O types. Thus, the specificity of O type itself should be controlled by alleles not at the *mt* locus but at some other loci. Genes controlling the specificities of E and O mating types were recently found by breeding analyses using intersyngenic hybrids of *P. caudatum.*

4.2 Three Genes Controlling the Specificities of E and O Types

In Mendelian genetics, we can find genes only when we have phenotypic differences of alleles. We identify the genes by crossbreeding analyses using such different phenotypes. In the species with two mating types, there are no phenotypic variations of type E or type O. Therefore, from breeding analyses, we only know genes controlling the difference of mating type, E or O (i.e., mt^+). However, if it is possible to introduce different mating types which belong to different syngens into a single breeding unit, these mating types can be used as phenotypic variants of type E or O and we may find genes controlling the specificity of each mating type.

In the *P. aurelia* complex and in *P. bursaria*, conjugation between certain sibling species (syngens) has been known to occur, but all these interspecific or intersyngenic conjugations were either lethal or sterile (Sonneborn and Dippell 1946; Bomford 1966; Haggard 1974). On the other hand, in *P. caudatum*, recent studies have shown that intersyngen hybrids are completely fertile (Tsukii and Hiwatashi 1983). Although the mating reaction occurs specifically between complementary mating types, subsequent pair formation is not mating-type specific (Hiwatashi 1951; Miyake 1960). Thus, by mixing mating reactive cells of four mating types belonging to two different syngens, clones of intersyngen hybrids were obtained. Extensive breeding analyses using these hybrids revealed three genes controlling the syngen specificities of E and O mating types (Table 1).

The syngen specificity of E types is controlled by multiple codominant alleles at the *Mt* locus (e.g., Mt^1, Mt^3, Mt^{12}; superscripts designate the number of syngen). *Mt* is evidently the same as mt^+. Clones heterozygous for the *Mt* locus express dual mating types (e.g., clones of Mt^1/Mt^3 express E^1E^3). On the other hand, the syngenic specificity of O types is controlled by interactions of codominant multiple alleles at two loci, *MA* and *MB*. *Mt* is epistatic to *MA* and*MB*, and O mating types can be expressed only in the recessive homozygote

Table 1. Mating types and corresponding genotypes of strains in each syngen and intersyngen hybrids of *P. caudatum*

Phenotypes	Genotypes	
Strains in each syngen		
E^1	$Mt^1/Mt^1; MA^1/MA^1; MB^1MB^1$	$Mt^1/mt; MA^1/MA^1; MB^1/MB^1$
O^1	$mt/mt; MA^1/MA^1; MB^1/MB^1$	
E^3	$Mt^3Mt^3; MA^3/MA^3; MB^3/MB^3$	$Mt^3/mt; MA^3/MA^3; MB^3/MB^3$
O^3	$mt/mt; MA^3/MA^3; MB^3/MB^3$	
Intersyngen hybrids		
E^1E^3	$Mt^1Mt^3; MA^a/MA^a; MB^a/MB^a$	
O^1O^3	$mt/mt; MA^1/MA^3; MB^1/MB^3$	
O^0 [b]	$mt/mt; MA^1/MA^1; MB^3/MB^3$	$mt/mt; MA^3/MA^3; MB^1/MB^1$

[a] Mt^1 and Mt^3 can suppress the expression of *MA* and *MB* regardless of the latter's syngen specificity.
[b] O^0: mating-typeless, for details see text.

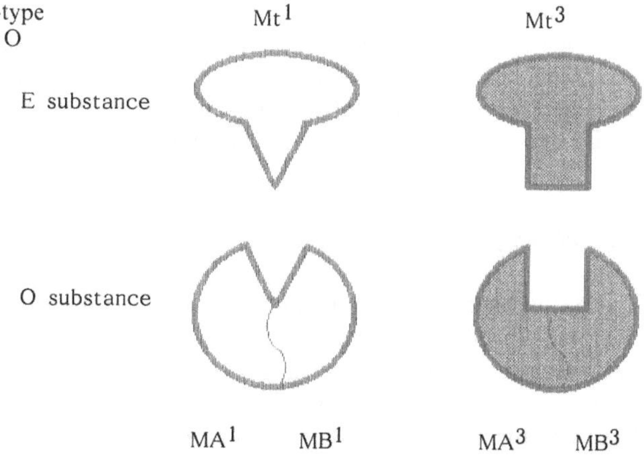

Fig. 1. Model for the mating-type substances of type E and type O

at the *Mt* locus (*mt*/*mt*). In addition, at least one allele each at the *MA* and *MB* loci must have a common syngen specificity for the expression of O types. Therefore, when *MA* and *MB* are homozygous for different syngen specificities and the *Mt* locus is the recessive homozygote (*mt*/*mt*) (e.g., *mt*/*mt*; MA^1/MA^1; MB^3/MB^3), no mating type is expressed (Fig. 1). Such mating-typeless clones, however, can be induced to conjugate chemically (Miyake 1958), so that we know that they are sexually mature. On the other hand, *Mt* is capable of expression even if the genotypes of *MA* and *MB* are such that they prevent the expression of any O type (Tsukii and Hiwatashi 1983).

Butzel (1953) proposed that there are not only similar genes in all syngens for O and E substances but, in addition, there were other genes which confer the syngen specificity on these substances. However, in the three-gene model for the determination of mating type in *P. caudatum* such an assumption of "similar" and additional "specificity" genes is not necessary.

From breeding analyses *Mt* is shown to control the syngen specificity of the E type or the "syngenic" difference of type E. However, it is evident that *Mt* also controls the specificity of the E type itself, because the "syngenic" difference of E type and the "type-E" specificity are always co-inherited (Tsukii and Hiwatashi 1983). This indicates that the reactive site of E substance is composed of a single protein. On the other hand, it is assumed that the reactive site of O substance consists of at least two subunits, each of which is controlled syngen-specifically by two genes, *MA* and *MB* (Fig. 1). According to this model, the mating-typeless clones should have aberrant type O reactive sites. Such reactive sites may be functional if appropriate E types were present. These predictions could be verified if we could reconstitute mating-reactive membrane vesicles from solubilized ciliary fractions of two mating-typeless clones of opposite genotypes (Kitamura and Hiwatashi 1980).

As to the mechanism of the epistasis between *Mt* and *MA* or *MB*, it is quite interesting that *Mt* can repress the activity of *MA* and *MB* even if their syngen specificities are different. For example, clones of a genotype, Mt^1/mt; MA^3/MA^3; MB^3/MB^3 express E^1. When these clones became unstable for the expression of mating type, they changed from E^1 to O^3. If the mechanism of epistasis did not work in some combinations of different syngens, i.e., if *Mt* of one syngen could not repress *MA* and *MB* of another syngen, hybrid clones might express heterotypic dual mating types, EO. Such dual mating types, if discovered, would make feasible the analysis of the mechanism of epistasis.

4.3 Polymorphism of Mating Type

The occurrence of intersyngenic (interspecies) mating reactions in the *P. aurelia* complex, *P. bursaria,* and *P. caudatum* (Sonneborn and Dippel 1946; Gilman 1949; Bomford 1966; Haggard 1974) suggests that the E types of most or all of the syngens, and also the various O types, are homologous and diverged from a common ancestor. Some syngens may still now be on the course of mating-type divergence. If so, it is highly possible that there are some variations of mating type within syngens, which may lead to a multiplication of syngens (speciation). Variations of mating type were recently found in two syngens of *P. caudatum* when checking mating activity of stocks by mixing with various strains of complementary mating types belonging to the same as well as different syngens.

In syngen 12, variations were found in both mating types, E and O, (E^{12a}, E^{12b} and O^{12a}, O^{12b}). E^{12a} is complementary both to type O^{12a} and O^{12b}, but E^{12b} is reactive only with type O^{12a} (Table 2). Furthermore, O^{12a} shows a pseudomating reaction with type E^3 of syngen 3. When O^{12a} is mixed with E^3, they immediately show mating agglutination, but this agglutination is not a real mating reaction, because within 5 or 10 min the agglutinative reaction gradually disappears and no pairs are induced. Crossbreeding analyses showed that the variation of type E^{12} is controlled by alleles Mt^{12a} and Mt^{12b} at the *Mt* locus, and that of type O^{12} by the alleles MA^{12a} and MA^{12b} at *MA* (Tsukii and Hiwatashi 1981; Tsukii, in prep.).

Table 2. Mating reactions among three different syngens of *P. caudatum*

Syngens	Mating types	Syngen 1			Syngen 3		Syngen 12			
		O^1	O^{1*}	E^1	O^3	E^3	O^{12a}	O^{12b}	E^{12a}	E^{12b}
Syngen 1	O^1	–	–	+	–	–	–	–	–	–
	O^{1*}		–	+	–	+	–	–	–	–
	E^1			–	–	–	–	–	–	–
Syngen 3	O^3				–	+	–	–	–	–
	E^3					–	(+)[a]	–	–	–
Syngen 12	O^{12a}						–	–	+	+
	O^{12b}							–	+	–
	E^{12a}								–	–
	E^{12b}									–

[a] (+) Indicate pseudoreaction, i.e., agglutinations disintegrate within several minutes and no conjugating pair is induced.

Though no mating reaction occurs between E^{12b} and O^{12b}, genes could be exchanged among these four mating types. Thus, all of them are included within a single breeding unit or a single syngen. However, if either O^{12a} or E^{12a} changes to the one incapable of cross-reaction, for example, if E^{12a} changes by mutation to a new mating type (e.g., E^{12c}) which cross-reacts only with O^{12b}, but not with O^{12a}, gene flow will stop between the two groups, the group of E^{12b} and O^{12a}, and that of O^{12b} and E^{12c}. Thus, the two groups would soon be separated into two new syngens.

On the other hand, in syngen 1, a new variant of O type, O^{1*} was found to cross-react with two E types, E^1 of syngen 1, and E^3 of syngen 3, and its mating reaction with E^3 is stronger than that with E^1 (Table 2). Thus, O^{1*} is phenotypically close to O^3 of syngen 3 rather than O^1 of syngen 1. However, by breeding analyses, this new variant was proved to be a variant of O^1 type controlled by an allele at the *MA* locus, designated as MA^{1*} (Tsukii and Hiwatashi 1982; Tsukii, in prep.).

Circumstantial evidence suggests that O^{1*} recently arose from O^1 and incidentally had a complementarity with type E^3. In *P. caudatum*, 16 syngens have already been reported (Gilman 1950). This means that the reactive sites of mating-type substances can exist at least in 16 different sets of conformations. We do not know how many sets of conformations are possible. It might depend, however, on the complexity of the reactive sites. The finding of O^{1*} suggests that the possible numbers of different mating types may not be so large, and therefore, the molecular structure of the reactive sites is not so complicated.

Acknowledgements. I thank Drs. Koichi Hiwatashi and Donald L. Cronkite for their invaluable help in improving the manuscript. This study was supported by grants 59740312 and 61740367 from the Ministry of Education, Science, and Culture.

References

Barnett A (1966) A circadian rhythm of mating type reversals in *Paramecium multimicronucleatum*, syngen 2, and its genetic control. J Cell Physiol 67:239–270

Beale GH (1954) The genetics of *Paramecium aurelia*. Cambridge Univ Press, New York

Bleyman L (1967) Selfing in *Paramecium aurelia*, syngen 5: persistent instability of mating type expression. J Exp Zool 165:139–146

Bomford R (1966) The syngens of *Paramecium bursaria:* new mating types and intersyngenic mating reactions. J Protozool 13 (3):497–501

Brygoo Y (1977) Genetic analysis of mating-type differentiation in *Paramecium tetraurelia*. Genetics 87:633–653

Brygoo Y, Keller AM (1981) Genetic analysis of mating type differentiation in *Paramecium tetraurelia*. III. A mutation restricted to mating type E and affecting the determination of mating type. Dev Genet 2:13–22

Brygoo Y, Sonneborn TM, Keller AM, Dippell RV, Schneller MV (1980) Genetic analysis of mating type differentiation in *Paramecium tetraurelia*. II. Role of the micronuclei in mating type determination. Genetics 94:951–959

Butzel HM, Jr (1953) The genic basis of mating type determination and development in the varieties of *Paramecium aurelia* belonging to group A. Doct Diss, Indiana Univ, Bloomington

Butzel HM, Jr (1955) Mating type mutations in variety I of *Paramecium aurelia*, and their bearing upon the problem of mating type determination. Genetics 40:321–330

Butzel HM, Jr (1973) Abnormalities in nuclear behavior and mating type determination in cytoplasmically bridged exconjugants of doublet *Paramecium aurelia*. J Protozool 20:140–143
Butzel HM, Jr (1974) Mating type determination and development in *Paramecium aurelia*. In: Van Wagtendonk WJ (ed) *Paramecium:* a current survey. Elsevier, Amsterdam, p 91
Byrne BC (1973) Mutational analysis of mating type inheritance in syngen 4 of *Paramecium aurelia*. Genetics 74:63–80
Cohen LW (1965) The basis for the circadian rhythm of mating in *Paramecium bursaria*. Exp Cell Res 37:360–367
Cohen LW, Siegel RW (1963) The mating type substances of *Paramecium bursaria*. Genet Res 4:143–150
Gilman LC (1949) Intervarietal mating reactions of *Paramecium caudatum*. Biol Bull 97:239
Gilman LC (1950) The position of Japanese varieties of *Paramecium caudatum* with respect to American varieties. Biol Bull 99:348–349
Haggard BW (1974) Interspecies crosses in *Paramecium aurelia* (syngen 4 by syngen 8). J Protozool 21:152–159
Hiwatashi K (1951) Studies on the conjugation of *Paramecium caudatum*. IV. Conjugating behavior of individuals of two mating types marked by a vital staining method. Sci Rep Tohoku Univ (Biol) 19:95–99
Hiwatashi K (1958) Inheritance of mating types in variety 12 of *Paramecium caudatum*. Sci Rep Tohoku Univ (Biol) 24:119–129
Hiwatashi K (1964) Mating type inheritance in *Paramecium caudatum* syngen 3. Genetics 50:255–256
Hiwatashi K (1968) Determination and inheritance of mating type in *Paramecium caudatum*. Genetics 58:373–386
Hiwatashi K (1969) Genetic and epigenetic control of mating type in *Paramecium caudatum*. Jpn J Genet 44:383–387
Hiwatashi K, Kitamura A (1985) Fertilization in *Paramecium*. In: Metz CB, Monroy A (eds) Biology of fertilization. Academic Press, London New York, p 57
Hiwatashi K, Myohara K (1976) A modifier gene involved in the expression of the dominant mating type allele in *Paramecium caudatum*. Genet Res 27:135–141
Karakashian MW (1968) The rhythm of mating in *Paramecium aurelia*, syngen 3. J Cell Physiol 71:197–210
Kitamura A, Hiwatashi K (1980) Reconstitution of mating active membrane vesicles in *Paramecium caudatum*. Exp Cell Res 125:486–489
Koizumi S (1971) The cytoplasmic factor that fixes macronuclear mating type determination in *Paramecium aurelia*, syngen 4. Genetics 68 (Suppl):34
Koizumi S, Kobayashi S (1984) A study of odd mating-type determination factor by transfer of macronuclear karyoplasm in *Paramecium tetraurelia*. Genetics 107:367–373
Metz CB (1954) Mating substances and the physiology of fertilization in ciliate. In: Wenrich DH (ed) Sex in microorganisms. Am Assoc Adv Sci, Washington, p 284
Mikami K, Koizumi S (1982) Nuclear transplant studies of the determination of mating type in germ nuclei of *Paramecium tetraurelia*. Exp Cell Res 137:397–402
Miwa I, Hiwatashi K (1970) Effect of mitomycin C on the expression of mating ability in *Paramecium caudatum*. Jpn J Genet 45:269–275
Miyake A (1958) Induction of conjugation by chemical agents in *Paramecium caudatum*. J Inst Polytech Osaka City Univ Ser D 9:251–296
Miyake A (1960) Artificial induction of conjugation by chemical agents in *P. aurelia, P. multimicronucleatum, P. caudatum* and between them. J Protozool 7 (Suppl):15 (Abstr)
Myohara K, Hiwatashi K (1975) Temporal patterns in the appearance of mating type instability in *Paramecium caudatum*. Jpn J Genet 50:133–140
Nanney DL (1954) Mating type determination in *Paramecium aurelia*. A study in cellular heredity. In: Wenrich IF, Lewis JR, Raper JR (eds) Sex in microorganisms. Am Assoc Adv Sci, Washington, p 266
Nanney DL (1977) Cell–cell interactions in ciliates: evolutionary and genetic constraints. In: Reissig JL (ed) Receptors and recognition, Ser B, vol 3. Microbial interactions. Halsted/ Wiley & Sons, New York; Chapman & Hall, London

Nanney DL (1980) Experimental ciliatology. An introduction to genetics and developmental analysis in ciliates. Wiley & Sons, New York

Preer JR, Jr (1968) Genetics of the Protozoa. In: Chen TT (ed) Research in protozoology, Pergamon, Oxford, p 129

Siegel RW, Larison LL (1960) The genetic control of mating types in *Paramecium bursaria.* Proc Natl Acad Sci USA 46:344–349

Sonneborn TM (1937) Sex, sex inheritance and sex determination in *Paramecium aurelia.* Proc Natl Acad Sci USA 23:378–395

Sonneborn TM (1938) Mating types in *Paramecium aurelia:* diverse conditions for mating in different stocks; occurrence, number and interrelations of the types. Proc Am Philos Soc 79:411–434

Sonneborn TM (1947) Recent advances in the genetics of *Paramecium* and *Euplotes.* Adv Genet 1:263–358

Sonneborn TM (1954) Pattern of nucleocytoplasmic integration in *Paramecium.* Caryologia 1 (Suppl):307–325

Sonneborn TM (1957) Breeding systems, reproductive methods and species problems in Protozoa. In: Mayr E (ed) The species problem. Assoc Adv Sci, Washington

Sonneborn TM (1974) *Paramecium aurelia.* In: King RC (ed) Handbook of genetics, vol 2. Plenum, New York, p 469

Sonneborn TM (1975) The *Paramecium aurelia* complex of 14 sibling species. Trans Am Microsc Soc 94:155–178

Sonneborn TM (1977) Genetics of cellular differentiation: stable nuclear differentiation in eukaryotic unicells. Annu Rev Genet 11:349–367

Sonneborn TM, Barnett A (1958) The mating type system in syngen 2 of *Paramecium multimicronucleatum.* J Protozool 5 (Suppl):18

Sonneborn TM, Dippel RV (1946) Mating reactions and conjugation between varieties of *Paramecium aurelia* in relation to conceptions of mating type and variety. Physiol Zool 19:1–18

Taub SR (1963) The genetic control of mating type differentiation in *Paramecium.* Genetics 48:815–834

Taub SR (1966a) Unidirectional mating type changes in individual cells from selfing cultures of *Paramecium aurelia.* J Exp Zool 163:141–150

Taub SR (1966b) Regular changes in mating type composition in selfing cultures and in mating type potentiality in selfing caryonides of *Paramecium aurelia.* Genetics 54:173–189

Tsukii Y, Hiwatashi K (1981) Polymorphism of mating type and its genetics in *Paramecium caudatum.* Zool Mag (Tokyo) 90:640 (in Japanese)

Tsukii Y, Hiwatashi K (1982) Mating-type variants in *Paramecium caudatum* and its evolutionary implications. Zool Mag (Tokyo) 91:608 (in Japanese)

Tsukii Y, Hiwatashi K (1983) Genes controlling mating-type specificity in *Paramecium caudatum:* three loci revealed by intersyngenic crosses. Genetics 104:41–62

Chapter 5 Conjugation

MASAHIRO FUJISHIMA [1]

CONTENTS

1 Introduction

Conjugation in *Paramecium* as in other ciliates is a sexual event in the life cycle. During this process, cells of complementary mating types mutually activate to acquire the capacity to unite and form conjugating pairs. In the pairs, a series of nuclear changes such as micronuclear migration, premeiotic DNA synthesis, meiosis, degradation of meiotic products, and formation of pronuclei, exchange and fusion of the pronuclei, and divisions of synkaryon, occurs. Furthermore, preexisting cortical structures including an oral apparatus are reconstructed. Then cells separate, but postzygotic nuclear divisions, differentiation of the new macro- and micronucleus, and degradation of the old macronucleus continue in the exconjugants. Thus, the conjugation process in *Paramecium* includes general

1 Biological Institute, Faculty of Science, Yamaguchi University, Yamaguchi 753, Japan

Paramecium. Ed. by H.-D. Görtz

features of fertilization and early development. In this chapter, various methods for induction of conjugation will be described first. Then, the discussion will be focused on cell surface events and nuclear events in conjugation, especially on control mechanism of meiosis.

2 Induction of Conjugation

2.1 Mixing of Complementary Mating-Type Cells

Since the discovery of mating types in *Paramecium* by Sonneborn (1937) several methods were developed to induce conjugating pairs. The simplest way for the induction of conjugation is the mixing of moderately starved cells of the complementary mating types. Immediately after mixing, cells stick to each other at their ventral surface cilia (Hiwatashi 1951, 1961; Miyake 1964), form large agglutinates, and eventually sink to the bottom of the vessel. This sexual cell agglutination is called the "mating reaction". In this process, no conjugation signals such as gamones in *Blepharisma* and *Euplotes* nor chemotactic attractants have been detected. Namely, the mating reaction occurs by accidental collisions among the ventral cilia of the two complementary mating types. Substances which exist on the ventral cilia and induce the mating reaction are called the "mating-type substances", which are not yet purified but almost certainly proteins (see review, Kitamura and Hiwatashi 1984 and Kitamura, Chap. 6, this Vol.). To express strong mating reactivity, cells must be neither too starved nor too fed. But nothing is known about how starvation induces the mating reactivity. Moderately starved cells in early stationary phase are in G_1 phase of the cell cycle (Woodard et al. 1966; Pieri et al. 1968; Golikova 1974; Fujishima 1983a). However, since *P. tetraurelia* cultured in a microcapillary with a small amount of culture medium can express mating reactivity through the whole cell cycle (Miwa and Umehara 1983), the expression of the mating reactivity may not depend on the cell cycle stage.

The clumps of cells in the mating reaction break down into individual pairs. In *P. caudatum* the first pair appears about 45 min after mixing of complementary mating types of 25° C. In the pairs cells loosely adhere at their deciliated anterior tips. Such an early pair is called the "holdfast union". About 1.5 h after the mixing, the cells of a pair adhere more intimately with deciliated surfaces just posterior to the oral apparatus. This pair is called the "paroral union". Because actinomycin, puromycin, and concanavalin A inhibit pair formation, protein synthesis and membrane fluidity might be involved in this process (Miyake 1969; Tsukii and Hiwatashi 1978).

The mating reaction is a mating-type specific reaction but pair formation can occur in any combination of mating types. Therefore, in the mixture of the two mating types, three kinds of pairs are formed: two kinds of homotypic pairs (selfing pairs) and heterotypic pairs (cross pairs). Meiosis and other nuclear changes of conjugation occur not only in the heterotypic pairs but also in the homotypic pairs. Discernment of the three kinds of pairs is especially important for genetical experiments by cross-analyses. These pairs are discernible if cells of one mating type are marked differently by brief incubation in India ink. However, as time passes, the India ink particles in the food vacuoles are discharged from the cells, so that the difference of pairs comes to be obscure. In *P. tetraurelia* and *P. caudatum* syngen 3, pawn and CNR mutations are used as genetic markers (see Ramanathan et al., Chap. 15; Takahashi, Chap. 17, this Vol.). When the mutants are mixed with wild-type cells of complementary mating type, mutant-mutant, wild-wild, and mutant-wild pairs are easily distinguishable throughout the period of conjugation on their swimming behavior, because the mutants have a defect in membrane excitability and cannot be induced to swim backward in high K^+ solution (Kung and Eckert 1972; Takahashi and Naitoh 1978). The duration of the conjugation process differs in *Paramecium* species, i.e., about 24 h in *P. bursaria,* about 15 h in *P. caudatum,* about 12 h in *P. multimicronucleatum,* about 6 h in *P. tetraurelia.*

2.2 Natural Selfing Pairs

In *P. caudatum,* clones of even mating type stably express the even mating type during the first month after conjugation (for explanation of the terms "even" and "odd mating type" see Tsukii, Chap. 4, this Vol.). Later, however, the clones express the even mating type only in the early stationary phase, but have the odd mating type in the late stationary phase. Consequently, during the transition period, mating reaction occurs in the clone and conjugation pairs are formed. These naturally formed homotypic pairs are called the "natural selfing pairs". The selfing due to the change of mating type is never observed in clones of the odd mating type. Generally, old clones of the even mating type tend to express the odd mating type in the early stationary phase.

2.3 Mating Reaction with Killed Cells

Several physical and chemical agents are known to be useful for obtaining mating reactive killed cells: lyophilization for *P. tetraurelia* and *P. calkinsi,* treatment with formalin for *P. tetraurelia, P. calkinsi,* and *P. caudatum,* picric acid and $HgCl_2$ for *P. calkinsi,* ammonium sulfate and glycerine for *P. caudatum,* $K_2Cr_2O_7$ for *P. multimicronucleatum* and *P. caudatum.* By means of mixing killed cells with mating reactive living cells of complementary mating type, mating reaction is induced between the killed cells and the living cells, and homotypic pairs consisting of the living cells are formed. For details to prepare the killed cells, see review of Hiwatashi (1969).

2.4 Mating Reaction with Isolated Cilia or Ciliary Components

The first crude fraction of mating reactive cilia was obtained by Cohen and Siegel (1963) in *P. bursaria* by differential centrifugations of cell brei. Later, the mating reactive cilia were isolated by different methods: with $K_2Cr_2O_7$ in *P. multimicronucleatum,* with $MnCl_2$, Triton X-100 and $CaCl_2$, and STEEP and $CaCl_2$ in *P. caudatum,* with $K_2Cr_2O_7$ in *P. octaurelia* (see review, Hiwatashi 1981). These cilia can induce mating agglutination and conjugation when applied to mating reactive living cells of the complementary mating type. Takahashi et al. (1974) succeeded in inducing mating agglutination between the isolated cilia from both mating types. Furthermore, Watanabe (1977) showed that pure isolated ciliary membranes have a strong mating reactivity, whereas axonemal and matrix proteins have no mating reactivity. Kitamura and Hiwatashi (see review, 1984) obtained membrane vesicles with high mating reactivity by treating isolated cilia of *P. caudatum* with urea-EDTA or lithium diiodosalicylate. For details on the mating-type substances, see Kitamura (Chap. 6, this Vol.).

2.5 Chemical Induction of Conjugation

Selfing can be induced by chemical treatment in certain *Paramecium* species: *P. caudatum, P. aurelia* species complex, *P. multimicronucleatum, P. jenningsi, P. traunsteineri,* and *P. bursaria.* This method was first developed by Miyake (1956) and named the "chemical induction of conjugation". The following cations and chemicals are known to be effective under Ca^{2+}-poor conditions: K^+, Mg^{2+}, Rb^+, Cs^+, Na^+, Li^+, Mn^{2+}, heparin, acriflavine, and EDTA (for review, see Miyake 1981). Ca^{2+} has an inhibitory effect to the chemical induction, although it has a weak inducing action at high concentrations. Acetamide, urea, biuret, and methylurea promote the inducing action of certain cations. During the chemical induction no mating reaction is seen, but the holdfast union appears almost without delay.

Generally, the capacity of cells for chemical induction of conjugation is intimately related with the capacity of cells for the expression of mating reactivity; chemical induction can be induced neither in immature cells nor in log phase cells of mature cells (Miyake 1968a). However, Takagi (1971) found that during the development of sexual maturity in *P. multimicronucleatum,* there is a period in which cells can form pairs by mating reaction but not by chemical treatment. The same was also observed in *P. caudatum* (Fujishima and Hiwatashi 1977). Furthermore, Cronkite (see review 1979) found two recessive mutants in *P. octaurelia* that were completely blocked in the chemical induction of conjugation but that could mate normally by the mating reaction. Tsukii and Hiwatashi (1983) obtained mating type-less clones by intersyngenic crosses in *P. caudatum,* in which only the expression of mating type is blocked but other mechanisms necessary for the chemical induction are normal. On the other hand, Miyake (1969) found that two kinds of stimulations for induction of conjugation, chemical treatment, and mating reaction, can work additively for the pair formation if the stimulations are sequentially applied to cells. From these results it is possible to assume that the receptor for the chemical stimulus is not the mating substance itself but the two kinds of stimulations operate through an identical regulation mechanism for the induction of holdfast unions.

Autogamy is known to occur naturally in certain *Paramecium* species (i.e., *P. aurelia* species complex. *P. polycaryum, P. jenningsi,* and *P. multimicro-*

nucleatum except syngen 2). The method of chemical induction of conjugation can be used with slight modifications for an artificial induction of autogamy. Miyake (1968b) first succeeded to do so with *P. multimicronucleatum* syngen 2. He suspended the cells in Ca^{2+}-poor conditioned medium containing KCl, acriflavine and ficin. Later, modified methods were succcessfully reported in *P. caudatum* (Tsukii and Hiwatashi 1979) and in *P. multimicronucleatum* syngen 2 (Shimomura and Takagi 1984).

2.6 *Mixing with Killers*

In *P. bursaria*, Chen (1945) observed conjugation-like pairing in sensitive stocks which were mixed with killer stocks. Later, a similar phenomenon was observed in the *P. aurelia* species complex (see review, Miyake 1981). The mode of conjugation induced by killers resembled chemical induction of conjugation. For example, it occurred without mating reaction. It did not need mixing of complementary mating types. It was inhibited by Ca^{2+} but enhanced by K^{+}. Homogenates of killer cells were also effective. Therefore, the conjugation-inducing activity seems to be associated with the symbiotic bacteria in the killer cytoplasm, but details for the induction mechanism are still unclear.

2.7 *Reconjugation*

In *Paramecium* species, exconjugants usually do not conjugate soon after the exconjugation. However, such early conjugation, the so-called reconjugation, is known to occur rarely in *P. caudatum* (Klitzke 1914; Diller 1942), *P. trichium* (Diller 1948; Jankowski 1966), and *P. triaurelia* (Kosćiuszko 1982). Because cells can enter the reconjugation process at various stages of the postzygotic nucleus this phenomenon provides us with an opportunity to examine this process when during the nuclear differentiation process the nuclei acquire the capacity needed for functional micro- and macronuclei. However, optimum conditions to gain a high induction ratio of the reconjugation have not been found.

3 Cell Surface Events in Conjugation

3.1 *Inactivation of Ciliary Movement*

Miyake (1958) reported that cells of *P. caudatum* which had been transferred into conjugation-inducing chemicals showed a special swimming pattern prior to conjugation. After mixing with chemicals, cells immediately began to spin and sank to the bottom of the culture vessel. Then, they whirled the anterior end in a wide circle, while holding the posterior end stationary. About 15 min after the mixing, they gradually became sluggish. Namely, their ciliary movement was temporarily inactivated. Thereafter, holdfast unions appeared simultaneously with recovery of the ciliary movement. Similar swimming behavior in the chemical induction was also reported in *P. octaurelia* (Cronkite 1979). On the other hand, such inactivation of the ciliary movement was found to be induced not only by conjugation-inducing chemicals but also by the mating reaction (Kitamura et al. 1979). These results indicate that the inactivation of ciliary movement is closely connected with the activation of cells to form conjugating pairs.

3.2 Local Degeneration of Cilia

In *P. caudatum*, Hiwatashi (1955b) observed that cells conjugated at the cilia-free area of the ventral surface. Later, detailed scanning, electron microscope observations were made by Watanabe (1978). Thirty minutes after the mixing of complementary mating types, their cilia began to degenerate at the anterior end of the ventral surface to form holdfast unions. Then, the deciliation proceeded to the posterior end to form paroral unions as time passed. However, no such deciliation occurred on the dorsal surface. A similar deciliation pattern was also observed in conjugating cells of *P. tetraurelia*, *P. multimicronucleatum*, *P. bursaria*, and *P. trichium*. Watanabe (1978) suggested that the deciliation of cells might be due to their resorption into the cell rather than their separation from the cell, because some cells had short cilia of various length during the deciliation. Furthermore, it is interesting to note that the deciliation also occurred in chemically induced autogamous cells of *P. caudatum*, though no deciliation was observed in natural autogamy of *P. tetraurelia*. Deciliation may occur only when cells are activated to form pairs.

3.3 Partial Cell Fusion

Ultrastructures of the united regions in the conjugation process were reported in several *Paramecium* species: *P. caudatum* (Vivier and André 1961; André and Vivier 1962; Watanabe 1977), *P. aurelia* (Schneider 1963), and *P. multimicronucleatum* (Inaba et al. 1966; Takagi et al. 1981). At the united regions of holdfast unions of *P. caudatum*, each alveolar vacuole of the mates is exactly facing an interspace of about 35 nm, so that cytoplasmic projections found between the alveolar vacuoles also face each other. No membrane fusions are observed, and kinetosomes remain intact. However, some trichocysts which were regularly positioned at the cytoplasmic projections with a right angle to the cell surface leave their location and become positioned irregularly near the cytoplasmic projections. At the united regions of paroral unions 2.5 – 3.0 h after the onset of the mating reaction, the number of trichocysts remarkably decreases at the united regions. Furthermore, cytoplasmic bridges of 0.1 – 0.5 µm widths appear at the cytoplasmic projections, and their number and width increases as time proceeds (Fujishima and Shinoda, unpubl. observations). Since their mitochondria are hardly transferred from one mate to the other in pair (see review, Beale and Tait 1981), the maximum width of the cytoplasmic bridges seems to be less than the size of a mitochondrion.

3.4 Cessation and Resumption of Feeding Activity

In *Paramecium*, the breakdown of the old oral apparatus and the reconstruction of the new one take place during conjugation, as well as during autogamy and every binary fission. During conjugation of *P. tetraurelia* the old buccal cavity collapses and new oral membranelles are formed at the stage of synkaryon, and a functional oral apparatus engaged in the formation of food vacuoles is completed about 40 min after exconjugation (Ng and Newmann 1984a). In *P. caudatum*, the exact times for the breakdown and the reconstruction of the oral apparatus are not yet clarified, but the capacity to form food vacuoles in exconjugants was found to recover just after the third postzygotic nuclear division (Fujishima and Görtz 1983). Furthermore, Fujishima (1983b) found that the loss of the feeding activity occurs prior to the beginning of meiotic prophase in the early stage of the conjugation process. At that stage, since the cells were tightly fused at the paroral region, the pairing itself might interfere with the uptake of particles into the buccal cavities. However, this possibility was excluded because cells which were surgically separated by a glass needle 2.0–2.5 h after the mixing of complementary mating types showed no feeding activity. In addition, when conjugating cells were separated 1.5–2.0 h after the mixing, they still had the feeding activity but about 50% of the cells soon ceased the activity (Fujishima 1983b). These results indicate that the cells are committed to a loss of feeding activity 1.5–2.0 h after the mixing (see Fig. 1 in Sect. 4.4). Therefore, if the breakdown of the old oral apparatus of *P. caudatum* begins at the synkaryon stage, as shown in *P. tetraurelia,* these results indicate that the oral apparatus has already lost its function before the breakdown. Similarly, the early loss of the feeding activity was also reported in *P. multimicronucleatum* (Takagi et al. 1981; Shimomura and Takagi 1984).

On the other hand, from studies of the stomatogenesis during conjugation and autogamy of amicronucleate cells, it was found that the micronucleus played an important role in the organization of a new oral apparatus, because in amicronucleate exconjugants and autogamonts new oral apparatuses could not be formed and the cells eventually perished (Mikami 1979; Ng and Newmann 1984a, b). For a discussion of this problem, see Mikami (Chap. 8, this Vol.).

4 Nuclear Events in Conjugation

4.1 Early Micronuclear Migration

In *Paramecium*, the micronucleus of a stationary phase cell lies in a concavity of the macronucleus, but within 10 min after the start of the mating reaction or a transfer of cells into conjugation-inducing chemicals, the micronucleus leaves the concavity and moves into the cytoplasm (Fujishima and Hiwatashi 1977). This phenomenon, called "early micronuclear migration (EMM)", is the earliest event so far described which follows the mating reaction or the mixing of mating reactive cells with chemicals, and is closely associated with the activation necessary for the formation of a holdfast union.

The discovery of EMM made possible the analysis of the activation-initiating mechanism of conjugation, especially of very early events in conjugation. Cronkite (1977) found that EMM could be induced when mating reactive cells were warmed after cold treatment or when mating reactive cells were treated with the calcium ionophore A23187. Interestingly, EMM did not occur in immature cells either after cold or ionophore treatment. By using $^{45}CaCl_2$, he also found that cells in the ice bath accumulated Ca^{2+} and then rapidly lost the ion upon warming. From these results he suggested that changes in internal Ca^{2+} concentration play a role in initiation of EMM. Recently, Fujishima and Hiwatashi (unpublished observation) found that the velocity of cytoplasmic streaming extremely increases within 5 min after the onset of the mating reaction in *P. caudatum*. Therefore, experiments are now in progress to clarify the relation between EMM, Ca^{2+}, and cytoplasmic streaming.

4.2 Premeiotic DNA Synthesis

In *Paramecium* cells enter the conjugation process with G_1 micronuclei (Woodard et al. 1966; Pieri et al. 1968; Golikova 1974; Fujishima 1983a). Therefore, premeiotic DNA synthesis takes place prior to meiotic prophase. In *P. caudatum*, the micronucleus is in G_1 during the first 1.5 h after the onset of the mating reaction at 25° C and in the premeiotic S-phase during the following 1 h; the G_2 phase is very short or lacking. Furthermore, in premitotic DNA synthesis [^{3}H]thymidine and [^{3}H]deoxyuridine monophosphate ([^{3}H]-dUMP) were well utilized, but in premeiotic DNA synthesis [^{3}H]thymidine was utilized very little even if the precursor had been taken into the cell (Fujishima 1983a). These results suggest that the activity of thymidine kinase is low in premeiotic S-phase cells and that deoxythymidylic acid is formed mainly by a de novo pathway in premeiotic S-phase.

In *Paramecium* the transition point from premeiotic interphase to meiotic prophase is at or very close to the end of premeiotic DNA synthesis. Therefore, up to the end of the premeiotic S-phase, cells have to be in a state ready to enter the meiotic prophase. Micronuclear transplantation experiments showed that the premeiotic S-micronucleus stopped proceeding into meiosis when it was transplanted into vegetative cells in the stationary phase, but entered the mitotic cycle when the recipient cells were grown after the transplantation (Fujishima and Hiwatashi 1978). The results showed that the premeiotic S-micronucleus is not yet committed to meiosis. On the other hand, when G_1 micronuclei were transplanted into cells in early meiotic prophase they did not enter meiosis, but when transplanted into cells in premeiotic S-phase they swelled in the same way as the host nucleus and entered meiosis. When premeiotic S-nuclei were transplanted into cells in early meiotic prophase they entered the meiotic prophase, but when transplanted into cells in meiotic metaphase I and II, they did not enter meiosis (Fujishima and Hiwatashi 1981). These results indicate that the readiness for entry into meiosis of the micronucleus is performed during the premeiotic S-phase. However, since the micronucleus could show nuclear changes typical for meiosis even if the premeiotic

DNA synthesis had been completely inhibited by aphidicolin treatment (Tsuda and Fujishima 1984), DNA duplication in the premeiotic S-phase is not required for the readiness for meiosis. To know what has been done for acquiring the readiness for meiosis is important for understanding the initiation mechanism of meiosis, but is not yet clear. The results of nuclear transplantation experiments also strongly suggest that the cytoplasm in meiotic prophase contains a factor that induces meiotic nuclear changes in a transplanted premeiotic S-nucleus. As to this cytoplasmic factor, see Section 4.5.

4.3 Meiosis

Detailed observations of nuclear events in the conjugation of *Paramecium* species have been carried out by Hamburger (1904), Calkins and Cull (1907), Wichterman (1940, 1948), Egelhaaf (1955), André and Vivier (1962), Schneider (1963), Kościuszko (1965), Woodard et al. (1966), Inaba et al. (1966), Jankowski (1966), Stevenson (1972), Mikami and Hiwatashi (1975), Jurand (1976), Fujishima and Hiwatashi (1977, 1981), Fujishima (1983a). Time schedules of the nuclear events have been reported in *P. caudatum* (Wichterman 1940; Mikami and Hiwatashi 1975), *P. bursaria* (Wichterman 1948), and *P. tetraurelia* (Woodard et al. 1966).

In *P. caudatum*, the micronucleus enters meiotic prophase 2.5 h after mixing of complementary mating types at 25 ° C. The micronucleus begins to elongate and becomes rod-shaped. At that time, condensed chromatin accumulates in granules on one side of the nucleus, and then chromatin threads arrange themselves parallel to the long axis of the nucleus. After a while the chromatin threads become a fine network structure in the elongated nucleus. About 5.5 h after mixing of complementary mating types, the elongated micronucleus curves and becomes crescent-shaped. Then, this crescent retracts in an ellipsoid and becomes prometaphase. In this stage, fine chromatin threads condense and become obvious chromosomes. Metaphases I and II occur about 8 and 10 h after the mixing.

Since the nuclear envelope remains intact during all stages of conjugation, the stages can be distinguished in living cells on the basis of the nuclear shape. However, the detailed correspondence of the rod and crescent-shaped micronuclei to those of the classical stages of meiotic prophase, i.e., leptotene, zygotene, pachtene, diplotene, and diakinesis, is not clear. In *P. primaurelia*, Stevenson (1972) has found synaptonemal complex (SC)-like structures in the stage prior to the crescent, but a typical SC has not yet been detected up to the crescent stage (Jurand 1976; Fujishima, unpublished observation). In *P. tetraurelia*, Jurand (1976) has found two electron-dense bodies at one end of the crescent, which may function as microtubule organizing centers. He has also found a single layer of intranuclear microtubules located close beneath the inner nuclear membrane of the elongated micronucleus running parallel to the long axis. This microtubular layer disappeared when bipolar meiotic spindles were formed at metaphase I. The significance of the nuclear elongation in meiotic prophase is not known, but these elongated nuclei are characteristic for the meiotic prophase of *Paramecium* and are never observed in mitotic prophase. Therefore, on the basis of morphological characters only, we can easily distinguish whether a nucleus has undergone meiotic nuclear division or not. As described in Sections 4.2 and 4.5, this character is utilized in studies for the induction mechanism of meiosis with micronuclear and cytoplasmic transplantation techniques.

After two meiotic divisions four haploid nuclei are formed. Only one of these nuclei, which was located in the paroral region, survives and three begin to degenerate about 11 h after the mixing. The surviving nucleus then divides mitotically to form two gamete pronuclei, a stationary and a migratory pronucleus. The migratory pronucleus moves into the partner and fuses with the stationary pronucleus to form the synkaryon.

4.4 Commitment Stages of Meiosis and Macronuclear Degradation

In the *P. aurelia* species complex, the mating reaction is sufficient to induce all the nuclear events of conjugation (Metz 1947). However, in *P. caudatum* and *P. multimicronucleatum*, a continuous pairing is necessary to induce the nuclear events (Hiwatashi 1955a; Ito 1972; Takagi et al. 1981). How long does the conjugation have to proceed for the initiation and the completion of the nuclear events? To examine this, conjugating pairs possessing specific stages of micronuclei were surgically separated by a glass needle at known times after mixing of complementary mating types of *P. caudatum:* (1) 1 – 1.5 h (G_1 phase), (2) 1.5 – 2 h (the first half of premeiotic S-phase), (3) 2 – 2.5 h (the latter half of premeiotic S-phase), and (4) 2.5 – 3 h (the beginning of meiotic prophase). Then, the cells were transferred into Dryl's solution and fixed and stained 6 – 6.5 h, 12 – 12.5 h, and 24 – 24.5 h after the mixing for observation of their nuclei (Fujishima 1983b). In (1) the micronuclei in the majority of cells remained in G_1 phase. Although a few cells entered premeiotic S-phase, they did not enter meiosis. In (2), the micronuclei maintained the premeiotic S-morphology, and could not enter meiosis. However, in many cells, macronuclear degradation was observed. In (3), the micronucleus of about half of the cells persisted to maintain the premeiotic S-morphology, whereas the nuclei of the remaining cells entered meiosis and completed it. Furthermore, macronuclear anlagen were also observed in some cells. In (4), all nuclear events occurred as in normal pairs. From these results it was found that cells are committed to meiosis 2 – 2.5 h, immediately prior to the beginning of meiosis, and committed to macronuclear degradation at 1.5 – 2 h. It should be noted that the commitment of macronuclear degradation occurs earlier than that of meiosis.

On the other hand, to examine whether these commitments are reversible phenomena, cells were transferred into culture medium immediately after the pair separation. The results were the same: the commitments were irreversible. Furthermore, from this experiment it was found that cells were committed to the loss of feeding activity at the same time with the commitment of macronuclear degradation. These results are summarized in Fig. 1.

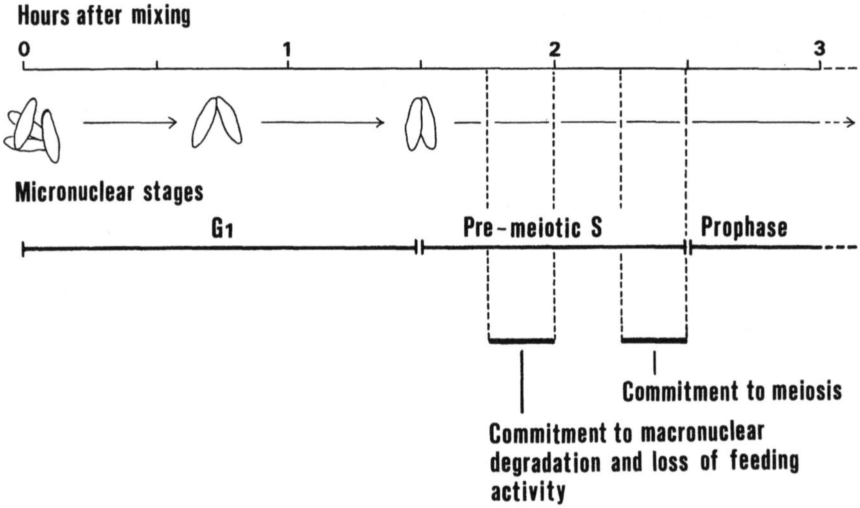

Fig. 1. Commitments to meiosis, macronuclear degradation, and loss of feeding activity in *P. caudatum*

In autogamous cells, Shimomura and Takagi (1985) showed that cells of *P. multimicronucleatum* were irreversibly committed to macronuclear degradation and cessation of feeding activity by incubating cells in autogamy-inducing medium for 1.5 h. Berger (1986) showed that cells of *P. tetraurelia* were also irreversibly committed to autogamy, immediately prior to initiation of meiosis.

4.5 Meiosis-Inducing Factor

As described in Section 4.2, Fujishima and Hiwatashi (1981) found by means of micronuclear transplantation that the cytoplasm in meiotic prophase contains a factor needed for the induction of meiosis in transplanted premeiotic S-micronuclei. To prove directly the existence of the meiosis-inducing factor, cytoplasm of meiotic prophase was injected into premeiotic S-arrested cells (see Sect 4.4) and it was observed whether meiosis was induced or not. However, this attempt has so far failed, probably due to the dilution of this factor because only 15% of the volume of the recipient cell can be injected. Therefore, for further biological and chemical characterization the factor has to be concentrated first and a bioassay system for the concentration has to be established.

Recently, Fujishima and Iwao (1984) found that occytes of amphibian *Bufo bufo japonicum* are useful for the bioassay system. In *Bufo* oocýtes which are arrested at diplotene of meiotic prophase, meiosis is reinitiated in vitro not only by progesterone treatment but also by injections of a cytoplasmic extract from meiosis-reinitiated oocytes. The reinitiation of meiosis involves nuclear membrane breakdown (germinal vesicle breakdown, GVBD) chromosome condensation, and progression through the first meiotic division. This meiosis-reinitiation factor called "maturation-promoting factor, MPF" of the oocyte is not yet purified. Recent reports suggest that MPF is a heat-sensitive protein generally produced in maturing oocytes of various animals and it has no species specificity among invertebrates and vertebrates (for review, see Ford 1985). We found that injected extracts of conjugating paramecia obtained by differential centrifugations can induce GVBD in *Bufo* oocytes.

Cells of *P. caudatum* were homogenized in ice-cold extraction medium containing 0.2 M NaCl, 0.25 M sucrose, 10 mM $MgSO_4$, 2 mM EGTA, and 10 mM Na-phosphate buffer (pH 6.8). GVBD-inducing activity of the homogenate was assayed by injecting 50 nl into each oocyte. Extracts of cells in stationary phase showed no GVBD-inducing activity. However, within 1 h after the mixing of complementary mating types, cells showed strong activities. Then the activity decreased at the crescent stage, and increased again at metaphases I and II. This fluctuation in the levels of GVBD-inducing activity seems to be associated with degrees of chromatin condensation of *Paramecium* micronucleus. The activity was present in the supernatants of centrifugations at 18,000 *g* for 20 min or at 126,700 *g* for 60 min, but absent in the pellets. Heat sensitivity of the activity showed that a 10-min incubation at 40 °C or higher resulted in a rapid loss of the activity. Furthermore, trypsin treatment resulted in a complete loss of the activity. These results indicate that the GVBD-inducing activity is closely associated with a soluble, heat-labile protein, and that the reinitiation of meiosis in amphibian *Bufo* oocytes and meiosis in *Paramecium* are controlled by the same or similar cytoplasmic factors. It was also found that cells in log phase have strong GVBD-inducing activity, suggesting that mitosis of *Paramecium* is also controlled by a cytoplasmic factor which is similar to MPF. The cytoplasmic factors from meiotic and mitotic cells of *Paramecium* are now being isolated and compared and their activity to induce meiosis and mitosis in *Paramecium* cells will be clarified soon.

5 Perspectives

Conjugation in *Paramecium* is an easily inducible developmental process. So far, the majority of the biological and physiological works on conjugation have been concentrated on the recognition mechanism of complementary mating types and the induction mechanism of conjugation, but almost nothing has been done on regulation mechanisms of the nuclear events. Recently, however, two general biological problems, the induction mechanism of meiosis and the differentiation mechanism of germinal and somatic nuclei are being analyzed. Details of the latter will be presented by Mikami (Chap. 8, this Vol.). As to the former, Fujishima and Iwao (1984) found that induction of reinitiation of meiosis of amphibian oocyte can be used as a good bioassay system for the isolation and characterization of meiosis and mitosis inducing factors from *Paramecium.* This work will contribute not only to the analysis of the induction mechanism of meiosis and mitosis of *Paramecium* at the molecular level, but may also give evidence that cytoplasmic factors which control nuclear divisions are common over a wide range of systematic taxa. Fujishima and Sakimura (in preparation) found that cells of *Tetrahymena thermophila* also have a cytoplasmic factor which can induce reinititation of meiosis of the oocytes. The comparison between the factors from *Paramecium* and *Tetrahymena* is in progress.

Furthermore, if mutants of meiosis are found in *Paramecium*, the mutants will be exploited not only for the analysis of genetic control of meiosis but also as good recipients for the bioassay system for detecting and isolating the gene products of the wild-type alleles.

References

André J, Vivier E (1962) Quelques aspects ultrastructuraux de l'échange micronucléaire lors de la conjugaison chez *Paramecium caudatum*. J Ultrastruct Res 6:390–406

Beale GH, Tait A (1981) Mitochondrial genetics of *Paramecium aurelia*. Int Rev Cytol 71:19–40

Berger JD (1986) Autogamy in *Paramecium*. Cell cycle stage-specific determination of the process. Exp Cell Res 116:475–486

Calkins GN, Cull SW (1907) The conjugation of *Paramecium aurelia (caudatum)*. Arch Protistenk 10:375–415

Chen TT (1945) Induction of conjugation in *Paramecium bursaria* among animals of one mating type by fluid from another mating type. Proc Natl Acad Sci USA 31:404–410

Cohen LW, Siegel RW (1963) The mating-type substances of *Paramecium bursaria*. Genet Res 4:143–150

Cronkite DL (1977) An analysis of the mechanism of activation of rapid micronuclear migration, a very early event in conjugation of *Paramecium caudatum*. Proc 5th Int Congr Protozool, 284 pp

Cronkite DL (1979) The genetics of swimming and mating behavior in *Paramecium*. In: Levandowsky M, Hutner SH (eds) Biochemistry and physiology of protozoa, vol 3, 2nd edn. Academic Press, London New York, pp 221–273

Diller WF (1942) Re-conjugation in *Paramecium caudatum*. J Morphol 70:229–256

Diller WF (1948) Nuclear behavior of *Paramecium trichium* during conjugation. J Morphol 82:1–52

Egelhaaf A (1955) Cytologisch-entwicklungsphysiologische Untersuchungen zur Konjugation von *Paramecium bursaria* Focke. Arch Protistenk 100:447–514

Ford CC (1985) Maturation promoting factor and cell cycle regulation. J Embryol Exp Morphol 89 (Suppl):271–284

Fujishima M (1983a) Microspectrophotometric and autoradiographic study of the timing and duration of pre-meiotic DNA synthesis in *Paramecium caudatum*. J Cell Sci 60:51–65

Fujishima M (1983b) Transition from mitosis to meiosis in *Paramecium* in relation to conjugation period. Zool Mag 92:603

Fujishima M, Görtz H-D (1983) Infection of macronuclear anlagen of *Paramecium caudatum* with the macronucleus-specific symbiont *Holospora obtusa*. J Cell Sci 64:137–146

Fujishima M, Hiwatashi K (1977) An early step in initiation of fertilization in *Paramecium*: early micronuclear migration. J Exp Zool 201:127–134

Fujishima M, Hiwatashi K (1978) Transplantation of germ nucleus in *Paramecium caudatum*. I. Nuclei in the pre-meiotic S phase can enter into mitotic cycle. Exp Cell Res 111:468–471

Fujishima M, Hiwatashi K (1981) Transplantation of germ nuclei in *Paramecium caudatum*. II. Induction of meiosis in transplanted interphase nucleus. Exp Cell Res 131:63–71

Fujishima M, Iwao Y (1984) Induction of reinitiation of meiosis in toad oocytes by injection of *Paramecium* extracts. 3rd Int Congr Cell Biol. Academic Press, Japan, p 391 (Abstr)

Golikova MN (1974) Cytophotometrical study of DNA amount in micronucleus of *Paramecium bursaria*. Acta Protozool 15:109–120

Hamburger C (1904) Die Konjugation von *Paramecium bursaria* Focke. Arch Protistenk 4:199–239

Hiwatashi K (1951) Studies on the conjugation of *Paramecium caudatum.* IV. Conjugation behavior of individuals of two mating types marked by a vital staining method. Sci Rep Tohoku Univ Ser 4 19:95–99

Hiwatashi (1955a) Studies on the conjugation of *Paramecium caudatum.* V. The time of the initiation of nuclear activation. Sci Rep Tohoku Univ Ser 4, 21:199–205

Hiwatashi K (1955b) Studies on the conjugation of *Paramecium caudatum.* VI. On the nature of the union of conjugation. Sci Rep Tohoku Univ Ser 4, 21:207–218

Hiwatashi K (1961) Locality of mating reactivity on the surface of *Paramecium caudatum.* Sci Rep Tohoku Univ Ser 4, 27:93–99

Hiwatashi K (1969) *Paramecium.* In: Metz CB, Monroy A (eds) Fertilization, vol 2. Academic Press, London New York, pp 255–293

Hiwatashi K (1981) Sexual interactions of the cell surface in *Paramecium.* In: O'Day DH, Horgen PA (eds) Sexual interactions in eukaryotic microbes. Academic Press, London New York, pp 351–378

Inaba F, Imamoto K, Suganuma Y (1966) Electron-microscopic observations on nuclear exchange during conjugation in *Paramecium multimicronucleatum.* Proc Jpn Acad Sci 42:394–398

Ito Y (1972) Studies on the nuclear reorganization in *Paramecium caudatum.* Bull Aichi Univ Ed 21:83–86

Jankowski AV (1966) The conjugation processes in *Paramecium putrinum* Clap. et Lachem. VIII. Nuclear phenomena during reconjugation. Zool Zh 45:818–829

Jurand A (1976) Some ultrastructural features of micronuclei during conjugation and autogamy in *Paramecium aurelia.* J Gen Microbiol 94:193–203

Kitamura A, Hiwatashi K (1984) Cell contact and the activation of conjugation in *Paramecium.* Zool Sci 1:161–168

Kitamura A, Onimal H, Naitoh Y, Hiwatashi K (1979) Relation between swimming and sexual behaviors in *Paramecium caudatum.* Dobutugaku Zasshi 88:528 (Abstr) (in Japanese)

Klitzke M (1914) Über Wiederconjuganten bei *Paramecium caudatum.* Arch Protistenk 33:1–20

Kościuszko H (1965) Karyologic and genetic investigations in syngen 1 of *Paramecium aurelia.* Folia Biol 13:339–368

Kościuszko H (1982) Reconjugation in *Paramecium triaurelia.* Folia Biol 30:9–13

Kung C, Eckert R (1972) Genetic modification of electric properties in an excitable membrane. Proc Natl Acad Sci USA 69:93–97

Metz CB (1947) Induction of "pseudoselfing" and meiosis in *Paramecium aurelia* by formalin killed animals of opposite mating type. J Exp Zool 105:115–140

Mikami K (1979) Stomatogenesis during sexual and asexual reproduction in an amicronucleate strain of *Paramecium caudatum.* J Exp Zool 208:121–128

Mikami K, Hiwatashi K (1975) Macronuclear regeneration and cell division in *Paramecium caudatum.* J Protozool 22:536–540

Miwa I, Umehara N (1983) Conjugation between G_1 and G_2 phase cells in *Paramecium tetraurelia.* J Protozool 30:271–274

Miyake A (1956) Physiological analysis of the life cycle of the protozoa. III. Artificial induction of selfing conjugation by chemical agents in *Paramecium caudatum.* Physiol Ecol 7:14–23

Miyake A (1958) Induction of conjugation by chemical agents in *Paramecium caudatum.* J Inst Polytech Osaka City Univ Ser D 9:251–296

Miyake A (1964) Induction of conjugation by cell-free preparations in *Paramecium multimicronucleatum.* Science 146:1583–1585

Miyake A (1968a) Induction of conjugation by chemical agents in *Paramecium.* J Exp Zool 167:359–380

Miyake A (1968b) Artificial induction of autogamy by chemical agents in *Paramecium multimicronucleatum* syngen 2 (Abstr). Jpn J Dev Biol 22:62–63 (in Japanese)

Miyake A (1969) Mechanism of initiation of sexual reproduction in *Paramecium multimicronucleatum.* Jpn J Genetics 44 (Suppl 1):388–395

Miyake A (1974) Cell interaction in conjugation of ciliates. Curr Top Microbiol Immunol 64:49–77

Miyake A (1978) Cell communication, cell union, and initiation of meiosis in ciliate conjugation. Curr Top Dev Biol 12:37–82

Miyake A (1981) Physiology and biochemistry of conjugation in ciliates. In: Levandowsky M, Hutner SH (eds) Biochemistry and physiology of protozoa, 2nd edn. Academic Press, London New York, pp 125–198

Ng SF, Newmann A (1984a) The role of the micronucleus in stomatogenesis in sexual reproduction of *Paramecium tetraurelia*: micronuclear and stomatogenetic events. Protistologica 20:43–64

Ng SF, Newmann A (1984b) The role of the micronucleus in stomatogenesis in sexual reproduction of *Paramecium tetraurelia*: conjugation of amicronucleates. Protistologica 20: 517–523

Pieri J, Vaugien C, Trouillier M (1968) Interprétation cytophotométrique des phénomènes micronucléaires au cours de la division binaire et des divisions prégamiques chez *Paramecium trichium*. J Cell Biol 36:664–668

Schneider L (1963) Elektronenmikroskopische Untersuchungen der Konjugation von *Paramecium*. Protoplasma 56:109–140

Shimomura F, Takagi Y (1984) Chemical induction of autogamy in *Paramecium multimicronucleatum*, syngen 2. J Protozool 31:360–362

Shimonura F, Takagi Y (1985) Commitment to sexual and asexual phases in *Paramecium multimicronucleatum*. J Exp Zool 233:317–326

Sonneborn TM (1937) Sex, sex inheritance and sex determination in *Paramecium aurelia*. Proc Natl Acad Sci USA 23:378–385

Stevenson I (1972) Ultrastructure of nuclear division in *Paramecium aurelia*. III. Meiosis in the micronucleus during conjugation. Aust J Biol Sci 25:775–789

Takagi Y (1971) Sequential expression of sex traits in the clonal development of *Paramecium multimicronucleatum*. Jpn J Genet 46:83–91

Takagi Y, Yoshiki Y, Baba C, Yamada S (1981) Determination of macronuclear fragmentation, stabilization of cell union, and cessation of feeding activity in triplet conjugants of *Paramecium*. J Protozool 28:99–102

Takahashi M, Naitoh Y (1978) Behavioral mutants of *Paramecium caudatum* with defective membrane electrogenesis. Nature (London) 271:656–659

Takahashi M, Takeuchi N, Hiwatashi K (1974) Mating agglutination of cilia detached from complementary mating types of *Paramecium*. Exp Cell Res 87:415–417

Tsuda M, Fujishima M (1984) Inhibition of pre-meiotic DNA synthesis by aphidicolin and its effect on meiosis of *Paramecium caudatum*. Jap J Protozool 17:9–10 (in Japanese)

Tsukii U, Hiwatashi K (1978) Inhibition of early events of sexual process in *Paramecium* by concanavalin A. J Exp Zool 205:439–446

Tsukii U, Hiwatashi K (1979) Artificial induction of autogamy in *Paramecium caudatum*. Genet Res Camb 34:163–172

Tsukii U, Hiwatashi K (1983) Genes controlling mating-type specificity in *Paramecium caudatum:* three loci revealed by intersyngenic crosses. Genetics 104:41–62

Vivier E, André J (1961) Données structurales et ultrastructurales nouvelles sur la conjugaison de *Paramecium caudatum*. J Protozool 8:416–426

Watanabe T (1977) Ciliary membranes and mating substances in *Paramecium caudatum*. J Protozool 24:426–429

Watanabe T (1978) A scanning electron-microscopic study of the local degeneration of cilia during sexual reproduction in *Paramecium caudatum*. J Cell Sci 32:55–66

Wichterman R (1940) Cytogamy: a sexual process occurring in living joined pairs of *Paramecium caudatum* and its relation to other sexual phenomena. J Morphol 66:423–451

Wichterman R (1948) The time schedule of mating and nuclear events in the conjugation of *Paramecium bursaria*. Turtox News 26:2–10

Woodard J, Woodard M, Gelber B, Swift H (1966) Cytochemical structures of conjugation in *Paramecium aurelia*. Exp Cell Res 41:55–63

Chapter 6 Mating-Type Substances*

Akio Kitamura †[1]

Contents

1 Introduction

Conjugation in *Paramecium* is initiated by a highly specific cell-cell recognition and adhesion event referred to as the mating reaction (Sonneborn 1937). The term "mating-type substances" was first introduced by Metz (1946, 1948, 1954) to define substances or molecular configurations on the cell surface that mediate the mating reaction. Even after 50 years of research on mating problems in *Paramecium*, there is still no convincing evidence that substances secreted from cells are involved in the mating reaction. This fact strongly confirms the supposition that sexual interactions of *Paramecium* are strictly cell surface events between complementary mating-type substances, although the existence of the latter still remains hypothetical. Thus, the mating reaction provides a simple and excellent system for studies on membrane–membrane interactions in general.

Here, I shall summarize recent progress in studying the mating-type substances in *Paramecium*. For reasons of space I will give only a short selection of references, although a number of recent reviews relating to sexual cell interactions in *Paramecium* are cited.

---

1 Deceased on January 1987. Biological Institute, Tohoku University, Aoba-yama, Sendai 980, Japan

* The author wishes to dedicate this work to Dr. Charles B. Metz, a great pioneer of the studies on the mating-type substances in *Paramecium*

Paramecium. Ed. by H.-D. Görtz

2 Role of Mating-Type Substances in Conjugation

The mating reaction can be instantaneously induced by mixing sexually mature cells of complementary types. Usually cells adhere to one another upon initial contact. Thus, unlike some other ciliates, sexual cell recognition and cell adhesion take place simultaneously in *Paramecium*. No diffusible substances are known to participate in the mating reaction (Sonneborn 1937; Larison and Siegel 1961). In view of these facts we still attribute the mating reaction entirely to the interactions of the mating-type substances as originally proposed by Metz (1954). Although possible involvement of mating-type nonspecific adhesion through hydrophobic interactions at the ciliary surface may occur in the mating reaction (Kitamura 1982), such interactions probably would not contribute much to the mating reaction because of the high mating-type specificity of the reaction.

Takahashi et al. (1974) found that detached cilia from both mating types of *P. caudatum*, when mixed together, agglutinate, but deciliated cells did not do so, and the detached, reactive cilia failed to adhere to deciliated cells of the opposite mating type. These observations suggest that the mating-type substances are located only on the cilia, and not on the cell surface.

Since the mating reactivity of *Paramecium* is known to be restricted to the ventral cilia (Hiwatashi 1961; Cohen and Siegel 1963; Byrne 1972), some control mechanism in intracellular differentiation of the mating-type substances must exist. However, no analytical investigations have been published to date concerning this problem, although potassium ions are known to regulate the transfer of the mating-type substances from the cytoplasm to the ciliary membranes (Takahashi and Hiwatashi 1974).

Another functional role of the mating-type substances is to trigger the entire series of activation during the conjugation process, as was originally suggested by Metz (1954). The earliest change observed in *P. caudatum* during the mating reaction is inactivation of ciliary movement (Kitamura and Hiwatashi 1984b). This ciliary inactivation was discovered in a model system of the mating reaction using ciliary membrane vesicles which had conjugation-inducing activity but lacked agglutination-inducing activity (Kitamura and Hiwatashi 1980). When mating reactive cells are treated with such vesicles from the opposite mating type, their swimming was inhibited within 1 min. Microscope observation revealed the perturbation of metachronal waves in the dorsal cilia as well as immobilization of the ventral cilia. These results suggest that interaction of the mating-type substances on ventral cilia affects the movement of dorsal cilia even though they lack mating reactivity. Since a positive correlation between membrane depolarization and change in ciliary beat frequency is known to operate in *Paramecium* (Machemer 1974), one may postulate that a change in membrane potential also known to occur during fertilization of various organisms (Hagiwara and Jaffe 1979, Jaffe and Cross 1986), occurs to inactivate ciliary movement at the start of the mating reaction. However, no change in membrane potential or resistance was detected in connection with ciliary inactivation (Kitamura et al. 1980). Some unknown change at the ciliary membrane surface is probably induced by mating-type substance interaction.

Other events tightly coupled with the trigger reaction of the mating-type substances are an increase in the hydrophobicity of the ciliary surface (Kitamura 1984), early micronuclear migration (Fujishima and Hiwatashi 1977), and ciliary degeneration (Hiwatashi 1955; Miyake 1966; Watanabe 1978). Probably the interaction between complementary mating-type substances transduces intracellular signals which in turn activate specific genes and/or initiate key biochemical reactions in the conjugation process. To date, however, no such signals, sometimes called second messages, have been detected. For detailed information on the problem of the activation initiating mechanism, readers are referred to recent reviews by Cronkite (1979), Hiwatashi (1981), Miyake (1981), Kitamura and Hiwatashi (1984a), and Hiwatashi and Kitamura (1985).

3 Chemical Nature of the Mating-Type Substances

3.1 Inactivation of the Mating-Type Substances in Vivo

Ordinarily the cellular adhesion in the mating reaction is sufficiently stable to allow mating agglutination to continue for about 1 h in most species of *Paramecium*. This phase is followed by holdfast union formation in which two cells are directly united at their anterior region. Unlike the mating reaction, the holdfast union is known to be mating-type nonspecific (Hiwatashi 1951; Metz 1954; Miyake 1958), indicating the involvement of substances other than the mating-type substances in this adhesion. The breakdown of agglutinated clumps into conjugating pairs is associated with a loss of mating reactivity (Metz 1954). However, this inactivation is not complete until paroral union formation (the final stage of cell contact) in *P. caudatum*. Most of the holdfast union pairs still retain weak mating reactivity (Miyake 1958) and complete inactivation takes place about 7 h after initiation of the mating reaction (A. Kitamura, unpublished data). Although the mechanism of this inactivation is as yet unknown, it is unlikely that secreted substances from cells are responsible, since cell-free fluids of mating mixtures are unable to destroy mating reactivity of cells (A. Kitamura, unpubl.). As will be described in the following section, the mating reactivity of isolated cilia is easily lost by treatment with proteolytic enzymes but not by nonproteolytic enzymes. Therefore, membrane-bound proteases may be the most probable candidates for in vivo inactivation of the mating-type substances, as such enzymes have been shown to be involved in the fertilization process of various types of organisms (McRorie and Williams 1974; Schleuning and Fritz 1976; Hoshi et al. 1981). In order to examine this possibility, Kitamura (1983) tested effects of 11 different protease inhibitors on the formation of holdfast unions in *P. caudatum*. If proteolytic enzymes participate in the loss of mating reactivity during the mating reaction, protease inhibitors should protect agglutinating cells from loss of mating reactivity, resulting in continued maintenance of agglutination or a marked delay in clump breakdown. However, clumps broke down in the normal manner to form holdfast unions in the presence of the inhibitors, which include chymostatin, pepstatin, antipain, bestatin, leupeptin, elastatinal, N-α-tosyl-L-lysyl-chloromethane

(TLCK), N-tosyl-L-phenylalanyl-chloromethane (TPCK), phenylmethane sulfonyl fluoride (PMSF), ovomucoid, and soybean trypsin inhibitors. These results indicate that proteolysis is not involved in the loss of mating reactivity during the mating reaction. Probably nonenzymatic mechanisms should be envisaged.

The effect of protein and RNA synthesis inhibitors on the mating reactivity of paramecia has also been reported. Both puromycin and actinomycin D inhibit agglutination in *P. pentaurelia* (Bleyman 1964) and *P. septaurelia* (Kang and Taub 1968). Nobili (1963) similarly found in the *P. aurelia* complex that the inhibitors actinomycin C, an amino acid analog, p-fluorophenylalanine, and a base analog, 6-azauracil reduced or totally inhibited the agglutinability of cells, depending on the dosages of those agents with which the cells were treated before expression of mating reactivity. These results suggest that de novo synthesis of RNA as well as protein is necessary for sexually mature paramecia to become reactive.

3.2 Inactivation of the Mating-Type Substances in Vitro

As mentioned in the preceding section, the mating-type substance activity is located at the ciliary surfaces. Thus, the mating agglutination can be induced specifically among cells of a single mating type by detached cilia from cells of the opposite mating type. Isolated mating reactive cilia can be obtained by treatment of reactive cells with $K_2Cr_2O_7$ (Miyake 1964), glycerol (Klayman 1969), $MnCl_2$ (Fukushi and Hiwatashi 1970). Triton X-100, EDTA and Ca (Takahashi et al. 1974), 2,4-dinitrophenol-Ca (Klimetzek 1974), and ethanol-EDTA and Ca (Watanabe 1977a). It should be emphasized that suspensions of detached cilia prepared by all of these methods contain not only reactive but also nonreactive cilia, since as already noted only cilia from the ventral surface of the cell possess mating reactivity. How many pure, mating reactive cilia are contained in the ciliary suspension? Kitamura (1978) tried to separate reactive from nonreactive cilia using a bioaffinity method. When cilia ($4 \sim 7$ mg as protein) were detached from mating reactive cells of *P. caudatum* by the $MnCl_2$ method and mixed with mating reactive cells ($0.5 \sim 1.1 \times 10^7$) of the opposite mating type, large mating clumps were formed. The remaining free cilia and unclumped living single cells were then removed by slow centrifugation, and the mating clumps were treated with 5% formalin to separate the cilia from the cells. Only 1% of the cilia used in the experiments was recovered. Although this value may not reflect the exact ratio of mating reactive to nonreactive cilia because of their possible loss or inactivation during these procedures, the proportion of reactive cilia would probably be one-tenth or less of the total cilia.

To elucidate the chemical nature of the mating-type substances, detached cilia of *P. caudatum* were used for inactivation experiments. The results obtained to date are summarized in Table 1. These studies reveal that both complementary mating-type substances V and VI in syngen 3 of *P. caudatum* are sensitive to digestion by proteolytic enzymes such as trypsin, chymotrypsin, and ficin, and to heating at 50 °C, but not to digestion by glycosidases or phospholipases. Neither high concentrations of salt (2 M NaCl) nor EDTA had any effect on mating reactivity of cilia, but many kinds of detergents and alcohol destroyed the reactivity, suggesting that the mating-type substances are tightly bound to the ciliary membranes. Since no clear differences were seen in the inactivation between cilia of complementary mating types V and VI, the chemical nature of mating-type substances V and VI must not be radically different. Basically, the results in Table 1 agree with the previous data with reactive killed cells of *P. tetraurelia, P. calkinsi,* and *P. caudatum* (Metz and Butterfield 1951; Metz 1954; Hiwatashi 1969) except for the effect of formalin. Hiwatashi (1950) reported differences in the stability of mating reactivity between formalin-fixed paramecia of different mating types in *P. caudatum.* However, such differential action of formalin was not seen in the inactivation test using detached cilia of the same syngen as he used. Additionally, formalin has no such differential action on the two mating types of *P. tetraurelia* (Metz 1947). The results of Table 1 confirm that the mating-type substances are tightly membrane-bound, intrinsic proteins of ciliary membranes and their active sites consist of simple proteins but not of sugar residues.

4 Recent Advances in Biochemical and Immunological Studies on the Mating-Type Substances

4.1 Biochemical Advances

Although many investigators have tried to isolate the mating-type substances in a molecular form, no one has yet succeeded. Based on the data of the inactivation experiments (Table 1), Kitamura (1977) tested for the presence of mating-substance activity in the supernatant of a ciliary suspension after mating reactivity of detached cilia was removed by each treatment. However, no positive results were obtained. For example, after treatment of reactive cilia with trypsin, the supernatant fluid (105000 *g*, for 1 h) was applied to a Sephadex G-50 column to fractionate trypsin and substances derived from cilia. At least two kinds of glycoproteins were isolated but no mating-substance activity was detected in the fractions when three types of mating-substance activity tests were made; agglutination, mating-type specific-pair formation, and inhibition of mating reactivity in cells of opposite type.

The simplest system exhibiting mating-substance activity so far obtained is that of ciliary membrane vesicles with a diameter of 100 to 150 nm (Kitamura and Hiwatashi 1976). The membrane vesicles can be isolated by treatment of reactive cilia of *P. caudatum* with a solution containing 2 M urea and 0.1 mM EDTA. They can induce the whole process of conjugation including mating agglutination in cells of the opposite mating type. Ciliary membrane vesicles with similar properties are also obtained by the same procedures from reactive cilia of *P. bursaria* and *P. duboscqui* (A. Kitamura, unpubl.).

A promising approach for identifying the mating-type substance is emerging, utilizing these membrane vesicles. This approach involves reconstituting

Table 1. Inactivation of detached cilia[a] by various agents

Agent or treatment	Conc.	Conditions			Inactivation[b]		Reference[c]
		Temp.	Time	pH	Mating type V	Mating type VI	
Enzymes							
α-Mannosidase	110 μg ml^{-1}	30° C	120 min	pH 5.8	–	–	(1)
α-L-Fucosidase	10 μg ml^{-1}		120 min		–	–	(1)
β-Galactosidase	110 μg ml^{-1}		120 min		–	–	(1)
β-Glucosidase	230 μg ml^{-1}		120 min		–	–	(1)
Neuraminidase	230 μg ml^{-1}		120 min		–	–	(1)
Lysozyme	5 mg ml^{-1}	25° C	20 h	pH 7.0	–	–	(3)
Galactose oxidase	110 μg ml^{-1}	30° C	60 min	pH 6.0	–	–	(4)
Phospholipase A	330 μg ml^{-1}	26° C	26 h	pH 7.4	–	–	(2)
Phospholipase C	2 mg ml^{-1}	30° C	120 min	pH 7.3	–	–	(2)
Snail digestive enzymes mixture	3 mg ml^{-1}	25° C	40 min	pH 5.1	–	–	(4)
Trypsin	0.1 μg ml^{-1}	25° C	60 min	pH 8.0	–	±	(1)
	10 μg ml^{-1}		60 min		±	+	(1)
	100 μg ml^{-1}		60 min		+	+	(1)
α-Chymotrypsin	100 μg ml^{-1}	25° C	60 min	pH 8.0	+	+	(2)
Thermolysin	1 mg ml^{-1}	25° C	15 min	pH 7.2	+	+	(3)
Ficin	1 μg ml^{-1}	25° C	120 min	pH 6.4	+	+	(2)
Lectins							
Concanavalin A	60 μg ml^{-1}	25° C	30 min	pH 7.3	–	–	(1)
	4 mg ml^{-1}		30 min		–	–	(1)
Phytohemagglutinin	250 μg ml^{-1}	25° C	30 min	pH 7.3	–	–	(1)
Sonication	5 A	2° C	10 s	pH 7.5	±	±	(4)
Heating	40° C		60 min	pH 7.3	–	–	(1)
	45° C		60 min		±	±	(1)
	50° C		60 min		+	+	(1)
pH	3	4° C	60 min		+	+	(1)
	4		60 min		±	±	(1)
	5~10		60 min		–	–	(1)
	11		60 min		±	±	(1)

Sodium chloride	2 M	4° C	23 h	pH 7.3	–	–	(1)
EDTA	10 mM	4° C	23 h	pH 7.3	–	–	(1)
Urea	2 M	4° C	60 min	pH 7.3	±	±	(4)
	4 M		60 min		+	+	(4)
Ethanol	30%	2° C	30 min	pH 7.3	±	±	(2)
	60%		30 min		+	+	(2)
n-Butanol	50%	2° C	30 min	pH 7.0	+	+	(2)
Formalin	5%	3° C	60 min	pH 6.9	–	–	(1)
Periodate	6.4 mM	2° C	60 min	pH 5.1	±[d]	±[d]	(1)
	64 mM		60 min		+[d]	+[d]	(1)
H_2O_2	5%	30° C	60 min	pH 6.0	+	+	(4)
$NaBH_4$	20 mM	25° C	30 min	pH 8.0	–	–	(4)
Detergents							
NP-40	0.5%	4° C	60 min	pH 9.0	±	±	(2)
Tween-20	0.1%	4° C	60 min	pH 8.6	+	+	(2)
Brij-58	0.1%	4° C	60 min	pH 7.3	+	+	(2)
DOC	0.5%	4° C	10 min	pH 8.4	±	±	(2)
LIS	0.1%	2° C	30 min	pH 7.5	+	+	(2)
SDS	0.1%	15° C	10 min	pH 7.5	+	+	(2)

[a] Detached cilia of *P. caudatum*, syngen 3, mating type V and VI. Cilia were treated at a final protein concentration of around 1 mg ml^{-1} in all experiments except for reference (3).

[b] + Complete inactivation; ± partial inactivation; – no inactivation.

[c] (1) Kitamura and Hiwatashi 1978; (2) Kitamura 1977; (3) Watanabe 1977b; (4) Kitamura, unpubl.

[d] This inactivation was caused by steric hindrance with mating-type substance activity due to cross-linking of nearby sugar residues (Kitamura and Hiwatashi 1978).

the functional organization of the vesicles from their components in vitro. Membrane vesicles can be reconstituted from the soluble fraction of the original membrane vesicles by dialysis after solubilization with the detergent lithium diiodosalicylate (LIS) (Kitamura and Hiwatashi 1980). The reconstitution was also successfully demonstrated by Sephadex G-200 gel filtration from a Tween-20 solubilized fraction of detached cilia (Hiwatashi and Kitamura 1977). In order to find constituents of ciliary membranes essential for mating activity, they compared polypeptide patterns on gel electrophoresis between the original membrane vesicles and partially LIS-solubilized fractions which were used for the reconstitution. Their results suggest that the agglutination-inducing activity is not contained in the major glycoproteins of about 180000 M_r since their presence does not coincide with the activity.

The reconstituted membrane vesicles have an interesting characteristic which differs from the original membrane vesicles obtained by the urea-EDTA method. When they are added to mating reactive cells of the opposite mating type, no mating agglutination of the latter was induced. However, most cells stopped their swimming within 1 min and then conjugating pairs were induced in about 1 h. This mating type specific ciliary inactivation or decrease in swimming velocity could be used as a more sensitive method for assaying for the mating-type substances than the three conventional types of assay (i.e., induction of agglutination, mating-type specific-pair formation and inhibition of mating reactivity in cells of opposite type), since higher mating-type activity is necessary for the induction of agglutination than induction of ciliary inactivation (Kitamura and Hiwatashi 1984b).

As stated previously, the mating-type substances in active form are believed to be localized only to the ciliary membranes. Biochemical comparisons of cilia or ciliary membranes between mating reactive and nonreactive cells or between cells of complementary mating types have been carried out in several laboratories. Using one-dimensional SDS-polyacrylamide gel electrophoresis, Klimetzek (1974, 1977) analyzed ciliary proteins in *P. multimicronucleatum.* He reported that cilia of mating reactive cells of both complementary mating type III and IV possess one major band on the gels which is a major component of ciliary membranes with a high molecular weight, but this band is virtually absent in cilia prepared from nonreactive cells of mating type III. On the other hand, when cilia of mating type IV were compared in the same way, no difference was detected in polypeptide bands between reactive and nonreactive cilia.

Adoutte et al. (1980) compared proteins of ciliary membranes in *P. tetraurelia* using one- and two-dimensional gel electrophoresis. They found an interesting polypeptide of 3100 daltons and pI 6.8 that varies in intensity on two-dimensional gels with the stage of cell culture. In addition, when proteins of ciliary membranes were analyzed on a one-dimensional gel system, this peptide was completely lacking in ciliary membranes from cells in log phase, became prominent in early stationary phase, and then decreased in starved, nonreactive cells in good correlation with cellular mating reactivity. However, the functions of this peptide and of that reported by Klimetzek are still unclear, although their possible involvement in the mating function was suggested by these authors.

Unexpectedly, however, in contrast to these above cited investigations, no recognizable difference was detected in the presence or intensity of polypeptides of mating reactive and non-reactive cilia or between cilia of complementary mating types in *P. caudatum.* These negative results were obtained when ciliary membrane vesicles prepared by the urea-EDTA method were compared by one-dimensional (Kitamura and Hiwatashi 1976) or two-dimensional gel electrophoresis (A. Kitamura, unpubl.). Similar negative results were also reported for *Tetrahymena* (Butzel and Decaprio 1978; van Bell and Williams 1983).

4.2 Immunological Advances

If antibodies which specifically prevent mating reactivity of cells are available, they might be useful to identify the mating-type substances. The immunological approach for analyzing the mechanism of the mating reaction began in the earlier period of the mating substance studies but has since made little progress. Metz and Fusco (1948) found a nonspecific inactivation of mating reactivity in formalin-killed cells of *P. aurelia* after treatment with some antisera against paramecia. Since antiserum prepared against nonreactive cells also had the same inhibiting effect and antiserum against one race sometimes failed to inhibit another race of the same mating type, they concluded that this inactivation was caused by a mechanical masking or steric hindrance of the mating-type substance through the inhibiting antibodies which combine with some neighboring antigens on the cilium.

Cilia of *Paramecium* carry surface antigens with a potent antigenicity (Preer 1959; Mott 1965). These antigens are called immobilization antigens (i-antigens, see Schmidt, Chap. 11, this Vol.) because the antigen-antibody reaction causes immobilization of ciliary movement (Beale 1954). The most important fact that emerged from the work of Metz and Fusco was that the immobilization antigen-antibody reaction interferes with the assay for the mating reaction-blocking antibodies. Therefore, for the assay, cells expressing the heterologous serotype to the immobilization antibody should be used. Hiwatashi and Takahashi (1967) first succeeded in obtaining antiserum which blocked the mating reaction without ciliary immobilization. The antiserum was raised by injecting frozen and thawed cells of *P. caudatum* into rabbits. However, the action of the antiserum had no mating-type specificity. Furthermore, this serum also inhibited the mating reaction of other species, *P. bursaria* and *P. multimicronucleatum.* Antibodies of a similar nature were obtained by injecting a brei of cells or frozen cilia of *P. multimicronucleatum* into rabbits (Barnett and Steers 1980). Of particular interest in the latter investigation is the finding that the mating reaction-blocking antibodies can be absorbed not only with living reactive cells but also with nonreactive cells such as sexually immature or nonstarved cells. These results led these workers to conclude that the antigenic site of the mating type substance is distinct from the mating recognition site and that the blocking antibodies are made against a common portion of the mating-type substance which might be exposed on the ciliary surface all the time, irrespective of the stages of cell growth and cell culture.

The only mating-type specific antibodies so far reported are those obtained by injecting into guinea pigs the mating reactive cilia from which i-antigens were completely removed by extraction with salt alcohol (Sasaki et al. 1972); however, further studies using such specific antibodies have not been reported. In spite of such a lack of success with immunological methods, trials involving production of monoclonal antibodies against specific recognition sites of the mating-type substance are currently being employed (G. H. Beale and A. Barnett, pers. commun.), encouraging the hope that such antibodies will eventually become available for the identification and purification of the mating-type substances.

5 Conclusion

The mating reaction of *Paramecium* seems to be an extremely simple system because only the mating-type substances are responsible for the reaction. Although evidence has accumulated showing that the substances are intrinsic proteins of ciliary membranes and their active site is made up of simple proteins, they have eluded all attempts at isolation, identification, or purification. One of the difficulties involved in those studies might be their extremely limited amounts in the ciliary membranes. Establishment of more sensitive assay systems than those conventionally used for detecting mating-type substance activity is required. An approach with a technique of reconstituting membrane vesicles seems to offer some promise for identifying the mating-type substances.

Acknowledgments. This work was carried out while the author was a fellow of the Alexander von Humboldt-Foundation in the laboratory of Dr. Heckmann of the University Münster. Thanks are due to Dr. R. K. Peck for his help with the English text and to Drs. G. H. Beale and A. Tait for reading the manuscript.

References

Adoutte A, Ramanathan R, Lewis RM, Dute RR, Ling K-Y, Kung C, Nelson DL (1980) Biochemical studies of the excitable membrane of *Paramecium tetraurelia.* III. Proteins of cilia and ciliary membranes. J Cell Biol 84:717–738

Barnett A, Steers E, Jr (1980) Antibodies blocking the mating reaction in *Paramecium multimicronucleatum* syngen 2. J Protozool 27:103–107

Beale GH (1954) The genetics of *Paramecium aurelia.* Cambridge Univ Press, London New York

Bell CT van, Williams NE (1983) Biochemical approaches to mating type expression in *Tetrahymena thermophila.* J Protozool 30:588–591

Bleyman L (1964) The inhibition of mating reactivity in *Paramecium aurelia* by inhibitors of protein and RNA synthesis. Genetics 50:236

Butzel HM, Jr, Decaprio A (1978) Ciliary membrane proteins of *Tetrahymena thermophila.* J Protozool 25:276–279

Byrne BC (1972) Mutagenic analysis of mating type and isolation of reactive cilia of both mating types in the ciliated protozoan, *Paramecium aurelia* syngen 4. Thes, Indiana Univ, Bloomington

Cohen LW, Siegel RW (1963) The mating-type substances of *Paramecium bursaria.* Genet Res 4:143–150

Cronkite DL (1979) The genetics of swimming and mating behavior in *Paramecium.* In: Levandowsky M, Hutner SH (eds) Biochemistry and physiology of protozoa, 2nd edn, vol 3. Academic Press, London New York, pp 351–378

Fujishima M, Hiwatashi K (1977) An early micronuclear migration. J Exp Zool 201:127–134

Fukushi T, Hiwatashi K (1970) Preparation of mating reactive cilia from *Paramecium caudatum* by $MnCl_2$. J Protozool 17:s21

Hagiwara S, Jaffe LA (1979) Electrical properties of egg membranes. Annu Rev Biophys Bioeng 8:385–416

Hiwatashi K (1950) Studies on the conjugation of *Paramecium caudatum.* III. Some properties of the mating type substances. Sci Rep Tohoku Univ Ser 4 (Biol) 18:270–275

Hiwatashi K (1951) Studies on the conjugation of *Paramecium caudatum.* IV. Conjugation behavior of individuals of two mating types marked by a vital staining method. Sci Rep Tohoku Univ Ser 4 (Biol) 19:95–99

Hiwatashi K (1955) Studies on the conjugation of *Paramecium caudatum.* VI. On the nature of the union of conjugation. Sci Rep Tohoku Univ Ser 4 (Biol) 21:207–218

Hiwatashi K (1961) Locality of mating reactivity on the surface of *Paramecium caudatum.* Sci Rep Tohoku Univ Ser 4 (Biol) 27:93–99

Hiwatashi K (1969) *Paramecium.* In: Metz CB, Monroy A (eds) Fertilization, vol 2. Academic Press, London New York, pp 255–293

Hiwatashi K (1981) Sexual interactions of the cell surface in *Paramecium.* In: O'Day DH, Horgen PA (eds) Sexual interactions in eukaryotic microbes. Academic Press, London New York, pp 351–378

Hiwatashi K, Kitamura A (1977) Mating reactive membrane vesicles and their reconstitution in *Paramecium caudatum.* In: Hutner SH (ed) Abstr Pap 5th Int Congr Protozoology, New York, p 275

Hiwatashi K, Kitamura A (1985) Fertilization in *Paramecium.* In: Metz CB, Monroy A (eds) Biology of fertilization, vol 1. Academic Press, London New York, pp 57–85

Hiwatashi K, Takahashi M (1967) Inhibition of mating reaction by antisera without ciliary immobilization in *Paramecium.* Sci Rep Tohoku Univ Ser 4 (Biol) 33:281–290

Hoshi M, Numakunai T, Sawada H (1981) Evidence for participation of sperm proteinases in fertilization of the solitary ascidian, *Halocynthia roretzi:* effects of protease inhibitors. Dev Biol 86:117–121

Jaffe LA, Cross NL (1986) Electrical regulations of sperm-egg fusion. Annu Rev Physiol 48:191–200

Kang HS, Taub SR (1968) Studies of the effects of actinomycin D and puromycin on mating type changes in cells of syngen 7, *Paramecium aurelia.* J Cell Biol 39:70a–71a

Kitamura A (1977) Studies on the mating type substances of *Paramecium caudatum.* Thes Tohoku Univ, Sendai

Kitamura A (1978) Attempts of bioaffinity separation of pure mating reactive cilia in *Paramecium caudatum.* Zool Mag (Tokyo) 87:498

Kitamura A (1982) Attachment of *Paramecium* to polystyrene surfaces: a model system for the analysis of sexual cell recognition and nuclear activation. J Cell Sci 58:185–199

Kitamura A (1983) Effects of protease inhibitors on early events in the conjugation of *Paramecium caudatum.* J Exp Zool 225:501–503

Kitamura A (1984) Evidence for an increase in the hydrophobicity of the cell surface during sexual interactions of *Paramecium.* Cell Struct Funct 9:91–95

Kitamura A, Hiwatashi K (1976) Mating-reactive membrane vesicles from cilia of *Paramecium caudatum.* J Cell Biol 69:736–740

Kitamura A, Hiwatashi K (1978) Are sugar residues involved in the specific cell recognition of mating in *Paramecium?* J Exp Zool 203:99–108

Kitamura A, Hiwatashi K (1980) Reconstitution of mating active membrane vesicles in *Paramecium.* Exp Cell Res 125:486–489

Kitamura A, Hiwatashi K (1984a) Cell contact and the activation of conjugation in *Paramecium.* Zool Sci 1:161–168

Kitamura A, Hiwatashi K (1984b) Inactivation of cell movement following sexual cell recognition in *Paramecium caudatum.* Biol Bull 166:156–166

Kitamura A, Onimaru H, Naitoh Y, Hiwatashi K (1980) Ciliary inactivation following sexual cell recognition in *Paramecium.* Dev Growth Differ 22:722

Klayman MB (1969) Clumping by isolated cilia of complementary mating types from axenically-grown *Paramecium aurelia.* Yale Sci Mag 43:32

Klimetzek V (1974) Isolierung und Untersuchung paarungsreaktiver Cilien von *Paramecium multimicronucleatum:* ein Beitrag zur Charakterisierung der Paarungstypsubstanzen von Ciliaten. Thes, Univ Münster

Klimetzek V (1977) Isolation and characterization of mating reactive cilia of *Paramecium multimicronucleatum.* In: Hutner SH (ed) Abstr Pap 5th Int Congr Protozoology, New York, p 275

Larison LL, Siegel RW (1961) Illegitimate mating in *Paramecium bursaria* and the basis for cell union. J Gen Microbiol 26:499–508

Machemer H (1974) Frequency and directional responses of cilia to membrane potential changes in *Paramecium.* J Comp Physiol 92:293–316

McRorie RA, Williams WL (1974) Biochemistry of mammalian fertilization. Annu Rev Biochem 43:777–803

Metz CB (1946) Effect of various agents on the mating type substances of *Paramecium aurelia* variety 4. Anat Rec 94:347

Metz CB (1947) Induction of "pseudoselfing" and meiosis in *Paramecium aurelia* by formalin killed animals of opposite mating type. J Exp Zool 105:115–139

Metz CB (1948) The nature and mode of action of the mating type substances. Am Nat 82:85–95

Metz CB (1954) Mating substances and the physiology of fertilization in ciliates. In: Wenrich DH (ed) Sex in microorganisms. Am Assoc Adv Sci, Washington DC, pp 284–334

Metz CB, Butterfield W (1951) Action of various enzymes on the mating type substances of *Paramecium calkinsi.* Biol Bull 101:99–105

Metz CB, Fusco EM (1948) Inhibition of the mating reaction in *Paramecium aurelia* with antiserum. Anat Rec 101:654–655

Miyake A (1958) Induction of conjugation by chemical agents in *Paramecium caudatum.* J Inst Polytech Osaka City Univ, Ser D 9:251–296

Miyake A (1964) Induction of conjugation by cell-free preparations in *Paramecium multimicronucleatum.* Science 146:1583–1585

Miyake A (1966) Local disappearance of cilia before the formation of holdfast union in conjugation of *Paramecium multimicronucleatum.* J Protozool 13:s28

Miyake A (1981) Physiology and biochemistry of conjugation in ciliates. In: Levandowsky M, Hutner SH (eds) Biochemistry and physiology of protozoa, 2nd edn, vol 4. Academic Press, London New York, pp 125–198

Mott MR (1965) Electron microscopy studies on the immobilization antigens of *Paramecium aurelia.* J Gen Microbiol 41:251–261

Nobili R (1963) Effects of antibiotics, base- and amino acid-analogues on mating reactivity of *Paramecium aurelia.* J Protozool 10:s24

Preer JR, Jr (1959) Studies on the immobilization antigens of *Paramecium:* IV. Properties of the different antigens. J Immunol 83:803–814

Sasaki S, Ito A, Kitamura A, Hiwatashi K (1972) Mating type specific antigen in *Paramecium caudatum.* Genetics 71:s55

Schleuning W-D, Fritz H (1976) Sperm acrosin. In: Lorand L (ed) Proteolytic enzymes, pt B. Academic Press, London New York, pp 330–341 (Methods in enzymology, vol 45)

Sonneborn TM (1937) Sex, sex inheritance and sex determination in *Paramecium aurelia.* Proc Natl Acad Sci USA 23:378–385

Takahashi M, Hiwatashi K (1974) Potassium: a factor necessary for the expression of mating reactivity in *Paramecium caudatum.* Exp Cell Res 85:23–30

Takahashi M, Takeuchi N, Hiwatashi K (1974) Mating agglutination of cilia detached from complementary mating types of *Paramecium.* Exp Cell Res 87:415–417

Watanabe T (1977a) Ciliary membranes and mating substances in *Paramecium caudatum.* J Protozool 24:426–429

Watanabe T (1977b) Chemical properties of mating substances in *Paramecium caudatum:* effect of various agents on mating reactivity of detached cilia. Cell Struct Funct 2:241–247

Watanabe T (1978) A scanning electron-microscopic study of the local degeneration of cilia during sexual reproduction in *Paramecium.* J Cell Sci 32:55–66

Chapter 7 The Cell Cycle and Regulation of Cell Mass and Macronuclear DNA Content

JAMES D. BERGER[1]

CONTENTS

1 Introduction

The chief developmental pathways in unicellular eukaryotes, such as *Paramecium*, are the vegetative replication pathway, and the sexual pathway. Both phenomena have as their basis the cell cycle with its recurrent pattern of growth, replication and division. The regulation of events within the cell cycle requires that cell components be elaborated in such a way as to maintain balanced growth. Perturbations of cell mass or gene dosage must also be attenuat-

1 Department of Zoology, University of British Columbia, Vancouver, B.C., Canada V6T 2A9

Paramecium. Ed. by H.-D. Görtz

ed, and cellular components must be properly allocated to daughter cells at division. Macronuclear DNA content must also be regulated so that a constant gene concentration (gene dosage per unit cell mass) is maintained over time. In ciliates the maintenance of a constant gene concentration is a significant problem because of the variable gene dosage of the polygenomic macronucleus and its relatively imprecise amitotic macronuclear division process. The mechanisms regulating gene dosage and cell mass must also be integrated into the processes controlling cell cycle events. In addition, it is necessary that the alternative pathways (e.g. meiosis, or stationary phase arrest) be entered under the appropriate conditions and with proper phasing with respect to the underlying vegetative replication pattern.

The style of virtually all subsequent cell cycle work in *Paramecium* was set by the initial studies of Kimball and co-workers (Kimball and Barka 1959; Kimball and Perdue 1962). They deduced the patterns of macromolecular synthesis and growth throughout the cell cycle using relatively small samples of highly synchronous cells obtained by hand selection of dividing cells from exponential growth-phase cultures. The absence of a system for obtaining synchrony in mass cultures precluded the use of large-scale biochemical procedures such as those used extensively in earlier work on the cell cycle of *Tetrahymena,* which relied on induced synchrony in mass cultures (reviews: Zeuthen 1971; Zeuthen and Rasmussen 1972). While the absence of a cell synchronizing system made it necessary to utilize procedures appropriate for single cells, the reliance on selection synchrony eliminated the problem of unbalanced growth which complicated much of the earlier *Tetrahymena* work. The second major development has been the use of gene mutations as physiological tools to perturb the cell cycle. The analysis of these experimentally induced perturbations has been essential for analysis of cell cycle control processes.

2 Patterns of Growth, Replication and Morphogenesis

The major events of the *Paramecium* cell cycle are shown in Fig. 1. These events form a number of semi-independent series involving the macronucleus, the micronucleus and the cell cortex (review: Berger 1984) and will be described briefly along with the patterns of cell growth and macromolecular synthesis.

2.1 Macronuclear Replication and Division

2.1.1 Macronuclear DNA Synthesis

Initiation of macronuclear DNA synthesis (IDS) is one of the earliest cell cycle landmarks and the first indication that cells have entered the vegetative replication pathway. Macronuclear DNA synthesis begins abruptly at 0.25 in the cell cycle in *P. tetraurelia* (Berger and Kimball 1964; Berger 1971, 1973; Smith-Sonneborn and Klass 1974; Rasmussen and Berger 1982), and continues until the

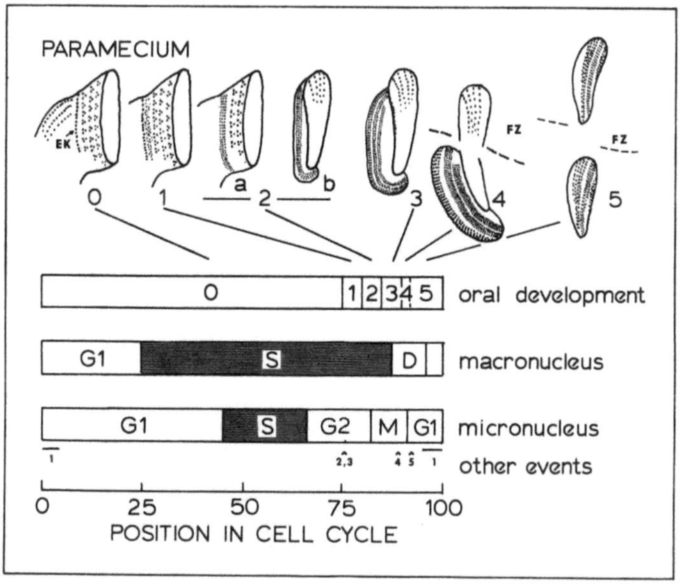

Fig. 1. Cell cycle landmarks. The drawings show stages of oral primordium development (Kaneda and Hanson 1974; Jones 1976). Initially, the oral primordium develops on the right wall of the vestibule between the endoral kinety (*EK*) and the first vestibular kinety. During stage *2* basal bodies form into rows and the primordium invaginates to form a saclike structure lying to the right of the buccal opening. The view in stage *2b* is looking into the overture, while earlier views are from the side. The kineties of the oral primordium differentiate into the oral membranelles (quadrulus and two peniculi) as the primordium moves posteriorly, away from the old oral apparatus. As the fission zone (*FZ*) forms, the buccal overture constricts to form two openings (stage *4*). Finally, in stage *5* the daughter oral structures are shown looking into the buccal openings. Macronuclear data are from Berger (1971) and Tucker et al. (1980); micronuclear data from Tucker et al. (1980) and Pasternak (1967). *Small numbers* indicate the following cell cycle events: *1* cytospindle (longitudinal microtubule bundles) present (Cohen et al. 1980); *2* point of commitment to cell division (last point at which blockage of DNA synthesis blocks cell division) (95); *3* new contractile vacuole pores appear (Kaneda and Hanson 1974); *4* endoral kinety is again visible (Jones 1976); and *5* oral primordium for the next cell cycle appears (Kaneda and Hanson 1974). (From Berger 1984)

beginning of macronuclear elongation at about 0.9 in the cell cycle (Kaneda and Hanson 1974; Tucker et al. 1980). There is no macronuclear G2 period. This information has been obtained both from autoradiographic studies of incorporation of [^{3}H]-thymidine, and from absorption microphotometry of Felgen-stained nuclei (Fig. 2).

The early autoradiographic and microphotometric studies were interpreted to indicate that DNA synthesis began about half way through the cell cycle (Woodard et al. 1961; Kimball and Barka 1959; Kimbal et al. 1960; Kimball and Perdue 1962). The failure of these studies to correctly estimate the beginning of the macronuclear S-period is likely a consequence of the small sample sizes used in the microphotometric experiments combined with the relatively high variability in macronuclear DNA content, and the lack of specificity in early DNA labelling experiments. Kimball and Vogt-Köhne (1962) correctly

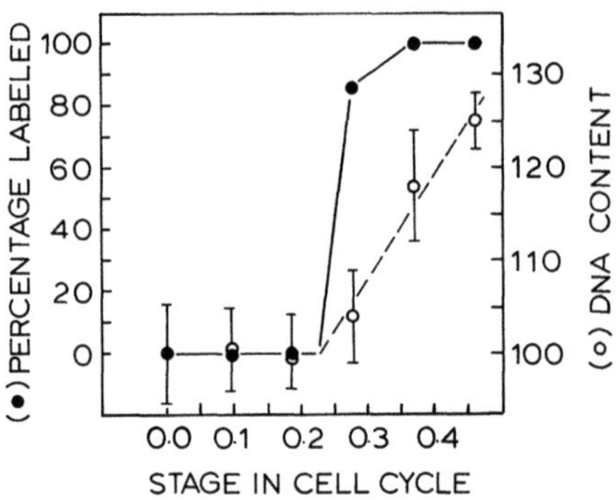

Fig. 2. Timing of initiation of macronuclear DNA synthesis in well-fed *P. tetraurelia* cells. *Open circles:* mean macronuclear DNA content as percentage of the mean post-fission value. *Filled circles:* percentage of cells incorporating [^{3}H]- thymidine. (Data from Rasmussen and Berger 1982). Both variables were determined from the same samples of synchronous cells

estimated the timing of IDS in *Paramecium.* Subsequent estimates of the timing of IDS were facilitated by the use of [^{3}H]-thymidine-labelled bacteria which results in increased specificity of labelling (Berger and Kimball 1964, Berger 1971).

The same relative duration of cell cycle stages is maintained when balanced (equilibrium) growth occurs under conditions which extend the cell cycle. For example, in doublet cells (Morton and Berger 1978) or in chemostat cultures (Ching and Berger 1986a) in which cell cycle duration may be as long as 30 h (Fig. 3a–c).

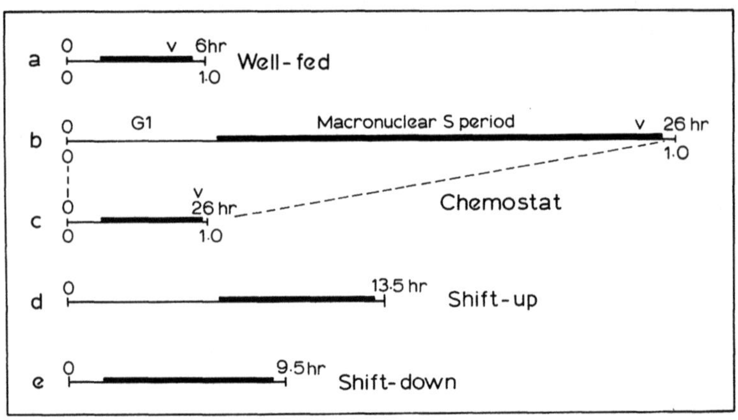

Fig. 3a–e. Effect of nutrient level and nutrient level shifts on the cell cycle of *Paramecium.* **a** Control cell cycle (well-fed cells). The period of macronuclear DNA synthesis is indicated by the *thicker line.* The point at which cells become committed to division is indicated by the downward *arrow head.* **b** The cell cycle in nutrient-limited cells (equilibrium chemostat culture with a generation time of 26 h). The horizontal scale is the same as in **a.** **c** The cell cycle of chemostat cells **b** reduced in scale so that the relative duration of the cell cycle is the same as that in well-fed cells **a** to show that IDS has the same relative location in the extended cell cycle of chemostat cells that it occupies in well-fed controls. **d** Effect of nutrient level shift-up at fission of chemostat cells on the subsequent cell cycle. Note that the timing of IDS is the same as that in the parental cell cycle (chemostat culture). **e** Effect of nutrient level shift-down at fission of well-fed cells on the timing of IDS. IDS occurs at the same time as in the preceding (well-fed) cell cycle. (Data from Ching and Berger 1986a, b)

The timing of macronuclear DNA synthesis in *P. caudatum* is similar to that of *P. tetraurelia.* The initial studies suggested that macronuclear DNA synthesis began about halfway through the cell cycle (Walker and Mitchison 1957; Blanc 1965). Rao and Prescott (1967) observed, however, that intense incorporation of [^{3}H]-thymidine begins about 2.5 h after fission (about 0.3 in the cell cycle). This suggests that the timing of IDS in *P. caudatum* is probably not greatly different than in *P. tetraurelia.* A similar pattern is observed in *P. putrinum* (Kudrjavtsev 1966). This pattern, however, does not extend throughout the genus for IDS occurs at the beginning of the cell cycle in *P. bursaria* (Just 1973).

Autoradiographic analysis indicates that the rate of macronuclear DNA accumulation over the course of the S-period in *P. tetraurelia* is approximately constant with a decrease at the end of the S-period (Berger 1971).

2.1.2 Macronuclear Growth

Nuclear growth occurs coordinately with DNA replication and continues in an approximately linear pattern throughout the macronuclear S-period. This has been shown by absorption cytophotometry (Woodard et al. 1961), microinterferometric measurements of dry mass in isolated macronuclei (Kimball et al. 1960) and isotope dilution studies (Berger 1971). The resolution of these studies is insufficient to decisively discriminate between linear and exponential patterns of nuclear mass doublings.

2.1.3 Macronuclear Division

Macronuclear amitosis begins at about 0.9 in the cell cycle. The process of macronuclear division and the morphology of the macronucleus during division has been examined by Tucker et al. (1980) and a variety of mutants affecting the amitotic macronuclear division process (*tam* or *amac* mutants) have been described (Ruiz et al. 1976; Berger and Schmidt 1978; Cohen and Beisson 1980; Cohen et al. 1980). The amitotic division process typically does not result in equal distribution of DNA to daughter cells. In *P. tetraurelia* the mean differences in DNA content between sister cells is 5–10% of the parental DNA content (Kimball and Barka 1959; Berger and Schmidt 1978).

2.2 Micronuclear Replication and Division

Woodard et al. (1961) reported that micronuclear DNA synthesis occurred between 0.5 and 0.6 in the cell cycle. In a careful study on the patterns of synthetic activity of micronuclei in *P. tetraurelia* Pasternak (1967) observed that DNA synthesis took place between approximately 0.35 and 0.65 in the cell cycle. In *P. caudatum* micronuclear DNA synthesis begins between 3 and 6.5 h after fission (0.35–0.6 in the cell cycle) (Rao and Prescott 1967). Micronuclei in both species also showed significant incorporation of [^{3}H]-uridine, especially during the period of DNA synthesis. Despite these observations of incorporation of RNA precursors (perhaps as primers for DNA synthesis), there is no indication of genetic activity in micronuclei in *P. tetraurelia* (Sonneborn 1974), although micronuclei play a role in the determination of mating type (Brygoo et al. 1980). The micronucleus also plays a significant role in oral morphogenesis (see Mikami, Chap. 8, this Vol.).

2.3 Cell Growth

The pattern of cell growth throughout the cell cycle is exponential as shown by the initial microinterferometric study of Kimball et al. (1959) and confirmed by microinterferometry (Kimball and Vogt-Köhne 1962) and isotope dilution (Kimball and Perdue 1965) studies. The specific rate of protein synthesis (rate of incorporation per unit mass) is constant throughout the cell cycle (Fig. 4) demonstrating that the pattern of growth is a uniform exponential pattern (Ching and Berger 1986a; Kimball and Perdue 1965). The cellular rate of protein synthesis increases throughout the cell cycle as expected (Fig. 4).

2.4 Cortical Morphogenesis

The morphogenetic events associated with fission have been reviewed elsewhere (Sonneborn 1970; Kaneda and Hanson 1974; Hanson and Ungerleider 1974; Jones 1976; Kaczanowska and Dubielecka 1983). The major observations pertinent to this discussion are that the morphogenetic processes associated with fission begin at about 0.75 in the cell cycle (90–100 min prior to fission in well-fed cells). At this time, the initial stages of oral primordium formation occur (Jones 1976) and new contractile vacuole pores first appear (Kaneda and Hanson 1974). These events are followed at about 0.8 by the beginning of micronuclear prophase. These early morphogenetic events occur at or soon after the point at which cells become committed to cell division as evidenced by the ability to complete division when DNA synthesis is blocked (see below). Throughout the remainder of the cell cycle the oral primordium differentiates

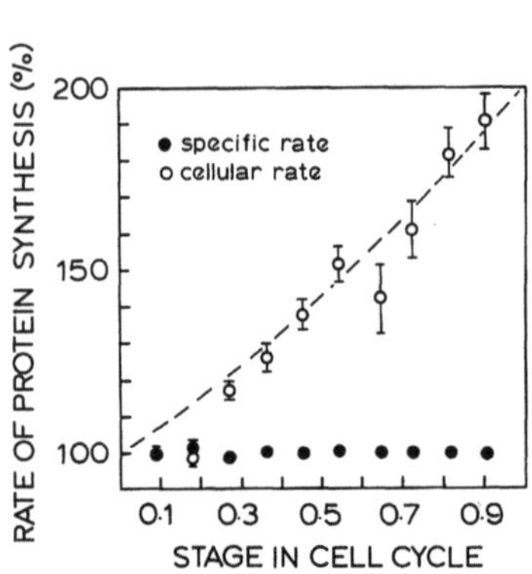

Fig. 4. Protein synthesis rates as a function of stage in the cell cycle. *Filled circles:* specific rate of protein synthesis (rate per unit mass). *Open circles:* cellular rate of protein synthesis (rate per cell). *Dashed line* indicates exponential doubling pattern (Data from Ching and Berger 1986a).

and separates from the parental oral apparatus as the buccal overture is divided by the fission zone (Fig. 1), which is underlain by the longitudinal microtubule bundles of the cytospindle that appear only during the latter part of the fission process (Cohen et al. 1980). There is a characteristic pattern of change in cell shape and dimensions during the course of the cell cycle (Kaneda and Hanson 1974; Jones and Berger 1982) with the greatest increase in cell length occurring during the last 10% of the cell cycle during the latter part of the prefission morphogenetic period.

3 Control of Cell Cycle Events

The cell cycle typically consists of two semi-independent pathways, the nuclear replication pathway and the cell growth and cytokinesis pathway (review: Mitchison 1971). These pathways must be coordinated to bring about proper regulation of cell mass and gene dosage. Experimental analysis of the cell cycle in *P. tetraurelia* has revealed that the coordination of cell growth with DNA replication occurs through the operation of two cell cycle control points. The first of these is associated with initiation of DNA synthesis and commitment of cells to the DNA replication pathway, and the second with commitment of cells to division. The approach used in the analysis was to perturb one or another of the physiological variables of the cell cycle and to examine the subsequent regulatory responses in small populations of synchronous cells.

3.1 Initiation of Macronuclear DNA Synthesis

The first major cell cycle control point in most eukaryotes is associated with initiation of DNA synthesis (IDS) (review: Yanishevsky and Stein 1981). The occurrence of IDS reflects the commitment of the cell to the vegetative replication pathway. For example, in yeasts commitment to vegetative replication and division is associated with the execution of the genetic function 'start'. Successful execution of 'start' is dependent on a number of conditions. These include sufficient nutrient concentrations, accumulation of sufficient cell mass and absence of mating pheromone if cells are haploid, or conditions for meiosis if cells are diploid (review: Pringle and Hartwell 1981). These requirements for execution of 'start' lead to the association of attainment of a 'critical' cell mass with IDS as observed in many organisms. It is therefore not surprising that the kinetics of IDS in *Paramecium* are influenced by changes in cell mass, macronuclear gene dosage and nutrient levels. Analysis of the effects of these variables on the timing of IDS allow some deductions to be made concerning the control of IDS and the initial stages of the cell cycle.

3.1.1 Change in Cell Mass

Experimental change in cell mass can be obtained through action of the temperature-sensitive conditional mutation *cc1* which reversibly blocks cell cycle

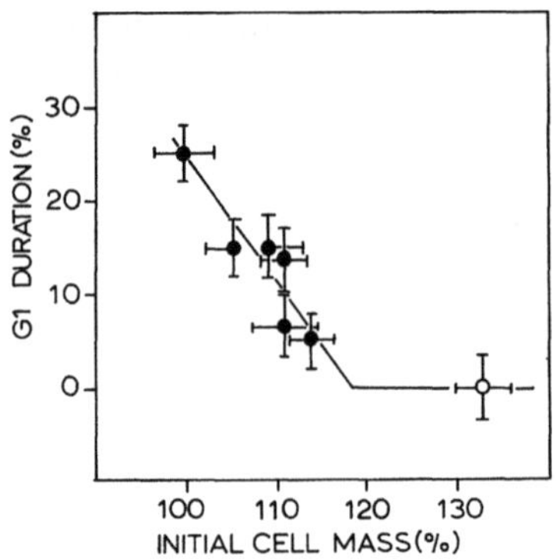

Fig. 5. Effect of cell mass at fission on the duration of the subsequent G1 interval. As cell mass increases, there is a progressive shortening of the G1 interval. IDS occurs when cell mass reaches or exceeds the value present at IDS in control cells (Rasmussen et al. 1986)

progression and macronuclear DNA synthesis, but not cell growth (Peterson and Berger 1976; Rasmussen and Berger 1984). This makes it possible to obtain newly divided cells with up to three times normal cell mass (Rasmussen and Berger 1982). In addition, it is possible to increase cell mass at specific points within the cell cycle, for example during the G1 interval.

When well-fed cells begin the cell cycle with increased cell mass, the duration of the G1 interval is reduced. The reduction in the duration of the G1 interval is proportional to the excess cell mass (Fig. 5). It reaches zero (the G1 period disappears) when the initial cell mass is equal to the normal cell mass at IDS in unperturbed cell cycles (119% of initial cell mass). When the initial cell mass is increased IDS occurs when the cell mass reaches the level normally present in unperturbed cells at the time of IDS. The timing of IDS is therefore strongly associated with cell mass increase. However, as described below, IDS is not dependent on cell mass increase, and can occur even in the presence of a net reduction in cell mass.

Although initial cell mass has a pronounced effect on the timing of IDS, increase in cell mass during the G1 interval does not lead to advancement of IDS. Blockage of the cell cycle during G1 with increase of cell mass to more than 170% of the initial level results in no decrease in the remaining portion of the G1 interval (Fig. 6; Rasmussen et al. 1986). This suggests that the timing of IDS

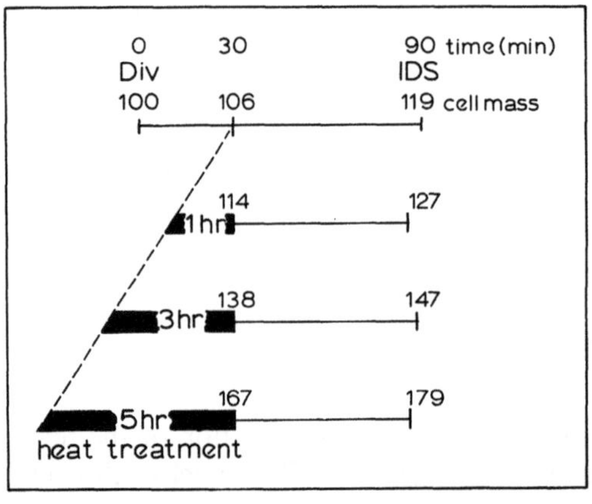

Fig. 6. Effect of increase in cell mass during the G1 interval on the timing of IDS. Increase in cell mass does not lead to advancement of IDS. (Data from Rasmussen et al. 1986)

is established during the preceding cell cycle. Consequently, as described below, changes in the rate of cell mass accumulation during the G1 interval have no effect on the timing of IDS (Ching and Berger 1986a). This does not mean that cell mass has no effect on the timing of IDS. Cell mass accumulated during the preceding cell cycle affects the timing of IDS. IDS, however, occurs independently of the cell mass present at the time of initiation.

3.1.2 Change in Macronuclear DNA Content

Similar observations have been made concerning the effect of variation in macronuclear gene dosage on the timing of IDS. Extensive variation in macronuclear DNA gene dosage (DNA content) can be obtained through the action of *tam* mutants which disturb the normal nearly equal segregation of macronuclei to daughter cells (Berger 1978, 1979; Berger and Schmidt 1978; Cohen and Beisson 1980; Cohen et al. 1980). Since changes in gene dosage occur at fission, it is possible to examine the consequences of changes in DNA content between sister cells which have an identical history prior to the immediately preceding fission. In such pairs of sister cells, differences in gene dosage of more than ten-fold do not lead to differences in the timing of IDS between the cells (Berger 1982b; Rasmussen et al. 1986). Further analysis of the data, however, reveals that differences in the gene dosage of their parents does have a slight effect on the timing of IDS. That is, daughter cells derived from a mother cell with less than average DNA content tend to initiate DNA synthesis slightly earlier than cells derived from mothers with greater gene dosage (Rasmussen et al. 1986). The magnitude of this effect is small relative to the dramatic effect of increase in cell mass during the parental cell cycle on the timing of IDS in the subsequent cell cycle. Despite the limited extent of the effect of parental DNA content on the timing of IDS, the observations support the hypothesis that the timing of IDS is established prior to fission: The mechanisms regulating the timing of IDS are unaffected by massive changes in gene dosage during the G1 interval.

3.1.3 Change in Nutrient Conditions

Changes in the nutrient level also affect the timing of IDS in *Paramecium,* as they do in other organisms. However, changes in the nutrient level at fission do not affect the timing of IDS in the subsequent cell cycle. For example, when cells are transferred at division from a chemostat culture with a generation time of 26 h to medium with excess food, IDS occurs about 6.5 h after fission (Fig. 3d). The duration of the G1 interval corresponds to 0.25 of the parental cell cycle (Fig. 3c), and not to either 0.25 of the cell cycle of well-fed control cells (Fig. 3a), or 0.25 of the present cell cycle (Fig. 3d). The next cell cycle, however, is normal. Similarly, if cells are transferred at fission from excess food to highly diluted medium, IDS occurs at a time corresponding to 0.25 of the preceding (control) cell cycle (Fig. 3a) and not at 0.25 in the extended experimental cell cycle (Fig. 3e).

The establishment of the timing of IDS prior to fission provides the basis for the uncoupling of cell mass increase and IDS observed in perturbed cell cycles. Changes in nutrient level at fission result in altered rates of cell mass accumulation. This combined with unaltered timing of IDS results in either greater or less cell mass at IDS than that present under pre-shift conditions. For example, when well-fed cells are subjected to severe nutrient down-shift at fission, IDS occurs with normal timing even though there may be a net decrease in cell mass between fission and IDS (Ching and Berger 1986a).

3.1.4 When is the Timing of IDS Established?

The observations summarized above indicate that the timing of IDS in *Paramecium* is established during the preceding cell cycle. If there is a 'start'-like control point in the *Paramecium* cell cycle, it occurs prior to fission. Further information concerning the location of this initial cell cycle control point can be obtained from the location of the point of initial commitment to meiosis. In yeasts, for example, meiosis is initiated from a point in the cell cycle just after start, but prior to initiation of the vegetative replication pathway (Hirschberg and Simchen 1977). Similarly in *Tetrahymena,* meiosis is entered from a point just prior to IDS, closely associated with the point of commitment to vegetative replication (Wolfe 1976). If this is also true for *Paramecium,* the point of initial commitment to meiosis might indicate the location of the point of commitment to IDS. In *Paramecium,* the initial point of commitment to meiosis occurs at about 0.75 in the cell cycle (Berger 1986), suggesting that commitment to IDS also occurs prior to fission, during the latter part of the preceding cell cycle (Fig. 7). This point is strongly associated with the point of commitment to cell division (see below).

Direct examination of the point at which the timing of IDS is established shows that the timing of IDS is set 90 – 100 min prior to fission. Nutrient down-shift at this point or later does not alter the timing of IDS in the next cell cycle, while down-shift at earlier points in the cell cycle results in delay of IDS in the subsequent cell cycle (Berger and Ching 1987).

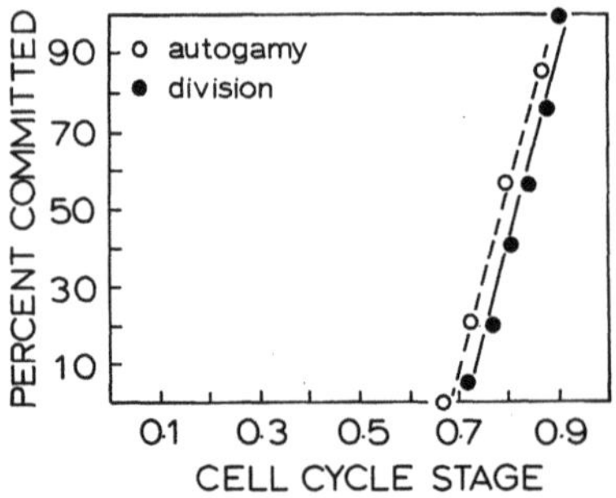

Fig. 7. Location within the cell cycle of transition points for commitment to cell division (*filled circles*) and for commitment to meiosis (autogamy) in the next cell cycle. (Data from Berger 1986)

3.2 Control of Cell Division

In *Paramecium*, as in yeasts (*Saccharomyces:* Piggott et al. 1982; *Schizosaccharomyces:* Nurse and Bissett 1981; Nurse and Fantes 1981, 1985), the second major cell cycle control point is associated with commitment to cell division.

3.2.1 Timing of Commitment to Division

In *Paramecium* the point of commitment to cell division can be determined by the use of temperature-sensitive mutations which block cell cycle progression. The point within the cell cycle at which transfer of cells to restrictive conditions no longer blocks fission is taken to indicate the point of commitment to cell division. In well-fed *Paramecium* cells this point occurs at approximately 0.75 in the cell cycle (Fig. 7) (Peterson and Berger 1976, Rasmussen et al. 1985). As two different gene mutations which block DNA synthesis provide very similar estimates of the location of the transition point for division (Peterson and Berger 1976) the location of the transition point seems to be a feature of the cell cycle and has little to do with the action of a particular mutation. The transition point for cell division occurs at a fixed interval (90 – 100 min) prior to fission rather than at a fixed relative position in the cell cycle, as does IDS. When the cell cycle is extended by growth at lower temperatures or by nutrient limitation, the point of commitment to division occurs at a later relative stage in the cell cycle, but at the same absolute time interval prior to fission. For example, in chemostat cultures with a 26-h generation time, the relative location of the point of commitment to division is at 0.96 ± 0.02 rather than at 0.73 ± 0.02 as in 27 °C well-fed controls (Ching and Berger 1986 b). The time interval between the transition point and division is similar in both cases (Fig. 3 a – c). The point of commitment to division in well-fed wild-type cells is closely associated with initiation of oral morphogenesis and micronuclear mitotic prophase (Kaneda and Hanson 1974).

3.2.2 Requirements for Division

Experiments which involve perturbation of the cell cycle have provided information concerning control of commitment to division, but the genetic and molecular mechanisms underlying this control system are unresolved. The observation that *Paramecium* cells synthesize a constant amount of macronuclear DNA, regardless of the initial DNA content (Berger and Schmidt 1978; Berger 1978, 1979) suggests that macronuclear DNA synthesis is an important precondition for initiation of division. The synthesis of a standard DNA increment in all cells implies that the DNA content does not double in cells with larger than average macronuclei, and in smaller than average macronuclei the DNA content increases more than two-fold. Therefore, initiation of cell division can not be dependent on the precise doubling of macronuclear DNA content.

Synthesis of the entire normal complement of macronuclear DNA is also not required for division. When *cc1* cells are placed under restrictive conditions just after the point of commitment to cell division they proceed to division, even though DNA synthesis does not occur under restrictive conditions and the mean DNA content of the resulting daughter cells is about 20% lower than that of controls (Rasmussen et al. 1985). If a fixed DNA increment is required prior to commitment to cell division, it must be substantially less than the full increment. These initial observations suggest that the DNA increment accumulated to the point of commitment to division is 50–60% of the normal increment (Ching and Berger 1986b). Similar, but slightly higher, estimates have been obtained from direct examination of the DNA content of cells in chemostat cultures (Ching and Berger 1986b). Direct examination of the DNA content of well-fed cells at the point of commitment to cell division suggests a somewhat higher DNA increment (Ching and Berger 1986b). The processes involved in establishing the DNA threshold increment require further study.

The mechanism regulating the magnitude of the DNA increment is rapidly reset after changes in nutrient level. When cells are transferred from nutrient-limited to well-fed conditions at fission, the DNA increment during the subsequent cell cycle is normal. Similarly, when cells are transferred from conditions of excess food to nutrient-limited conditions, the DNA increment drops to about 58% of normal (Ching and Berger, 1986b). Unlike the timing of IDS, the magnitude of the DNA increment is adjusted during the first cell cycle after a nutritional shift.

These observations indicate that cell mass, like initial DNA content, has little to do with the magnitude of the DNA increment. Following nutritional up-shift cells with small macronuclei and low initial cell mass synthesized the normal DNA increment, while cells with normal initial cell mass and DNA content produced a much smaller increment after nutrient down-shift (Ching and Berger 1986b).

Although increases in cell size or macronuclear DNA content alone lead to no change in the DNA increment, an increased DNA increment is obtained when both DNA and cell mass are increased (Rasmussen and Berger 1982). In cells with greatly increased cell mass, lesser increases in initial DNA content lead to the production of proportionately greater rates of DNA accumulation and final DNA increments (Berger 1982, Rasmussen and Berger 1982a). Similarly, when cells are subjected to nutrient limitation, the rate of DNA accumulation is drastically reduced (Ching and Berger 1986a).

3.3 Cell Cycle Model

These observations suggest a general model for cell cycle control in *Paramecium*. The cell cycle is divided into three segments by the two major control points (Fig. 8). The first segment extends from the point of commitment to IDS until IDS. This segment begins prior to fission. The location of the point of commitment to meiosis at 0.75 in the normal cell cycle suggests, by analogy with other eukaryotes, that the point of commitment to vegetative replication is located at or just prior to this point. The duration of this segment of the cell cycle is variable, and is influenced by cell mass, nutritional conditions and gene

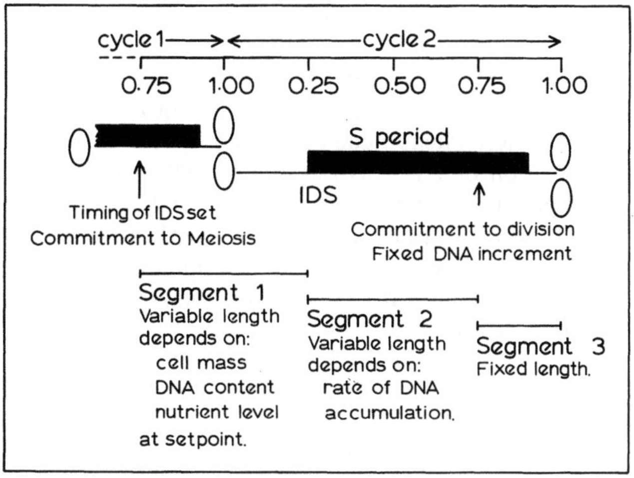

Fig. 8. Cell cycle model for *P. tetraurelia* (Ching and Berger 1986b)

dosage present at or before the point of commitment. The observations suggest that the duration of this first segment of the cell cycle is fixed when the point of commitment to IDS is reached. Perturbation of the variables which alter the timing of IDS in the parental cycle during the following G1 interval does not alter the timing of IDS. Consequently, the mechanism(s) controlling the duration of the G1 interval must be largely independent of drastic changes in gene dosage, cell mass or nutrient level.

The second segment of the cell cycle is of variable duration and appears to be primarily dependent on the rate of DNA accumulation. It is greatly extended by decreases in macronuclear gene dosage (Berger 1978, 1979), or nutrient limitation. For example in chemostat cultures the rate of DNA synthesis is reduced to less than 20% of the normal rate and the second segment of the cell cycle is extended from 3 to 18 h (Ching and Berger 1986b). The second segment of the cell cycle ends with the attainment of the threshold DNA increment and commitment to division.

The third segment of the cell cycle is of fixed duration and coincides with the period of prefission morphogenesis and nuclear division. The relative portion of the cell cycle occupied by this segment decreases with decreasing growth rate.

There are two unusual features of this scheme: First, a particular cell cycle operationally begins prior to fission at the point of commitment to IDS, and second, the temporal locations of the two major control points, commitment to IDS, or meiosis, and commitment to division are close together, rather than widely separated within the cell cycle. Although the two control functions act in close temporal proximity to each other, one affects commitment to division in the present cell cycle and the other commitment to IDS in the next cell cycle. From the standpoint of control, the cell cycles of *Paramecium* are overlapped.

4 Quantitative Regulation of Cell Components

The foregoing analysis of cell cycle control provides a framework for understanding how regulation of DNA content and cell mass occur in *Paramecium* and how these phenomena relate to other cell cycle processes. In every organism coordination of DNA synthesis and cell growth is required to maintain balanced growth. To maintain a constant cell size and gene dosage it is necessary that at least one point in the DNA replication and division pathway be dependent on cell growth, and that one point in the cell growth and division pathway be dependent on DNA replication. The first check ensures that cell size does not fall drastically below normal levels and the second that DNA replication does not lag behind cell growth. Such mechanisms ensure that the variation in cell mass and DNA content introduced at each fission through cytokinesis and macronuclear amitosis are attenuated over the course of subsequent cell cycles. A further feature of the regulatory system of an organism such as *Paramecium* is to allow adjustment of cell mass and gene dosage so that the appropriate gene concentration (gene dosage per unit mass of cytoplasm) is maintained over a wide range of physiological conditions.

4.1 Coordination of DNA Synthesis and Cell Growth

In the analysis presented above the most variable portion of the cell cycle is the second segment extending from IDS to the point of commitment to division. During this interval a fixed DNA increment is accumulated. The rate of DNA accumulation appears to be the major variable controlling the duration of this segment of the cell cycle. Both cell mass and DNA content influence the rate of DNA accumulation. These variables interact through their effect on the cellular rate of protein synthesis. The nature of this interaction was revealed by two series of experiments. The first deals with the effect of variation in DNA content and cell mass on the rate of protein synthesis, and the second with the effects of changes in cell mass and DNA content on the rates of DNA accumulation and cell growth.

The occurrence of increased variation in DNA content in *tam* mutants made it possible to examine the effects of variation in gene dosage on the rate of protein synthesis in cells of different sizes (Berger 1982a). The effect of cell size was assessed by examination of the rate of protein synthesis at different stages within the cell cycle in cells with widely varying DNA contents.

The rate of protein synthesis decreases in a manner directly proportional to gene dosage when gene dosage is experimentally decreased below the normal level (Fig. 9a). This occurs throughout the cell cycle and suggests that at normal or lower gene concentrations (gene dosage per unit mass) the rate of protein synthesis is transcriptionally limited. However, increases in gene dosage do not produce increases in the rate of protein synthesis above the control level. This plateau rate of protein synthesis, however, increases throughout the cell cycle and is strongly correlated with mean cell mass (Fig. 9b) (Berger 1982, 1984). This correlation suggests that the rate of protein synthesis is limited by

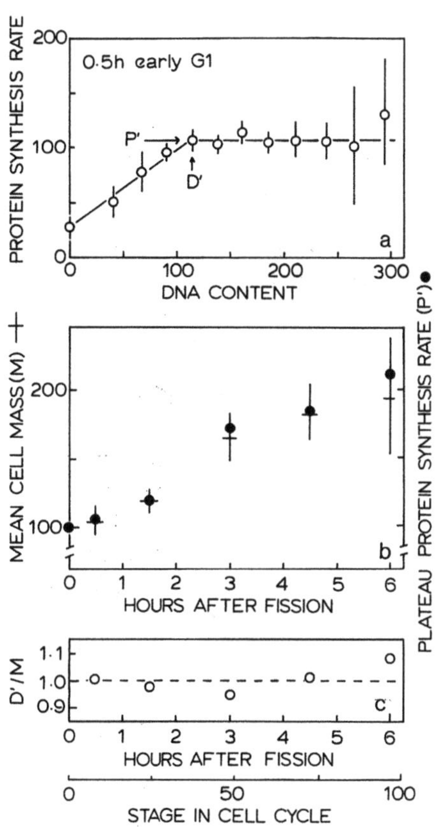

Fig. 9a–c. Effects of variation in macronuclear gene dosage and cell size on the rate of protein synthesis in *Paramecium.* **a** At any stage in the cell cycle the rate of protein synthesis is proportional to DNA content at low DNA contents, but reaches a maximum, plateau, level as DNA content increases; **b** the relative value of the plateau rate of protein synthesis is strongly correlated with the mean cell mass when cells of different sizes (different stages of the cell cycle) are compared; **c** the gene concentration (gene dosage per unit cell mass) at which the plateau rate of protein synthesis is reached remains constant throughout the cell cycle (Berger 1984)

cell mass when gene concentration is higher than normal and that the protein synthesizing system is saturated at normal gene concentrations. The gene concentration at which the plateau rate of protein synthesis is reached does not change over the course of the cell cycle (Fig. 9c), suggesting that the concentration of ribosomes remains constant throughout the cell cycle. These observations suggest that increases in gene dosage or cell mass alone do not result in increases in the rate of protein synthesis, while decreases in either variable result in directly proportional decreases in the cellular rate of protein synthesis.

The maintenance of a constant or nearly constant gene concentration throughout the cell cycle is presumably responsible for the maintenance of a constant rate of protein synthesis per unit mass of cytoplasm (specific rate of protein synthesis) throughout the cell cycle, and a constantly increasing cellular rate of protein synthesis which results in the observed exponential pattern of cell growth (Fig. 4). Saturation of the protein synthesizing system at normal gene concentrations is also suggested by the constant rate of protein synthesis per unit mass of cytoplasm observed throughout the cell cycle (Ching and Berger 1986a).

These results are correlated with observations that increases in DNA content alone or cell mass alone do not result in increases in either the rate of DNA accumulation or cell growth. On the other hand, coordinated increases in DNA content and cell mass result in increases in both the rate of DNA accumulation and cell growth which are proportional to the relative magnitude of the lower of the two variables. For example, when both DNA content and cell mass are increased, the rates of growth and DNA accumulation are proportional to the relative DNA content which was relatively lower than the increase in cell mass (Berger 1979; Rasmussen and Berger 1982). Reduction in cell size in the *sm2* mutant which interferes with cell surface proliferation (Jones and Berger 1982) leads to reduction in both the DNA and cell mass increments (Berger 1979). Similarly, reduction in gene dosage leads to marked decreases in the rates of growth and DNA accumulation (Berger 1978, 1979).

4.2 Quantitative Regulation of Cell Components

Growth related processes in *Paramecium* interact with the cell cycle control system in three ways; first, the association of cell mass increase with IDS; second, the dynamic interaction of DNA content and cell mass involved in the regulation of the rate of protein synthesis; and third, the apparent requirement for a fixed DNA increment as a precondition for initiation of cell division. All three processes are involved in regulation of cell mass, and the latter two in regulation of DNA content.

4.2.1 Regulation of DNA Content

Regulation of DNA content in *Paramecium* occurs through an incremental mechanism: All cells synthesize a standard amount (increment) of DNA during each cell cycle. The general properties of incremental regulatory mechanisms have been discussed by Fantes and Nurse (1981). Regulation of macronuclear DNA content in *Paramecium* occurs as a consequence of the DNA increment requirement for initiation of cell division. Because of this, half of the variance in DNA content introduced at each fission is removed during the course of the subsequent cell cycle (Berger and Schmidt 1978) and individual perturbations of DNA content are reduced by half during each subsequent cell cycle. This occurs following either increases or decreases in DNA content (Berger 1978, 1979).

Several variables affect the magnitude of the DNA increment. If both cell mass and gene dosage are coordinately increased there is a proportional increase in the size of the DNA increment (Rasmussen and Berger 1982). Thus, doublet cells with twice the normal DNA content and cell mass produce a DNA increment which is twice the normal value (Morton and Berger 1978). Factors which reduce the rate of DNA accumulation reduce the DNA increment by reducing the amount of DNA accumulated between the point of commitment to division and the end of the S-period (Ching and Berger 1986b). The number of macronuclei per cell has no effect on the total DNA increment (Berger 1979).

The size of the macronucleus, and thus the mean DNA increment, changes under different conditions. As described above, the DNA increment in nutritionally limited cells is substantially smaller than that in well-fed cells. Similarly, the mean DNA content (and therefore the mean DNA increment) in aged clones of *P. tetraurelia* is substantially lower than that in young well-fed cells. In aged clones there is a progressive decrease in DNA content with a minimum of about 50% of the level for young well-fed clones (Schwartz and Meister 1975; Klass and Smith-Sonneborn 1976). The decrease in macronuclear DNA content occurs in conjunction with decrease in the rate of DNA synthesis (Smith-Sonneborn and Klass 1974) and with decrease in the rate of feeding in aged cells (Smith-Sonneborn and Rodermel 1976). These observations are consistent with the cell cycle model and the DNA content regulation scheme proposed above. However, this scheme may not apply without modification to other *Paramecium* species. In *P. caudatum*, the mean macronuclear DNA content of aged cells also decreases, but to a much lower value, about 20% of the mean level for young cells (Takagi and Kanazawa 1982).

4.2.2 Regulation of Cell Mass

The downward regulation of cell mass is chiefly the result of two processes. First, the association of cell mass with IDS results in shortening or elimination of the G1 period when the initial cell mass is increased. This results in a shortening of the cell cycle. Shortening of the cell cycle results in a smaller than normal mass increment because there is less time before division for cell growth. The second process contributing to regulation of cell mass is the lack of effect of increased cell mass on the rate of cell growth. The rate of cell mass (total protein) accumulation is not increased unless *both* DNA content and cell mass are increased. Increase in cell mass alone does not change the rate of growth (Berger 1979; Rasmussen and Berger 1982). Consequently, the rate of cell mass increase is not increased in experimentally enlarged cells and their shorter cell cycle (0.75 of normal) results in a cell mass increment which is also 0.75 of the normal value (Rasmussen and Berger 1982). Downward regulation of cell mass thus occurs more rapidly than would be expected from a purely incremental mechanism alone.

Upward regulation of cell mass has not been as well studied because it has not been possible to generate cells with decreased cell mass without concomitant decreases in gene dosage. When small cells from chemostat cultures are returned to excess food at fission, the cell mass increment produced in the next cell cycle is somewhat greater than normal (Ching and Berger 1986b). This is a result of the extended cell cycle; more time is available for growth. The increased cell cycle duration is, in turn, a consequence of the long G1 interval and the lower than normal rate of DNA accumulation. The long G1 interval is a consequence of the fact that the timing of IDS is established during the preceding cell cycle prior to the nutrient shift at fission. The lower than normal rate of DNA accumulation is primarily a consequence of the initial low values for cell mass and DNA content. Consequently, a longer than normal interval is required to accumulate the threshold DNA increment that is associated with commitment to cell division. The specific rate of protein synthesis returns to normal levels within a couple of hours after the nutrient shift, well before the end of the G1 interval (Ching and Berger 1986b).

Examination of control of cell cycle processes in *Paramecium* suggests that cell mass is not closely controlled, but that the interaction between DNA content and cell mass in determining the overall rate of growth and DNA synthesis assures that there is a substantial correlation between increase in cell mass and increase in DNA content. Despite the apparent lack of direct control of cell mass increase, there is usually a strong correlation between cell mass and macronuclear DNA content, both in mutant cells with defects in cytokinesis or in cell surface proliferation (Berger 1984) or in cells grown under different nutrient conditions (Kimball 1967).

Since both cell mass and macronuclear DNA content directly effect the rates of growth and DNA accumulation we must ask how the set point for regulation of cell mass and DNA content is determined. The answer may lie in the size of the cell cortex. Newly divided singlet cells with twice the normal cell mass and normal DNA content can be generated by heat treating *cc1* cells. These cells regulate so that cell mass is reduced and DNA content remains constant. A newly divided doublet cell with the same initial cell mass and DNA content is generated under some conditions when a bimacronucleate doublet parent gives rise to a unimacronucleate daughter cell through abnormalities in macronuclear division and segregation. In this case the cell regulates in such a way that the cell mass remains unchanged while DNA content increases to the normal doublet level (Morton and Berger 1978). These observations are also consistent with informal observations that incomplete doublets are larger than normal singlet cells and have proportionately enlarged macronuclei (Sonneborn 1963). The observations of Kimball (1967) and Dupy-Blanc and Genermont (1979) showing persistent clonal differences in cell mass and macronuclear DNA content may be based on differences in the corticotype of the cells.

5 Comparison with Other Organisms

The greatest difference in the control of the cell cycle between *Paramecium* and *Tetrahymena* lies in the control of IDS. This leads to significant differences in the pattern of DNA replication in the two organisms and in the manner of regulation of DNA content. In *Tetrahymena* the point of commitment to IDS occurs early in the cell cycle, about 15 min prior to IDS (Wolfe 1976). Unlike *Paramecium,* DNA synthesis occurs as complete doublings in *Tetrahymena* (Andersen and Zeuthen 1971; Cleffmann 1975). The main factor associated with IDS is reduced gene concentration (Worthington et al. 1976; Seyfert 1977; Cleffmann et al. 1979). A variable number of complete doublings of DNA content occur within a single cycle, depending on the initial gene dosage. If gene dosage is sufficiently low, an extra round of replication occurs, and if gene dosage is sufficiently high no replication takes place (Cleffmann 1968; Doerder and DeBault, 1978; Doerder 1979).

Regulation of DNA content does not occur through an incremental mechanism acting in each cell during each cell cycle as in *Paramecium.* The regulatory system acts intermittently and in individuals at the extremes of the range of DNA content. Typically, about 5% of the cells undergo either an extra round of DNA replication or skip replication in any cell cycle (Cleffmann 1968; Doerder 1979). This pattern of regulation occurs in many tetrahymenids (Doerder et al. 1981) and similar patterns of regulation occur in *Colpidium* (Morat 1980) and in *Colpoda* (Morat et al. 1982).

Initiation of cell division in *Tetrahymena*, as in *Paramecium*, occurs prior to or at the onset of the morphogenetic period. This typically occurs during the latter part of the macronuclear S-period as shown by the observation that blockage of DNA synthesis at this point in the cell cycle does not block fission while earlier blockage of DNA synthesis does (Jeffery et al. 1970; Andersen 1972). Worthington et al. (1976) suggested that initiation of cell division is associated with a high gene concentration. In contrast to *Paramecium*, macronuclear DNA synthesis is not a precondition for initiation of division as cell cycles without replication of macronuclear DNA occur regularly in cells with sufficiently high initial DNA content (Cleffmann 1968, Doerder and DeBault 1978; Doerder 1979). Under conditions of rapid growth the morphogenetic period begins very early in the cell cycle, prior to virtually all DNA synthesis (Nelson et al. 1981). The morphogenetic period occupies a fixed interval, as in *Paramecium*, which does not change with cell cycle duration or with varying nutritive conditions (Nelson et al. 1981; Suhr-Jessen et al. 1977). The beginning of the morphogenetic period is associated with a standard relative cell size increment. Under a range of nutritive conditions which lead to cell cycles of different lengths, the relative cell size at the mid-point of stage 1 of oral morphogenesis is about 1.8 times the initial value. The absolute size of the cells, however, varies with nutrient conditions (Nelson et al. 1981).

As suggested above, regulation of cell mass in *Tetrahymena* occurs through an incremental mechanism (Schaefer and Cleffmann 1982; Berger 1984). Observations on the rates of protein synthesis as a function of macronuclear DNA content (Cleffmann et al. 1979) suggest that mechanisms similar to those in *Paramecium* (Berger 1982a) may be regulating the rate of protein synthesis.

The *Paramecium* cell cycle differs from that of many eukaryotes in which the duration of the interval from IDS to cytokinesis is relatively constant over a range of conditions (review: Yanishevsky and Stein 1981). In *Paramecium* the duration of the S-period is variable and depends on gene dosage and nutrient level. This is a consequence of the incremental nature of macronuclear DNA replication in this organism.

Cell cycle control has been analyzed more extensively in yeasts than in any other organism (Nurse and Fantes 1981; Pringle and Hartwell 1981; Fantes 1984; Reed 1984; Nurse 1985) and shows some strong parallels to the cell cycle of *Paramecium*. The cell cycles of both fission and budding yeasts, like that of *Paramecium*, have two major control points (Fig. 10). The 'start' function, which is the major cell cycle control system, acts twice during the course of the cell cycle, once in controlling entry to DNA synthesis or alternative pathways of mating or meiosis, and again later at commitment to mitosis (Nurse and Bissett 1981; Piggott et al. 1982). The main difference between these organisms and *Paramecium* is that the initial cell cycle control point in *Paramecium* is located

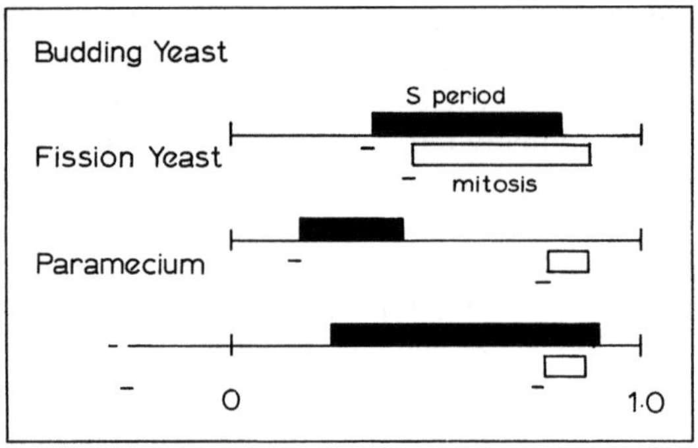

Fig. 10. Comparison of the cell cycle regulation schemes for budding yeast (*Saccharomyces*), fission yeast (*Schizosaccharomyces*) and *Paramecium*. The *short horizontal lines* beneath the cell cycle diagrams indicate the location of the major control points regulating initiation of DNA synthesis and cell division respectively. (Yeast data from Nurse 1985)

prior to fission during the preceding cell cycle. The cell cycles of *Paramecium* are thus overlapped, with the functional beginning of the next cell cycle occurring prior to the end of the first cycle.

References

Andersen HA (1972) Requirements for DNA replication preceding cell division in *Tetrahymena pyriformis*. Exp Cell Res 75:89–94

Andersen HA, Zeuthen E (1971) DNA replication sequence in *Tetrahymena* is not repeated from generation to generation. Exp Cell Res 68:309–314

Berger JD (1971) Kinetics of incorporation of DNA precursors from ingested bacteria into macronuclear DNA of *Paramecium aurelia*. J Protozool 18:419–429

Berger JD (1973) Nuclear differentiation and nucleic acid synthesis in well-fed exconjugants of *Paramecium aurelia*. Chromosoma 42:247–268

Berger JD (1978) Downward regulation of macronuclear DNA content in *Paramecium tetraurelia:* the effects of excess macronuclear DNA content on subsequent DNA and protein content and the cell cycle. Exp Cell Res 114:253–261

Berger JD (1979) Regulation of macronuclear DNA content in *Paramecium tetraurelia*. J Protozool 26:18–28

Berger JD (1982a) Effects of gene dosage on protein synthesis rate in *Paramecium tetraurelia:* implications for regulation of cell mass, DNA content and the cell cycle. Exp Cell Res 142:261–275

Berger JD (1982b) Initiation and termination of macronuclear DNA synthesis in *Paramecium:* effect of variation in macronuclear DNA content. Can J Zool 60:2501–2506

Berger JD (1984) The ciliate cell cycle. In: Nurse P, Streiblova E (eds) The microbial cell cycle. CRC, Boca Raton, Fla, pp 191–208

Berger JD (1986) Autogamy in *Paramecium tetraurelia:* cell cycle stage specific determination of the process. Exp Cell Res 166:475–486

Berger JD, Ching AS-L (1987) The timing of initiation of DNA synthesis in *Paramecium tetraurelia* is established during the preceding cell cycle as cells become committed to cell division. Exp Cell Res

Berger JD, Kimball RF (1964) Specific incorporation of precursors into DNA by feeding labeled bacteria to *Paramecium aurelia*. J Protozool 11:534–537

Berger JD, Schmidt HJ (1978) Regulation of macronuclear DNA content in *Paramecium tetraurelia*. J Cell Biol 76:116–126

Blanc J (1965) Etude cytophotométrique des périodes de duplication de l'acide désoxyribonucléique dans l'appareil nucléaire de *Paramecium caudatum*. Protistologica 1:11–15

Brygoo Y, Sonneborn TM, Keller AM, Dippell RV, Schneller MV (1980) Genetic analysis of mating type differentiation in *Paramecium tetraurelia*. II. Role of the micronuclei in mating type determination. Genetics 94:951–959

Ching AS-L, Berger JD (1986 a) The timing of initiation of macronuclear DNA synthesis is set during the preceding cell cycle in *Paramecium tetraurelia:* analysis of the effects of abrupt changes in nutrient level. Exp Cell Res 167: 177 – 190

Ching AS-L, Berger JD (1986 b) Control of cell division in *Paramecium tetraurelia:* effects of abrupt changes in nutrient level on accumulation of macronuclear DNA and initiation of cell division. Exp Cell Res 167: 191 – 202

Cleffmann G (1968) Regulierung der DNS-Menge im Makronukleus von *Tetrahymena.* Exp Cell Res 50: 193 – 207

Cleffmann G (1975) Amount of DNA produced during extra S phases in *Tetrahymena.* Chromosoma 78: 313 – 325

Cleffmann G, Reuther W-O, Seyfert H-M (1979) Increase in macromolecular amounts during the cell cycle of *Tetrahymena:* a contribution of cell cycle control. J Cell Sci 37: 117 – 124

Cohen J, Beisson J (1980) Genetic analysis of the relationships between the cell surface and the nuclei in *Paramecium tetraurelia.* Genetics 95: 797 – 818

Cohen J, Beisson J, Tucker JB (1980) Abnormal microtubule deployment during defective macronuclear division in a *Paramecium* mutant. J Cell Sci 44: 153 – 167

Cohen J, Adoutte A, Grandchamp S, Houdebine LM, Beisson J (1982) Immunocytochemical study of microtubular structures throughout the cell cycle of *Paramecium tetraurelia.* Biol Cell 44: 35 – 44

Doerder FP (1979) Regulation of macronuclear DNA content in *Tetrahymena.* J Protozool 26: 28 – 35

Doerder FP, DeBault LE (1978) Life cycle variation and regulation of macronuclear DNA content in *Tetrahymena thermophila.* Chromosoma 69: 1 – 19

Doerder FP, Ditaranto J, DeBault LE (1981) Evolutionary constraints on quantitative variation and regulation of macronuclear DNA content in the genus *Tetrahymena.* J Cell Sci 49: 177 – 193

Dupy-Blanc J, Genermont J (1979) The hereditary transmission of the amount of macronuclear DNA in *Paramecium caudatum.* Protistologica 15: 177 – 182

Fantes P (1984) Cell cycle control in *Schizosaccharomyces pombe.* In: Nurse P, Streiblova E (eds) The microbiol cell cycle. CRC, Boca Raton, Fla, pp 109 – 125

Fantes P, Nurse P (1981) Division timing: controls, models and mechanisms. In: John PCL (ed) The cell cycle. Univ Press, Cambridge, p 11

Hanson ED, Ungerleider RM (1974) The formation of the feeding organelle in *Paramecium aurelia.* J Exp Zool 185: 175 – 187

Hirschberg J, Simchen G (1977) Commitment to the mitotic cell cycle in yeast in relation to meiosis. Exp Cell Res 105: 245 – 252

Jeffery WR (1972) Evidence for temporal incompatibility between deoxyribonucleic acid synthesis and cell division in *Tetrahymena.* J Cell Biol 53: 624 – 634

Jeffery WR, Stewart KD, Frankel J (1970) The relationship between deoxyribonucleic acid replication and cell division in heat-synchronized *Tetrahymena.* J Cell Biol 533 – 543

Jones D, Berger JD (1982) Temperature-sensitive mutations affecting cortical morphogenesis and cell division in *Paramecium tetraurelia.* Can J Zool 60: 2296 – 2308

Jones WR (1976) Oral morphogenesis during asexual reproduction in *Paramecium tetraurelia.* Genet Res 27: 187 – 204

Just E (1973) Untersuchungen zur zeitlichen Ordnung der Syntheseleistung in der Interphase von *Paramecium bursaria.* Arch Protistenk 115: 22 – 68

Kaczanowska J, Dubielecka B (1983) Pattern determination and pattern regulation in *Paramecium tetraurelia.* J Embryol Exp Morphol 74: 47 – 68

Kaneda M, Hanson ED (1974) Growth patterns and morphogenetic events in the cell cycle of *Paramecium aurelia.* In: Wagtendonk WJ van (ed) *Paramecium,* a current survey. Elsevier/North Holland Biomedical Press, New York Amsterdam, pp 219 – 262

Kimball RF (1967) Persistent intraclonal variation in cell dry mass and DNA content in *Paramecium aurelia.* Exp Cell Res 48: 378 – 394

Kimball RF, Barka T (1959) Quantitative cytochemical studies on *Paramecium aurelia.* II. Feulgen microspectrophotometry of the macronucleus during exponential growth. Exp Cell Res 17: 173 – 182

Kimball RF, Perdue SW (1962) Quantitative cytochemical studies on *Paramecium aurelia.* V. Autoradiographic studies of nucleic acid synthesis. Exp Cell Res 27: 405 – 415

Kimball RF, Perdue SW (1965) Autoradiographic studies on the conservation of label from ribonucleosides and amino acids in *Paramecium aurelia.* Exp cell Res 38:660–669

Kimball RF, Vogt-Köhne L (1961) Quantitative cytochemical studies on *Paramecium aurelia.* IV. The effect of limited food and starvation on the macronucleus. Exp Cell Res 23:479–487

Kimball RF, Vogt-Köhne L (1962) Effects of radiation on cell and nuclear growth in *Paramecium aurelia.* Exp Cell Res 28:228–238

Kimball RF, Caspersson TO, Svensson G, Carlson L (1959) Quantitative cytochemical studies on *Paramecium aurelia.* I. Growth in total dry weight measured by scanning interference microscope and X-ray absorption methods. Exp Cell Res 17:160–172

Kimball RF, Vogt-Köhne L, Casparsson TO (1960) Quantitative cytochemical studies on *Paramecium aurelia.* III. Dry weight and ultraviolet absorption of isolated macronuclei during various stages of the interdivision interval. Exp Cell Res 20:368–377

Klass MR, Smith-Sonneborn J (1976) Studies on DNA content, RNA synthesis and DNA template activity in ageing cells of *Paramecium aurelia.* Exp Cell Res 98:63–72

Kudrjavtsev BN (1966) Changes in the DNA content in macro- and micronucleus of *Paramecium putrinum* in the interdivision phase. Acta Protozool 4:51–57

Mitchison JM (1971) The biology of the cell cycle. Univ Press, Cambridge

Morat G (1980) Régulation de la teneur en ADN macronucléaire et variation du rapport nucléo-cytoplasmique au cours de la crossance exponentielle chez le cilié *Colpidium campylum.* J Protozool 27:85A

Morat G, Chessa G, Crippa-Franceschi T (1982) Étude de la régulation des teneurs en ADN nucléaire au cours de la réproduction végétative et de l'enkystment d'attente chez le cilié *Colpoda cucullus.* Protistologica 17:313–329

Morton GT, Berger JD (1978) Comparison of singlet and doublet *Paramecium tetraurelia:* DNA content, protein content and the cell cycle. J Protozool 25:203–207

Nelson EM, Frankel J, Martel E (1981) Development of the ciliature of *Tetrahymena thermophila.* I. Temporal coordination with oral development. Dev Biol 88:27–38

Nurse P (1985) Cell cycle control genes in yeast. Trends Genet 1:51–55

Nurse P, Bissett Y (1981) Gene required for G1 commitment to cell cycle and in G2 for control of mitosis in fission yeast *Schizosaccharomyces pombe.* Nature (London) 292:558–560

Nurse P, Fantes PA (1981) Cell cycle controls in fission yeast: a genetic analysis. In: John PCL (ed) The Cell Cycle. Univ Press, Cambridge, pp 85–105

Pasternak JJ (1967) Differential genic activity in *Paramecium aurelia.* J Exp Zool 165:395–418

Peterson EL, Berger JD (1976) Mutational blockage of DNA synthesis in *Paramecium tetraurelia.* Can J Zool 54:2089–2097

Piggott JR, Rai R, Carter BLA (1982) A bifunctional gene product involved in two phases of the yeast cell cycle. Nature (London) 298:391–393

Pringle JR, Hartwell LH (1981) The *Saccharomyces* cell cycle. In: Strathern JN, Jones EW, Broach J (eds) The Molecular Biology of the Yeast *Saccharomyces.* Cold Spring Harbor Lab, New York, p 91

Rao MVN, Prescott DM (1967) Micronuclear DNA synthesis in *Paramecium caudatum.* Exp Cell Res 33:281–285

Rasmussen CD, Berger JD (1982) Downward regulation of cell size in *Paramecium tetraurelia:* effects of increased cell size, with or without increased DNA content, on the cell cycle. J Cell Sci 57:315–329

Rasmussen CD, Berger JD (1984) A gene function required for cell cycle progression during the G1 portion of the cell cycle and for maintenance of macronuclear DNA synthesis in *Paramecium tetraurelia.* Exp Cell Res 155:593–597

Rasmussen CD, Ching AS-L, Berger JD (1985) The full schedule of macronuclear DNA synthesis is not required for cell division in *Paramecium tetraurelia.* J Protozool 32:366–368

Rasmussen CD, Berger JD, Ching AS-L (1986) Effects of increased cell mass and altered gene dosage on the timing of initiation of macronuclear DNA synthesis in *Paramecium tetraurelia.* Exp Cell Res 165:53–62

Reed S (1984) Genetic and molecular analysis of division control in *Saccharomyces.* In: Nurse P, Streiblova E (eds) The microbial cell cycle. CRC, Boca Raton, Fla, pp 89–107

Ruiz F, Adoutte A, Rossignol M, Beisson J (1976) Genetic analysis of morphogenetic processes in *Paramecium.* Genet Res Camb 27:109–122

Schaefer E, Cleffmann G (1982) Division and growth kinetics of the division mutant conical of *Tetrahymena.* Exp Cell Res 137:277–284

Schwartz V, Meister H (1975) Einige quantitative Daten zum Problem des Alterns bei *Paramecium.* Arch Protistenk 117:85–109

Seyfert H-M (1977) A short G1 period is correlated with low macronuclear DNA content in *Tetrahymena.* Exp Cell Res 108:456–459

Seyfert H-M, Cleffmann G (1982) Mean macronuclear DNA contents are variable in the ciliate *Tetrahymena pyriformis.* J Cell Sci 58:211–223

Smith-Sonneborn J, Klass MR (1974) Changes in the DNA synthesis pattern of *Paramecium* with increased clonal age and interfission time. J Cell Biol 61:591–611

Smith-Sonneborn J, Rodermel S (1976) Loss of endocytotic capacity in ageing *Paramecium:* the importance of cytoplasmic organelles. J Cell Biol 71:575–588

Sonneborn TM (1963) Does preformed cell structure play an essential role in cell heredity? In: Allen JM (ed) The nature of biological diversity. McGraw-Hill, New York, pp 165–211

Sonneborn TM (1970) Gene action in development. Proc R Soc London. Ser B 176:347–366

Sonneborn TM (1974) *Paramecium aurelia.* In: King RC (ed) Handbook of genetics II. Plenum, New York, pp 469–594

Suhr-Jessen P, Stewart JM, Rasmussen L (1977) Timing and regulation of nuclear and cortical events in the cell cycle. J Protozool 24:299–303

Takagi Y, Kanazawa N (1982) Age-associated change in macronuclear DNA content in *Paramecium caudatum.* J Cell Sci 54:137–147

Tucker JB, Beisson J, Roche DLJ, Cohen J (1980) Microtubules and control of macronuclear 'amitosis' in Paramecium. J Cell Sci 44:135–151

Walker PMB, Mitchison JM (1957) DNA synthesis in two ciliates. Exp Cell Res 13:167–170

Wolfe J (1976) G1 arrest and the division/conjugation decision in *Tetrahymena.* Dev Biol 54:116–126

Woodard JB, Gelber B, Swift H (1961) Nucleoprotein changes during the course of the mitotic cycle in *Paramecium aurelia.* Exp Cell Res 23:258–264

Worthington DH, Salamone M, Nachtwey DS (1976) Nucleocytoplasmic ratio requirements for the initiation of DNA replication and fission in *Tetrahymena.* Cell Tissue Kinet 9:119–130

Yanishevsky RM, Stein GH (1981) Regulation of the cell cycle in eukaryotic cells. Int Rev Cytol 69:223–259

Zeuthen E (1971) Synchronization of the *Tetrahymena* cell cycle. In: Prescott DM, McConkey E, Goldstein L (eds) Advances in biology, vol 2. Appleton-Century-Crofts, New York, p 111

Zeuthen E, Rasmussen L (1972) Synchronous cell division in Protozoea. In: Chen TT (ed) Research in protozoology, vol 4. Pergamon, New York, p 9

Chapter 8 Nuclear Dimorphism and Function

KAZUYUKI MIKAMI[1]

CONTENTS

1 Introduction

Ciliates have two kinds of nuclei, macronuclei and micronuclei, in a common cytoplasm. They differ both in structure and in function. The micronucleus is an ordinary diploid and plays the role of a germinal nucleus which undergoes meiosis and gives rise to gametic nuclei which form the zygotic nucleus (synkaryon) at fertilization. After fertilization the synkaryon gives rise to new micro- and macronuclei. Both micronuclei and macronuclei have typical nuclear envelopes, however, the chromatin of the micronucleus is densely packed and nucleoli are absent. The macronucleus is transcriptionally very active and is considered to control cytoplasmic and nuclear events. Does the micronucleus act only as a puppet of the macronucleus throughout the life cycle? The somatic function of the micronucleus has been debated for a long time. Recent work on *Paramecium* shows that the micronucleus plays an important role in stomatogenesis during both sexual and asexual phases of the life cycle. Nuclear dualism also provides unique and interesting problems for the study of nuclear function and differentiation.

---

1 Research Institute for Science Education, Miyagi University of Education, Sendai 980, Japan

Paramecium. Ed. by H.-D. Görtz

2 Nuclear Function During Vegetative Phase

The macronucleus is physiologically very active; intense RNA synthesis was detected by autoradiography and cytochemical techniques (for review, see Raikov 1969, 1982; Wichtermann 1986). Amacronucleate cells cannot survive long, and undergo no more than one or two fissions, when the macronucleus is lost by unequal distribution at cell division in the *P. aurelia* complex (Sonneborn 1947). This was ascertained also in *P. caudatum* after elimination of the macronucleus by micropipetting (Mikami, unpubl.). Thus, the macronucleus is indispensable for survival and reproduction of the cell. The questions are whether the micronucleus is transcriptionally active, and whether it is indispensable for cell reproduction.

2.1 Phenotypic Expression in Heterokaryons

Heterokaryons make it possible to assess the contribution of the micronucleus during asexual reproduction. For this purpose, the macronucleus should carry recessive mutant genes and the micronucleus, the corresponding dominant alleles. Heterokaryons were produced by regeneration of the prezygotic macronucleus after conjugation, with regard to the *K* gene (for maintenance of the endosymbiont kappa), serotype genes (Sonneborn 1954), *pwA* in *P. tetraurelia* (Berger 1974), and the *M* gene (for maintenance of the endosymbiont mu) in *P. primaurelia* (Pasternak 1967). Advances in nuclear transplantation have made it possible to prepare heterokaryons in *P. tetraurelia* by whole macronucleus transplantation in the case of mating type (Mikami and Koizumi 1982), or by micronuclear implantation, in the case of the markers $+^{nd3b}$ (trichocyst discharge, Ng 1981; Ng and Mikami 1981) and $+^{pwA}$ (ciliary reversal, Ng and Mikami 1981). In *P. caudatum*, heterokaryons were obtained by micronuclear implantation for $+^{tnd\text{-}1}$, $+^{tnd\text{-}2}$ (trichocyst discharge), $+^{cnrA}$, $+^{cnrB}$, $+^{cnrD}$ (for ciliary reversal) (Mikami 1987), and $+^{cnrC}$ (Mikami, unpubl.). In all of these cases, the heterokaryons expressed exclusively the phenotypes according to the macronucleus and no micronuclear gene activity was detected.

However, the gene-dosage effect must be considered before arriving at a conclusion. The DNA content of the macronucleus is about 430 times (860 C) that of the micronucleus in *P. tetraurelia* (reviewed by Sonneborn 1974) and about 40 times (80 C) in *P. caudatum* (Mikami 1980, etc.). When conjugation was induced between heterokaryons with the wild-type micro- and the recessive macronucleus in *P. caudatum*, the phenotypic expression of the genes $+^{tnd\text{-}1}$ and $+^{cnrB}$ was found to occur at quite early stages after conjugation, probably before the first round of DNA synthesis. At that time, total gene dosage of the macronuclear anlagen in a cell seemed to be 4 to 8 C (each anlage is diploid at the beginning and there were four per exconjugant). Moreover, the timing of the expression of *tnd-1* did not differ significantly between homozygotes and heterozygotes. Therefore, only a few gene copies may be necessary for expression of the wild-type phenotype. Nevertheless, the heterokaryons with the wild-type micro- and the recessive (*tnd-1* or *cnrB*) macronucleus expressed the macronuclear genotype throughout the cell cycle (Mikami 1987). These observations indicate that the micronucleus is genetically silent during vegetative growth.

2.2 *Effect of Micronuclear Elimination*

Although the somatic function of the micronucleus has been debated for nearly a century, the somatic function of the micronucleus has been controversial. Amicronucleates of *Paramecium* have been collected from nature and also artificially generated since Lewin (1910), though such cells and their subsequent asexual progenies often show marked loss of vigor (review: Ng 1986). Schwartz (1946) first reported that amicronucleate clones of *P. caudatum* have a longer interfission time than micronucleates. Later, artificial elimination of the micronucleus resulted in the depression of the cell division rate and nonviability in extreme cases (Ng, 1986).

In *P. caudatum*, almost all cells divided at least once after micronuclear elimination and both daughter cells formed food vacuoles normally. Reduction in the rate of food vacuole formation and the expression of the interfission period began two or three cell cycles after micronuclear elimination (Mikami, unpubl.). Thereafter, some cells in each amicronucleate clone divided with normal interfission time but others did not (Mikami 1979 a). In extreme cases, all of the cells in an amicronucleate clone failed to form food vacuoles so that the clone perished. The survival rate of amicronucleate clones was zero or very low when they were derived by UV irradiation (Fokin and Ossipov 1981; Fokin 1983 b), and about 60% (Fujishima and Watanabe 1981) or more (Mikami, unpubl.) when they were obtained by micropipetting.

Absence of the micronucleus frequently leads to abnormalities in stomatogenesis as explained below. The abnormalities of amicronucleate cells are unrelated to damage produced at the time of micronuclear elimination, and are consequences of the absence of the micronucleus. In *P. tetraurelia*, stomatogenic abnormalities were not observed when one of the two micronuclei was removed (Ng and Mikami 1981). In addition, reimplantation of a micronucleus into amicronucleate cells restored normal stomatogenesis and food vacuole formation (Fujishima and Watanabe 1981; Ng and Mikami 1981). Amicronucleates, obtained by treatment with colchicine prior to cell division showed almost the same viability and vigor as those obtained by micropipetting in *P. caudatum* (Mikami, unpubl.).

2.3 *Stomatogenic Role of the Micronucleus*

Reduction or loss of the buccal cavity was first reported in amicronucleate *P. caudatum* by Schwartz (1934). Thereafter, aberrant oral structures were observed in amicronucleates during vegetative growth (Mikami 1979 a; Fujishima and Watanabe 1981; Fokin 1983 a, b; etc.). The abnormality was expressed as a reduction in the rate of food vacuole formation, in the length of the buccal cavity, or its absence. In some amicronucleate *P. caudatum,* the rate of food vacuole formation was extremely low and each vacuole was often very large. This was considered to be caused by elevation of the inner wall of the buccal cavity (Mikami 1979a). When the defect in oral apparatus is serious and few food vacuoles are formed, the cell cannot recover following reimplantation of a micronucleus (Mikami, unpubl.).

A study of silver impregnation in *P. tetraurelia* revealed that amicronucleate cells in the depressed period of growth rate following loss of the micronucleus often possessed abnormal oral apparatuses (Ng and Mikami 1981). The oral membranelles were frequently disorganized and the buccal cavity was markedly reduced in size. Food vacuoles were few and frequently of abnormal size. In extreme cases the cells did not feed and eventually perished (Ng and Mikami 1981). Stomatogenic defects observed in silver-impregnated *P. caudatum* amicronucleates were very similar to those in *P. tetraurelia* (Mikami, unpubl.).

A new oral apparatus develops at each cell division. In some amicronucleate cells, cytokinesis was abnormal with the newly formed oral membranelles of the opisthe (the posterior cell), blocking the fission zone. In this case, the buccal cavity was absent and cytokinesis was arrested to form a V-shape chain (Ng and Mikami 1981). Also in some amicronucleate dividers of *P. bursaria,* the new oral apparatus remains near the fission zone (Fokin 1983a). We conclude that the micronucleus functions in stomatogenesis during asexual reproduction and that the defect in stomatogenesis in amicronucleates is the primary cause of other phenomena such as reduced rate of food vacuole formation and longer interfission time. Amicronucleate clones include cells which have mild defects in the oral apparatus and these can reproduce to sustain the clone.

2.4 Recovery from the Depression

Amicronucleate cells tended to recover in fission rate gradually after a period of growth rate depression in *P. caudatum* (Mikami 1979a; Fujishima and Watanabe 1981). In *P. tetraurelia,* recovery of amicronucleate cell lines was nearly, but not fully complete before the onset of clonal senescence. Oral abnormalities became modest and eventually the oral apparatus became rather normal if only somewhat shorter (Ng and Mikami 1981). In *P. caudatum,* the length of the buccal cavity later increased and subsequently returned to normal (Fujishima and Watanabe 1981) or not to fully normal dimension in 2 years (Fokin 1983b).

There is no evidence that micronuclear function can be replaced by the macronucleus. However, the recovery from the depression is very lasting in *P. caudatum.* Cells which had partially but not completely recovered from effects of micronuclear loss were implanted with micronuclei and cultivated for about 1 month. Then, the micronuclei were eliminated from them again. But the amicronucleate did not show an apparent depression of the fission rate (Mikami, unpubl.).

2.5 *Is the Function of the Micronucleus Genic?*

Autoradiography revealed labeling of [^{3}H]-uridine associated with the micronucleus preceding the S-period and at the S-period in *P. tetraurelia* (Pasternak 1967), and in the S-period in *P. caudatum* (Rao and Prescott 1967). Significant amounts RNA were extracted from micronuclei in *P. primaurelia, P. tetraurelia,* and *P. tredecaurelia* (Cummings 1972). However, Gorovsky (1980) pointed out the possibility that RNA synthesized during the S-period might serve as a primer for DNA synthesis in the micronucleus and may not represent genic activity.

Polyploid micronuclei have been reported in *Paramecium* but influenced the phenotype only slightly (reviews Raikov 1982). Moreover, haploid micronuclei appear to function in *P. caudatum* (Mikami, unpubl.). When micronuclei, of which the euploid condition was estimated to be broken by unequal distribution of chromosomes at binary fission after infection with the bacteria *Holospora elegans,* were removed in *P. caudatum,* cells went into depression (Mikami and Görtz 1985). Therefore, euploidy may not be necessary for micronuclear vegetative function. Furthermore, cell lines with achromatic and mini-micronuclei could propagate for many cell cycles and were much more vigorous than amicronucleates in *P. tetraurelia* (review: Ng 1986). Is the function of the micronucleus genic? If ghost micronuclei lack chromosomes and retain their vegetative function, the nongenic theory could be acceptable. However, the presence of DNA in achromatic micronuclei cannot be ruled out completely (Borchsenius and Fokin 1982).

3 Nuclear Function During Sexual Cycle

The process of nuclear reorganization during sexual phase is complicated because of nuclear dualism, but it remains well regulated. After the onset of conjugation or autogamy, the micronucleus undergoes meiosis and subsequently gives rise to gametic nuclei (see Fujishima, Chap. 5, this Vol.). Following exchange of gametic nuclei between the mates, the gametic nuclei fuse to form a zygotic nucleus (synkaryon), which divides mitotically two to four times depending upon species. Among the daughter nuclei, a certain number (typically half) differentiate into macronuclear anlagen. In the mean time, the prezygotic (old) macronucleus breaks down. These nuclear events of the sexual cycle have been reviewed by many authors (Grell 1973; Raikov 1982; Wichtermann 1986; etc.). How the micronucleus and the macronucleus take respective parts in the nuclear reorganization is an interesting problem from the standpoint of internuclear or nucleo-cytoplasmic interaction.

3.1 *Effects of Micro- and Macronuclear Elimination*

It is well known that amicronucleate cells show mating reactivity under appropriate conditions and form tightly conjugating pairs after mixing with amicronucleates of complementary mating types. In *P. caudatum*, these mating pairs separate after a certain length of time as do micronucleates. In these amicronucleate "exconjugants", the breakdown of the prezygotic macronucleus (skein formation and fragmentation) occurred normally (Mikami 1979a), although Ossipov and Skovlo (1973) reported incomplete fragmentation in the same species (Ossipov and Skovlo 1973). The temporary reduction in the length of exconjugants, which occurs normally after conjugation of micronucleates coinciding with the determination of macronuclear anlagen, was also observed in amicronucleates (Mikami 1979a). In the amicronucleate exconjugants, however, food vacuole formation was not resumed, whereas it was in normal micronucleate exconjugants soon after the reduction of cell length (Mikami 1979a, 1980). Therefore, the events of macronucleus and the cytoplasm, except for stomatogenesis, took place normally in the absence of the micronucleus. Transcriptional activity of the micronucleus has not been reported during sexual phase of *Paramecium*.

On the other hand, the macronucleus is indispensable for the commitment and the progress of the sexual phase. Amacronucleate *P. caudatum* cells retained mating reactivity for a while after the nuclear elimination, but never proceeded to paroral unions (Mikami, unpubl.). The macronucleus is also required during reciprocal fertilization for transfer of a gametic nucleus to the partner. However, it is not required for receipt of a nucleus prior to fertilization as amacronucleate cells can receive a gametic nucleus from the partner following removal of the macronucleus at early meiotic stages (Mikami, unpubl.).

3.2 Determination of Nuclear Differentiation

Maupas (1889) pointed out the possibility of the nuclear differentiation being correlated with nuclear localization in the cell. Long afterwards, the hypothesis was proven in *P. caudatum* (Mikami 1980) and in *P. tetraurelia* (Grandchamp and Beisson 1981). The determination of the nuclear differentiation occurs immediately after the last division of synkaryon (third in *P. caudatum* and second in *P. tetraurelia*) and is closely associated with a brief (15–30 min in *P. caudatum*) localization of the nuclei near the opposite ends of the cell, with the macronuclear anlagen in the posterior region and the micronuclei in the anterior. The antero-posterior localization of the daughter nuclei is due to the orientation of the nuclear spindles which lie parallel to the longitudinal axis of the cell and their extention to opposite ends of the cell, and the temporary decrease in cell length (Mikami 1980; Grandchamp and Beisson 1981). The different fates of the nuclei may depend on their antero-posterior position in the cytoplasm/cortex. When the vegetative micronucleus was transplanted into an early exconjugant in *P. tetraurelia*, it occasionally differentiated into a new macronucleus (Mikami and Ng 1983).

There is no intrinsic difference between nuclei localized anteriorly and those situated posteriorly. Both were retained as germinal micronuclei when implanted into vegetative cells (Mikami 1985). The "micronuclei" during conjugation may not fundamentally differ from the micronucleus of a vegetative cell

(Harumoto and Hiwatashi 1982) until it is determined as a macronuclear anlage. Irreversible differentiation of macronuclear anlagen occurs later than the antero-posterior localization of nuclei. When the prezygotic macronucleus was eliminated soon after the first division of synkaryon, the postzygotic micronuclei divided once more but did not show any sign of further differentiation into macronuclear anlagen in *P. caudatum* (Mikami 1980). The prezygotic macronucleus must play an important role in the determination of the macronuclear anlagen.

3.3 Development of Macronuclear Anlagen

In the postzygotic micronuclei, determined to be macronuclear anlagen in *P. caudatum* and *P. bursaria,* parts of the chromosomes despiralize, and other parts condense and partly fuse, producing heterochromatic aggregates (review: Raikov 1982). In *P. bursaria,* the pericentromeric regions of the chromosomes are preserved in the anlage, while the chromosome arms detach and disintegrate (Schwartz 1978). After the disappearance of chromatic aggregates, the anlage is faintly stained by the Feulgen reaction in most *Paramecium* species (review: Raikov 1982). In the next stage, macronuclear anlagen increase their DNA content. Cytophotometric studies indicate that DNA increases progressively in the developing macronuclear anlagen of *P. caudatum* (Dupy-Blank 1969) and *P. tetraurelia* (Berger 1973a). In the latter species, four or more discontinuous rounds of DNA synthesis occur in the macronuclear anlage during the first cell cycle after conjugation (Berger 1973a). Discontinuous rounds of DNA synthesis seem to occur also in *P. caudatum* (Mikami 1987).

During the development of the macronuclear anlagen, partial elimination of the micronuclear genome has been shown in other ciliates such as *Euplotes, Stylonychia, Oxytricha,* and *Tetrahymena.* The macronucleus of *P. bursaria* contains shortened chromosomal fragments rather than whole chromosomes (Schwartz 1978). In this species, the DNA content of the macronuclear anlage increases initially up to sixfold, and then decreases by half temporarily at the "achromatic" stage prior to resumption of DNA synthesis (Schwartz and Meister 1975). At vegetative G1 the macronucleus of *P. caudatum* contains about 40 times (80C) as much DNA as the micronucleus. If the postzygotic micronucleus (2C) duplicates its entire genome at each round of DNA synthesis, the content should be 64C after five rounds and 128C after six rounds but should not be 80C (Mikami 1987). The hypothesis of underreplication or partial elimination is attractive to rationalize the DNA content. According to microspectrophotometric results, the DNA content was estimated to be lower than 4C after the first replication. It is likely that underreplication or partial elimination of micronuclear DNA occurs during the first few rounds of DNA synthesis during macronuclear development in *P. caudatum* (Mikami 1987). In a mutant of *P. tetraurelia,* the 51A i-antigen gene, which is present in the micronucleus, is deleted during the development of macronuclear anlagen and absent from the macronucleus (Espstein and Forney 1984).

When do the chromosomes of the macronuclear anlage gain transcriptional ability? In the case of the $+^{pwA}$ gene of *P. tetraurelia*, the phenotypic expression of macronuclear anlagen occurred late in the first cell cycle or during the first part of the second cell cycle (Berger, 1976). In *P. caudatum*, the phenotypes of genes $+^{cnrB}$ and $+^{tnd\text{-}1}$ are expressed very early, during the first round of DNA synthesis. Considering the time lag necessary for the accumulation of gene products, the genes can be transcribed even before the first round of DNA synthesis. Hence, with regard to at least a few genes, the chromosomal or intranuclear reorganization from the micronuclear type to the macronuclear type may occur by the first round of DNA synthesis (Mikami 1987).

3.4 Transition of the Macronuclear Function from the Old Macronucleus to the Anlagen

Exconjugant cells have two kinds of macronuclei, new macronuclei (anlagen) and prezygotic (old) macronuclei which form several dozen fragments in the species except for *P. bursaria.* The old macronucleus eventually ceases function and is absorbed in the cytoplasm. However, nucleoli are actively formed in it soon after conjugation (see review by Raikov 1978), and the rate of RNA synthesis per unit volume of nucleus is much greater in the old macronucleus than in the anlagen in *P. aurelia* (Berger 1973a). The time of macronuclear disintegration varies among species. In *P. tetraurelia*, autolysis of the fragments was more rapid in starved exconjugants than in well-fed ones (Berger 1973a, 1974), whereas in *P. caudatum* the fragments persisted for a long time even in starved exconjugants (Mikami 1979b).

In *P. caudatum*, DNA synthesis in the macronuclear fragments continues until the third cell cycle after conjugation and selective suppression of the DNA synthesis takes place at the fourth cell cycle. The selective suppression of DNA synthesis in macronuclear fragments is due to the presence of the new macronucleus (anlage) in the common cytoplasm, as demonstrated in *P. tetraurelia* by Sonneborn (1947) and Berger (1973b). The selective inhibition does not occur when the postzygotic macronucleus (anlage) is removed by micropipetting (Mikami 1980) or as a result of mis-distribution of the anlagen at postconjugation fissions (Mikami and Hiwatashi 1975). However, once the suppression of DNA synthesis occurs, macronuclear fragments cannot regenerate if the anlagen are removed after the initiation of the suppression (Mikami 1980). Berger (1973b) reported that the selective suppression of the DNA synthesis in the old macronucleus was the result of a specific inhibitor of DNA synthesis released from the developing macronuclear anlage in *P. tetraurelia.* In *P. caudatum,* competition for the limited amount of DNA precursors, which occurred also in *P. tetraurelia* (Berger 1973b), was concluded to be playing an important role in the onset of selective suppression of DNA synthesis (Mikami 1979b).

3.5 Function of the Micronucleus in Stomatogenesis Following Sexual Reorganization

The micronucleus is absolutely indispensable for stomatogenesis during the sexual phase. In the *P. aurelia* complex, the old oral apparatus is also replaced in the sexual phase (Ng and Newman 1984a; etc.); the preexisting oral apparatus is resorbed and a new oral primordium is developed on the posterior-right of the preexisting oral apparatus. Diller (1965) mentioned that amicronucleate exconjugants of *P. multimicronucleatum* or *P. trichium* lacked a buccal cavity. In amicronucleate *P. caudatum*, the inner wall of the oral apparatus is elevated and lacks a typical buccal cavity after conjugation or autogamy (Mikami 1979a). The literature on stomatogenesis in exautogamous or exconjugant cells of amicronucleate ciliates were reviewed comprehensively by Ng (1986). In a detailed study of *P. tetraurelia* amicronucleates, resorption of the preexisting oral apparatus occurred, but the development of the oral primordium was arrested at the early stage of alignment of the oral membranellar basal bodies. Therefore, the following stomatogenic steps were missing (Ng and Newman 1984b). Thus, the exautogamous cells or exconjugants were astomatous and did not feed and eventually died. Ossipov and Skovlo (1973) reported in their discussion the possibility that the migrating pronucleus secreted a factor into the cytoplasm which induced stomatogenesis. Recent results of micronuclear elimination and implantation during conjugation indicate that the gametic nuclei at the stage between division of the gametic nuclei and fertilization could function in stomatogenesis after conjugation (Mikami, unpubl.). The micronuclear function at similar stages was also concluded in *P. tetraurelia* (Ng and Newman 1984a; Tam and Ng 1986). However, whether the function of the micronucleus is genic or nongenic is still an open question.

Acknowledgment. The author would like to thank Dr. J.D. Berger and Dr. S.F. Ng for their critical reading of the manuscript and constructive comments.

References

Berger JD (1973a) Nuclear differentiation and nucleic acid synthesis in well-fed exconjugants of *Paramecium aurelia.* Chromosoma 42:247–268

Berger JD (1973b) Selective inhibition of DNA synthesis in macronuclear fragments in *Paramecium aurelia* exconjugants and its reversal during macronuclear regeneration. Chromosoma 44:33–48

Berger JD (1974) Selective autolysis of nuclei as a source of DNA precursors in *Paramecium aurelia* exconjugants. J Protozool 21:145–152

Berger JD (1976) Gene expression and phenotypic change in *Paramecium tetraurelia* exconjugants. Genet Res Cambr 27:123–134

Borchsenius ON, Fokin SI (1982) Ultrastructural changes of the UV-irradiated micronucleus in vegetative cells of *Paramecium bursaria* and its functional importance. Tsitologiya 24:635–640

Cummings DJ (1972) Isolation and partial characterization of macro- and micronuclei from *Paramecium aurelia.* J Cell Biol 53:105–115

Diller WF (1965) The relation of amicronuclearity to stomatogenic activity during conjugation in several ciliates. J Morphol 116:51–64

Dupy-Blank J (1969) Étude cytophotomérique des teneus en ADN des micronucléus de *Para-*

mecium caudatum au cours de la conjugaison et pendant la différenciation des "Anlages" en macronucléus. Protistologica 5:239–248
Epstein LM, Forney JD (1984) Mendelian and non-Mendelian mutations affecting surface antigen expression in *Paramecium tetraurelia.* Mol Cell Biol 4:1583–1590
Fokin SI (1981) The damage of the nuclear apparatus of *Paramecium caudatum* and the vitality of cells after UV irradiation. Tsitologiya 23:448–453
Fokin SI (1982) The alteration of *Paramecium caudatum* germinal nucleus morphology after UV irradiation. Tsitologiya 24:1061–1066
Fokin SI (1983a) Morphology of the *Paramecium bursaria* cells after UV irradiation of the germ nuclei. Tsitologiya 25:328–334
Fokin SI (1983b) Morphology and viability of cells in the amicronucleate clones of *Paramecium caudatum.* Tsitologiya 25:811–817
Fokin SI, Ossipov DV (1981) Generative nucleus control over cell vegetative functions in *Paramecium.* Acta Protozool 20:51–73
Fujishima M, Watanabe T (1981) Transplantation of germ nuclei in *Paramecium caudatum.* III. Role of germinal micronucleus in vegetative growth. Exp Cell Res 132:47–56
Gorovsky MA (1980) Genome organization and reorganization in *Tetrahymena.* Annu Rev Genet 14:203–239
Grandchamp S, Beisson J (1981) Positional control of nuclear differentiation in *Paramecium.* Dev Biol 81:336–341
Grell KG (1973) Protozoology. Springer, Berlin Heidelberg New York
Harumoto T, Hiwatashi K (1982) Transplantation of synkaryon in *Paramecium caudatum:* analysis of its competence as germ nucleus. Exp Cell Res 137:476–481
Lewin KR (1910) Nuclear relations of *Paramecium caudatum* during the sexual period. Proc Cambridge Philos Soc 16:39–41
Maupas E (1889) Le rajeunissement karyogamique chez les ciliés. Arch Zool Exp Gen Ser 27:149–157
Mikami K (1979a) Stomatogenesis during sexual and asexual reproduction in an amicronucleate strain of *Paramecium caudatum.* J Exp Zool 208:121–128
Mikami K (1979b) Internuclear control of DNA synthesis in exconjugant cells of *Paramecium caudatum.* Chromosoma 73:131–142
Mikami K (1980) Differentiation of somatic and germinal nuclei correlated with intracellular localization in *Paramecium caudatum* exconjugants. Dev Biol 80:46–55
Mikami K (1985) Nuclear differentiation in exconjugants of *Paramecium caudatum:* role of nuclear division in differentiation. Dev Growth Differ 27:21–27
Mikami K (1987) Macronuclear development and gene expression in the exconjugants of *Paramecium caudatum.* Dev Biol 123:161–168
Mikami K, Görtz H-D (1985) *Holospora elegans*-infected micronuclei of *Paramecium caudatum* still serve a function during vegetative life. J Protozool 32:736–737
Mikami K, Hiwatashi K (1975) Macronuclear regeneration and cell division in *Paramecium caudatum.* J Protozool 22:536–540
Mikami K, Koizumi S (1982) Nuclear transplant studies of the determination of mating type in germ nuclei of *Paramecium tetraurelia.* Exp Cell Res 137:397–402
Mikami K, Ng SF (1983) Nuclear differentiation in *Paramecium tetraurelia.* Exp Cell Res 144:25–30
Nanney DL (1980) Experimental ciliatology. John Wiley & Sons, New York
Ng SF (1981) Transplantation of the germ nucleus in *Paramecium tetraurelia* and its genetic consequences. Protistologica 17:489–496
Ng SF (1986) The somatic function of the micronucleus in ciliated protozoa. In: Corliss JO, Patterson DJ (eds) Progress in protistology, vol 1, Biopress, Bristol, pp 215–286
Ng SF, Mikami K (1981) Morphogenetic role of the germ nucleus in *Paramecium tetraurelia.* Protistologica 17:497–509
Ng SF, Newman A (1984a) The role of the micronucleus in stomatogenesis in sexual reproduction of *Paramecium tetraurelia:* micronuclear and stomatogenetic events. Protistologica 20:43–64
Ng SF, Newman A (1984b) The role of the micronucleus in stomatogenesis in sexual reproduction of *Paramecium tetraurelia:* conjugation of amicronucleates. Protistologica 20:517–523

Ossipov DV, Skovlo II (1973) Morphogenetic activity of generative nucleus in the sexual cell reorganization of ciliates. Acta Protozool 12:151–164
Pasternak J (1967) Differential genic activity in *Paramecium aurelia*. J Exp Zool 165:395–418
Raikov IB (1969) The macronucleus of ciliates. In: Chen TT (ed) Research in protozoology, vol 3. Pergamon, Oxford New York, pp 4–128
Raikov IB (1972) Nuclear phenomena during conjugation and autogamy in ciliates. In: Chen TT (ed) Research in protozoology, vol 4. Pergamon, Oxford New York, p 147
Raikov IB (1978) The protozoan nucleus. Springer-Verlag, Wien New York
Raikov IB (1982) The protozoan nucleus, 2nd edn. Springer, Berlin Heidelberg New York
Rao M, Prescott DM (1967) Micronuclear RNA synthesis in *Paramecium caudatum*. J Cell Biol 33:281–286
Schwartz V (1934) Versuche über Regeneration und Kerndimorphismus der Ciliaten. Nachr Ges Wiss Göttingen Math Phys Kl Fachgr VI NF 1:143–155
Schwartz V (1946) Der Formwechsel des Makronucleus in der Konjugation micronucleusloser Paramecien. Biol Zentralbl 65:89–94
Schwartz V (1978) Struktur und Entwicklung des Makronucleus von *Paramecium bursaria*. Arch Protistenk 120:255–277
Schwartz V, Meister H (1975) Die Extinktion der feulgengefärbten Makronucleusanlage von *Paramecium bursaria* in der DNS-armen Phase. Arch Protistenk 117:60–64
Sonneborn TM (1947) Recent advances in the genetics of *Paramecium* and *Euplotes*. Adv Genet 1:263–358
Sonneborn TM (1954) Gene-controlled, aberrant nuclear behavior in *Paramecium aurelia*. Microb Gen Bull 11:24–25
Sonneborn TM (1974) *Paramecium aurelia*. In: King C (ed) Handbook of genetics. Plenum, New York, p 469
Tam LW, Ng SF (1986) The role of the micronucleus in stomatogenesis in sexual reproduction of *Paramecium tetraurelia:* laser microbeam irradiation of the micronucleus. J Cell Sci (in press)
Wichtermann R (1986) The biology of *Paramecium*, 2nd edn. Plenum, New York

Chapter 9 Aging

YOSHIOMI TAKAGI[1]

CONTENTS

1 Mortal or Immortal

Is *Paramecium* mortal or immortal? This had been controversial for decades around the turn of the century (for review, see Wichterman 1953). Maupas, Hertwig, Calkins, and Jennings were those who claimed that *Paramecium* had a life cycle with a definite potential for cell divisions and thus clonal death would be inevitable unless clones were rejuvenated by means of conjugation. However, the hypothesis of clonal immortality appeared unambiguous. Theoretically, protozoan cells should consist of immortal germ plasm according to Weismann's germ plasm theory. Experimentally, Woodruff, for example, presented data to show that *P. biaurelia* had indefinite division potential. Also, the data showing clonal death may be ascribable to poor techniques or inadequate culture medium.

The controversy came to an end when Sonneborn (1954) demonstrated that (1) the endomixis, which had been described as an asexual nuclear process, was erroneous; instead, autogamy, a true sexual process in a single cell, including meiosis and fertilization, occurred regularly in Woodruff's immortal strain; (2) clones without autogamy died after a definite number of cell divisions, while cell lines having recurrent autogamies lived indefinitely; and (3) the clonal life cycle consisted of sequential stages called immaturity, maturity, and senescence.

So far, clonal mortality has been demonstrated in *P. primaurelia, P. biaurelia, P. tetraurelia, P. jenningsi,* and *P. caudatum,* with maximal life spans of 350, 303, 250–330, 250–300, and 600–750 fissions, respectively (Sonneborn 1954; Miyake and Miyake 1967; Smith-Sonneborn 1979, 1981; Takagi and Yoshida 1980; Karino and Hiwatashi 1981, 1984; Takagi et al. 1987a). Less in-

1 Department of Biology, Nara Women's University, Nara 630, Japan

Paramecium. Ed. by H.-D. Görtz

formation is available on the division potential of *P. bursaria* and *P. multimicronucleatum.* Although the clonal life cycle and definite division potential in *Paramecium* are generally accepted, it still remains open whether or not there are some strains capable of indefinite cell divisions (Wichterman 1986). For example, some of our stocks of *P. multimicronucleatum* have been maintained for more than 20 years and have never produced viable progeny after conjugation, and yet they show vigorous cell divisions. Such reports, however, should not be overrated. According to our estimate of the number of fissions from growth kinetics, our stock cultures, maintained by replacing half of the culture with fresh culture medium every 3 weeks, should have experienced not more than 1200 fissions in 20 years. If this were the sum of two sexual generations caused by an oversight of only one viable exconjugant, then the fission life spans would be reduced to about 600 fissions.

However, I am in favor of the idea that cells are basically immortal and have acquired during the evolutionary process a mechanism to restrict the division potential when they acquired the sexual process. If so, we may expect an emergence of strains capable of indefinite divisions in mortal species by means of loss or damage of such a mechanism. The immortal strains of *Tetrahymena pyriformis* may be taken as such an example.

2 Life Cycle Stages

The clonal life cycle of *Paramecium* begins with fertilization (conjugation and autogamy) and terminates in clonal death. Immaturity is the first life cycle stage, during which cells are unable to conjugate. Immaturity follows after conjugation but not after autogamy. During a certain fission period after autogamy, cells are able to conjugate but are unable to undergo autogamy. This period is called autogamous immaturity. Maturity is primarily defined as the stage during which cells are able to conjugate. Siegel (1967) marked maturity by the readiness for conjugation, by the high frequency of viability among progeny clones, by phenotypic stability, and by the virtual absence of cell death. He pointed out that in both protozoans and higher organisms the two invariable signs of senescence are phenotypic instability and a progressive loss of reproductive function.

The number of fissions spanning each stage is different among species. Smith-Sonneborn (1981) found that among ciliates the length of immaturity is roughly proportional to the length of the life span. Transition from immaturity to maturity is not sudden but stepwise. In *P. bursaria* a transitional stage called adolescence intervenes between immaturity and maturity. Cells in this stage can mate with only one of the other multiple mating types (Siegel 1967). Since conjugation is a long, complex process, its initiation does not necessarily mean completion (Shimomura and Takagi 1985). In transition from immaturity to maturity in *P. multimicronucleatum*, cells are first able to undergo only the mating reaction, then the holdfast union, and finally the paroral union (Takagi 1971). Besides qualitative changes, the transition accompanies quantitative changes as exemplified by a fluctuating, but gradually increasing mating reactivity (Takagi 1970).

Many parameters have been monitored to characterize senescence (for review, see Siegel 1967; Smith-Sonneborn 1981, 1985). In the *Paramecium aurelia* complex where two sexual processes, conjugation and autogamy, are optional and inducible under identical environmental conditions, i.e., food depletion, maturity is the stage in which cells respond to food depletion by conjugation and senescence by autogamy (Sonneborn 1974). However, generally, no parameters are clear enough to indicate the inauguration of senescence.

Parameters that increase in senescence include a micronuclear anomaly in its number and chromosomal morphology (Dippell 1955; Sonneborn and Dippell 1960), an expression of the gene *am,* which is responsible for unequal macronuclear division (Nobili 1959), an anomaly in the cell shape and macronuclear morphology and location (Sonneborn and Dippell 1960; Sundararaman and Cummings 1976a; Takagi and Yoshida 1980; Karino and Hiwatashi 1981), DNA strand breaks (Holmes and Holmes 1986a, 1986b), UV sensitivity (Smith-Sonneborn 1971, 1979), X-ray sensitivity (Tixador et al. 1972), the duration of the cell cycle especially of the S-period (Smith-Sonneborn and Klass 1974), the number of mitochondria, lysosomal bodies, large dense bodies (age pigments), and autophagous vacuoles (Sundararaman and Cummings 1976a), proneness to undergo autogamy under excess food conditions (Smith-Sonneborn and Rodermel 1976; Takagi et al. 1987a), unequal distribution of macronuclear DNA at cell division (Takagi and Kanazawa 1982), antigenic variation (Sundararaman and Cummings 1983), and catalase activity (Croute et al. 1985).

Those that decrease in senescence include the progeny viability after sexual reproduction (Mitchison 1955; Sonneborn and Schneller 1960; Fukushima 1975; Rodermel and Smith-Sonneborn 1977; Karino and Hiwatashi 1981, 1984), fission rate (Sonneborn 1954; Smith-Sonneborn and Klass 1974; Takagi and Yoshida 1980; Karino and Hiwatashi 1981, 1984), DNA template activity and RNA synthesis (Klass and Smith-Sonneborn 1976), macronuclear DNA content (Schwartz and Meister 1973, 1975; Klass and Smith-Sonneborn 1976; Takagi and Kanazawa 1982), the size and number of macronuclear chromatin bodies (Sundararaman and Cummings 1976b), the rate of food-vacuole formation and excretion (Smith-Sonneborn and Rodermel 1976), UV-induced repair capacity (Smith-Sonneborn 1979), nucleolar volume (Heifetz and Smith-Sonneborn 1981), and the proportion of cells dividing synchronously (Takagi and Yoshida 1980).

Those that remain unchanged in senescence include the duration of G_1 period (Smith-Sonneborn and Klass 1974), copper tolerance (Nyberg 1978), DNA polymerase activity (Williams and Smith-Sonneborn 1980), morphological appearance of nucleoli, nuclear envelopes, and chromatin bodies (Heifetz and Smith-Sonneborn 1981), the size ratio of the macronucleus to the cytoplasm (Takagi and Kanazawa 1982), and enzyme activities of glutathione peroxidase and superoxidismutase (Croute et al. 1985).

Although expressions of these parameters are neither simultaneous nor inclusive in every member of the clone, cells considered senescent in some of these parameters would be highly expected to be senescent also in other parameters.

3 Fission Life Span or Calendar Life Span

Paramecia cells should have a cellular clock to measure their species-specific clonal life spans. The question thus arises whether the time-measuring mechanism of the clock is associated with an internal signal represented by the number of fissions or with an external one represented by solar movements. The experimental design employed to answer this question was a comparison of mean clonal life spans measured in fissions and in days between subclones grown at different fission rates caused by different temperatures. Smith-Sonneborn and Reed (1976) showed that the time-measuring mechanism of *P. tetraurelia* was associated with fissions rather than calendar time within a physiological temperature range of 24° – 27° C. It should be noted that they used the temperature difference of only 3° ± 1° C around the optimal temperature in order not to cause a physiological imbalance resulting in dissociation of the life span from the normal mechanism.

Takagi and Yoshida (1980) observed that fission-associated life spans have persisted in *P. caudatum* for 468 days, during which subclones were kept as stock cultures at 17° C under such conditions that cells were overfeeding and starving alternately. In contrast, the clonal life span of *P. tetraurelia* was easily dissociated with fissions under nutritionally poor conditions (Takagi et al. 1987a). Subclones grown in two culture media prepared with different concentrations of Cerophyl powder (1×-group and 4×-group) showed a significant difference in their mean life spans in fissions, while those grown at 20° C and 25° C in the same culture medium showed similar fission life spans. However, even with uniform temperatures and nutrition, intraclonal variations of fission life spans are not small.

Cells may not count the number of fissions but the number of DNA replications. Explanation for the intraclonal variation of life spans is easier in this case. Two daughter cells could have a different life expectancy in terms of fissions if they had different amounts of DNA and if DNA replications in more than one cycle were occurring to restore the parental level without cell division. Takagi and Kanazawa (1982) reported that aged cells of *P. caudatum* were apt to produce two daughter cells with different amounts of macronuclear DNA. Berger and Schmidt (1978) found that *P. tetraurelia* cells can correct the consequence of unequal macronuclear division by extra DNA replications. The *am* mutants, which distribute the macronucleus unequally, had in fact short clonal life spans measured in fissions (Aufderheide and Schneller 1985). Further, cells reached autogamous maturity with fewer fissions when their macronucleus was partly eliminated microsurgically (Mikami and Koizumi 1983).

Nothing is known about the mechanism by which ages are counted by the cycles of DNA replication. However, we can speculate a mechanism (Kuroda 1976). At the replication of linear DNA double strands, newly formed strands have small gaps at the 5′ ends where primer RNAs were located. Thus, DNA strands become shorter through successive replications, unless these gaps are filled by the 5′ → 3′ synthesis, which is possible only when special molecular forms such as a concatemer or a circularization occur. If this DNA shortening were the case in the macronucleus (but not in the micronucleus) and if some critical gene were located at a certain distance from either end, a phenotypic expression can occur at any time. Depending on the kinds of the critical gene, we can expect different types of phenotypic expressions: cell death in the case of the essential gene; derepression in the case of the repressor gene. Further, depending on the gene location, we can expect different spreading rates of the phenotypic expression among the members of a clone. This model is speculative but testable.

Factors that shorten the clonal life span include irradiation by UV, X-ray, and γ-ray (Smith-Sonneborn 1971; Fukushima 1974; Tixador et al. 1981), natural ionizing radiations (Tixador et al. 1981), fertilization at older parental ages (Smith-Sonneborn et al. 1974), caffeine (Smith-Sonneborn 1974), and novobiocin (Smith-Sonneborn 1981). Of great interest is the development of methods to lengthen the clonal life span (Smith-Sonneborn 1979, 1983). Although UV-irradiated cells had shorter life spans than nonirradiated controls, they had longer life spans if UV irradiation was followed by photoreactivation. This was interpreted as an activation of a dark repair mechanism as well as a light repair mechanism by photoreactivation. Another method to extend the clonal life span was to put cells in an electromagnetic field. Membrane changes activating the cellular function were argued as a possible mechanism. It should be noted that these methods are not to extend the maximal potential itself but to realize it, because still a longer life span can be seen without any special treatment (Takagi et al. 1987a).

4 Nuclear or Cytoplasmic

Senescence-associated changes include those in the macronucleus, in the micronucleus, and in the cytoplasm. Are there primary changes? And, if so, what are they? In conjugation, two cells of a conjugating pair produce two exconjugant clones or a synclone that is genotypically identical but cytoplasmically different unless gross cytoplasmic exchanges occur. In autogamy, exautogamous clones become homozygous in all of the alleles with the cytoplasm identical to the parental one. Synclonal cell death after conjugation indicates dominant lethal mutations in the micronuclei. Clonal segregation with one abortive and another normal exconjugant indicates cytoplasmic malfunctions in one of the pair members. Recessive lethals will be segregated in F_2 exautogamous clones.

Sonneborn and Schneller (1960) emphasized that the mutations were not occurring at a uniform rate as cells aged, but were specifically "induced" in the aged cytoplasm. In other words, primary senescent changes occur in the cytoplasm, not in the nucleus (Sonneborn 1974). Their basic observation was that the percentage of progeny survival was kept practically 100% in the course of 1500 fissions if autogamy was permitted every 25 fissions, while the survival rate decreased from 100% at the age of 80 fissions to 0% at the age of 200 fissions if autogamy was inhibited. The aged cytoplasm-induced mutations were shown by the crossings of young with old clones, which yielded pairs of isogenic exconjugants, one producing a viable clone derived from the young parent cell and the other producing a nonviable clone derived from the old parent cell. An extreme case was the crossing of young cells to "merogones", i.e., old cells whose own nuclei fail to survive meiosis, but into which nuclei from young cells can be introduced by mating. The general consequence was uniparental deaths of exconjugants from the merogones. The action of the aged cytoplasm to induce lethal micronuclear mutations had been suggested by Mitchison (1955), but his arguments were considered to be speculative (Rodermel and Smith- Sonneborn, 1977).

Karino and Hiwatashi (1981, 1984), examined the nuclear versus cytoplasmic problem in *P. caudatum* with a micronuclear transplantation technique. They showed that there were two classes of micronuclei in aged clones: those that function normally if transplanted into young cytoplasm, and those that cannot function normally even in a young cytoplasmic environment. They argued that the micronuclei were basically kept normal unless they were induced to become defective by the aged cytoplasm. When micronuclei were transplanted from young cells to such old cells, producing few viable progeny, the transplanted clones restored their progeny survival to the level of young clones; but the survival rate decreased abruptly between 14 and 42 fissions after the transplantation. This mutation-inducing effect of the aged cytoplasm was, however, not demonstrated by cytoplasmic transplantation.

Aufderheide (1984) studied the effect of cytoplasmic transplantation in detail. His assumption was that if senescence is the result of progressive loss of some critical cytoplasmic material, old cells will be rescued from senescence by injections of cytoplasm from young cells, and thus have longer life spans than uninjected controls. And if senescence is the result of progressive accumulation of some factor or state in the cytoplasm, young cells will become prematurely senescent by injections of cytoplasm from old cells, and thus have shorter life spans than uninjected controls. The results were, however, all negative: mean clonal life spans were not affected by cytoplasmic injections. The injected material, however, even if effective itself and quantitatively sufficient at that time, will be readily diluted through successive cell divisions, unless it is self-reproductive. Therefore, the possibility of the cytoplasm as the site of primary senescent changes cannot be ruled out. However, his investigations showed that the term "cytoplasm" in this context does not refer to the self-reproductive components such as mitochondria, endosymbionts, or unknown plasmids.

5 Programmed or Error Accumulated

The question whether the aging process is a programmed event or a result of error accumulation has long been discussed and is still a subject of argument. The term "aging" is sometimes regarded as the counterpart of "development". According to this view, the aging process is a disorder as a result of cessation of genic regulations that are subject to error accumulations. Alternatively, the aging process as well as the developmental process is considered programmed. Predictability or species specificity of the life span may be the only supporting evidence. Sonneborn (1978) pointed out that programmed aging and death might have evolved in diploids in which sex and fertilization became mandatory for survival of the species, as explained by the law: Fertilize and recombine genes when you can, or tomorrow you will die.

Without postulating intrinsic temporal regulations, we cannot understand the timing and sequence involved in some events during early clonal development (Siegel 1967; Bleyman 1971; Nanney 1974, 1980). In *P. bursaria,* the transition pattern from adolescence to maturity, i.e., mating-type expression of both the A and B loci in maturity shifting from the B locus expression in adolescence, is determined genetically. Thus, a mutation in this shift will result in permanent adolescence (Siegel and Cole 1967). In *P. caudatum* and *P. multimicronucleatum,* the timing of maturation is predictable in terms of the number of fissions since the last fertilization (Kroll and Barnett 1968; Takagi 1970, 1974; Miwa and Hiwatashi 1970). Genic control of the length of immaturity is shown by Mendelian segregation in *P. primaurelia* (Siegel 1961), and is shown more directly by mutants with reduced lengths of immaturity in *P. caudatum* (Myohara and Hiwatashi 1978). A material basis for controlling the length of immaturity was obtained by microinjections of the cytoplasm from cells of immaturity to those of maturity (Miwa et al. 1975; Miwa 1979). The key material is identified as immaturin, a single protein of molecular weight 10000 (Haga and Hiwatashi 1981; Miwa 1984).

Regarding the events of later life cycle stages, however, the programmed hypothesis is no longer decisive. Fukushima (1975) reasoned that if mutations were occurring at random in two micronuclei of *P. tetraurelia,* then the proportion of defective progeny after autogamy would begin to increase at any clonal age and would increase stepwise from 25% (one mutation) to 37.5 or 50% (two mutations) and then to 44, 50, 62.5, or 75% (three mutations) and so on. The experimental results were consistent with his idea and presented a striking contrast to the previous report (Sonneborn and Schneller 1960), which indicated no mutations in young cells and an abrupt increase in old cells.

Rodermel and Smith-Sonneborn (1977) reconciled these conflicting reports by large-scale experiments. They showed that an abrupt increase in percent death and reduced viability occurred around 50–80 fissions. Yet, lethal micronuclear mutations were occurring randomly as shown by the fact that death after autogamy was detected in some young clones and not at all in some old clones. They claimed that the abrupt increase would be a consequence of the imbalance between the mutation rate and the repair capacity and suggested that stochastic hits mutated the micronucleus and the organism was programmed to lose error-free repair.

Since the macronucleus supports most, if not all, of the cellular functions, its defects will certainly affect clonal aging and life span. However, it is difficult to tell whether alterations in the macronucleus with age are due to stochastic events or programmed events. Nanney (1974) demonstrated elegantly how precise timing could be set by random events. Isolations of mutants with respect to aging and life span will be a fruitful way to understand the underlying mechanism. Aufderheide and Schneller (1985) found that some *nd* mutants of *P. tetraurelia,* which are incapable of discharging trichocysts, were also characterized by a very short clonal life span. They further discovered that the short clonal life span was closely associated with the inability to properly divide the macronucleus. By chemical mutagenesis, we recently isolated a *P. tetraurelia* mutant which had a clonal life span of about one-tenth of the wild type (Takagi et al. 1987b). This mutant was associated with a low fission rate, intraclonal heterogeneity of division potential, and a proneness to undergo autogamy in excess food; but was not associated with inabilities to discharge trichocysts, to properly divide the macronucleus, to swim backward, and to normally form and excrete food vacuoles. The questions whether these couplings are separable or not and how many genes are involved (cytoplasmic contribution is, if any, negligible) remain to be clarified. We can now expect to isolate more mutants with altered life cycle features and to undertake breeding analyses with the hope that we will gain important insights on how cells age and how the life span is limited.

Acknowledgments. This research was partly supported by a grant from Itoh Science Foundation. I am greatly indebted to Drs. K. Hiwatashi and S. Fukushima for their critical reading of the manuscript. This work is dedicated to the memory of Sadako Miyake.

References

Aufderheide KJ (1984) Clonal aging in *Paramecium tetraurelia*. Absence of evidence for a cytoplasmic factor. Mech Ageing Dev 28:57–66

Aufderheide KJ, Schneller M (1985) Phenotypes associated with early clonal death in *Paramecium tetraurelia*. Mech Ageing Dev 32:299–309

Berger JD, Schmidt HJ (1978) Regulation of macronuclear DNA content in *Paramecium tetraurelia*. J Cell Biol 76:116–126

Bleyman LK (1971) Temporal patterns in the ciliated protozoa. In: Cameron IL, Padilla GM, Zimmer AM (eds) Developmental aspects of the cell cycle. Academic Press, London New York, p 67

Croute F, Vidal S, Dupouy D, Soleilhavoup JP, Serre G (1985) Studies on catalase, glutathione peroxidase and superoxidismutase activities in aging cells of *Paramecium tetraurelia*. Mech Ageing Dev 29:53–62

Dippell RV (1955) Some cytological aspects of aging in variety 4 of *Paramecium aurelia*. J Protozool 2 (Suppl):7

Fukushima S (1974) Effect of X-irradiations on the clonal life-span and fission rate in *Paramecium aurelia*. Exp Cell Res 84:267–270

Fukushima S (1975) Clonal age and the proportion of defective progeny after autogamy in *Paramecium aurelia*. Genetics 79:377–381

Haga N, Hiwatashi K (1981) A protein called immaturin controlling sexual maturity in *Paramecium*. Nature (London) 289:177–179

Heifetz SR, Smith-Sonneborn J (1981) Nucleolar changes in aging and autogamous *Paramecium tetraurelia*. Mech Ageing Dev 16:255–263

Holmes GE, Holmes NR (1986a) Accumulation of DNA damages in aging *Paramecium tetraurelia*. Mol Gen Genet 204:108–114

Holmes GE, Holmes NR (1986b) Accumulation of DNA fragments in aging *Paramecium tetraurelia* in axenic and nonaxenic media. Gerontol 32:252–260

Karino S, Hiwatashi K (1981) Analysis of germinal aging in *Paramecium caudatum* by micronuclear transplantation. Exp Cell Res 136:407–415

Karino S, Hiwatashi K (1984) Resistance of germinal nucleus to aging in *Paramecium:* evidence obtained by micronuclear transplantation. Mech Ageing Dev 26:51–66

Klass MR, Smith-Sonneborn J (1976) Studies on DNA content, RNA synthesis, and DNA template activity in aging cells of *Paramecium aurelia*. Exp Cell Res 98:63–72

Kroll RJ, Barnett A (1968) The effect of different fission rates on the onset of maturity in *Paramecium multimicronucleatum.* J Protozool 15 (Suppl): 10
Kuroda Y (1976) Concerning Hayflick's hypothesis. Igakuno-Ayumi 97:472–478 (in Japanese)
Mikami K, Koizumi S (1983) Microsurgical analysis of the clonal age and the cell-cycle stage required for the onset of autogamy in *Paramecium tetraurelia.* Dev Biol 100:127–132
Mitchison NA (1955) Evidence against micronuclear mutations as the sole basis for death at fertilization in aged, and in the progeny of ultraviolet irradiated, *Paramecium aurelia.* Genetics 40:61–75
Miwa I (1979) Immaturity substances in *Paramecium primaurelia* and their specificity. J Cell Sci 38:193–199
Miwa I (1984) Destruction of immaturin activity in early mature mutants of *Paramecium caudatum.* J Cell Sci 72:111–120
Miwa I, Hiwatashi K (1970) Effect of mitomycin C on the expression of mating ability in *Paramecium caudatum.* Jpn J Genet 45:269–275
Miwa I, Haga N, Hiwatashi K (1975) Immaturity substances: material basis for immaturity in *Paramecium.* J Cell Sci 19:369–378
Miyake A, Miyake S (1967) Immaturity, maturity and senescence in exautogamous clones of *Paramecium jenningsi.* Jpn J Genet 42 (Suppl): 424 (in Japanese)
Myohara K, Hiwatashi K (1978) Mutants of sexual maturity in *Paramecium caudatum* selected by erythromycin resistance. Genetics 90:227–241
Nanney DL (1974) Aging and long-term temporal regulation in ciliated protozoa: a critical review. Mech Ageing Dev 3:81–105
Nanney DL (1980) Experimental ciliatology. John Wiley & Sons, New York Chichester Brisbane Toronto
Nobili R (1959) The effects of aging and temperature on the expression of the gene *am* in variety 4, stock 51, of *Paramecium aurelia.* J Protozool 6 (Suppl): 29
Nyberg D (1978) Copper tolerance and segreation distortion in aged *Paramecium.* Exp Gerontol 13:431–437
Rodermel SR, Smith-Sonneborn J (1977) Age-correlated changes in expression of micronuclear damage and repair in *Paramecium tetraurelia.* Genetics 87:259–274
Schwartz V, Meister H (1973) Eine Altersveränderung des Makronucleus von *Paramecium.* Z Naturforsch 28 c: 232
Schwartz V, Meister H (1975) Einige quantitative Daten zum Problem des Alterns bei *Paramecium.* Arch Protistenk Biol 117:85–109
Shimomura F, Takagi Y (1985) Commitment to sexual and asexual phases in *Paramecium multimicronucleatum.* J Exp Zool 233:317–326
Siegel RW (1961) Nuclear differentiation and transitional cellular phenotypes in the life cycle of *Paramecium.* Exp Cell Res 24:6–20
Siegel RW (1967) Genetics of ageing and the life cycle in ciliates. Symp Soc Exp Biol 21:127–148
Siegel RW, Cole J (1967) The nature and origin of mutations which block a temporal sequence for genic expression in *Paramecium.* Genetics 55:607–617
Smith-Sonneborn J (1971) Age correlated sensitivity to ultraviolet radiation in *Paramecium.* Radiat Res 46:64–69
Smith-Sonneborn J (1974) Age-correlated effects of caffeine on non-irradiated and UV-irradiated *Paramecium aurelia.* J Gerontol 29:256–260
Smith-Sonneborn J (1979) DNA repair and longevity assurance in *Paramecium tetraurelia.* Science 203:1115–1117
Smith-Sonneborn J (1981) Genetics and aging in protozoa. Int Rev Cytol 73:319–354
Smith-Sonneborn J (1983) Programmed increased longevity induced by weak pulsating current in *Paramecium.* J Electroanal Chem 156:373–382
Smith-Sonneborn J (1985) Aging in unicellular organisms. In: Finch CE, Schneider EL (eds) Handbook of the biology of aging, 2nd edn. Van Nostrand Reinhold, New York, p 79
Smith-Sonneborn J, Klass M (1974) Changes in the DNA synthesis pattern of *Paramecium* with increased clonal age and interfission time. J Cell Biol 61:591–598
Smith-Sonneborn J, Reed JC (1976) Calendar life-span versus fission life-span of *Paramecium aurelia.* J Gerontol 31:2–7

Smith-Sonneborn J, Rodermel SR (1976) Loss of endocytic capacity in aging *Paramecium.* J Cell Biol 71:575–588
Smith-Sonneborn J, Klass M, Cotton D (1974) Parental age and life-span versus progeny life-span in *Paramecium.* J Cell Sci 14:691–699
Sonneborn TM (1954) The relation of autogamy to senescence and rejuvenescence in *Paramecium aurelia.* J Protozool 1:38–53
Sonneborn TM (1974) *Paramecium aurelia.* In: King RC (ed) Handbook of genetics, vol 2. Plenum, New York, p 469
Sonneborn TM (1978) The origin, evolution, nature and causes of aging. In Behnke JA, Finch CE, Moment CB (eds) The biology of aging. Waverly, Baltimore, p 286
Sonneborn TM, Dippell RV (1960) Cellular changes with age in *Paramecium.* In: Strehler BL (ed) The biology of aging. Waverly, Baltimore, p 279
Sonneborn TM, Schneller MV (1960) Age-induced mutations in *Paramecium.* In: Strehler BL (ed) The biology of aging. Waverly, Baltimore, p 285
Sundararaman V, Cummings DJ (1976a) Morphological changes in aging cell lines of *Paramecium aurelia.* I. Alterations in the cytoplasm. Mech Ageing Dev 5:139–154
Sundararaman V, Cummings DJ (1976b) Morphological changes in aging cell lines of *Paramecium aurelia.* II. Macronuclear alterations. Mech Ageing Dev 5:325–338
Sundararaman V, Cummings DJ (1983) Antigenic variation during cellular aging in *Paramecium tetraurelia.* Mech Ageing Dev 22:89–96
Takagi Y (1970) Expression of the mating-type trait in the clonal life history after conjugation in *Paramecium multimicronucleatum* and *Paramecium caudatum.* Jpn J Genet 45:11–21
Takagi Y (1971) Sequential expression of sex-traits in the clonal development of *Paramecium multimicronucleatum.* Jpn J Genet 46:83–91
Takagi Y (1974) The effect of ultraviolet irradiation on the period of immaturity in *Paramecium caudatum.* Zool Mag (Tokyo) 83:96–98 (in Japanese)
Takagi Y, Kanazawa N (1982) Age-associated change in macronuclear DNA content in *Paramecium caudatum.* J Cell Sci 54:137–147
Takagi Y, Yoshida M (1980) Clonal death associated with the number of fissions in *Paramecium caudatum.* J Cell Sci 41:177–191
Takagi Y, Nobuoka T, Doi M (1987a) Clonal lifespan of *Paramecium tetraurelia:* effect of selection on its extension and use of fissions for its determination. J Cell Sci 88:129–138
Takagi Y, Suzuki T, Shimada C (1987b) Isolation of a *Paramecium tetraurelia* mutant with short clonal life-span and with novel life-cycle features. Zool Sci 4:73–80
Tixador R, Richoilley G, Planel H (1972) Radiosensitivity variation of *Paramecium aurelia* as a function of the clonal age. J Protozool 19 (Suppl):73
Tixador R, Richoilley G, Monrozies E, Planel H, Tap G (1981) Effects of very low doses of ionizing radiation on the clonal life-span in *Paramecium tetraurelia.* Int J Radiat Biol 39:47–54
Williams TJ, Smith-Sonneborn J (1980) DNA polymerase activity in aged clones of *Paramecium tetraurelia.* Exp Gerontol 15:353–357
Wichterman R (1953) The biology of *Paramecium.* Blakiston, New York Toronto
Wichterman R (1986) The biology of *Paramecium,* 2nd edn. Plenum, New York London

Chapter 10 Organization and Expression of the Nuclear Genome

MANFRED FREIBURG [1]

CONTENTS

1 Introduction

During the last decade enormous progress was made in the investigation of the structure of the ciliate genome. Whereas for a long time ciliates have been thought to be rather burlesque relicts of evolution of interest only as an ecological niche for a few enthusiasts, they are now becoming more and more important for the research on basic mechanisms of the molecular biology of eukaryotic genes. Most of this progress was made on a few genera of ciliates: among the hypotrichs *Oxytricha, Stylonychia* and *Euplotes* and among the holotrichs essentially on *Tetrahymena* (see Gall 1984). *Paramecium*, most important in the past, seems to be somewhat ousted from the centre of interest in this field of research. Nevertheless, as it was demonstrated very recently by the discovery of its deviation in the usage of the "universal" genetic code, *Paramecium* and its molecular genetics remain most important for our understanding of the organization and evolution of eukaryotic genes.

1 Zoologisches Institut der Westfälischen Wilhelms-Universität Münster, D-4400 Münster, FRG

Paramecium. Ed. by H.-D. Görtz

2 Nuclear Dimorphism

Like other ciliates paramecia contain two different types of nuclei within a single cell: one large macronucleus and one or more micronuclei (for review and references see Nanney 1980; Raikov 1982). Both types of nuclei exhibit different functions during the life cycle of a cell. The macronucleus serves as the somatic nucleus governing the vegetative functions and determining thereby the phenotype of the cell. The micronucleus represents the "germ line" of the organism serving mainly the maintenance of the genetic continuity from one generation to the next and, on the other hand, the recombination of the genome and the development of a new macronucleus during the process of conjugation (see Fujishima, Chap. 5, this Vol.). The existence of two functionally different types of nuclei within the same cytoplasmic environment has raised questions concerning the organization of their genomes, on possible genome reorganizations during macronuclear development and the mechanisms by which macronuclear genes are expressed while corresponding micronuclear genes remain repressed.

2.1 Structure and Function of the Micro- and the Macronucleus

Micronuclei from paramecia are small, their diameters ranging from 1 to 10 µm. During cell division the generally diploid micronucleus divides mitotically without disintegration of its envelope. During the course of mitosis and during the meiotic processes in conjugation or autogamy small chromosomes become visible. Estimates of the diploid numbers of chromosomes in different species range from approximately 20 to several hundreds (summarized by Raikov 1982; for review of the number of chromosomes in the *P. aurelia* complex see Sonneborn 1974). In addition, the number of chromosomes can vary within a given species from one strain to the other. Polyploid micronuclei have been found in different species, e.g. in *P. bursaria, P. caudatum* and *P. putrinum* (see Raikov 1982). Cells can lose their micronuclei spontaneously or under certain experimental conditions. These cells are viable and able to divide for long periods of time, indicating that micronuclear functions are not essential for vegetative life. The view of transcriptionally inert micronuclei is supported by the finding that under normal conditions very little micronuclear RNA synthesis can be measured, occurring largely in the micronuclear S-phase (Pasternak 1967; Rao and Prescott 1967). Nearly the same amount of micronuclear RNA synthesis as in normal cells has been observed by Pasternak (1967) in amacronucleate cells. For possible functions of the micronucleus during the vegetative life of the cell see Mikami (Chap. 8, this Vol.).

Macronuclei of paramecia are rather large and oval or kidney-shaped. In general, the size of the macronucleus is approximately proportional to the cell size. The macronuclear DNA content is many times that of the micronucleus. Chromosomes have never been observed in macronuclei. During cell division the macronucleus divides without any visible mitosis. The mode of the regular distribution of the genetic material to the two daughter macronuclei is mostly unknown. Small nucleoli have been observed containing presumably the genes coding for rRNA. Amacronucleate cells are unable to survive and die within few days.

2.2 Isolation of Micro- and Macronuclei

One prerequisite for investigating the different functions and the organization of the two types of nuclei in vitro is the development of isolation procedures leading to separated fractions of pure macronuclei and pure micronuclei. In addition, the isolated nuclei should retain their in vivo composition and function as far as possible. To achieve this goal, cells are usually lysed by the addition of detergents or octanol in the presence of spermine, spermidine, Ca^{2+}, Mg^{2+}, gum arabic, etc. to protect the nuclear membranes. Separation of the nuclei is more or less achieved by successive steps of centrifugation of the cell lysate through continuous or discontinuous sucrose gradients (for review see Cummings and Tait 1975; Cummings 1977). Especially the preparation of pure micronuclei is hampered by contaminating macronuclear fragments generated during the process of autogamy in certain species and/or artificially during lysis of the cells. To overcome this problem Shiomi et al. (1980) separated macro- and micronuclei from *P. bursaria* and *P. trichium* by membrane filtration and centrifugation in Metrizamide gradients. However, sensitive controls to exclude the presence of macronuclear components in micronucleus preparations are still pending.

3 DNA Structure

The existence of two functionally and structurally different nuclei within a single *Paramecium* cell raises basic questions according to the amount and structure of their DNA: Is the macronucleus a simple polyploid copy of the micronucleus, or are there any differences in genome organization between both types of nuclei resulting presumably from a process of "genome differentiation" that may take place during the development of the new macronucleus from the micronucleus? Does the macronucleus contain chromosome-sized DNA molecules or is the DNA fragmented into subchromosomal pieces? Since no chromosomes can be detected at any time in the macronucleus, what is the structural unit of genome organization?

3.1 Genome Size and Complexity in Micro- and Macronuclei

The amounts of DNA in the micro- and macronucleus differ significantly in diverse species of *Paramecium.* The largest differences have been found in the *P. aurelia* complex showing ratios of macro-/micronuclear DNA content of 400–500, whereas the lowest values of approximately 10–20 were obtained in *P. bursaria.* Especially in those cases where low values of the macro-/micronuclear DNA ratio had been found, polyploidy of the micronucleus seems likely (for review and references see Raikov 1982).

In various stocks of the *P. aurelia* complex cytophotometric measurements of G1 cells using bull sperm nuclei as standard led to estimates of approximately 0.6 pg DNA in the micronucleus and 268 – 346 pg in the macronucleus (Gibson and Martin 1971; Allen and Gibson 1972). In contrast, a more direct determination of the DNA content of macronuclei isolated from unsynchronized cultures of *P. primaurelia* revealed lower values of approximately 100 pg/macronucleus corresponding to 70 pg/macronucleus in G1 (Cummings and Tait 1975). A similar value has been obtained by Skoczylas and Soldo (1975) and Soldo et al. (1981) for macronuclei isolated from exponentially growing cultures of *P. octaurelia*. In addition, the latter reported a significant decrease of approximately 50% in DNA content of macronuclei isolated from cultures starved for 48 h. Higher amounts of DNA were measured in isolated macronuclei derived from *P. caudatum*, i.e. 348 pg in the mid-log phase and 169 pg in starved cultures (Soldo et al. 1981). In *P. bursaria*, the DNA content of the macronucleus was determined cytophotometrically to be 150 or 277 pg respectively (Cullis 1973; Steinbrück et al. 1981), whereas the value for the 2c micronucleus was approximately 7.5 pg.

As in other ciliates the guanine and cytosine (GC) content of macronuclear DNAs of different paramecia species is low, comprising 26 – 29% in the *P. aurelia* complex (Gibson et al. 1971; Allen and Gibson 1972; McTavish and Sommerville 1980; Soldo et al. 1981), 26.2% in *P. caudatum* (Soldo et al. 1981; Allen and Gibson 1972) and 29 and 26.3% respectively in *P. bursaria* (Steinbrück et al. 1981; Cullis 1972).

Various attempts have been made to determine the sequence complexity of the macronuclear DNA by measurements of its reassociation rate after denaturation compared to a DNA sample of known complexity. Soldo and Godoy (1972) reported that the renaturation kinetics of macronuclear DNA from *P. octaurelia* indicates the presence of two different fractions. The main fraction, comprising approximately 96% of the DNA, reassociates slowly. The minor fraction (4%) reassociates ca. 3000 times faster than the main fraction. The kinetic complexity of the main fraction corrected for the GC content was estimated to be 3.8×10^{10} daltons and that of the minor fraction to be 1.45×10^{7} daltons. Similar values have been obtained from two different stocks of *P. primaurelia* leading to estimates of their sequence complexities of 3×10^{10} and 5×10^{10} daltons respectively (Cummings 1975; McTavish and Sommerville 1980). In contrast, significantly higher kinetic complexities have been measured by Allen and Gibson (1972) in different stocks of *P. primaurelia* and *P. octaurelia*. In *P. bursaria* Cullis (1972) reported a kinetic complexity of 9.4×10^{10} daltons (corrected for the GC content) and the occurrence of repetitive DNA sequences comprising approximately 13% of the total, whereas Steinbrück et al. (1981) measured a kinetic complexity of 1.1×10^{10} (corrected for GC content) and less than 2% of repetitive DNA. As pointed out by Steinbrück et al. (1981), the estimation of kinetic complexities depends very much on the experimental conditions. Therefore, the comparison of data derived by different groups is faced with high uncertainties.

From cytophotometric data Gibson and Martin (1971) concluded that in species of the *P. aurelia* complex the macronucleus contains approximately 400 times as much DNA as the micronucleus. Supposing that no drastic differences of the genome organization of the micro- and macronucleus exist, this value corresponds to 800 copies of the haploid genome in the macronucleus. These calculations are supported by data derived from renaturation kinetics leading to estimates of 860 haploid genomes/macronucleus for *P. primaurelia* (McTavish and Sommerville 1980) and 730 haploid genomes/macronucleus for *P. octaurelia* (Soldo et al. 1981). Conflicting estimates have been made by Cummings (1975) for *P. primaurelia* (1400 copies of the haploid genome/ macronucleus) and by Soldo and Godoy (1972) in *P. octaurelia* (1680 haploid copies/macronucleus of the slowly renaturing component, see above). Most interestingly, the data obtained from *P. caudatum* (which is the largest species investigated) led to an estimation of 6780 haploid copies of the genome/ macronucleus and to the assumption that the DNA content of the macronucleus might be somehow correlated with the cell size (Soldo et al. 1981).

3.2 Genome Reorganization During Macronuclear Development?

Chromatin diminution during the differentiation of soma and germ line was discovered cytologically in the late 19th century in the nematode *Ascaris* by Boveri (reviewed by Boveri 1899), but lost attention thereafter. Now it has become of immediate interest again by the discovery of the, in part, dramatic genome reorganization processes occurring in hypotrichous, and, to a lesser extent, in holotrichous ciliates during the macronuclear development from the new micronucleus after conjugation (for review see Brunk 1986; Kraut et al. 1986).

As mentioned above, in some species of *Paramecium* a tremendous quantitative difference in the DNA content between the micro- and macronucleus can be measured. It is one of the central questions of the molecular biology of ciliates whether besides this quantitative difference, qualitative differences can arise during macronuclear development due to rearrangements, selective amplification, fragmentation and selective elimination of micronuclear DNA sequences. In the case of the genes coding for rRNA it has been clearly shown that they are highly amplified in the macronucleus of *P. tetraurelia* as compared with those of the micronucleus and are of subchromosomal size (see below).

The question of whether bulk DNA of the macronucleus occurs in micronucleuslike chromosome-sized pieces or in smaller fragments has been investigated by different techniques. Cummings (1975) analyzed the size of DNA molecules isolated from *P. primaurelia* in the electron microscope. The average length was found to depend largely on the experimental conditions and was on average smaller than 150 kilobases (kb). The longest molecules he observed were of 105 μm corresponding to approximately 300 kb. In addition, using viscoelastic techniques, he and his collaborators measured a fraction of molecules reaching 12000 kb (Cummings 1975). Preer and Preer (1979) sedimented radioactively labelled DNA derived from carefully lysed cells of *P. tetraurelia* in low-speed neutral sucrose gradients and calculated from its sedimentation behaviour an average size of 300 kb. This value was significantly lower than the figure of 600 kb they obtained for *Tetrahymena*-DNA (for discussion of DNA length measurements using different techniques see Preer and Preer 1979). However, in contrast to the "gene-sized" DNA fragments (on the average 2–3 kb) in the macronucleus of hypotrichs such as *Oxytricha, Euplotes* or *Stylonychia* (see Kraut et al. 1986), no discrete bands of different size classes can be resolved by gel electrophoretic separation of *Paramecium*-DNA (McTavish and Sommerville 1980; Steinbrück et al. 1981).

From measurements of the kinetic complexity of macronuclear DNA of *P. primaurelia* a size (corrected for the GC content) of approximately 45000–75000 kb has been calculated (Cummings 1975; McTavish and Sommerville 1980). Assuming a haploid chromosome number of 37–63, depending on the stock investigated (for references see Sonneborn 1974), and understanding that one chromosome contains one DNA molecule and that all chromosomes may have approximately the same size, one can roughly calculate that one chromosome-sized DNA molecule might be 700–2000 kb long. Making very similar assumptions Preer and Preer (1979) calculated an average length of a chromosome-sized molecule to be 2100–2700 kb. On the basis of these calculations, it can be assumed that at least in *P. primaurelia,* no extensive fragmentation of the micronucleus genome occurs during macronuclear development. The identical conclusion was drawn by McTavish and Sommerville (1980) comparing the DNA content of the macronucleus (as measured by Cummings, 1975) with their estimation of the macronuclear sequence complexity. However, using these methodologic approaches, small reorganizations and elimination of DNA sequences to lesser degrees cannot be detected with certainty.

More direct experimental indications on genome reorganization events have been obtained in *P. bursaria:* By cytophotometrical measurements Schwartz and Meister (1975) found a rise followed by a significant decrease of the DNA content during the development of early macronuclear anlagen. Thereafter the remaining DNA was multiplied again building up the genome of the mature macronucleus. These findings resemble the events occurring during macronuclear development of hypotrichous ciliates. Steinbrück et al. (1981), however, determined the size of macronuclear DNA of *P. bursaria* to be on the average more than 20 kb, which excludes a fragmentation of the micronuclear genome as extensive as it is found in the hypotrichs.

In *Tetrahymena,* using methods which permit more direct comparisons of micro- and macronuclear DNA, it has been concluded that 10–20% of the complexity of the micronuclear genome is lost during the development of macronuclear anlagen (Brunk 1986). In the course of genome reorganization each micronuclear chromosome most likely becomes fragmented leading to macronuclear DNA molecules of 21 kb up to 1100 kb and more (Altschuler and Yao 1985). Despite the relatively close evolutionary relationship of *Paramecium* to *Tetrahymena* these data cannot simply be transferred to the situation in *Paramecium,* since in another tetrahymenid, *Glaucoma chattoni,* much smaller macronuclear DNA molecules in the range of 2.1 kb to more than 100 kb have been measured (Katzen et al. 1981).

As mentioned above, in contrast to the micronucleus, no chromosomes sensu stricto are visible at any time during macronuclear division. However, from the size of DNA molecules in the macronucleus of *Paramecium* a physical linkage of genes, although subchromosomal in size, presumably exists. These "linkage groups" must be maintained and replicated from one cell division to the other, whereby specific DNA sequences at the ends of the molecules (telomeres) may play an important role (for review see Blackburn and Szostak 1984). In *Tetrahymena* telomeres have been found in the macronucleus at the ends of rDNA as well as of non-rDNA and have been shown to consist of tandem repeats of the hexanucleotide C_4A_2 (reviewed by Blackburn 1986). Yao and Yao (1981) have identified this repeat in *P. tetraurelia* near the ends of macronuclear DNA. In addition, Epstein and Forney (1984) obtained a *P. tetraurelia* mutant called D48 which is unable to express the surface antigen A despite its presence in the micronucleus. Like the surface antigen G gene of *P. primaurelia* (Meyer et al. 1985), the A gene lies near the end of a macronuclear chromosome. Presumably, as pointed out by Blackburn (1986), the inability of its expression in the macronucleus originated from a wrong addition of telomere sequences into the 5′-region of that gene removing the remaining downstream sequences.

From the data available up to now one can conclude that in *Paramecium* genome reorganization occurs during the development of a new macronucleus but not as extensive as in hypotrichous ciliates. The extent of the reorganization, however, might be species specific. (For a comprehensive review of nuclear differentiation in ciliates see Steinbrück 1986.)

3.3 Organization of the Genes Coding for Ribosomal RNA

Reassociation kinetics of macronuclear DNA of *P. primaurelia* and *P. octaurelia* revealed two different fractions, a slowly reassociating component presumably representing bulk DNA and a faster reassociating component assumed to represent repetitive genes of the macronuclear genome, i.e. essentially the genes for ribosomal RNA (rDNA) and tRNA. This fraction comprises 2–4% of the total DNA (Soldo and Godoy 1972; Cummings 1975; McTavish and Sommerville 1980).

In a more direct approach to identify one of these genes Findley and Gall (1978) sedimented whole cell DNA isolated from *P. tetraurelia* in alkaline sucrose gradients (due to the large differences in the amounts of micro- and macronuclear DNA it must have been essentially of macronuclear origin). They identified the position of rDNA by its hybridization with radioactively labelled 17 and 25S rRNA from the same species. The data indicated that the rDNA comprises approximately 0.2% of total DNA (a similar value has been determined by Cummings, 1975) and occurs as multicopy extrachromosomal molecules heterogeneous in size. Analysis of the isolated rDNA in the electron microscope showed both linear and circular molecules. After partial denaturation a regular repeating pattern of denatured regions could be observed. In each molecule 2–13 repeats of approximately 5.5×10^6 daltons were found. In some cases the repeat was heterogeneous in size. Digestion of the rDNA with restriction endonucleases supported both the concept of a tandemly repeated basic unit and its size heterogeneity. In denaturation-renaturation experiments no half-sized, double-stranded molecules could be identified. This excludes a palindromic structure of rDNA as it was found in *Tetrahymena* (Engberg et al. 1976; Karrer and Gall 1976), *Dictyostelium* (Cockburn et al. 1978) and *Physarum* (Vogt and Braun 1976). According to a restriction map Findley and Gall (1980) proposed the following structure of the rRNA genes:

– 25 S – (– spacer – 17 S – 5.8 S – 25 S –) (– spacer – 17 S –

Each repeat (given in parentheses) includes a spacer region and the sequences for 17 S, 5.8 S and 25 S rRNA. The limited size heterogeneity of the repeat obtained in the previous experiments could be restricted to the spacer region. No indication has been found for the presence of 5 S rRNA genes in the rDNA molecule. Since no defined ends could be identified in the isolated rDNA molecules, they seem to exist in the macronucleus as covalently closed circles which break randomly during the isolation procedure leading to heterogeneous linear molecules.

A comparison of the rDNA structure of *P. tetraurelia* with that of the other well-analyzed hypotrichous ciliates *Tetrahymena* and *Glaucoma* indicates differences in organization: Although, in all cases investigated the genes coding for the large rRNAs are extrachromosomal and do not contain 5 S rRNA; in *Tetrahymena* these molecules are found to be linear palindromes containing one gene on each half (Engberg et al. 1976; Karrer and Gall 1976), whereas *Glaucoma* contains linear, free monomers of the gene (Katzen et al. 1981). In *Paramecium* the genes are arranged as tandem repeats on covalently closed circles and resemble in this respect the organization of ribosomal RNA genes in *Xenopus* (for review see Blackburn 1982). Due to the size heterogeneity found in the spacer regions, Findley and Gall (1980) proposed that the micronucleus from which at least the macronuclear genes originate contains more than one rRNA gene. Presumably these genes are integrated into the micronuclear genome and tandemly arranged. In contrast, in *Tetrahymena* there is good evidence for the existence of a single, integrated, non-palindromic rRNA gene per haploid genome in the micronucleus (Yao and Gall 1977).

3.4 Methylation of Micro- and Macronuclear DNA

In eukaryotes there is convincing evidence that DNA methylation is involved in the regulation of gene activity. DNA in sperm cells has been found to be highly methylated, whereas somatic cells in general show significantly lower contents of methyl groups along their DNA. This and other findings have led to the assumption that the degree and pattern of DNA methylation might be involved in the activation or inactivation of transcription (for review see Doerfler 1983; Razin and Szyf 1984).

Ciliates might be an excellent experimental system to prove this assumption (if what is true for higher eukaryotes is true for ciliates) due to the fact that they possess two types of genomes within one cell differing in their state of activity: the transcriptionally rather inactive micronucleus and the highly active macronuclear genome. Cummings et al. (1974) investigated this question looking for the presence of methylated bases in micro- and macronuclear DNA in *P. primaurelia.* In addition, they analyzed the macronuclear DNA of some other species of the *P. aurelia* complex and of *P. caudatum.* They found that the DNA of paramecia does not contain 5-methylcytosine in contrast to higher eukaryotes where it is the sole modified base (Razin and Szyf 1984). Instead, they found as a modified base exclusively N^6-methyladenine. The same has been reported for *Tetrahymena thermophila* (reviewed by Gorovsky 1980). However, in *Tetrahymena* a striking difference in the N^6-methyladenine content between the micronucleus (less than 0.1 mol%) and the macronucleus (0.6–0.8 mol%) has been found, whereas Cummings et al. (1974) reported very similar amounts of N^6-methyladenine in micro- (3.3 mol%) and macronuclear DNA (2.5 mol%) of *P. primaurelia.* Since during the isolation of micronuclei a contamination with macronuclear fragments can hardly be avoided, it cannot be excluded that at least part of the N^6-methyladenine content of the micronuclear fraction from *Paramecium* might be of macronuclear origin.

4 Chromatin Structure and Chromosomal Proteins

Eukaryotic chromatin in general seems to be organized in a fundamental packing unit called the nucleosome. One nucleosome monomer contains approximately 200 base pairs (bp) of DNA. It consists of a nucleosome bead made up of eight histone molecules around which a 146-bp segment of double-stranded DNA is wound and a "linker" of about 60 bp connecting the nucleosome beads (reviewed by McGhee and Felsenfeld 1980). Micrococcal nuclease digestion preferentially cleaves the linker DNA, whereas the DNA in the nucleosome beads remains fairly protected by its association with the histone core. Limited digestion thus produces regular series of fragment lengths indicating the nucleosomal structure of the chromatin.

Using this method Prince et al. (1977) investigated the macronuclear chromatin from *P. primaurelia.* As in other eukaryotes, the chromatin of *Paramecium* showed a nucleosomal structure. However, compared with the DNA fragments obtained from isolated HeLa cell nuclei and from macronuclei of *Tetrahymena pyriformis* digested in parallel, the DNA repeats from *Paramecium* were smaller which indicates differences in the size of the DNA linker. In addition, a repeat size of 153 ± 7 bp has been found in logarithmically growing cultures, whereas stationary cells showed a repeat size of about 178 bp. The question of whether these differences between growing and growth-arrested cells may be due to differences in the transcriptional activity of the genome, as discussed by the authors, is still undecided.

Within the nucleus nucleosomes are packed normally in complex, higher order chromatin structures. In *Paramecium* three different chromatin components have been observed with the light and electron microscope in thin sections of macronuclei: (1) small chromatin bodies (approximately 10^5/macronucleus) having diameters of 0.1 – 0.2 μm; (2) larger chromatin bodies with diameters of 0.5 – 1 μm, considered to be nucleoli (approximately 1000/macronucleus), and (3) a fibrillar matrix (for references and review see Samuel et al. 1981; Raikov 1982).

After spreading macronuclear chromatin clearly discernible structures were observed (Samuel et al. 1981): (1) nucleosomal filaments; they consist of 10 – 15 nm beaded filaments corresponding to the "beads on a string" structure of nucleosomes; (2) supranucleosomal globules; these are densely packed chromatin aggregates with 30 – 40 nm fibres and correspond presumably to the 30-nm chromatin fibre ("superbeads"); from these aggregates nucleosomal filaments seem to be pulled out in well-spread preparations; and (3) convoluted fibrillar elements; these structures presumably arise from aggregates of supranucleosomal globules to knobby 40 – 60 nm fibres. The small chromatin bodies observed in thin sections of macronuclei presumably correspond to the superbead structure observed in spread chromatin. It is perhaps due to technical reasons that nucleolar chromatin could not be identified in the course of this study.

Chromatin spreads, especially from growing cultures, show lateral 20-nm fibrils attached to nucleosomal filaments and representing transcription products organized as RNP complexes (Samuel et al. 1981; Sommerville and McTavish 1982). These RNP fibrils are normally found as densely packed nodules emerging from four to eight nucleofilaments lying in parallel. This arrangement perhaps, may indicate something like polyteny in the macronucleus. In addition, actively transcribed chromatin seems to be spatially separated from non-transcribed regions, as indicated by separate positions in chromatin spreads.

Only little information is presently available on the biochemical characterization of chromosomal proteins of paramecia. Isaacks and Santos (1973) extracted histones from isolated macronuclei of *P. tetraurelia.* This protein fraction amounted to approximately 40% of the total macronuclear protein and could be separated electrophoretically into the five major fractions known from other eukaryotes. Among the non-histone proteins three DNA polymerase activities have been identified and partially purified from macronuclei of *P. primaurelia* (Tait and Cummings 1975). The different forms are similar in size (90 – 110 kDa) and show some peculiarities distinct from those known from higher eukaryotes. RNA polymerase activities have been measured in vitro in isolated macronuclei, but the enzymes themselves have not been isolated so far (Stevenson 1967; Skoczylas and Soldo 1975; Shiomi et al. 1980; Freiburg 1985).

5 Transcription of the Macronuclear Genome

Considering the multitude of publications dealing with the DNA structure in ciliate nuclei, comparatively little is known about transcription in ciliates. Hruby et al. (1977) isolated total polyadenylated RNA (polyA$^+$ RNA) from *P. primaurelia.* The fraction obtained by binding to oligo (dT)-cellulose comprised 2 – 3% of the total cellular RNA and appeared to have a 5′-capped structure and a polyA-tail at the 3′-end. Further analysis of the polyA$^+$ RNA fraction indicated that it contained molecules of 100–7000 nucleotides length most of them ranging between 1000 and 2000 nucleotides and presumably transcribed from 5–8% of the macronuclear DNA (McTavish and Sommerville 1980). In addition, among the polyA$^+$ RNAs three different classes have been discerned: one being transcribed with low frequency from the bulk of DNA, the others being transcribed at medium or high frequencies from 100–200 or from 1–3 genes respectively. The number of expressed genes can be induced (and thereby the synthesis of polyA$^+$ RNA) by shifting the *Paramecium* cultures from 24 to 32° C. Throughout this temperature shift a partial redistribution among the different classes of polyA$^+$ RNA occurs (Sommerville and McTavish 1982).

So far, in *Paramecium* no DNA sequences have been identified that could serve as regulatory elements in gene transcription (Prat et al. 1986). In addition, no introns have been found in the genes coding for cell surface antigens (see Schmidt, Chap. 11, this Vol.)

6 The Genetic Code

Heterologous in vitro systems which are well established in the translation of mRNA from other eukaryotes gave insufficient results in the translation of mRNA isolated from *Paramecium* (or other ciliates) (Hruby et al. 1977; David and Smith 1981; Preer et al. 1981). The in vitro translation products were much smaller than expected from the size of mRNA used in the experiments. Meyer et al. (1984), using the mRNA of the G-surface antigen of *P. primaurelia,* first reported evidence indicating that this failure was due to the primary structure of the mRNA itself. Recently, the reason for insufficient in vitro translation of the *Paramecium* RNA was clarified: *Paramecium* uses two of the three codons functioning in higher eukaryotes as termination signals to code for amino acids (Caron and Meyer 1985; Preer et al. 1985).

So far, deviations from the "universal" genetic code have been known from the mitochondrial genome. Now it seems very likely that ciliates in general use the termination codons TAA and TAG as codons for glutamine or glutamic acid respectively (Helftenbein 1985; Horowitz and Gorovsky 1985; Kuchino et al. 1985). This deviation from the general genetic code may be a relic of a less accurately, primitive translation machinery or might have been acquired during early evolution of ciliates (for discussion see Horowitz and Gorovsky 1985; Caron 1986; Hanyu et al. 1986).

7 Conclusions and Summary

Despite the fact that the organization of the micro- and macronuclear genome of *Paramecium* is by far not as well investigated as that of the closely related holotrichous ciliate *Tetrahymena,* drastic genome reorganization and elimination events as those found in the hypotrichous ciliates can be excluded. Minor qualitative differences, however, between the genomes of the micro- and the macronucleus seem to be likely, due to chromosome fragmentation during the development of the macronuclear anlagen. The genes coding for rRNA are found to be extrachromosomal and arranged as tandem repeats on covalently closed circles. The macronuclear chromatin seems to be organized in a typical eukaryotic mode but showing some peculiarities presumably due to the specific macronuclear architecture. Like other ciliates, *Paramecium* deviates from the "universal" genetic code using the codons TAA and TAG to code for glutamine or glutamic acid respectively.

Acknowledgments. I wish to thank Drs. K. Mueller and G. Steinbrück für critically reading the manuscript.

References

Allen S, Gibson I (1972) Genome amplification and gene expression in the ciliate macronucleus. Biochem Gen 6:293–313

Altschuler MI, Yao MC (1985) Macronuclear DNA of *Tetrahymena thermophila* exists as defined subchromosomal-sized molecules. Nucl Acids Res 13:5817–5831

Blackburn EH (1982) Characterization and species differences of rDNA: protozoans. In: Busch H, Rothblum L (eds) The cell nucleus, vol 10. Academic Press, London New York, pp 145–170

Blackburn EH (1986) Structure and formation of telomeres in holotrichous ciliates. Int Rev Cytol 99:29–47

Blackburn EH, Szostak JW (1984) The molecular structure of centromeres and telomeres. Annu Rev Biochem 53:163–194

Boveri Th (1899) Die Entwicklung von *Ascaris megalocephala* mit besonderer Rücksicht auf die Kernverhältnisse. Kupfer-Festschrift. Fischer, Jena

Brunk CF (1986) Genome reorganization in *Tetrahymena.* Int Rev Cytol 99:49–83

Caron F (1986) Deviations from the "universal" genetic code. Microbiol Sci 3:36–40

Caron F, Meyer E (1985) Does *Paramecium primaurelia* use a different genetic code in its macronucleus? Nature (London) 314:185–188

Cockburn AF, Taylor WC, Firtel RA (1978) *Dictyostelium* rDNA consists of non-chromosomal palindromic dimers containing 5 S and 36 S coding regions. Chromosoma 70:19–29

Cullis CA (1972) The basis of cell-to-cell transformation in *Paramecium bursaria.* II. Investigation into the molecular nature of the transforming agent. J Cell Sci 11:611–619

Cullis CA (1973) DNA amounts in the nuclei of *Paramecium bursaria.* Chromosoma 40:127–133

Cummings DJ (1972) Isolation and partial characterization of macro- and micronuclei from *Paramecium aurelia.* J Cell Biol 53:105–115

Cummings DJ (1975) Studies on macronuclear DNA from *Paramecium aurelia.* Chromosoma 53:191–208

Cummings DJ (1977) Methods for the isolation of nuclei from ciliated protozoans. In: Stein G, Stein J, Kleinsmith L (eds) Methods in cell physiology, vol 16. Academic Press, London New York, pp 97–112

Cummings DJ, Tait A (1975) The isolation of nuclei from *Paramecium aurelia.* In: Prescott DM (ed) Methods in cell biology, vol 9. Academic Press, London New York, pp 281–309

Cummings DJ, Tait A, Goddard JM (1974) Methylated bases in DNA from *Paramecium aurelia.* Biochim Biophys Acta 374:1–11

David ET, Smith KE (1981) Preparation and partial characterization of cell-free proteinsynthesizing extracts from *Tetrahymena pyriformis.* Biochem J 194:761–770

Doerfler W (1983) DNA methylation and gene activity. Annu Rev Biochem 52:93–124

Engberg J, Andersson P, Leick V, Collins J (1976) Free ribosomal DNA molecules from *Tetrahymena pyriformis* GL are giant palindromes. J Mol Biol 104:455–470

Epstein LM, Forney JD (1984) Mendelian and non-Mendelian mutations affecting surface antigen expression in *Paramecium tetraurelia.* Mol Cell Biol 4:1583–1590

Findley RC, Gall JG (1978) Free ribosomal RNA genes in *Paramecium* are tandemly repeated. Proc Natl Acad Sci USA 75:3312–3316

Findley RC, Gall JG (1980) Organization of ribosomal genes in *Paramecium tetraurelia.* J Cell Biol 84:547–559

Freiburg M (1985) Isolation and characterization of macronuclei of *Paramecium caudatum* infected with the macronucleus-specific bacterium *Holospora obtusa.* J Cell Sci 73:389–398

Gall JG (1984) Ciliates come of age. Nature (London) 310:453–454

Gibson I, Martin N (1971) DNA amounts in the nuclei of *Paramecium aurelia* and *Tetrahymena pyriformis.* Chromosoma 35:374–382

Gibson I, Chance ML, Williams JM (1971) Extranuclear DNA and the endosymbionts of *P. aurelia.* Nature New Biol 234:75–77

Gorovsky MA (1980) Genome organization and reorganization in *Tetrahymena.* Annu Rev Genet 14:203–239

Hanyu N, Kuchino Y, Nishimura S, Beier H (1986) Dramatic events in ciliate evolution: alteration of UAA and UAG termination codons to glutamine codons due to anticodon mutations in two *Tetrahymena* $tRNAs^{Gln}$. EMBO J 5:1307–1311

Helftenbein E (1985) Nucleotide sequence of a macronuclear DNA molecule coding for α-tubulin from the ciliate *Stylonychia lemnae.* Special codon usage: TAA is not a translation termination codon. Nucl Acids Res 13:415–433

Horowitz S, Gorovsky MA (1985) An unusual genetic code in nuclear genes of *Tetrahymena.* Proc Natl Acad Sci USA 82:2452–2455

Hruby DE, Maki RA, Cummings DJ (1977) Isolation and characterization of mRNA from *Paramecium aurelia.* Biochim Biophys Acta 477:89–96

Isaacks RE, Santos BG (1973) Studies on nuclei of *Paramecium aurelia.* II. Amino acid composition and electrophoretic properties of the chromosomal basic proteins. J Protozool 20:482–489

Karrer KM, Gall JG (1976) The macronuclear ribosomal DNA of *Tetrahymena pyriformis* is a palindrome. J Mol Biol 104:421–453

Katzen AL, Cann GM, Blackburn EH (1981) Sequence-specific fragmentation of macronuclear DNA in a holotrichous ciliate. Cell 24:313–320

Kraut H, Lipps HJ, Prescott DM (1986) The genome of hypotrichous ciliates. Int Rev Cytol 99:1–28

Kuchino Y, Hanyu N, Tashiro F, Nishimura S (1985) *Tetrahymena thermophila* glutamine tRNA and its gene that corresponds to UAA termination codon. Proc Natl Acad Sci USA 82:4758–4762

McGhee JH, Felsenfeld D (1980) Nucleosome structure. Annu Rev Biochem 49:1115–1156

McTavish C, Sommerville J (1980) Macronuclear DNA organization and transcription in *Paramecium aurelia.* Chromosoma 78:147–164

Meyer E, Caron F, Guiard B (1984) Blocking of *in vitro* translation of *Paramecium* messenger RNAs is due to messenger RNA primary structure. Biochimie 66:403–412

Meyer E, Caron F, Baroin A (1985) Macronuclear structure of the G surface antigen gene of *Paramecium primaurelia* and direct expression of its repeated epitopes in *Escherichia coli.* Mol Cell Biol 5:2414–2422

Nanney DL (1980) Experimental ciliatology. An introduction to genetic and developmental analysis in ciliates. Wiley & Sons, New York

Pasternak J (1967) Differential genic activity in *Paramecium aurelia.* J Exp Zool 165:395–418

Prat A, Katinka M, Caron F, Meyer E (1986) Nucleotide sequence of the *Paramecium primaurelia* G surface protein. A huge protein with a highly periodic structure. J Mol Biol 189:47–60

Preer JR, Jr, Preer LB (1979) The size of macronuclear DNA and its relationship to models for maintaining genic balance. J Protozool 26:14–18

Preer JR, Jr, Preer LB, Rudman BM (1981) mRNAs for the immobilization antigens of *Paramecium.* Proc Natl Acad Sci USA 78:6776–6778

Preer JR, Jr, Preer LB, Rudman BM, Barnett AJ (1985) Deviation from the universal code shown by the gene for surface protein 51 A in *Paramecium.* Nature (London) 314:188–190

Prince DJ, Cummings DJ, Seale RL (1977) Analysis of chromatin repeat units in logarithmically and stationary growing cells of *Paramecium aurelia* and *Tetrahymena pyriformis*. Biochem Biophys Res Commun 79:190–197

Rao MVN, Prescott DM (1967) Micronuclear RNA synthesis in *Paramecium caudatum*. J Cell Biol 33:281–285

Raikov IB (1982) The protozoan nucleus. Morphology and evolution. Springer, Berlin Heidelberg New York

Razin A, Szyf M (1984) DNA methylation patterns. Formation and function. Biochim Biophys Acta 782:331–342

Samuel C, Mackie J, Sommerville J (1981) Macronuclear chromatin organization in *Paramecium primaurelia*. Chromosoma 83:481–492

Schwartz V, Meister H (1975) Die Extinktion der feulgengefärbten Makronucleusanlage von *Paramecium bursaria* in der DNS-armen Phase. Arch Protistenk 117:60–64

Shiomi Y, Higashinakagawa T, Saiga H, Mita T (1980) Metrizamide isopycnic centrifugation for the isolation of macro- and micronuclei from *Paramecium*. J UOEH 2:323–330

Skoczylas B, Soldo AT (1975) Separation and purification of macronuclei from macronuclear fragments and micronuclei in the ciliate *Paramecium aurelia*. Exp Cell Res 90:143–152

Soldo AT, Godoy GA (1972) The kinetic complexity of *Paramecium* macronuclear deoxyribonucleic acid. J Protozool 19:673–678

Soldo AT, Brickson SA, Larin F (1981) The kinetic and analytical complexities of the DNA genomes of certain marine and fresh-water ciliates. J Protozool 28:377–383

Sommerville J, McTavish C (1982) The effect of temperature change on gene expression in *Paramecium primaurelia*. Biochim Biophys Acta 698:158–166

Sonneborn TM (1974) *Paramecium aurelia*. In: King RC (ed) Handbook of genetics, vol 2. Plenum, New York, pp 469–594

Steinbrück G (1986) Molecular reorganization during nuclear differentiation in ciliates. In: Hennig W (ed) Results and problems in cell differentiation, vol 13. Springer, Berlin Heidelberg New York Tokyo, pp 105–174

Steinbrück G, Haas I, Hellmer K-H, Ammermann D (1981) Characterization of macronuclear DNA in five species of ciliates. Chromosoma 83:199–208

Stevenson I (1967) A method for the isolation of macronuclei from *Paramecium aurelia*. J Protozool 14:412–414

Tait A, Cummings DJ (1975) DNA-dependent DNA polymerase activities from *Paramecia* macronuclei. Biochim Biophys Acta 378:282–295

Vogt VM, Braun R (1976) Structure of ribosomal DNA in *Physarum polycephalum*. J Mol Biol 106:567–587

Yao MC, Gall JG (1977) A single integrated gene for ribosomal RNA in a eucaryote, *Tetrahymena pyriformis*. Cell 12:121–132

Yao MC, Yao CH (1981) Repeated hexanucleotide C-C-C-C-A-A is present near free ends of macronuclear DNA of *Tetrahymena*. Proc Natl Acad Sci USA 78:7436–7439

Chapter 11 Immobilization Antigens

HELMUT J. SCHMIDT[1]

CONTENTS

1 Introduction

When Rössle (1905) discovered, more than 80 years ago, that homologous antiserum from a rabbit or guinea pig immobilizes and kills *Paramecium* cells he did not know that large proteins on the surface of paramecia cause this reaction. The proteins became later known as immobilization, i-antigens or surface proteins, and paramecia with specific i-antigens were called serotypes. Today, an explanation for the serotype system must be sought in terms of one of the central problems of modern cell biology, the control of phenotypic expression. *Paramecium* offers distinct advantages for the study of this problem. The cells exhibit a range of easily distinguishable, alternative types, which are generally mutually exclusive and can be made to change reversibly. Furthermore, biochemical compounds which are important for such studies (proteins, mRNAs and DNAs of structural genes) can be readily isolated.

The phenomenon has been examined extensively since the 1940s (Sonneborn 1943). Much of the early work was devoted chiefly to a description of the systems and a study of their genetics. Later, the emphasis of the investigations shifted to a characterization of the chemistry and molecular biology of the i-antigen molecules. Within the past few years attention has focussed on a detailed analysis of serotype proteins, their mRNAs and genes. Important new insights were gained, but many questions remain to be answered.

1 Zoologisches Institut der Westfälischen Wilhelms-Universität Münster, D-4400 Münster, FRG

Paramecium. Ed. by H.-D. Görtz

This chapter covers and summarizes our present knowledge about serotypes and the genetics of their expression, about transformation of serotypes and the localization of i-antigens in the cell, about their chemistry and possible functions, and about their molecular biology and what is known regarding the control of i-antigen expression.

Beale (1957), Finger (1974), Preer (1968, 1986), Sommerville (1970) and Sonneborn (1975) have written thorough reviews dealing with surface antigens of paramecia, sometimes with detailed and extensive discussions of aspects that can only be summarized here.

2 Serotypes

Immobilization serotypes have been found in all *Paramecium* species that were investigated for them such as *P. primaurelia, P. biaurelia, P. triaurelia, P. tetraurelia, P. octaurelia, P. nonaurelia, P. caudatum* and *P. multimicronucleatum* (Hiwatashi 1967; Sonneborn 1975; Steers and Barnett 1982). Except for *Tetrahymena* (for references on *Tetrahymena* surface proteins, see Williams et al. 1985) i-antigens have not been studied in other ciliates.

There is no single serotype for a given *Paramecium* stock and usually only few animals survive the treatment with antiserum; such resistant lines, when injected into rabbits, can be used to obtain antibodies specifically directed against their own antigens. All isogenic stocks of *Paramecium* display a spectrum of possible types. Up to a dozen different ones have been found for some stocks of *P. tetraurelia,* but no systematic search has ever been made to determine the complete array of types that can be produced. Perhaps the serologically most thoroughly investigated ciliate stock is stock 51 of *P. tetraurelia* which produces serotypes A, B, C, D, E, G, H, I, J, N and Q.

Sonneborn and Beale (see Beale, 1954) investigated the genetic basis for the ability of a stock to express a given serotype and clearly showed that alternative forms were due to differential expression of genes at different loci. Thus far, no two i-antigen loci have been found to be linked. The study of null alleles, such as F (Sonneborn 1950) and A (Epstein and Forney 1984) in stock 51 or H (Reisner 1955) in stock 169 confirmed the conclusion that the ability of a stock to establish a certain serotype is determined by Mendelian genes, one gene for each serotype. Within a species the antigens coded by genes at different loci are given different letters; those coded by different alleles have the same letter together with a stock number. The number of possible alleles at a given locus is unknown, but allelic variation is very common and greater than originally thought as many peptide mapping experiments, especially with *P. primaurelia* serotypes have shown (Jones and Beale 1963). Allelic differences are obvious for stocks isolated from different geographical locations.

Antisera can cross-react. The polymorphic locus for antigen D in *P. primaurelia,* for example, is serologically related to D of *P. tetraurelia,* and A, B and G are related in all species of the *P. aurelia* complex. Within *P. tetraurelia* serotypes A, B, G and Q as well as types D, J and M form related groups. The cross-reactions between different serotypes within a stock, however, are usually much weaker than the cross-reactions between homologous types in different stocks. *P. primaurelia,* where allelic differences are often very great, is an exception. In such cases genetic tests for allelism are sometimes required to prove that two types in different stocks are allelic (Beale 1957).

Capdeville et al. (1985) concluded for i-antigen G of *P. primaurelia* that allelic antigens share a great number of molecular surface epitopes even when they do not cross-react "in vivo" or only very weakly. Cross-reactions between antigens coded by different genes imply similar sequence determinants in their view. This supports earlier findings by Finger (1964) that allelic types share the same reactive sites on their i-antigen molecules, and by Preer (1959d), who showed that cross-reactions, unless caused by contaminating animals when sera are prepared, are often due to chemical relatedness. Another possible explanation for certain cross-reactions was proposed by Sonneborn (1947, 1948) when he postulated the presence of secondary antigens. Such antigens should be able to induce antibody formation, but not participate in the immobilization reaction. Secondary antigens occur sporadically in several stocks of *P. biaurelia* (Seed et al. 1964; see review by Finger 1974). Thus far they have never been detected in *P. tetraurelia* or *P. primaurelia.*

Mutual exclusion is a characteristic feature of the serotype system. A classical exception is stock 172 with serotypes D and M which can be expressed simultaneously for long periods of time (Margolin 1956). The same is true in clones heterozygous for two specificity alleles which generally do not display dominance; usually both alleles are expressed, although mutual exclusion between alleles in heterozygotes has been observed and named "allelic repression" (Finger and Heller 1964; Capdeville 1969). It was further found (Finger and Heller 1963) that some heterozygotes can produce hybrid antigens with specificities of both parental types. These antigens were truly hybrid molecules and not simply a mixture of parental molecules. Since i-antigens are single polypeptides, hybrid antigens are difficult to explain and remain an unresolved problem of serotype inheritance.

Serotype inheritance is clearly influenced cytoplasmically. After conjugation, with both exconjugants alike in genotype, the exconjugant clones are serologically like their parents. This has been shown for *P. tetraurelia* with crosses between paramecia expressing serotypes A and B (Preer 1968). Cells that have switched to the type of their partner can only be found in the exconjugant clones, when broad cytoplasmic bridges are allowed to form during conjugation. A strong cytoplasmic influence on serotype inheritance was also found for *P. primaurelia.* Beale (1952) demonstrated this in a classical analysis involving crosses of stocks 90 and 60 with serotypes G and D, which unlike A and B of *P. tetraurelia* cannot be stabilized under identical environmental conditions. The mechanism for this cytoplasmic influence on serotype expression is not understood.

3 Serotype Transformation

Serotype stability is influenced by the environment, but spontaneous transformation from one serotype to another is a common occurrence. In addition, such a change can be induced. Transformation induced in a culture generally does not act upon the selection of a few spontaneous transformants, but influences most of the individuals of the culture (Sonneborn 1943). Transformation is also reversible and does not involve gene mutations. Many inducing agents have been examined for numerous stocks and serotypes. It was discovered that the following factors induce transformation, but the list is doubtlessly incomplete: temperature, pH, salt concentration, type of culture medium, rate of feeding, antiserum, proteolytic enzymes, UV and X-irradiation, patulin and other antibiotics (for respective references, see Finger 1974).

One of the best known and most important factors is temperature. Certain types have temperature optima at high temperatures, such as A of *P. tetraurelia,* D of *P. primaurelia* or E of *P. biaurelia.* Other types are stable at lower temperatures, such as B and H of *P. tetraurelia* and G of *P. biaurelia.* When temperature is used to induce transformation, Beale (1957) noted that a sudden change is much more effective than a gradual change.

Another widely used transforming agent is homologous antiserum, which has long been known as a powerful inducer of serotype change. Even non-immobilizing antiserum prepared against fluid in which paramecia have lived is a good inducer of change (Finger 1974). Finger and Audi (1985) isolated certain proteins with a molecular weight of 70000 from such media as transforming agents which possibly transform serotypes by inhibiting antigen expression. On the other hand, it may be the actual removal of the old antigen by antiserum which plays a role in the serotype change, but, despite many theories, the action of the inducing agent in changing the serotype is still unknown for all of them.

Normally serotype transformation only becomes apparent after a certain amount of metabolism has produced enough i-antigen for an immobilization reaction, but with sensitive autoradiography techniques it was shown (for types 168G, D of *P. primaurelia*) that the new type i-antigen appears at an early stage (15 min) and the old type continues to be synthesized for at least 4 h (for a detailed discussion, see Sommerville 1970). Similar investigations in which the amounts of old and new i-antigens were measured during transformation showed that the original antigen generally decreases in quantity with its replacement increasing nearly proportionately. Once started, this process lasts from two to three fissions. Although cell division generally occurs during transformation, the synthesis of the new antigen does not necessarily require cell division. A patulin-induced change from 51D to 51B, for example, can take place without it (Austin et al. 1956).

4 Localization and Possible Functions

Since the earliest observations of the immobilization reaction, it was assumed that the antigens are localized on the surface of the cells. Indeed, most of the antigen appears as a thick coat of about 30 nm that covers the surface of the outer membrane of both cilia and the cell body. This was shown by Beale and co-workers (Beale and Mott 1962; Mott 1963, 1965) using labelled antibodies, conjugated with fluorescein or ferritin, and observed with the light microscope and, later, in sectioned material with the electron microscope. In addition, very low concentrations of i-antigen are found on internal membranes, on ribosomes (Sinden 1971, 1973) and in the medium in which paramecia have lived (Wyroba 1977). Thirty percent of the total ciliary proteins consist of i-antigen, which agrees with a layer of 20–30 nm thickness as found with the electron microscope (Preer 1968).

I-antigen was also estimated to represent 3.5% of total cellular protein, and more than 90% of the i-antigen can be easily sedimented by centrifugation, since it is apparently bound to cilia and outer cell membranes (Macindoe and Reisner 1967). A large portion of this is highly soluble and can be extracted with a dilute salt solution (Preer 1959 a–c). The remainder could only be extracted with detergent. Both forms of the antigen have similar mobilities in Triton X-100 polyacrylamide gel electrophoresis, and could be distinguished by precipitation with antiserum. Their roles are unclear (Hansma and Kung 1975). Similar proteins were studied in *P. primaurelia* by Capdeville et al. (1985, 1986). They concluded that the membranal form which requires detergents for its isolation is the physiological one, and that it can be converted into the soluble form by an enzymatic conversion also yielding two cross-reacting light chains of 40–45 kD. These light chains do not correspond to the 40-kD proteins found in ciliary membranes (see below).

Capdeville et al. (1986) also discovered that the soluble forms of surface proteins as well as the cross-reacting light chains share immunological homologies with *Trypanosoma* variant surface antigens, whereas such a homology could not be demonstrated for the membranal form.

I-antigen proteins from different serotypes vary in size and therefore in position on SDS-polyacrylamide gels, whereas usually all the other peptides seen in such gels have constant positions and intensities. In a study on ciliary proteins, however, it was discovered that three to five bands (molecular weights 40000 to 46000) vary in intensity but not in position from one serotype to the other (Adoutte et al. 1980; Merkel et al. 1981; Capdeville et al. 1985). Whether these peptides are related to or are important for i-antigen proteins is unknown.

Recent studies (Barnett and Steers 1984) showed that immobilization can only occur when, after adding an antiserum, a cross-linking of antigen-antibody complexes takes place; the mere binding of antibodies to antigen molecules is not sufficient. Barnett and Steers (1984) also observed the fusing of ciliary membranes at the tips of the cilia in the presence of homologous serum. Perhaps it is this membrane fusion that causes sticking and, later, the death of the cells rather than a cross-linking of adjacent cilia by divalent antibodies.

The biological function of the i-antigens remains obscure, but they appear to be indispensable. Paramecia lacking these surface proteins have never been found. Their importance is further suggested by the large number of alternate forms available to a cell, and by the observation that high concentrations of homologous antiserum not only immobilize the cells but subsequently kill them. Even dilute antiserum kills when protein synthesis is inhibited at the same time (Finger 1974). A possible function is that i-antigens regulate the interaction of certain substances with the membrane by providing a protective buffer between the cell membrane and the environment (see Preer 1986).

5 Chemistry and Molecular Biology of i-Antigens

5.1 Proteins

Surface antigens are high molecular weight proteins consisting of large, single polypeptides on the order of 300000 daltons. They have been isolated from *P. primaurelia, P. tetraurelia, P. caudatum* (Koizumi 1966) and *P. multimicronucleatum* (Steers and Barnett 1982). These highly soluble molecules can be extracted from whole cells in dilute salt alcohol. They are usually further purified by fractional ammonium sulphate precipitation after differential centrifugation at low and neutral pH (Preer 1959b), by ion exchange chromatography (Jones 1965) or on antibody affinity columns (Davis and Steers 1978). Isolated proteins can be identified with specific bands in immunodiffusion gels (Finger 1956). The proteins are not simply different conformations of similar polypeptides, because tryptic peptide digestions resulted in clearly different peptide maps for i-antigens from different serotypes (Steers 1962; Jones 1965). The same was observed when the proteins were treated with cyanogen bromide. The obtained banding pattern on SDS-polyacrylamide gels for i-antigens of *P. primaurelia* and *P. tetraurelia* was characteristic for each i-antigen (Schmidt 1983). This work confirmed the early findings from amino acid analysis that surface proteins are unusually low in methionine. On the other hand, they are very rich in cysteine residues with apparently all the –SH groups bound in cystine (Reisner et al. 1969).

For many years several groups discussed a subunit structure for i-antigens (see review by Sommerville 1970), but such subunits do not exist. The i-antigen preparations often contain a contaminant protease that can be easily activated by mercaptoethanol, the very agent used to study the subunits. This protease has been isolated (Davis and Steers 1978). It can be inactivated by boiling, and, if it is, no subunits are seen.

Merkel et al. (1981) found that lectins recognizing glucose and mannose sugars bind specifically to surface proteins of *P. tetraurelia,* stock 51. This result substantiates the early discovery of a small carbohydrate content in the i-antigen molecules (Preer 1959c).

5.2 Nucleic Acids

Preer and co-workers (Preer et al. 1981; Forney et al. 1983) investigated the mRNAs for 11 serotypes from stock 51 of *P. tetraurelia.* They consistently found a single, very large mRNA whose size is large enough to synthesize the respective i-antigen. Their evidence that these RNAs are the mRNAs for the i-antigens is still indirect, but very strong. Similar RNAs were studied for G and D serotypes from *P. primaurelia* (Sommerville 1983; Meyer et al. 1984). Several structural genes for surface proteins (51A, 51C, 51G, 51H, 156G) (Preer 1986) have been isolated from genomic libraries. This was either done by isolating the mRNAs, labelling them and using them immediately as probes for screening the genomic libraries or by preparing a cDNA library first (with mRNA from a given serotype) and screening the genomic libraries later with isolated cDNA clones (Meyer et al. 1985). A verification of a clone as an i-antigen gene was accomplished by showing that it hybridizes on a Northern blot to an RNA band from cells of the proper serotype and not to RNA from cells of other serotypes (Forney et al. 1983). Schmidt (1987) used the same method to study clones with putative D genes. He found two types of "genes", entirely different in their restriction and transcription maps, but both with a strong hybridization reaction to poly A^+ RNA from serotype 51D. It seems likely that one of the clones contains the 51D gene. Whether the other clone belongs to a pseudogene or a related, strongly cross-reacting i-antigen gene is still unclear.

The following general conclusions can be drawn from the present characterization of i-antigen genes:

1. The genetic and molecular analysis of deletion mutations provides strong evidence that the cloned genes are really the i-antigen genes.
   One of the deletions, where the DNA beginning at the 5′ end of the gene was missing, is incapable of expressing serotype A (Epstein and Forney 1984). This deletion is allelic with the specificity locus for serotype A in *P. tetraurelia* (Preer et al. 1985).
2. The transcribed segments are of the right size to code for the mRNAs. Introns have never been found, and large introns are certainly absent.
3. There is probably only one copy of each i-antigen gene per genome (Forney et al. 1983).
4. The two most extensively studied genes 51A and 156G are located in the vicinity of the end of a macronuclear "chromosome" oriented with their 3′ ends towards the terminus (Preer 1986).
   This is an interesting parallel to surface antigens from trypanosomes where expression linked copies of antigen genes have been found at chromosomal ends (Myler et al. 1984). Preer (1986), admitting that only speculations are possible on the likely function of such a terminal position in paramecia, nevertheless deduces that this location cannot be attributed to chance.
5. Serotype expression does not seem to involve DNA rearrangements.
   Rearrangements of DNA have been shown to occur concomitantly with shifts in gene expression in trypanosomes (Myler et al. 1984), *Salmonella* (Zieg et al. 1978), *Neisseria gonorrhoeae* (Meyer et al. 1982) and yeast (Strathern et al. 1980). Rearrangements were postulated (Brown 1981) for *Paramecium*, too, but are doubtful due to distinct peculiarities of the serotype system. The speed of serotype transformation, the high level of ploidy in the macronucleus and the fact that nearly all cells in a culture participate in the transformation process are difficult to reconcile with a regulation of gene expression by DNA rearrangements. Indeed, rearrangements of the sizes typical for the other organisms have never been found for *Paramecium*, despite numerous

experiments designed to detect them (Forney et al. 1983; Meyer et al. 1985; Schmidt 1987).

6. Serotype expression can be influenced by genetic loci outside of the i-antigen genes.
   It was believed until recently that the serotype genes themselves were the main, perhaps only, genes affecting serotype stability (Sonneborn 1975). Then, Epstein and Forney (1984) isolated X-ray induced mutations with a decreased stability for serotype A, and increased frequencies of serotypes C and E. The mutations were at separate loci, unlinked with the A locus, and the A gene was completely present. These were the first identified loci outside of the i-antigen genes which affect serotype expression.

Finally, work on i-antigen genes led to two important discoveries which are interesting for our understanding of serotypes, but very likely of more general importance.

One mutant with a deletion of the A gene (d48) produced a remarkable result when crossed to stock 51, when Epstein and Forney (1984) demonstrated that only the macronucleus of this mutant lacks the A gene, while the micronucleus contains it. Such a pattern of inheritance parallels precisely other cytoplasmically inherited traits in *Paramecium,* such as the killer trait or mitochondrially determined traits, or the inheritance of mating types in *P. tetraurelia* or the trichocyst non-discharge character (Sonneborn and Schneller 1979). The discovery of this mutation is important for the general understanding of macronuclear differentiation in *Paramecium.* Further evidence for the presence of the normal A gene in the micronucleus of mutant d48 was provided by Harumoto (1986). She showed that transfer of macronucleoplasm or, though less effective, of cytoplasm from wild type to d48 caused d48 to revert to wild type after autogamy. Therefore, the hereditary determinants responsible for the d48 trait seem to be located in the macronucleus, and act by producing cytoplasmic factors determining the differentiation of the new macronucleus.

The first sequences available for any DNA of *Paramecium* were obtained for two antigen genes (Caron and Meyer 1985; Preer et al. 1985). The surprising observation was the absence of open reading frames because TAA codons were found approximately every 120 bases. TAG, though more rarely, was also found. Both, TAA and TAG are "universal" translation termination codons. Similar observations were made at the same time for a tubulin gene of another ciliate, *Stylonychia* (Helftenbein 1985). It is now generally assumed that ciliates deviate from the universal genetic code by using TAA and TAG to code for glutamine instead of being stop signals. Thus far, UGA is the only termination codon that has been found in any ciliate (see also Chap. 10 by Freiburg, this Vol.).

For a number of years several groups have reported difficulties with "in vitro" translation systems for ciliate mRNAs. With one exception (Sommerville 1983) a high percentage of polypeptides synthesized "in vitro" is of discrete but very small sizes. Meyer et al. (1984) were the first to exclude simple explanations for this result, and concluded that the primary structure of the ciliate RNA itself is responsible for this finding. The unexpected DNA sequencing results strongly support this conclusion (Caron 1986). Recently, the complete sequence of the G-antigen of *P. primaurelia* has been determined (Prat et al. 1986). It is free of introns, contains an open reading frame of 8145 nucleotides and codes for a protein of 329000 daltons. Its 5′-coding sequence is very homologous to a sequenced region of the A gene from *P. tetraurelia*. On the other hand, surprisingly, neither eukaryotic transcription (TATA, CAAT boxes) nor polyadenylation signals were found. The deduced amino acid sequence reveals a monotonous amino acid distribution which could be expected from a protein that has a structural role and no known enzymatic function. This putative protein has a high degree of internal homology, and also shares a homology with a segment near the carboxy terminus of the mature protein from surface antigens of trypanosomes.

6 Conclusion

Several theories have been discussed to explain the expression of serotypes in *Paramecium* (Finger 1974; Capdeville 1979). A common feature of these models is the involvement of regulatory molecules specific for the i-antigen that is being expressed, and perhaps the i-antigens themselves function as such regulatory molecules. Nevertheless, none of these models is conclusively supported by data. Our knowledge about the molecular biology of serotype mRNAs and genes contributes two main perceptions to the perhaps most important question concerning the i-antigens, namely the mechanism of their mutual exclusion. Serotype control does not seem to use rearrangements of DNA, and occurs at the level of transcription or RNA processing, but no further insights were gained. The same applies to the question on the function of these surface proteins.

Since both of these questions still need to be answered, and encouraged by the rapid progress made within the past few years, it is certain that further investigation of immobilization antigens will continue at a high pace. Presently, *Paramecium* appears to be the best-suited ciliate for such studies.

Acknowledgements. The author wishes to thank Prof. J. R. Preer, Jr. for his helpful comments on the manuscript. This work was supported by the Deutsche Forschungsgemeinschaft (DFG No. Schm 544/2-1,2,3).

References

Adoutte A, Ramanathan R, Lewis RM, Dute RR, Ling K-Y, Kung C, Nelson DL (1980) Biochemical studies of the excitable membrane of *Paramecium tetraurelia*. III. Proteins of cilia and ciliary membranes. J Cell Biol 84:717–738

Austin ML, Widmayer D, Walker LM (1956) Antigenic transformation as adaptive response of *Paramecium aurelia* to patulin; relation to cell division. Physiol Zool 29:261–287

Barnett A, Steers E, Jr (1984) Antibody-induced membrane fusion in *Paramecium*. J Cell Sci 65:153–162

Beale GH (1952) Antigen variation in *Paramecium aurelia*, variety 1. Genetics 37:62–74

Beale GH (1954) The genetics of *Paramecium aurelia*. Cambridge University Press, New York

Beale GH (1957) The antigen system of *Paramecium aurelia*. Int Rev Cytol 6:1–23

Beale GH, Mott MR (1962) Further studies on the antigens of *Paramecium aurelia* with the aid of fluorescent antibodies. J Gen Microbiol 17:68–74
Brown DB (1981) Gene expression in eukaryotes. Science 211:667–674
Capdeville Y (1969) Sur les interactions entre allèles controlant le type antigénique G chez *Paramecium aurelia.* CR Acad Sci Paris 269:1213–1215
Capdeville Y (1979) Intergenic and interallelic exclusion in *Paramecium primaurelia:* immunological comparisons between allelic and non-allelic surface antigens. Immunogenetics 9:77–95
Capdeville Y, Deregnaucourt C, Keller A-M (1985) Surface antigens of *Paramecium primaurelia:* membrane-bound and soluble forms. Exp Cell Res 161:495–508
Capdeville Y, Baltz T, Deregnaucourt C, Keller A-M (1986) Immunological evidence of a common structure between *Paramecium* surface antigens and *Trypanosoma* variant surface glycoproteins. Exp Cell Res 167–175
Caron F (1986) Deviations from the "universal" genetic code. Microbiol Sci 3/2:36–40
Caron F, Meyer E (1985) Does *Paramecium primaurelia* use a different genetic code in its macronucleus? Nature (London) 314:185–188
Davis RH, Jr and Steers E, Jr (1978) Purification of the i-antigen 51A from *Paramecium tetraurelia* by immunoaffinity chromatography. Immunochemistry 15:371–378
Epstein LM, Forney JD (1984) Mendelian and non-mendelian mutations affecting surface antigen expression in *Paramecium tetraurelia.* Mol Cell Biol 4:1583–1590
Finger I (1956) Immobilizing and precipitating antigens of *Paramecium.* Biol Bull 111:358–363
Finger I (1964) Use of simple gel-diffusion techniques to assign antigenic markers to native proteins. Nature (London) 203:1035–1039
Finger I (1974) Surface antigens of *Paramecium aurelia.* In: Van Wagtendonk WJ (ed) *Paramecium,* a current survey. Elsevier, New York, pp 131–164
Finger I, Audi D (1985) Isolation of extracellular inhibitors of *Paramecium* surface antigen expression. J Cell Biol 101:76a
Finger I, Heller C (1963) Immunogenetic analysis of proteins of *Paramecium.* IV. Evidence for presence of hybrid antigens in heterozygotes. J Mol Biol 6:190–202
Finger I, Heller C (1964) Cytoplasmic control of gene expression in *Paramecium* I. Preferential expression of a single allele in heterozygotes. Genetics 49:485–498
Forney JD, Epstein LM, Preer LB, Rudman BM, Widmayer DJ, Klein WH, Preer JR Jr (1983) Structure and expression of genes for surface proteins in *Paramecium.* Mol Cell Biol 3:466–474
Hansma HG, Kung C (1975) Studies on the cell surface of *Paramecium.* Ciliary membrane proteins and immobilization antigens. Biochem J 152:523–528
Harumoto T (1986) Induced change in a non-Mendelian determinant by transplantation of macronucleoplasm in *Paramecium tetraurelia.* Cell Mol Biol 6:3498–3501
Helftenbein E (1985) Nucleotide sequence of a macronuclear DNA molecule coding for alpha-tubulin from the ciliate *Stylonychia lemnae.* Special codon usage: TAA is not a translation termination codon. Nucl Acids Res 13:415–423
Hiwatashi K (1967) Serotype inheritance and serotypic alleles in *Paramecium caudatum.* Genetics 57:711–717
Jones IG (1965) Studies on the characterization and structure of the immobilization antigens of *Paramecium aurelia.* Biochem J 96:17–23
Jones IG, Beale GH (1963) Chemical and immunological comparisons of allelic immobilization antigens of *Paramecium aurelia.* Nature (London) 197:205–206
Koizumi S (1966) Serotypes and immobilization antigens in *Paramecium caudatum.* J Protozool 13:73–76
Macindoe H, Reisner AH (1967) Adsorption titration as a specific semi-quantitative assay for soluble and bound *Paramecium* serotypic antigen. Aust J Biol Sci 20:141–152
Margolin P (1956) An exception to mutual exclusion of the ciliary antigens in *Paramecium aurelia.* Genetics 41:685–699
Merkel SJ, Kaneshiro ES, Gruenstein EI (1981) Characterization of the cilia and ciliary membrane proteins of wild-type *Paramecium tetraurelia* and a pawn mutant. J Cell Biol 189:206–215

Meyer TF, Mlawer N, So M (1982) Pilus expression in *Neisseria gonorrhoeae* involves chromosomal rearrangement. Cell 30:45–52

Meyer E, Caron F, Guiard B (1984) Blocking of in vitro translation of *Paramecium* messenger RNAs is due to messenger RNA primary structure. Biochimie 66:403–432

Meyer E, Caron F, Barion A (1985) Macronuclear structure of the G surface antigen gene of *Paramecium primaurelia* and direct expression of its repeated epitopes in *Escherichia coli.* Mol Cell Biol 5:2414–2422

Mott MR (1963) Cytochemical localization of antigens of *Paramecium* by ferritin-conjugated antibody and by counterstaining the resultant adsorbed globin. J R Microsc Soc 81:159–162

Mott MR (1965) Electron microscopy studies on the immobilization antigens of *Paramecium aurelia.* J Gen Microbiol 41:251–261

Myler PJ, Allison J, Agabian N, Stuart K (1984) Antigenic variation in African trypanosomes by gene replacement or activation of alternate telomeres. Cell 39:203–211

Prat A, Katinka M, Caron F, Meyer E (1986) Nucleotide sequence of the *Paramecium primaurelia* G surface protein. A huge protein with a highly periodic structure. J Mol Biol 189:47–60

Preer JR, Jr (1959a) Studies on the immobilization antigens of *Paramecium.* I. Assay methods. J Immunol 83:276–283

Preer JR, Jr (1959b) Studies on the immobilization antigens of *Paramecium.* II. Isolation. J Immunol 83:378–384

Preer JR, Jr (1959c) Studies on the immobilization antigens of *Paramecium.* III. Properties. J Immunol 83:385–391

Preer JR, Jr (1959d) Studies on the immobilization antigens of *Paramecium.* IV. Properties of the different antigens. Genetics 44:803–814

Preer JR, Jr (1968) Genetics of the protozoa. In: Chen TT (ed) Research in protozoology 3. Pergamon, Oxford, pp 129–278

Preer JR, Jr (1986) Surface antigens of *Paramecium.* In: Gall J (ed) Molecular biology of the ciliated Protozoa. Academic Press, London New York, pp 301–339

Preer JR, Jr, Preer LB, Rudman BM (1981) mRNAs for the immobilization antigens of *Paramecium.* Proc Natl Acad Sci USA 78:6776–6778

Preer JR, Jr, Preer LB, Rudman BM, Barnett AJ (1985) Deviation from the universal code shown by the gene for surface protein 51A in *Paramecium.* Nature (London) 314:188–190

Reisner AH (1955) A method for obtaining specific serotype mutants in *Paramecium aurelia* stock 169, var. 4. Genetics 40:591–592

Reisner AH, Rowe J, Sleigh R (1969) Concerning the tertiary structure of the soluble surface proteins of *Paramecium.* Biochemistry 8:4637–4644

Rössle R (1905) Spezifische Seren gegen Infusorien. Arch Hyg Bakteriol 54:1–31

Schmidt HJ (1983) Identification of the immobilization antigens of *Paramecium* by their cyanogen bromide cleavage patterns. J Protozool 30:577–580

Schmidt HJ (1987) Characterization and comparison of genomic DNA clones containing complementary sequences to mRNA from serotype 51D of *Paramecium tetraurelia.* Mol Gen Genet 208:450–456

Seed JR, Shafer S, Finger I, Heller C (1964) Immunogenetic analysis of proteins of *Paramecium.* VI. Additional evidence for the expression of several loci in animals of a single antigenic type. Genet Res 5:137–140

Sinden RE (1971) The synthesis of immobilization antigen in *Paramecium aurelia:* in situ localization of immobilization antigen using fluorescein- or ferritin-conjugated antibodies. J Microsc 93:129–144

Sinden RE (1973) The synthesis of the immobilization antigen in *Paramecium aurelia:* in vitro localization of antigen in ribosomal cell fractions. J Protozool 20:307–315

Sommerville J (1970) Serotype expression in *Paramecium.* Adv Microbiol Physiol 4:131–178

Sommerville J (1983) Isolation and translation of mRNA coding for the variant surface antigens of *Paramecium.* Nucl Acids Res 11:7375–7385

Sonneborn TM (1943) Acquired immunity to specific antibodies and its inheritance in *P. aurelia.* Proc Indiana Acad Sci 52:190–191

Sonneborn TM (1947) Developmental mechanisms in *Paramecium*. Growth Symp 11: 291–307

Sonneborn TM (1948) The determination of hereditary antigenic differences in genetically identical *Paramecium* cells. Proc Natl Acad Sci USA 34:413–418

Sonneborn TM (1950) The cytoplasm in heredity. Heredity 4:11–36

Sonneborn TM (1975) *Tetrahymena pyriformis. Paramecium aurelia.* In: King RC (ed) Handbook of genetics 2. Plenum, New York, pp 433–594

Sonneborn TM, Schneller MV (1979) A genetic system for alternative stable characteristics in genomically identical homozygous clones. Dev Genet 1:21–46

Steers E, Jr (1962) A comparison of the tryptic peptides obtained from immobilization antigens of *Paramecium aurelia.* Proc Natl Acad Sci USA 48:867–874

Steers E, Jr, Barnett A (1982) Isolation and characterization of an immobilization antigen (C) from *Paramecium multimicronucleatum.* Comp Biochem Physiol 71 B:217–222

Strathern JN, Spatola E, McGill C, Hicks JB (1980) Structure and organization of transposable mating type cassettes in *Saccharomyces* yeasts. Proc Natl Acad Sci USA 77:2839–2843

Williams NE, Doerder FP, Ron A (1985) Expression of a cell surface immobilization antigen during serotype transformation in *Tetrahymena thermophila.* Mol Cell Biol 5:1925–1931

Wyroba E (1977) Studies on the surface coat of *Paramecium aurelia.* II. Relationship to the immobilization antigen. Cell Tissue Res 181:245–253

Zieg J, Hilmen M, Simon M (1978) Regulation of gene expression by site-specific inversion. Cell 15:237–244

Chapter 12 Mitochondria

ANNIE SAINSARD-CHANET[1] and DONALD CUMMINGS[2]

CONTENTS

1 Introduction

Studies of protozoa have made an important contribution to the early history of mitochondrial biogenesis (see Swift and Wolstenholme 1969). Mitochondrial DNAs of *Tetrahymena* and *Paramecium* were among the first mitochondrial DNAs isolated (Suyama and Preer 1965) and the study of the mitochondrial genetics of *Paramecium* started as early as 1969 with the isolation of drug-resistant mutations (Beale 1969).

In recent years the knowledge of the yeast mitochondrial genome has reached a high degree of sophistication. Admittedly the data from *Paramecium* are far less extensive but these data as those from other orgamisms (filamentous fungi, animal cells, etc.) have revealed substantial differences in regard both to the organization and the functions of the mitochondrial genome. This review shall draw attention to what distinguishes *Paramecium* from other organisms. It is subdivided into three sections dealing successively with the genetics, the respiratory chain, and the molecular biology of the genome. It must be noted that

1 Centre de Génétique Moléculaire, F-91190 Gif-sur-Yvette, France
2 Department of Microbiology, University of Colorado Medical Center, Denver, CO 80262, USA

Paramecium. Ed. by H.-D. Görtz

a very good review on mitochondrial genetics of *Paramecium* had already been published in 1981 by Beale and Tait.

2 Genetics

With respect to mitochondrial genetics, *Paramecium* offers several advantages. (1) During the process of conjugation, haploid gamete nuclei pass from one conjugating partner into another, while the cytoplasm of the conjugants usually does not mix. However, in some cases, cytoplasmic exchanges occur through bridges between the two conjugants as shown in Fig. 1. This permits us to easily follow the transmission of mitochondrial markers and to study situations of nucleomitochondrial interactions. (2) The large size of the cells has enabled us to develop microinjection techniques of mitochondria (Knowles 1974).

2.1 *Mitochondrial Mutations*

Most of the mitochondrial mutations have been selected in *Paramecium aurelia.* They can be classified into two groups: (1) drug-resistant mutations; (2) suppressor mutations of a nuclear gene (*c11*).

Drug-Resistant Mutations. Mutants resistant to inhibitors of protein synthesis on mitochondrial ribosomes such as erythromycin (E), chloramphenicol (C), and spiramycin (S) have been obtained both in species 1 and 4 (Beale 1969; Adoutte and Beisson 1970; Beale et al. 1972; Adoutte 1974, 1977). Erythromycin-resistant mutants have also been selected in species 5 and 7 (Beale and Knowles, 1976) and mikamycin-resistant mutants in species 1 (Beale 1973). Mutants simultaneously resistant to two antibiotics (C^RE^R and M^RE^R) have been obtained by two-step selection procedures. Tests with some 20 other inhibitors of mitochondrial functions have failed to give any mutant (Adoutte 1977).

Besides the resistance itself, these mutations may have pleiotropic effects bearing mainly on cross-resistance with other antibiotics, thermosensitivity, growth rate, and decreased level of cytochrome oxidase (Beale 1973; Adoutte 1974; Adoutte and Doussière 1978).

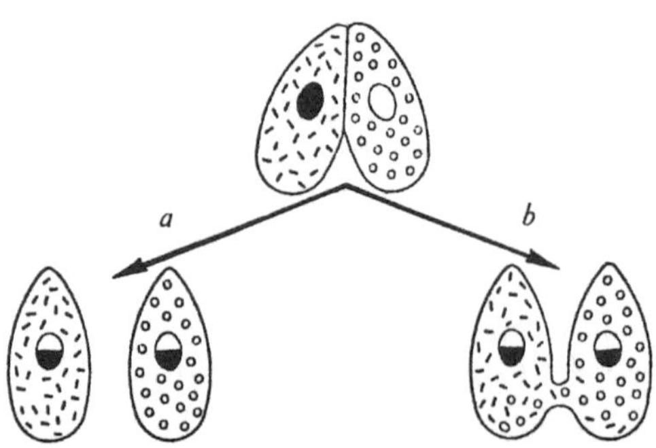

Fig. 1. Diagram showing conjugation with (*b*) and without (*a*) cytoplasmic exchange. The cytoplasmic properties of the two strains are symbolized by *lines* and *circles*

Genetic analysis has shown that all these mutations are cytoplasmically inherited (Beale 1969, 1973; Adoutte and Beisson 1970) and microinjection experiments demonstrated that they are mitochondrial. Injection of mitochondria from a resistant strain into sensitive cells and exposure of the recipients to the drug results in "transformation" of the sensitive cells into resistant ones (Beale et al. 1972; Knowles 1972, 1974; Beale 1973). This process of "transformation" has been studied cytologically and genetically both on microinjected cells (Knowles 1972) and on erythromycin-sensitive cells which have received erythromycin-resistant mitochondria by a cytoplasmic bridge at conjugation (Perasso and Adoutte 1974). Under the selective pressure of the antibiotic, the recipient paramecia pass through an intermediate stage containing a mixture of the two types of mitochondria and after a few fissions, become pure for resistant mitochondria. Direct proof that the cytoplasmic factors controlling drug resistance are mitochondrial mutations was obtained by Maki and Cummings (1977), who showed after injection of resistant mitochondria from one species into sensitive cells of another species that the mitochondrial DNA of transformed cells has the same restriction pattern as the donor species. Furthermore, this result allows us to conclude that the transformation of the recipient strain to antibiotic resistance does not involve a genetic recombination between the mitochondrial genomes of the donor and the recipient, but an elimination of the sensitive mitochondria and replacement by the resistant ones.

At least some of these antibiotic resistance mutations seem to affect mitoribosomal properties since mitoribosomes of some erythromycin-resistant strains isolated in species 1 fail to bind radioactive erythromycin in vitro (Tait 1972). In addition, most of the mutants display a decreased amount of cytochrome oxidase (Adoutte and Doussière 1978) that can be attributed to a decreased efficiency of mitochondrial protein synthesis. In terms of their sensitivity to inhibitors, mitoribosomes of *Paramecium* resemble those of yeast (Tait and Knowles 1977) and by analogy with the antibiotic-resistant mutants of yeast (Dujon 1983), the antibiotic-resistant mutations of *Paramecium* could be located in the large mitoribosomal RNA gene. It must be noted, however, that the restriction map of the mitochondrial DNA of at least one E^R mutant differs from the wild type in a region different from the rRNA gene (Cummings and Laping, 1981). If the erythromycin-resistant phenotype is correlated with this modification, this would represent the single example of a mitochondrial E^R mutation located in a gene other than an rRNA gene.

Suppressors of the cl1 Nuclear Mutation. A nuclear mutation, *cl1*, responsible for a severe disorganization of mitochondria, thermosensitivity, and a very reduced growth rate has been isolated in *Paramecium tetraurelia* (Sainsard et al. 1974). This mutation is interesting in that the incompatibility between the nuclear mutant genome and mitochondria is not final. During growth, all the cells improve their growth rate. The genetic analysis of this nucleomitochondrial readjustment showed that it results from two types of mitochondrial modification: reversible mitochondrial changes whose nature has not been elucidated and mitochondrial mutations partially suppressing the deleterious effect of gene *cl1* (Sainsard-Chanet and Knowles 1979). Several different mitochondrial mutations (M^{cl}, M^{su}, M^*), differing in their degree of suppression have been described (Sainsard 1975, 1978, 1979). They arise spontaneously with a frequency of the order of that of spontaneous mitochondrial erythromycin-resistant mutations (Queiroz and Beale 1974) and are subject to positive selection in the presence of the nuclear *cl1* genome.

The cells carrying these mutations (M^{cl}, M^{su}) in combination with a mutated *cl1* nucleus have a growth rate depending on the suppressor. Their mitochondria are completely devoid of cytochrome *aa3* during the exponential phase of growth and they display a high level of cyanide-insensitive respiration (80% instead of 10–20% for wild type; see Sect. 3, Doussière et al. 1979a, b). In combination with a wild-type nuclear genome, these mutations slightly reduce the cellular growth rate and are responsible of a partial deficiency in cytochrome oxidase. Furthermore, these mutants show a polymorphism in the initiation region of the mitochondrial DNA (Cummings and Laping 1981). This polymorphism is not due to a modified sequence of the initiation region, but to variations in the number of repeat units in this region (Pritchard et al. 1983, see Sect. 4).

It has been suggested that both the nuclear *cl1* mutation and its mitochondrial suppressors could alter mitochondrial products of dual nucleomitochondrial origin such as the mitoribosome or the mitochondrial membrane and that the suppression could result from functional interactions between the nuclear and the mitochondrial modified products (Sainsard-Chanet and Knowles 1979). However, no molecular data have allowed us to verify this hypothesis.

2.2 Interactions Between Nucleus and Mitochondria

It is well known that the biogenesis of mitochondria is the result of an intricate interplay between the nuclear and the mitochondrial genomes. A specifically clear illustration of this interplay is furnished by the analysis of the nuclear mutation *cl1* and its mitochondrial suppressors which restore nucleomitochondrial harmony as described above. The role of the nuclear genome on mitochondrial multiplication is also shown by the study of the evolution of mixtures of mutated (M^{cl}) and wild-type (M^+) mitochondria. Such mixtures obtained by cytoplasmic bridges in *cl1/cl1*$^+$ heterozygotes evolve quite differently according to the nuclear segregation in the second division.

The same mixed cytoplasm becomes rapidly pure for M^{cl} or M^+ mitochondria, whereby a *cl1/cl1* or a *cl1/cl1*$^+$ nucleus is established (Sainsard-Chanet 1976).

The role of the nucleus on mitochondrial multiplication has also been illustrated by microinjecting mitochondria from one species into another. In this way, Beale and Knowles (1976) showed that E^R mitochondria from species 1 and 5 can be transferred into species 1, 5, and 7. In contrast, mitochondria from species 7 usually failed to develop in cells containing species 1 or 5 nuclei. Moreover, mitochondria from hybrid cells containing species 7 nuclei and, species 1 or 5 mitochondria were not transferable to species 1 or 5 cells as if mitochondria were modified by the nucleus of species 7.

A very great diversity in the initiation region of the mitochondrial DNA has been observed between these species (Pritchard et al. 1983). It has been postulated that this diversity could be the crucial factor for successful interspecies transfer mitochondria; the nuclear enzyme involved in replication may not be able to recognize the initiation origin of the interspecies mitochondrial DNA.

Finally, a number of clear positive and negative interactions between nuclear mutations conferring resistance to tetrazolium (Ruiz and Adoutte 1978) and mitochondrial mutations ranging from effects on the replication rate or selective advantage of particular mitochondrial mutant types to suppression have been found in *P. tetraurelia* (Ruiz and Knowles 1980).

2.3 Autonomy of Mitochondria, Absence of Complementation, and Intermolecular Recombination

A characteristic trait of the mitochondrial genetic system of *Paramecium* is its very great functional autonomy. This can be shown by the following facts: (1) In a wild-type nuclear context, when cells containing two genetically distinct types of mitochondria are grown in normal medium, one always observes the progressive loss of one of the two types. The kinetics of loss strictly depends on the markers used. It can vary from ten to several hundred cellular generations (Adoutte and Beisson 1972; Adoutte and Doussière 1978). Furthermore, the replication rate of mutated mitochondria in mixed cytoplasm is very similar to that of these mitochondria in pure mutant cytoplasm indicating a complete absence of interaction in the mixed cytoplasm (Adoutte and Doussière 1978). (2) Cells containing both erythromycin- and chloramphenicol-resistant mitochondria are resistant to either antibiotic added separately but cannot grow when both antibiotics are present, indicating that there is no complementation between mitochondria at least for these mutations (Adoutte and Beisson 1972; Beale et al. 1972). (3) Recombination has never been observed between mitochondrial markers. Thirty-two combinations involving 24 distinct mitochondrial markers in several genetic configurations (mutant × mutant, double mutant × wild type) were studied (Adoutte et al. 1979). The markers were either antibiotic-resistant mutations displaying or not displaying thermosensitivity or slow growth and the suppressor mutations of gene *c11*. None of the combinations yielded any evidence for the occurrence of recombined genotypes. The absence (or the extreme rarity) of recombination in the mitochondria of *Paramecium* is in marked contrast to what happens in a set of organisms such as yeast and filamentous fungi. This property is interesting from an evolutionary point of view but unfortunately imposes serious limits to the use of *Paramecium* for mitochondrial genetics. The reasons for this lack of recombination are not known. It could be due to the fact that no "hybrid" mitochondria containing the two types of DNA are formed. Indeed, *Paramecium* mitochondria appear at all stages of their life cycle as discrete, independent entities as opposed to the formation of extensive networks described in yeast (Perasso and Beisson 1978). One other hypothesis could be that "hybrid" mitochondria are formed but no recombination occurs. Recently, a novel activity of intron-encoded proteins, concerning the induction of genetic recombination and referred to as "DNA recombinase", has been found in yeast (Kotylak et al. 1985). These authors suggest that only genomes which contain introns and intron-encoded proteins can evolve, while genomes which have lost these introns during evolution, have consequently lost the property of genetic recombination and are now frozen in evolution. Now, although *Paramecium* contains intervening sequences in its mitochondrial genome, these do not seem to code for intron-related proteins (see Sect. 4).

Finally, it must be noted that although intermitochondrial recombination seems to be rare or nonexistent in *Paramecium,* intramitochondrial recombination may well occur. Such a recombination mechanism has been proposed by Pritchard et al. (1983) to explain the generation of the repeat units present in the region of the mitochondrial DNA which contains the replication origin.

3 Respiratory Chain

Although the mitochondrial chain of protozoa carries out oxidative phosphorylation with the same efficiency as the respiratory chain of mammalian mitochondria, marked differences between them appear to exist. The protozoa which have been studied in this respect include *Tetrahymena pyriformis* (Lloyd and Chance 1972; Kilpatrick and Erecinska 1977), a number of trypanosomatids (Hill 1978), and *Acanthamoeba castellani* (Edwards et al. 1977).

P. tetraurelia is another example of a protozoan with unusual properties of the components involved in mitochondrial respiration. Its respiration is partially insensitive to cyanide and it possesses in addition to a cytochromic standard respiratory chain, two alternative pathways to molecular oxygen: a cyanide-resistant oxidase like many higher plants and microorganisms (for review: Henry and Nyns 1975; Lloyd and Edwards 1978) and an autoxidizable *b*-type cytochrome (Doussière et al. 1979a, b). *Paramecium* offers an advantage over higher systems because its alternative and standard pathways can be manipulated genetically or environmentally and the selection of mutants has greatly helped in the understanding of its respiratory chain (Doussière et al. 1976, 1979a, b).

3.1 Spectral Properties

At low temperature, 77 K, the difference spectrum of wild-type mitochondria from *P. tetraurelia* reveals peaks at 608, 450, and 441 nm attributable to cytochrome *aa3*, at 557, 553, 526, and 426 nm assigned to *b*-type cytochromes and at 549, 518, and 419 nm assigned to cytochrome *c* (Fig. 2). The spectrum of mitochondria depleted for cytochrome *c* reveals a peak at 551 nm which probably belongs to cytochrome *c1*(Doussière et al. 1979a).

In common with animal and plant mitochondria, *Paramecium* cytochrome *c* can be selectively extracted by saline solutions, indicating that it is loosely bound to the mitochondrial membrane. The cytochrome *aa3* consists of two

Fig. 2. Difference spectra obtained at 77 K from wild-type mitochondria from *P. tetraurelia* reduced with dithionite minus oxygenated reference

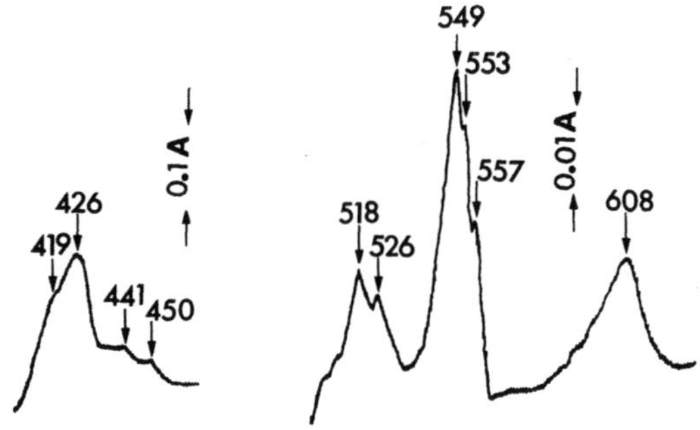

components which closely resemble those of mammalian cytochrome *aa3*. However, it differs from those of higher organisms in that it specifically reacts with its own cytochrome *c* but not with cytochrome *c* extracted from beef heart or *S. cerevisiae*. A similar specificity has also been observed for *T. pyriformis* (Wikström et al. 1976). *Paramecium* possesses two cytochromes *b* with α peaks at 553 and 557 nm though only one low-potential component (Eo = 120 mV) can be detected. These *b*-type cytochromes have a number of original properties. First, they lack antimycin-binding sites (though they can bind HQNO), which explains the absence of an inhibitory effect of this drug on respiration.

Second, one of them (cytochrome *b*557) is able to combine CO and the resulting CO-cytochrome *b* is photodissociable: The fact that light can restore CO-inhibited respiration of mitochondria from the *c11* mutant completely devoid of cytochrome *aa3*, strongly suggests that in *Paramecium* cytochrome *b*557 acts as a functional oxidase (Doussière et al., 1979a and b). Cytochrome *b* oxidases referred to as cytochromes *o* have been reported in a number of bacteria and some trypanosomatids (for review: Lloyd and Edwards 1978). In *Paramecium*, the autoxidizable cytochrome *b* is sensitive to high concentrations of cyanide and hydroxamic acids.

3.2 Cyanide-Insensitive Respiration

Respiration in *P. tetraurelia* is partially resistant to cyanide. Under standard conditions of growth, about 20% of respiration is insensitive to 1 mM cyanide. This respiration is located in mitochondria and branched to the main cytochromic chain. As in plants and a number of microorganisms (review: Henry and Nyns 1975), the cyanide-resistant respiration is fully sensitive to salicylhydroxamic acid (SHAM). The localization of the branching point is not clearly established. In wild-type mitochondria, P/O ratios are about 3 for NAD-linked substrates; in mutants devoid of cytochrome *aa3*, it is about 2 as if the second phosphorylation site was present in the cyanide-insensitive pathway (Doussière et al. 1976), in contrast to the situation in plants where only one phosphorylation site is retained in the alternative pathway, the branching point being located at the ubiquinone level (Solomos 1977). However, the difference between the P/O ratios in plants and *Paramecium* could be due to the presence

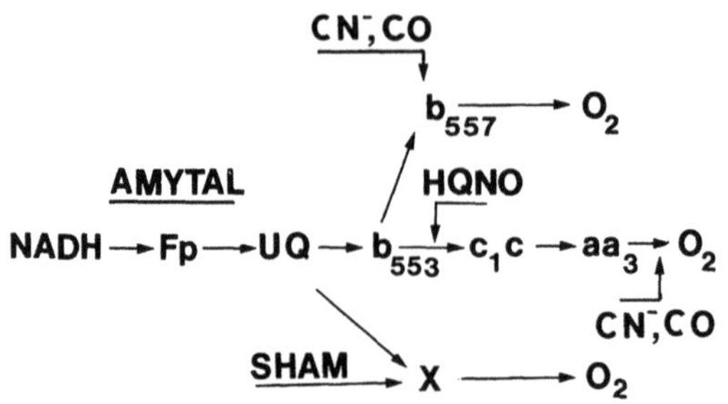

Fig. 3. Model of pathways of electron transport in wild-type mitochondria from *P. tetraurelia*. The sites of action of inhibitors are indicated

of cytochrome *b*557. *Paramecium* mutants devoid of cytochrome *aa3* use the SHAM-sensitive oxidase and also the autoxidizable cytochrome *b*, thus retaining the second phosphorylation site. The branching point has therefore been tentatively located as in plants at the ubiquinone level (Fig. 3) (Doussière et al. 1979b; Doussière 1982). The SHAM-sensitive pathway is not electrogenic and presents a lower affinity to oxygen than the main cyanide-sensitive pathway.

In *Paramecium*, the alternative SHAM-sensitive pathway has the peculiarity to be specifically stimulated by AMP (Doussière 1982; Doussière and Vignais 1984). This property has also been observed in *Euglena gracilis* (Sharpless and Butow 1970), *Moniliella tomentosa* (Hansens and Verachtert 1976; Vanderleyden et al. 1980a), *A. castellani* (Edwards and Lloyd 1978), and *Neurospora crassa* (Vanderleyden et al. 1980b). However, in these organisms, the mechanism has so far remained elusive. In *Paramecium*, Doussière (1982) and Doussière and Vignais (1984) demonstrated that AMP activates the cyanide-insensitive respiration by binding to a specific component of the alternative pathway that they have partially purified. They showed that a photoactivated derivative of AMP, NAP_4-AMP, added to *Paramecium* mitochondria in the absence of light, inhibits competitively the AMP-stimulated cyanide-resistant respiration. Photoirradiation of mitochondria treated with $[^3H]NAP_4$-AMP revealed a specific protein component as the target site for NAP_4-AMP which probably represents the AMP-binding protein. On the other hand, they have extracted and partially purified from wild-type mitochondria a protein fraction able to oxidize ubiquinol with the characteristics of the cyano-resistant oxidase. The addition of this protein fraction to submitochondrial particles from beef heart (completely cyanide-sensitive) allows the reconstitution of a cyanide-resistant, SHAM-sensitive, AMP-stimulatable oxidase activity strongly suggesting that this fraction contains the alternative oxidase.

3.3 Regulation of the Respiratory Pathways in P. tetraurelia

As depicted in Fig. 3, the respiratory chain of *Paramecium* comprises three pathways: a classical cytochromic chain with cytochrome *aa3* inhibited by low concentrations of cyanide (Ki $\simeq$ 10 μM), a cyanide-resistant oxidase sensitive to salicylhydroxamic acid (Ki > 120 μM), activated by AMP and probably branched at the ubiquinone level, and a *b*-type cytochrome oxidase slightly sensitive to cyanide (Ki > 50 μM) and to hydroxamic acids (Ki 700 μM). The main pathway is electrogenic (three phosphorylation sites), while the two alternative ones are not.

The extent of these different pathways depend on growth conditions (phase of growth, temperature, presence of erythromycin or cyanide in the medium; Adoutte and Doussière 1978; Doussière et al. 1979b) and on the genetic context. Most of the mitochondrial antibiotic-resistant mutants assumed to have an impaired mitochondrial protein-synthesizing machinery (Adoutte and Doussière 1978) as nuclear mutants affected in their cytochromic content (Ruiz and Beisson 1980) show an increased cyanide-insensitive respiration.

The regulation of the electron flux in these different pathways in wild type is not clearly understood, but the fact that the cyanide-resistant, SHAM-sensitive, nonelectrogenic pathway is specifically activated by AMP, whereas the main cytochromic electrogenic pathway is activated by ADP, suggests that the adjustment of the activities of the two pathways depends on the relative concentrations of AMP and ADP.

4 Molecular Biology of the Genome

4.1 Linear Structure and DNA Replication Origin

Several years ago we showed that the mitochondrial (mt) genome of *P. aurelia* is a linear 41-kbp structure (Goddard and Cummings 1975, 1977). Except for *Tetrahymena* and *Chlamydomonas* mt DNA (Ryan et al. 1970; Borst 1972), this is in sharp contrast to other mitochondrial DNAs which are circular. Electron microscopic and denaturation/renaturation experiments revealed that replication of this linear molecule was initiated from a unique single-stranded DNA cross-linked end (Goddard and Cummings 1975, 1977) and that the mode of replication was semiconservative (Goddard and Cummings 1977). A model depicting the general mechanism of replication is shown in Fig. 4.

Here, we illustrate the replication mechanism in two ways. First (Fig. 4A–D), electron microscope data indicated that replication was initiated from the unique cross-linked end and proceeded unidirectionally via lariat (circle and trail) and dimer-length intermediates, terminating in two monomer-length molecules (Goddard and Cummings 1975, 1977). Second, DNA sequence analysis of a cloned dimer initiation fragment (see Fig. 4C) revealed that the initiation region in five different species consisted of a series of AT-rich repeats (Pritchard et al. 1983). This is illustrated in Fig. 4 E–H, where the initiation region is enlarged to show the repeats. We postulated that processing of the dimer intermediate (Fig. 4F,G) occurs via staggered cuts, generating two "flip-flop" complementary monomers (Fig. 4H). Direct evidence for this type of processing was obtained by isolating the two monomer species from a mitochondrial DNA population and determining their sequence (Pritchard and Cummings 1984).

The DNA sequence of these AT-rich repeats for the five species is shown in Fig. 5. As can be noted, while all are AT-rich, little identity can be detected in

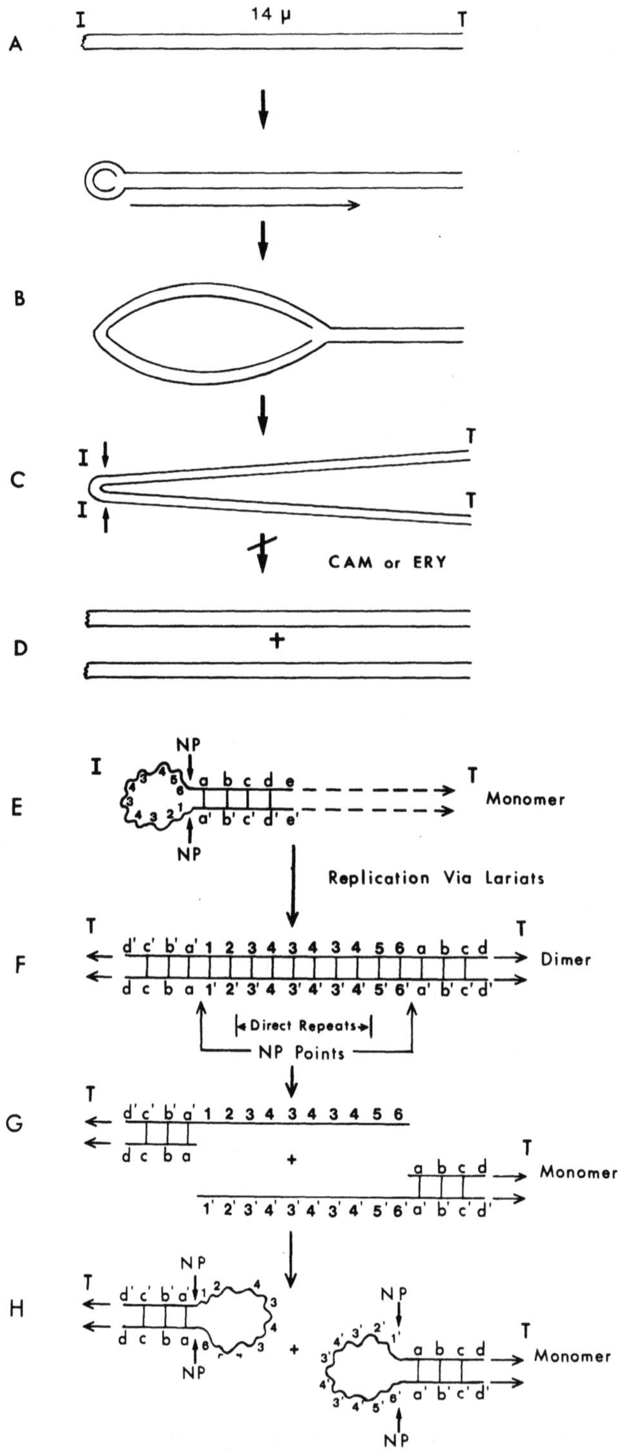

Fig. 4. Replication scheme of *Paramecium* mitochondrial DNA

	Sequence	Length bp	#
Sp. 1	ATATAATTAATTATAAATTTAATTTTTTAATAAATTATATTATTTATATACATATATTATAATTATTTTT	72	6
Sp. 5	ATATATTATAAATAATTTAATATTAAAATATATTTATATAATTTAAATATTATTATTTTTAAATAC	66	8
Sp. 7	ATATTATAATTAAATATTTATAATAATTTATTTTTACAATTAAATTATTTTTATTTTATTTATATAAAT	69	7
Sp. 4	TAATATATTTATATTTTTTTATTTTAATAAATA	34	6–14
Sp. 8	AATAATAATTTATTATTTATATTTATAAT	28	2
	TAA	3	42

Fig. 5. Interspecies sequence comparison of replication origin. *Starred underline* shows the TATA box common to all species and the *underline without stars* that common to odd species. Also indicated are the length and number of each repeat

an interspecies comparison. Except for an ATATTTATA sequence, the repeats are not conserved. Clearly, the sets can be separated into two classes, even and odd species. The odd species have another sequence, ATTATTTTT, in common, but the significance of this, or that of the species' common sequence is not known. Very little difference in four species (1, 4, 5, or 7) was noted in an intraspecies comparison. We concluded from this study that sequence conservation is not necessary for the replication origin, but that a secondary or structural conservation is all that is required. In retrospect, this should have been expected for such an integral function. An alternate explanation for this structural, and not sequential, conservation is that these repeats are not the actual replication origin. This was ruled out by experiments which showed that subsets of this replication origin functioned as autonomously replicating sequences only when the repeats were present (Lazdins and Cummings 1984).

Finally, another feature of this replication origin is that polymorphisms occur, especially in species 4. Restriction enzyme analysis of several mutants showed a wide range in molecular weight of the dimer initiation region. These differences were entirely accounted for by a variation in the number of repeats, which ranged from 6 to 14. We proposed an unequal recombination mechanism resulting in the deletion or tandem duplication of this region. This was based on an earlier model that Smith (1976) put forth to explain unequal recombination between sister chromatids.

4.2 Genetic Map and Codon Usage

Localizing genes on the *Paramecium* mitochondrial genome has not been easy. Heterologous hybridization with other known mitochondrial genes was unsuccessful, as was also the case for *Tetrahymena* mitochondrial genes (Suyama et al. 1985). Rather, we had to rely on computer-generated translation sequences. Excluding the ribosomal RNA genes and tRNA genes, which will be dealt with in the next section, we have located five genes. The first of them, labeled P1 in Fig. 6, was identified with a 2-Kb sequence, common to all five species, immediately adjacent to the replication origin. For this gene, as well as

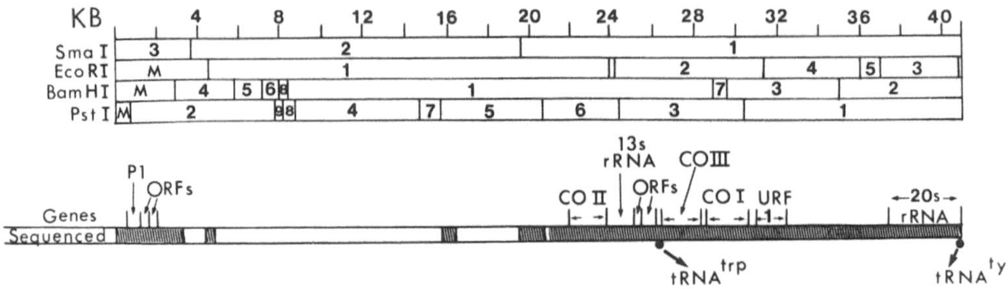

Fig. 6. Genetic and restriction map of *Paramecium tetraurelia. Hatched areas* indicate regions which have been sequenced. *P1* refers to that protein immediately 3′ of the replication origin. *ORFs* show open reading frames, as yet unidentified

a cytochrome oxidase gene, we have demonstrated its presence in mitochondrial extracts by Western analysis with an antibody raised against an in vitro produced polypeptide (Mahalingam et al. 1986). Because of its placement, we feel that it may be involved in DNA replication. We have also localized the three genes of the cytochrome oxidase complex, CO I, II, and III, as well as URFI, the first gene in the respiratory chain NADH dehydrogenase complex (Chomyn et al. 1985). All three were identified solely from their computer-generated amino acid sequences (Pritchard et al. 1986). Even here, the homology with other homologous genes was only 30 to 40%. The DNA sequences themselves were less than 10% homologous, thus accounting for our failure to identify these genes by heterologous hybridization.

No introns have yet been found in the 22 kbp sequenced. However, for CO I, the deduced amino acid sequence of 645 amino acids (Pritchard et al. 1986) contains a stretch of 108 amino acids not found in this gene in other organisms. This "insert" does not conform to the usual sequence properties of mt introns (Davies et al. 1982; Michel and Dujon 1983) and is probably not an intron, since it contains no stop codons in the CO I reading frame and S1 protection experiments reveal no evidence for a spliced RNA. Hydropathy pattern comparisons with other CO I sequences show quite good (de la Cruz et al. 1984) agreement except for this stretch.

The gene order displayed in Fig. 6 is quite unusual. Most mitochondrial genomes have URFI upstream of the small ribosomal gene (13s rRNA) whereas in *Paramecium* it is located on its 3′ side. Similarly, CO III is commonly on the 3′ flank of the large ribosomal gene (20s rRNA) and here it is shown well upstream of the 5′ side. CO II is often found on the 3′ side of another CO III or CO I and here we find it on the 5′ side, upstream of the small rRNA gene. This location is probably unique to ciliates, since we have detected it also in sequences published for *Tetrahymena* mitochondrial DNA (Schnare et al. 1986).

Having identified five coding sequences, as well as several unidentified open reading frames (ORFs), we are able to specify codon usage (Table 1). As can be noted, few surprises are found. Like other mitochondrial codes, UGA

	P1	COI	COII	URF1	ORF	ORF		P1	COI	COII	URF1	ORF	ORF
Phe- TTT	9	41	8	23	6	11	Tyr- TAT	2	16	2	6	0	1
TTC	13	36	4	10	14	13	TAC	7	19	6	9	2	5
Leu- TTA	3	9	2	7	0	1	Ter- TAA	0	1	0	0	0	1
TTG	1	12	3	7	3	2	TAG	1	0	1	1	1	0
Leu- CTT	3	25	5	7	1	1	His- CAT	0	8	3	1	1	0
CTC	16	11	5	13	18	11	CAC	3	14	6	1	0	2
CTA	1	14	3	8	1	0	Gln- CAA	3	5	0	0	0	0
CTG	3	2	3	4	3	1	CAG	3	2	3	3	0	5
ILe- ATT	1	18	7	4	0	3	Asn- AAT	2	13	6	6	2	1
ATC	1	15	6	11	5	4	AAC	10	14	3	4	4	7
ATA	2	13	3	2	0	1	Lys- AAA	4	12	7	3	3	1
Met- ATG	2	19	1	7	3	4	AAG	12	19	9	4	4	3
Val- GTT	2	19	3	12	1	2	Asp- GAT	0	9	1	2	0	1
GTC	9	10	0	5	3	7	GAC	4	8	6	1	2	1
GTA	1	10	2	5	1	0	Glu- GAA	1	2	3	6	0	0
GTG	2	6	2	1	0	3	GAG	11	10	3	6	2	5
Ser- TCT	1	17	4	5	1	2	Cys- TGT	0	4	2	0	1	0
TCC	7	7	2	7	3	1	TGC	3	2	5	4	0	3
TCA	0	5	2	2	1	1	Trp- TGA	0	7	2	2	0	1
TCG	4	3	0	2	5	1	TGG	2	7	2	2	2	1
Pro- CCT	1	6	4	2	0	1	Arg- CGT	0	1	0	1	0	0
CCC	5	12	4	3	0	0	CGC	5	1	0	1	0	0
CCA	0	7	0	1	0	0	CGA	0	1	1	2	0	0
CCG	0	0	0	0	0	0	CGG	1	0	0	0	0	0
Thr- ACT	1	11	4	3	0	0	Ser- AGT	0	5	1	2	0	1
ACC	4	13	1	3	3	1	AGC	8	10	4	2	1	1
ACA	0	10	2	2	1	1	Arg- AGA	0	5	5	0	0	0
ACG	5	8	3	1	2	0	AGG	1	16	4	2	0	1
Ala- GCT	2	23	3	5	2	1	Gly- GGT	0	7	1	3	0	2
GCC	9	8	3	5	5	3	GGC	7	11	2	6	4	0
GCA	1	10	1	6	1	0	GGA	0	16	4	3	0	1
GCG	2	2	0	2	1	1	GGG	9	8	0	4	1	0

serves to recognize tryptophan rather than as a stop codon. The only striking difference with the universal and mitochondrial code is the lack of use of CCG as a proline codon. With all the unique features of the *Paramecium* mitochondrial DNA, i.e., structure, insertion sequences and, as we will see, ribosomal RNA genes, it is, in some respects, comforting to find that its codon usage is apparently normal.

4.3 Ribosomal and tRNA Genes

The sequences of the small and large mitochondrial rRNA genes from two divergent species, *P. primaurelia* and *P. tetraurelia*, were determined (Seilhamer et al. 1984a, b). In general, the sequences resemble their counterparts in cytoplasmic, prokaryotic, and other mitochondrial sources but most closely resemble prokaryotic genes. The small rDNA covers about 1680 nucleotides, including an extraneous 83-bp sequence near the 3′ end which is unique to *Paramecium* mitochondria. According to proposed models for the small subunit rRNA secondary structure, this extraneous sequence occurs at the apex of the penultimate helix, which is highly variable in length. Other ciliate mitochondrial DNAs may lack this insert since it has not been found in *Tetrahymena* (Schnare et al. 1986). A more significant difference with *Tetrahymena* small rDNA is the presence of a 5′-5s like RNA. All other small subunit rDNAs, cytoplasmic and mitochondrial, show continuous sequences. Schnare et al. (1986) reported that in *Tetrahymena* mitochondrial DNA, the small rDNA was discontinuous, being preceded by a 200-bp sequence on its 5′ flank. This was based on hybridization of the small rDNA sequence to the 14s RNA, as expected, as well as to a 200-bp RNA. However, no precursor relationship was noted. *Paramecium* RNA also contains this 200-bp RNA species but we were unable to detect hybridization to the small rDNA region. Its presence as a 5′ component of the small rDNA gene in *Paramecium* remains an open question.

There is no doubt that we detected a 5′-5.8s RNA-like component of the large subunit rDNA in *Paramecium*. In prokaryotes, the 5′ end of 23s rRNA consists of a "fusion" of a 5.8s and a 20s component (Nazar 1980, 1982). In eukaryotes the 5.8s rRNA gene is separated from the 5′ end of the large rRNA gene by a spacer which is excised during posttranscriptional processing. In total *Paramecium* mitochondrial RNA there is a 283-base RNA species which hybridizes to the large subunit rDNA. This rDNA region consists of this 5.8s-like RNA gene separated from the 20s rDNA by a 19- to 26-bp excised region. Moreover, a precursor relationship was demonstrated. The large subunit rRNA gene complex spans about 2654 bp and is preceded by a 15-base poly (T) tract and terminates randomly within a 20-base A- + T-rich segment immediately preceding a tRNA-Tyr gene. We postulated that this tRNA-Tyr gene may be a contiguous part of the large subunit rRNA gene complex, similar to the contiguous gene organization found in mammalian mitochondria (Battey and Clayton 1980; Ojala et al. 1981), but there is no evidence here to support this view.

The only other tRNA found thus far in *Paramecium* mitochondrial DNA is that for tryptophan (UGA) (Seilhamer and Cummings 1982). Suyama et al.

(1985) have proposed that *Tetrahymena* and other ciliates may have an abnormally low complement of tRNA genes. Even though we have sequenced more than half the genome of *Paramecium* mitochondrial DNA, thus far we have found only two tRNA genes. Two other structures have been detected (Seilhamer and Cummings 1981), but these are integral parts of the 20s rDNA gene and participate in the secondary structure folding of this gene. Clearly, we must await the completion of the entire sequence to be certain of the tRNA components.

4.4 Overview and Evolution

Nanney (1980) has put forth the notion that ciliates may have branched off from other eukaryotes at a very early stage in evolution and, thus, may be closer to our ancestral progeny than mammals and fungi. Some features of *Paramecium* DNA might serve to support this view. For example, in surface antigen nuclear genes, TAA and TAG encode glutamine rather than act as stop codons (Caron and Meyer 1985; Preer et al. 1985). As we have seen here, mitochondrial rRNA genes more closely resemble prokaryotic counterparts and, in addition, the large subunit rRNA gene contains an eukaryotic-like 5.8s RNA component. DNA sequence comparisons between two divergent species provide a good assessment for microevolution. First, the same general novel gene order and structure has been maintained. This gene order appears to be also true for *Tetrahymena* mitochondrial DNA genes. With regard to the two rDNA genes, it is interesting to note, however, that the relative rate of divergence between the large and small rRNA genes is different (6.4% for the small and 4.1% for the large). This might be due to fewer constraints on the secondary structure of the small rRNA gene. The tRNA-Tyr gene is 100% conserved, whereas mRNA and nontranscribed regions are only 80% conserved, so the degree of conservation may well reflect the relative functional importance of the genes or gene segments. Finally, the structure and replication mode of the mitochondrial DNA appears to be unique to the ciliates. In many respects, it resembles viruses more than any other organism (see Pritchard et al. 1983).

While several features of *Paramecium* mitochondrial DNA appear to argue that ciliates may have a unique and important niche in eukaryotic evolution, it is important to note that when other genes are used to evaluate the evolutionary position, a different answer may be reached. For example, Sogin and Elwood (1986) have recently found that when the sequence and structure of the small subunit rRNA genes of the nuclear genome are compared with that for other lower eukaryotes, *Paramecium* (and *Tetrahymena*) occupy a higher place in the evolutionary pathway than *D. Discoideum* and some fungi. It is clear that in spite of the many interesting properties of ciliate nuclear and mitochondrial DNA, some caution is necessary in defining their evolutionary significance.

References

Adoutte A (1974) Mitochondrial mutations in *Paramecium:* phenotypical characterization and recombination. In: Kroon A, Saccone C (eds) The biogenesis of mitochondria. Academic Press, London New York, pp 263–271

Adoutte A (1977) La génétique des mitochondries chez la paramécie. Thesis, Univ Paris Sud
Adoutte A, Beisson J (1970) Cytoplasmic inheritance of erythromycin resistant mutations in *Paramecium aurelia*. Mol Gen Genet 108:70–77
Adoutte A, Beisson J (1972) Evolution of mixed populations of genetically different mitochondria in *Paramecium aurelia*. Nature (London) 235:393–396
Adoutte A, Doussière J (1978) Physiological consequences of mitochondrial antibiotic-resistant mutations in *Paramecium*. Mol Gen Genet 161:121–134
Adoutte A, Knowles JK, Sainsard-Chanet A (1979) Absence of detectable mitochondrial recombination in *Paramecium*. Genetics 93:797–831
Battey J, Clayton D (1980) The transcription map of human mitochondrial DNA implicates transfer RNA excision as a major processing event. J Biol Chem 255:11599–11606
Beale GH (1969) A note on the inheritance of erythromycin resistance in *Paramecium aurelia*. Genet Res 14:341–342
Beale GH (1973) Genetic studies on mitochondrially inherited mikamycin-resistance in *Paramecium aurelia*. Mol Gen Genet 127:241–248
Beale GH, Knowles JKC (1976) Interspecies transfer of mitochondria in *Paramecium aurelia*. Mol Gen Genet 143:197–201
Beale GH, Tait A (1981) Mitochondrial genetics of *Paramecium aurelia*. Int Rev Cytol 71:19–40
Beale GH, Knowles JKC, Tait A (1972) Mitochondrial genetics in *Paramecium*. Nature (London) 235:396–397
Borst P (1972) Mitochondria nucleic acids. Annu Rev Biochem 41:333–376
Caron I, Meyer E (1985) Does *Paramecium primaurelia* use a different genetic code in its macronucleus. Nature (London) 314:185–188
Chomyn A, Mariottini P, Chester MWJ, Ragan CI, Matsunoyagi A, Hatefi Y, Doolittle RF, Attardi G (1985) Six unidentified reading frames of human mitochondrial DNA encode components of the respiratory-chain NADH dehydrogenase. Nature (London) 314:592–597
Cruz VF de la, Neckelmann N, Simpson L (1984) Sequences of six genes and several open reading frames in the kinetoplast maxicircle DNA of *Leishmania tarentolae*. J Biol Chem 259:15136–15147
Cummings DJ, Laping JL (1981) Organization and cloning of mitochondrial deoxyribonucleic acid from *Paramecium tetraurelia* and *Paramecium primaurelia*. Mol Cell Biol 1:972–982
Davies RW, Waring RB, Ray JA, Brown TA, Scazzocchio C (1982) Making ends meet: a model for RNA in fungal mitochondria. Nature (London) 300:719–724
Doussière J (1982) Respiration cyano-resistante des mitochondries de *Paramecium tetraurelia*. Thesis, Univ Grenoble
Doussière J, Vignais PV (1984) AMP-dependence of the cyanide-insensitive pathway in the respiratory chain of *Paramecium tetraurelia*. Biochem J 220:787–794
Doussière J, Adoutte A, Sainsard A, Ruiz F, Beisson J, Vignais P (1976) Physiological and genetical analysis of the respiratory chain of *Paramecium*. In: Genetics and biogenesis of chloroplasts and mitochondria. Elsevier/North Holland Biomedical Press, Amsterdam New York, pp 873–879
Doussière J, Sainsard-Chanet A, Vignais PV (1979a) The respiratory chain of *Paramecium tetraurelia* in wild type and the mutant cl1. Biochim Biophys Acta 548:224–235
Doussière J, Sainsard-Chanet A, Vignais PV (1979b) Cyanide-insensitive respiration, function and regulation. Biochim Biophys Acta 548:236–252
Dujon B (1983) Mitochondrial genes, mutants and maps: a review. In: Schweyen RJ, Wolf Y, Kaudewitz F (eds) Mitochondria 1983. de Gruyter, Berlin, pp 1–24
Edwards SW, Chagla AH, Griffiths AJ, Lloyd D (1977) The cytochromes of *Acanthamoetra castellanii*. Biochem J 168:113–121
Goddard JM, Cummings DJ (1975) Structure and replication of mitochondrial DNA from *Paramecium aurelia*. J Mol Biol 97:593–609
Goddard JM, Cummings DJ (1977) Mitochondrial DNA replication in *Paramecium aurelia*. Cross-linking of the initiation end. J Mol Biol 109:327–344
Hansens L, Verachtert H (1976) Adenosine 5′-monophosphate-stimulated cyanide-insensitive respiration in mitochondria of *Monilliela tomentosa*. J Bacteriol 125:829–836

Henry MF, Nyns EJ (1975) Cyanide-insensitive respiration, an alternative mitochondrial pathway. Subcell Biochem 4:1–65
Hill GC (1978) Characterization of the electron transport systems present during differentiation of African *Trypanosomas.* In: Functions of alternative terminal oxidase, vol 49. 11th FEBS-Meet, Copenhagen 1977, pp 67–77
Kilpatrick L, Erecinska M (1977) Mitochondrial respiratory chain of *Tetrahymena pyriformis.* Biochem Biophys Acta 460:346–363
Knowles JKC (1972) Observations of two mitochondrial phenotypes in single *Paramecium* cells. Exp Cell Res 70:223–226
Knowles JKC (1974) An improved microinjection technique in *Paramecium aurelia.* Transfer of mitochondria conferring erythromycin-resistance. Exp Cell Res 88:79–87
Kotylak Z, Lazowska J, Hawthorne DC, Slonimski PP (1985) Intron encoded proteins of mitochondria: key elements of gene expression and genomic evolution. In: Quagliariello E, Slater EC, Palmieri F, Saccone C, Kroon AM (eds) Achievements and perspective of mitochondrial research, vol 2. Biogenesis. Elsevier, New York, pp 1–20
Lazdins I, Cummings DJ (1984) Structural and functional analysis of replication of mitochondrial DNA from *Paramecium aurelia.* Curr Genet 8:483–487
Lloyd D, Chance B (1972) The cytochromes of mitochondria from *Tetrahymena pyriformis* strain ST. Biochem J 128:1171–1182
Lloyd D, Edwards SW (1978) Electron transport pathways alternative to the main phosphorylating respiratory chain. In: Functions of alternative terminal oxidase, vol 49. 11th FEBS Meet, Copenhagen 1977, pp 1–10
Mahalingam R, Seilhamer JJ, Pritchard AE, Cummings DJ (1986) Identification of *Paramecium* mitochondrial proteins using antibodies raised against fused mitochondrial gene products. Gene 49:129–138
Maki RA, Cummings DJ (1977) Characterization of mitochondrial DNA from *Paramecium aurelia* with EcoRI and HaeII restriction endonuclease. Plasmid 1:106–114
Michel F, Dujon B (1983) Conservation of RNA secondary structures in two intron families including mitochondrial-, chloroplast- and nuclear-encoded members. EMBO J 2:33–38
Nanney DL (1980) Experimental ciliatology. John Wiley and Sons, New York
Nazar RN (1980) A 5.8 rRNA-like sequence in prokaryotic 23 S rRNA. FEBS Lett 119:212–214
Nazar RN (1982) Evolutionary relationship between eukaryotic 29–32 S nucleolar rRNA precursors and the prokaryotic 23 S rRNA. FEBS Lett 143:161–162
Ojala D, Montoya J, Attardi G (1981) tRNA punctuation model of RNA processing in human mitochondria. Nature (London) 290:470–474
Perasso R, Adoutte A (1974) The process of selection of erythromycin-resistant mitochondria by erythromycin in *Paramecium.* J Cell Sci 14:475–497
Perasso R, Beisson J (1978) Temporal pattern of mitochondrial multiplication during the cell cycle of *Paramecium.* Biol Cell 32:275–290
Preer JR, Jr, Preer LB, Rudman BM, Barnett AJ (1985) Deviation from the universal code shown by the gene for surface protein 51A in *Paramecium.* Nature (London) 314:188–190
Pritchard AE, Cummings DJ (1984) Structural and functional analysis of the origin of replication of mitochondrial DNA from *Paramecium aurelia.* Curr Genet 8:477–482
Pritchard AE, Laping JL, Seilhamer JJ, Cummings DJ (1983) Inter-species sequence diversity in the replication initiation region of *Paramecium* mitochondrial DNA. J Mol Biol 164:1–15
Pritchard AE, Seilhamer JJ, Cummings DJ (1986) *Paramecium* mitochondrial DNA sequences and RNA transcripts for cytochrome oxidase subunit I, URFI, and three ORFs adjacent to the replication origin. Gene 49:243–253
Queiroz C, Beale GH (1974) Mitochondrial mutations affecting resistance to erythromycin and mikamycin in *Paramecium aurelia:* provisional results with a new method. Genet Res 23:233–238
Ruiz F, Adoutte A (1978) Selection and characterization of nuclear mutations affecting mitochondria in *Paramecium.* Mol Gen Genet 162:1–8
Ruiz F, Beisson J (1980) Genetic interactions in the control of mitochondrial functions in *Paramecium.* I. Interactions between nuclear genes. Mol Gen Genetics 180:553–561

Ruiz F, Knowles J (1980) Genetic interactions in the control of mitochondrial function in *Paramecium.* II. Interactions between nuclear and mitochondrial genomes. Mol Gen Genetics 180:563–572

Ryan R, Grant D, Chiang K-S, Swift H (1970) Isolation and characterization of mitochondrial DNA from *Chlamydomonas reinhardtii.* Proc Natl Acad Sci USA 75:3268–3272

Sainsard A (1975) Mitochondrial suppressor of a nuclear gene in *Paramecium.* Nature (London) 257:312–314

Sainsard-Chanet A (1976) Gene-controlled selection of mitochondria in *Paramecium.* Mol Gen Genet 145:23–30

Sainsard-Chanet A (1978) A new type of mitochondrial mutation in *Paramecium.* Mol Gen Genet 159:117–123

Sainsard-Chanet A (1979) Etude génétique et physiologique des interactions entre une mutation nucléaire et ses suppresseurs mitochondriaux chez la paramecie. Thesis, Université Paris Sud

Sainsard-Chanet A, Knowles J (1979) Restoration of nucleo-mitochondrial compatibility in *Paramecium.* Genetics 93:833–859

Sainsard A, Claisse M, Balmefrezol M (1974) A nuclear mutation affecting structure and function of mitochondria in *Paramecium.* Mol Gen Genet 130:113–125

Schnare MW, Heinonen TYK, Young PG, Gray MW (1986) A discontinuous small subunit ribosomal RNA in *Tetrahymena pyriformis* mitochondria. J Biol Chem 261:5187–5193

Sharpless TK, Butow RA (1970) An inducible alternate terminal oxidase in *Euglena gracilis* mitochondria. J Biol Chem 245:58–70

Seilhamer JJ, Cummings DJ (1981) Structure and sequence of the mitochondrial 20s rDNA and tRNA tyr gene of *Paramecium primaurelia.* Nucl Acids Res 9:6391–6406

Seilhamer JJ, Cummings DJ (1982) Altered genetic code in *Paramecium* mitochondria: possible evolutionary trends. Mol Gen Genet 187:236–239

Seilhamer JJ, Olsen GJ, Cummings DJ (1984a) *Paramecium* mitochondrial genes I. Small subunit rRNA gene sequence and microevolution. J Biol Chem 259:5167–5172

Seilhamer JJ, Gutell RR, Cummings DJ (1984b) *Paramecium* mitochondrial genes II. Large subunit rRNA gene sequence and microevolution. J Biol Chem 259:5173–5181

Smith GP (1976) Evolution of repeated DNA sequences by unequal crossover. Science 191:528–535

Sogin ML, Elwood HJ (1986) Primary structure of the *Paramecium tetraurelia* small-subunit rRNA coding region: phylogenetic relationships with the ciliophora. J Mol Evol 23:53–60

Solomos T (1977) Cyanide-resistant respiration in higher plants. Annu Rev Plant Physiol 28:279–297

Suyama Y, Preer JR, Jr (1965) Mitochondrial DNA from protozoa. Genetics 52:1051–1058

Suyama Y, Fukuhara H, Sor F (1985) A fine restriction map of the linear mitochondrial DNA of *Tetrahymena pyriformis:* genome size, map locations of rRNA and rRNA genes, terminal inversion repeat, and restriction polymorphism. Curr Genet 9:473–493

Swift H, Wolstenholme DR (1969) Mitochondria and chloroplasts: nucleic acids and the problem of biogenesis. In: Lima of Faria A (ed) Handbook of molecular cytology. North Holland, Amsterdam, pp 972–1046

Tait A (1972) Altered mitochondrial ribosomes in an erythromycin resistant mutant of *Paramecium.* FEBS Lett 24:117–120

Tait A, Knowles JKC (1977) Characterization of mitochondrial and cytoplasmic ribosomes from *Paramecium aurelia.* J Cell Biol 73:139–148

Vanderleyden J, Van den Eynde E, Verachtert H (1980a) Nature of the effect of adanosine 5′-monophosphate on the cyanide-insensitive respiration in mitochondria of *Moniliella tomentosa.* Biochem J 186:309–316

Vanderleyden J, Peeters C, Verachtert H, Bertrand H (1980b) Stimulation of the alternative oxidase of *Neurospora crassa* by nucleoside phosphates. Biochem J 188:141–144

Wikström MKF, Harmon HJ, Ingledew WJ, Chance B (1976) A reevaluation of the spectral, potentiometric and energy-linked properties of cytochrome oxidase in mitochondria. FEBS Lett 65:259–276

Chapter 13 Electrophysiology *

HANS MACHEMER [1]

CONTENTS

* In memoriam Roger Eckert (1934–1986)

1 Arbeitsgruppe Rezeptoren, Fakultät für Biologie, Ruhr-Universität, D-4630 Bochum, FRG

Paramecium. Ed. by H.-D. Görtz

1 Introduction

Bioelectricity is a characteristic of living matter. It occurs in conjunction with the existence of lipid bilayers or "membranes" which enclose cells and some intracellular organelles. A membrane potential, i.e. the difference in voltage between the external and internal faces of the membrane, results from an uneven ion distribution at both sides of the membrane. This applies in particular to Ca^{2+} which is kept at concentrations near 10^{-7} M in living cells. Because of the low intracellular concentration of Ca^{2+}, a minor entry of Ca^{2+} modifies the number of free cytosolic Ca ions so that this cation can serve as a potent second messenger of the cell. Bioelectricity, cell membranes and the unique intracellular Ca milieu are closely interrelated in function and may have a common evolutionary origin.

A "resting" cell membrane has a high specific resistance[1] and, due to a gauge of about 5 nm, it is an excellent electric condensor approximating a capacitance of near 1 $\mu F\ cm^{-2}$. The membrane of *Paramecium* includes, as other cell membranes do, various types of channel proteins which penetrate the hydrophobic lipid bilayer and can "open" and "close", thereby regulating the passage of ions and molecules. Electrophysiology is a discipline dealing with mechanisms and functions of cellular electricity. It is common use to separate passive electrical properties of the membrane from those requiring active regulation, in particular of the membrane conductance. The present survey deals with some passive properties (resistance, capacitance, cable properties) in a section on the resting membrane. Two more sections are devoted to active responses of the membrane following stimuli and shifts in voltage. These central sections will be placed between notes characterizing the electrical organization of *Paramecium* as being simple in design and sophisticated in performance.

2 A Historical Note

In 1934 T. Kamada, following his studies in J. Gray's laboratory at Cambridge, published a paper "on potential differences across the ectoplasm membrane of *Paramecium*" in a well-known British journal. This first report of reproducible measurements of a cation-dependent membrane potential in an animal cell is not well known in the physiological literature. Kamada used intracellular, electrolyte-filled quartz microelectrodes in modification of a method developed by Ettisch and Peterfi (1925). Successful attempts of membrane potential recording by microelectrodes had been previously performed in large plant cells (*Nitella* and *Chara;* Umrath 1930). After the war, electrophysiological studies of *Paramecium* were successfully continued in the laboratory of the late Kamada[2] (Yamaguchi 1960; Kinosita et al. 1964) and have become the basis of the now well-established branch of animal cell physiology.

---

[1] In the order of 10^{10} Ωcm; this value of specific resistance ranges between that of Bakelite and glass.

[2] Zoological Institute, The University of Tokyo.

3 Ion Batteries and Membrane Channels

3.1 *The Ca-K Membrane Circuit*

Various species of *Paramecium* employ two essential ion concentration batteries: a calcium battery with an inward-directed concentration gradient, and a potassium battery with the gradient directed outward (Fig. 1). These gradients cause Ca ions to move inward, and K ions to move outward across the membrane respectively. Removal of Ca^{2+} from the environment leads to deterioration of the cell membrane. Unorthodox membrane behaviour of *Paramecium* occurs in the absence of K^+ from the external solution (see Sect. 6.3.6).

The diffusion of a particular species of ion is impeded by electrostatic repulsion built up during the migration of charges across the membrane. According to the Nernst equation, the equilibrium potentials for Ca^{2+} and K^+ are the potentials of perfect balance of the mutually opposing diffusion work and electrical work. Neither Ca nor K ions will be in electrochemical equilibrium under physiological conditions, because the Ca and K batteries are electrically connected in series, thereby establishing the membrane potential by ohmic division of voltage. Because the membrane potential is unequal to the equilibrium potentials of Ca as well as of K, Ca ions flow inward, and K ions flow outward in a closed electric circuit[3]. These permanent ion fluxes tend to degrade the concentration gradients which would eventually lead to cell death. Mechanisms of uphill transport of Ca and K ions secure a long-term steady- state condition of the ion gradients. Poisoning of the aerobic and anaerobic metabolic pathway can rapidly deplete the ATP pool which leads to a breakdown of the bioelectric homeostasis of the cell.

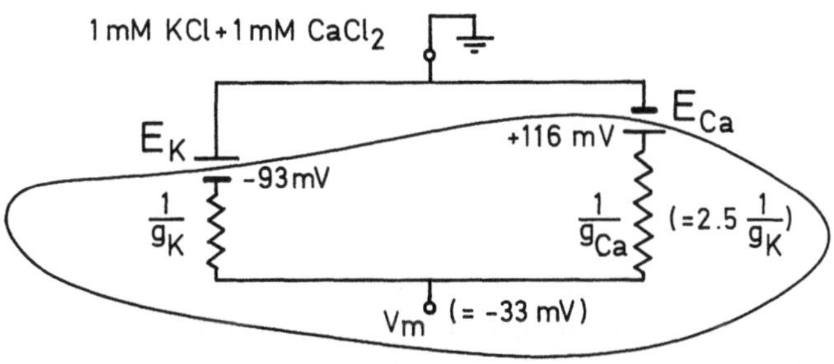

Fig. 1. Electric circuit diagram of *Paramecium.* Two ion batteries, as represented by the equilibrium potentials, E_{Ca} and E_K, occur across the plasma membrane. The assumed inward Ca concentration gradient is 10^4:1 in a solution of 1 mM Ca^{2+}, the outward K-concentration gradient is near 40:1 in a solution of 1 mM K^+. The batteries and their internal resistances ($1/g_{Ca}$; $1/g_K$) are connected in series so that the membrane potential results conventionally due to simple voltage division. With $E_{Ca} = +116$ mV, $E_K = -93$ mV, and the conductance ratio, g_{Ca}/g_K, equal to 0.4, the membrane potential, V_m, is calculated as -33 mV [Eq. (1)].

[3] It should be remembered that for an undisturbed cell the number of charges entering the cell equals the number of charges (of same sign) leaving the cell. The size of the resting current is limited by the sum of series resistances, as represented by the membrane conductances. This basic circuit applies also to events of rapid recharging of the membrane, when ionic currents are partly replaced by capacitive currents.

3.2 Other Ion Batteries

The existence of a negative membrane potential leads other cations and anions to unevenly distribute at both sides of the cell membrane. On the assumption that this distribution is fully passive, a particular ion species will, in theory, approach an equilibrium potential equal to the membrane resting potential. This may, in fact, apply to chloride which was found to affect neither the resting potential nor the peak of the action potential in *Paramecium* and other ciliates (Naitoh et al. 1972; De Peyer and Machemer 1977; Pape and Machemer 1986).

Conversely, an ion battery may be ineffective due to an extremely low membrane conductance for that ion. This appears to apply to magnesium ions which occur at millimolar concentrations in the cytosol of *Paramecium.* Although Mg^{2+} is vital to the ciliary function, *Paramecium* can survive for days in solutions devoid of Mg^{2+}. The recorded membrane potential is little affected by Mg ions at concentrations up to 2 mM[4].

Paramecium may establish weak batteries for some ions which can play a role in modulating the behaviour of the cell (see Sect. 6.2.8). Sodium is commonly more abundant than potassium in freshwater biotopes (Machemer and Deitmer 1985) and approximates half-molar strength in a marine environment. Although the classical voltage-dependent channel is impermeable to Na^+ even for brackish- water paramecia (Deitmer and Machemer 1982), a minor slow inward Na current occurs in the depolarized state of *Paramecium* immersed in solutions with raised Na^+ (Saimi and Kung 1980). These data prove the existence of an inward driving force for Na^+ so that E_{Na} is more positive than the potential of the resting and physiologically depolarized membrane.

3.3 The Membrane Channels

The electric membrane behaviour is, in the first place, determined by the conductances of various channels (Fig. 1). A membrane "channel" or "pore" is a conductance of particular ion specificity and kinetics. Preliminary single-channel recordings from the *Paramecium* membrane (Boheim et al. 1981; Ehrlich et al. 1984; Martinac et al. 1986) suggest that more than one open and one closed state of a channel can exist; these states may show, in addition, complex temporal properties. The structural and biochemical properties of ion channels in *Paramecium* are still unknown. Table 1, in giving a summary of ion conductances (or channels) in *Paramecium,* does not attempt to relate the physical entities of ion channels to these conductances. Table 1 shows that most ion channels in ciliates regulate the passage across the membrane of either Ca^{2+} or K^+.

---

[4] Depolarizing effects of Mg^{2+} on the membrane bulk potential in the presence of low Ca are similar to those of Ca^{2+} (Machemer 1976) and are now believed to involve partial neutralization of negative surface charges on the external side of the cell membrane (see Sect. 4.2).

Table 1. Simplified survey of 11 ion conductances in the ciliate membrane

Type[a, b]	Membrane site	Permeant ions[b]	Activation[b]	Inactivation[b]
"Leakage"				
Ca	Soma	Ca, ?	Depol.	Hyperpol.
K	Soma	K, ?	?	?
Mechanoreceptor				
Ca	Anterior soma	Ca, Mg*, Sr* Ba*, Mn	Rate of mechan. deformation*	Zero deform. rate*
K	Posterior soma	K, Rb*	Rate of mechan. deformation*	Zero deform. rate*
Voltage-dependent				
Ca_{fast}	Cilia	Ca, Sr, Ba	Depol. ≧5 mV	Ca_i
$Ca_{l.t.}$*	Cilia	Ca, Sr, Ba Mg, Na	Depol. ≦10 mV	Voltage
Ca_{slow}	Cilia	Ca, Sr, Ba	Depol. ≧? mV	Voltage
K_{fast}	Soma	K, ?	Depol. ≧15 mV	Repol., ?
K_{slow}	Soma	K, ?	Ca_i; depol. ≧20 mV	Repol., ?
$K_{i.r.}$	Soma	K, ?	Hyperpol. ≧10 mV; Ca_i	$V_m \rightarrow V_{rest}$
Na_{slow}	Soma?	Na, Li	Ca_i	Repol., $\downarrow Ca_i$?

[a] Ca_{fast} = classical Ca current; $Ca_{l.t.}$ = low-threshold current identified in *Stylonychia;* Ca_{slow} = persistent Ca current; K_{fast} = voltage-dependent K current; K_{slow} = delayed = outward rectifier current; $K_{i.r.}$ = inward rectifier current; Na_{slow} = Ca-activated Na current (see text for references).

[b] Properties identified in cells other than *Paramecium* are marked by an asterisk.

4 Properties of the Resting Membrane

4.1 Membrane Potential

Freshwater paramecia in the standard solution of 1 mM $CaCl_2$ and 1 mM KCl show resting potentials (V_{mr} between −30 and −40 mV depending on recording techniques, and it is expected from the Nernst equation that the membrane is less negative at increased concentrations of either Ca^{2+} or K^+. Conversely, the negativity of the membrane potential should be increased with reduction in concentration of these ions. The equation of the resting potential[5]:

$$V_{mr} = E_{Ca}\, g_{Ca}/(g_{Ca} + g_K) + E_K\, g_K/(g_{Ca} + g_K), \qquad (1)$$

allows the determination of the conductance ratio, g_{Ca}/g_K, using the Nernst equation and observed shifts in the resting potential (ΔV_{mr}) determined at

[5] In this equation, the value of the transmembrane potential, E_m, may differ from the membrane potential, V_m, as picked up by the voltage recording electrodes (see Fig. 2). In those cases which involve electrochemical equilibria it is appropriate to use the expression V_m.

20 °C in different concentrations of external Ca^{2+} ($[Ca]_{o1}$, $[Ca]_{o2}$) and/or K^+ ($[K]_{o1}$, $[K]_{o2}$):

$$g_{Ca}/g_K = \Delta V_{mr}/\{29\ \text{mV}\ (\log [Ca]_{o2} - \log [Ca]_{o1}) - \Delta V_{mr}\}$$
$$= \{58\ \text{mV}\ (\log [K]_{o2} - \log [K]_{o1}) - \Delta V_{mr}\}/\Delta V_{mr}. \quad (2)$$

Since the driving force for Ca^{2+}, $E_{Ca} - V_{mr}$, exceeds the driving force for K^+, $E_K - V_{mr}$, it is reasonable to expect a conductance ratio of ≦1 of the resting potential. Equation (2) suggests that slopes of half-logarithmic plots of observed V_{mr} against K^+ or Ca^{2+} concentration yield similar conductance ratios. Unfortunately, published slopes of the resting potential lead to discrepancies in calculated values of the conductance ratio. Errors in reading of shifts in membrane potential are produced by cation-dependent changes in the surface potential[6]. Observed unorthodox Ca-dependent shifts of the threshold for the transient Ca current in *Paramecium* (Satow and Kung 1979) suggest that the "read" (V_m) and the "real" membrane potential (E_m) can substantially differ due to membrane surface-charge neutralization.

4.2 Surface Potentials and the "Ca-Paradox"

Changing the external ionic environment produces complex effects upon the electric membrane properties which are insufficiently explained by the Hodgkin-Horowicz equation of the membrane potential (Eq. 1). Raising Ca_o depolarizes the observed membrane potential and increases the input resistance of the membrane (Naitoh and Eckert 1968a; Machemer 1976; Machemer and Ogura 1979). A membrane depolarization activates depolarization-dependent motor responses of the cilia (see Chap. 14, this Vol.). However, upon raising the external Ca concentration, the reverse is observed: the cells speed up forward swimming, i.e. they exhibit a hyperpolarization-dependent behavior. Because the cilia respond to the transmembrane potential (E_m), increases in external Ca appear to affect, besides equilibrium potentials and conductances, the surface potential of the membrane (Fig. 2). Surface potentials occur at various salinities and are substantial in freshwater solutions (McLaughlin 1977). The observed membrane bulk potential (V_m) is the sum of the outer and inner surface potentials and the transmembrane potential (E_m). Negative surface charge neutralization by Ca ions more than compensates the Nernstian positive shift in Ca equilibrium potential (Fig. 2, left column, steps 1 and 2). In addition, Ca binding to, and screening of, membrane channels raises the input resistance and lowers the g_{Ca}/g_K conductance ratio which tends to hyperpolarize the transmembrane potential even more (Fig. 2, step 3). In general, external Ca has been

[6] The inferred conductance ratio of the resting membrane of *Paramecium* in standard solution, using data from Naitoh and Eckert (1968a) and from Machemer and Ogura (1979), is between 3.2 and 5.5 with K^+ varied, and between 1.8 and 2.7 with Ca^{2+} varied. High and/or incongruent ratios are also calculated for other ciliates (*Stylonychia:* 18, 0.8; De Peyer and Machemer 1977; *Didinium:* 6.6, 3.1; Pape and Machemer 1986). A real conductance ratio approximating 0.4 may apply to *Paramecium* (see legend to Fig. 1).

known to decrease the excitability of tissue cells ("stabilizing effect"; Frankenhaeuser and Hodgkin 1957), i.e. to hyperpolarize E_m due to surface charge neutralization and screening.

An equivalence of the "Ca-paradox", or divergence of changes in E_m and V_m, does not apply to K-dependent shifts in membrane potential. Raising external K implicates comparatively small changes in surface potentials (McLaughlin 1977), so that the major depolarizing effect of positive shifts in E_K predominates (Fig. 2, right column).

While the majority of monovalent and divalent cations can depolarize V_m, their effects on E_m are more difficult to assess. In addition, after changing solutions or imposing steady-state shifts of the membrane potential, the membranes of ciliates gradually "accommodate", i.e. they restore the previous resting E_m. Consequently, the electric and the motor responses relax in the course of tens of seconds. It has been suggested that this potentiostatic membrane response is due to ion redistribution across the membrane (Eckert and Brehm 1979).

4.3 Leakage Conductance

Ideally, the membrane resting conductance, $g_{Car} + g_{Kr} + g_{Xr}$[7], is a passive, i.e. voltage-independent, property of the membrane. In *Paramecium*, the Ca resting conductance, as deduced from Ca currents recorded under voltage clamp, decreases with hyperpolarization and increases with depolarization of the membrane from the resting voltage level (Table 1; Naitoh and Machemer, unpubl.). This voltage dependence of the resting Ca conductance is unrelated to the ciliary Ca channel, because it persists in a cnr-mutant with a non-functional Ca channel. Figure 3 diagrammatically illustrates experiments revealing the variable Ca resting conductance.

4.4 Resistance

The input resting resistance of the membrane in *Paramecium* varies depending on the size of the cell and thus the membrane surface area. Typical values of freshwater paramecia range between 40 and 80 MΩ. It is interesting to note that the resistance of the ciliary membrane highly exceeds the somatic resistance, as deduced from unchanged input resistance after ethanol deciliation (Machemer and Ogura 1979). Similar observations have been made in ciliated and deciliated *Didinium* (Pape and Machemer 1986). Since the removal of the cilia in *Paramecium* or *Didinium* affected neither the resistance nor the resting potential, the leakage conductances of the membrane appear to be limited to somatic areas. Deciliation techniques using chloral hydrate led to a doubling of the input resistance (Dunlap 1977); a possible interference of the drug with the resting conductance was not tested in that case.

---

[7] g_{Xr} represents an unspecific conductance for ions other than Ca^{2+} and K^+.

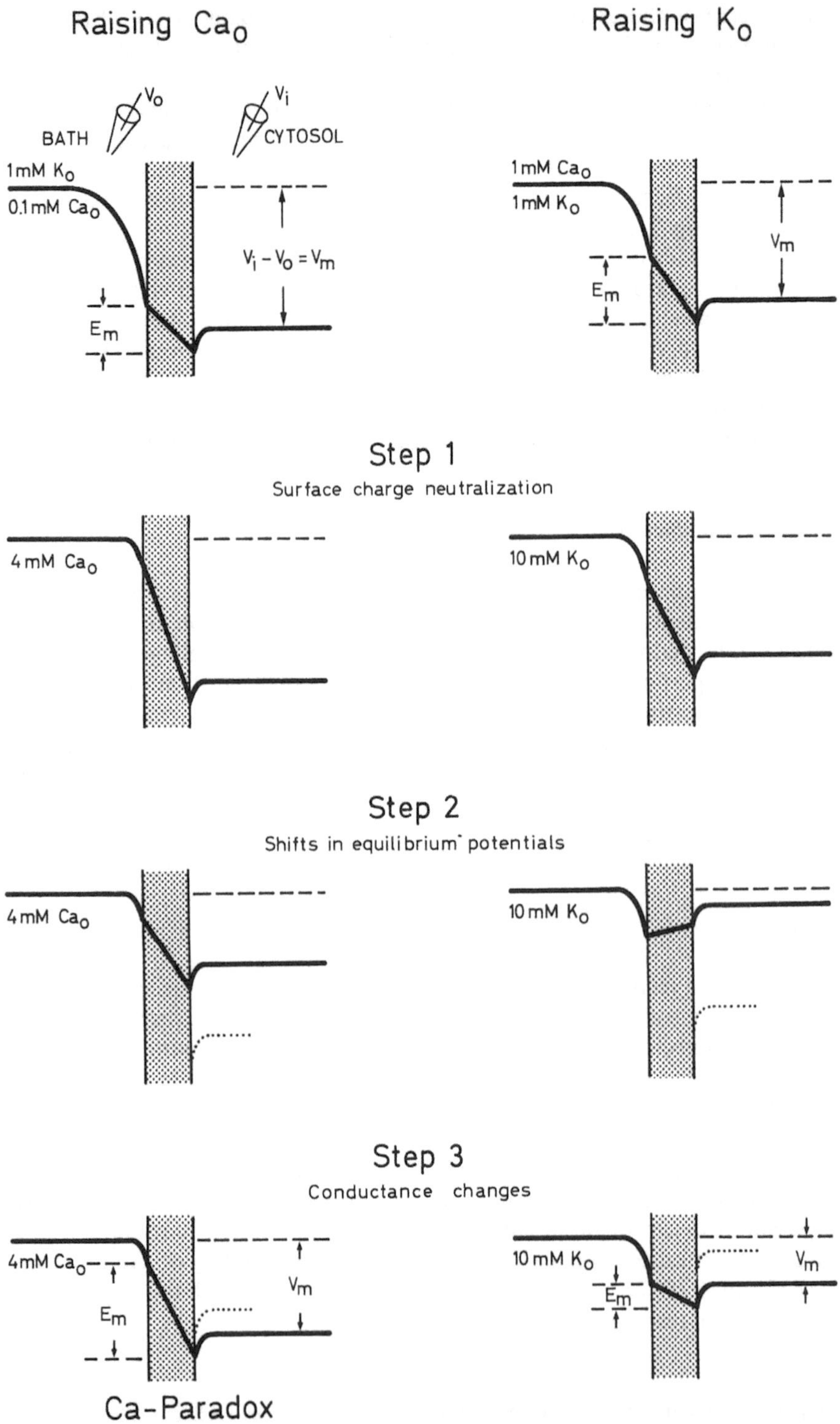
Raising Ca$_o$
Raising K$_o$
V$_o$
V$_i$
BATH
CYTOSOL
1mM K$_o$
0.1mM Ca$_o$
V$_i$ − V$_o$ = V$_m$
E$_m$
1mM Ca$_o$
1mM K$_o$
V$_m$
E$_m$
Step 1
Surface charge neutralization
4mM Ca$_o$
10mM K$_o$
Step 2
Shifts in equilibrium potentials
4mM Ca$_o$
10mM K$_o$
Step 3
Conductance changes
4mM Ca$_o$
E$_m$
V$_m$
10mM K$_o$
E$_m$
V$_m$
Ca-Paradox

◀ **Fig. 2.** How external cations affect the membrane potential. Three types of parameters: surface potentials, Nernstian equilibrium potentials and ion conductances superimpose in generating the "real" transmembrane potential, E_m. The bulk membrane potential, as "read" from the internal (V_i) and reference voltage electrodes (V_o), is the arithmetic sum of E_m and the extermal and internal surface potentials. Cations, in particular Ca^{2+}, neutralize the negative surface charges, so that the voltage offset between E_m and V_m changes together with the external ion concentration. This discrepancy between "real" and "read" membrane potential may lead to paradoxical observations in the case of changing external Ca (e.g. from 0.1 to 4 mM; *left column*). If V_m and E_m change largely in proportion, such as upon raising external K (e.g. from 1 to 10 mM), the voltage offset between these parameters in less meaningful. For the purpose of analysis, the figure attempts to separate the instantaneous electric events (*top diagrams → bottom diagrams*) into three steps.
Left column: step 1: Partial surface charge neutralization hyperpolarizes E_m but leaves V_m unmodified; *step 2*: a positive shift in E_{Ca} depolarizes the membrane potential (E_m, V_m); *step 3:* Ca binding and screening decreases the leakage conductances, in particular for Ca, so that g_{Ca}/g_K drops, and the membrane potential (E_m, V_m) hyperpolarizes. *Right column: step 1:* Comparatively less effective surface charge neutralization by K; *step 2:* with g_K as the leading conductance, a ten-fold increase in K_o strongly shifts E_m and V_m in the positive direction; *step 3:* the depolarization raises g_K, lowers g_{Ca}/g_K and therefore induces a negative shift in E_m and V_m. Consult Section 4.2 for conclusions

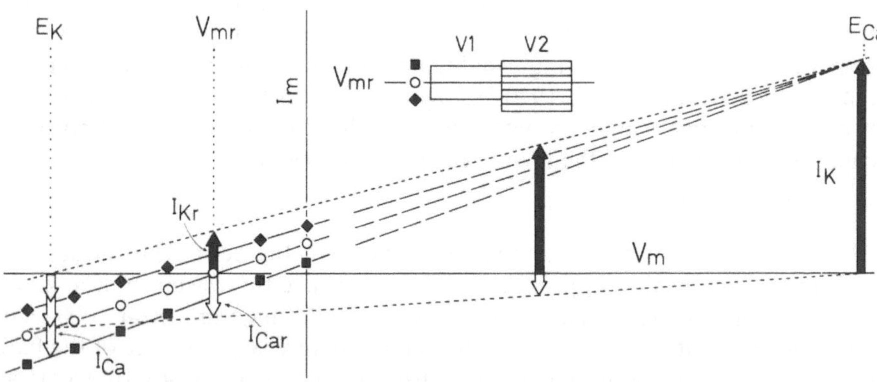

Fig. 3. Diagrammatic summary of the instantaneous current-voltage relationship under voltage clamp in *Paramecium*. In Ca-K solution, the net membrane current is the arithmetic sum of an outward K current (I_K; *filled arrows*) and an inward Ca current (I_{Ca}; *open arrows*) which add up to a zero current at the membrane resting potential (V_{mr}). A pure Ca current flows at the K-equilibrium potential, E_K, and a K current at the Ca-equilibrium potential, E_{Ca}. Sizes of the current components are indicated by *dashed lines* for any membrane potential between E_K and E_{Ca}. Weak prepulses within ±10 mV (*inset, V1*) do not affect the K conductance, but they alter the conductance for Ca, as indicated by the observed near-parallel shifts of the instantaneous I−V relationship of the main pulse (*inset, V2*) between E_K and 0 mV. A predepolarization (*squares*) raises, a prehyperpolarization (*diamonds*) diminishes a persistent Ca leakage current, as indicated by the size of Ca currents in E_K. With prepulses beyond ±20 mV the parallel shift of the I−V relation is disturbed due to activation of an additional K conductance (Naitoh and Machemer, unpublished data)

4.5 Capacitance and Membrane Area

The input capacitance (C_{inp}) of *Paramecium* is calculated from the time constant divided by the input resistance:

$$\tau = R_{inp} C_{inp}. \quad (3)$$

A 200-μm cell has a typical input capacitance of 700 pF which is reduced to about 50% upon cell deciliation. Regrowth of the cilia goes along with a rise in capacitance and voltage-dependent Ca current (Dunlap 1977; Machemer and Ogura 1979). Assuming a specific membrane capacitance of 1 μF cm^{-2}, the ciliary and somatic areas are calculated (near 3.5×10^{-4} cm^2 for each in a 200-μm cell). Similar values result using the geometric data of approximately 5000 cilia and the sculptured cell cortex. The large ciliary membrane area, which is in *Didinium* near 60% (Pape and Machemer 1986) and in *Stylonychia* 70% of the total surface area (Machemer and Deitmer 1987), explains why ciliates like to shed their ciliary organelles under various conditions of stress. Lines of scission pre-exist distal to the ciliary axosome in *Paramecium.*

4.6 Cable Properties

4.6.1 Soma

The large cell soma of *Paramecium* is a good conductor of currents originating from local stimulation (Sect. 5). Point injection of current pulses led to voltage responses which were indistinguishable near the anterior and posterior ends of the cell (Eckert and Naitoh 1970; Satow and Kung 1979). The length constant (λ) of a "biological cable" is determined by the specific resistance (R_i) of the core (the cytoplasm) and the membrane resistance (R_m):

$$\lambda = \sqrt{r R_m / 2 R_i} \quad [\mathrm{cm}], \quad (4)$$

where r denotes the radius of the cable (here, of the cell). The intracellular resistivity (200 Ωcm) is approximated from the specific resistance of a 40 mM KCl solution (190 Ωcm). The resistivity of the outer medium (near 1000 Ωcm) does not contribute to λ because of the comparatively large volume of the bath. R_m of the resting membrane in standard solution is about 4×10^4 Ωcm^2 assuming a total surface of 7×10^{-4} cm^2 and an input resistance of 6×10^7 Ω. With the diameter of a *Paramecium* "cable" estimated at 45 μm, a length constant of 4.7 mm results, which is equivalent to a 5% decay of a voltage signal along a 200-μm body length. This rough estimate of the body length constant corresponds to observations.

4.6.2 Cilia

The spread of voltage signals along the cilia is more critical because these organelles are narrow, long cables (diameter 0.25 μm; length 10 μm). A stimulus-regulated behaviour of the cell requires that ciliary activity is coupled to the locally arising receptor response (see Chap. 14, this Vol.). Although the total input resistance of the ciliary membrane has been beyond direct measurement (see Sect. 4.4), reasonable assumptions of the shunting effect of the ciliary resistance suggest that this resistance is in the order of 1 GΩ corresponding to a specific ciliary membrane resistance of about 4×10^5 Ωcm²; this leads to a ciliary length constant of 1.1 mm using Eq. (4). Assuming that only 50% of the ciliary volume are occupied by electrolytes, the calculated value of λ is slightly reduced (0.94 mm). This rough approximation shows that cilia are presumably excellent cables. A signal, such as a receptor potential, suffers no significant loss in amplitude from the base to the ciliary tip.

As the ciliary conductance rises due to depolarization-dependent channel activation, the length constant drops. With a calcium driving force of 120 mV passing an 8-nA peak Ca current across the total ciliary surface of *Paramecium*, λ will be reduced to a value slightly above 100 µm; this length constant still supplies at least 90% of a basal signal amplitude at the tip of the cilium. In summary, the cable properties of *Paramecium* are such as to provide virtual isopotentiality throughout the extension of the membrane. This conclusion corresponds to observations of the rapid and synchronous activation of stimulus-induced ciliary motor responses.

5 Responses to Stimuli

5.1 Sensory Transduction

Cells which are permanently exposed to the outside world have commonly developed the ability to sense "stimuli" which represent meaningful alterations in physical or chemical properties of the normal environment. The process of conversion of the stimulus into a proper signal of the cell is called sensory transduction. A characteristic of sensory transduction is that the impinging energy of the stimulus is a minor fraction of the intrinsic signal energy released. In unicellular organisms, such as ciliates, this rule does not apply throughout. Some type of stimulus may be meaningful only at raised energy levels; here, mechanisms of attenuation serve to restrict the sensory response to adequate intensities of the stimulus.

Sensory transduction appears to go along, in most cases, with bioelectric events called receptor potentials. These potentials are, in essence, a negative or positive transient alteration of the resting membrane potential. Some kinds of signals can directly and rapidly pass the cell membrane, such as CO_2 (Roos and Boron 1981), thus affecting the intracellular balance of pH. In hormone action and chemosensation a signal molecule is thought to bind to a receptor molecule of the membrane, thus inducing the production of second messenger molecules (e.g. cyclic AMP) at the inner face of the membrane (cf. Kaissling 1982).

5.2 Receptor Potentials

Receptor potentials occur whenever a stimulus interferes with the membrane conductance of a particular ion or class of ions. In theory, either the entrance of an ion or the electric consequence of the conductance change may be relevant to sensory transduction. Experiments in ciliates have indicated that the receptor potential is the signal, whereas the nature of the charge carrier has no meaning (see Chap. 14, this Vol.). Among various modes of stimuli processed by *Paramecium*, mechanosensation is highly developed and is, at present, comparatively well understood. A separate chapter of this book is devoted to chemotransduction. Photo- and thermoresponses of *Paramecium* will be reviewed in Sections 5.6 and 5.7.

5.3 Anterior Mechanoreceptor Response

5.3.1 Ion Mechanism

The existence in *Paramecium* of a responsiveness to a mechanical disturbance has been well known to observers, because the bumping of the anterior cell end into a mechanical obstacle induces an eye-catching behaviour of the cell, including a transient reversal in swimming direction. This response has been termed "avoiding reaction" by Jennings (1906). Mechanical-pulse stimulation and recording of the electric membrane events showed that deformation of the anterior cell end leads to a depolarizing receptor potential which triggers an action potential (Naitoh and Eckert 1969a). The ionic flux was presumed to be related to Ca^{2+} although this event was difficult to separate from of the Ca-dependent action potential. Artificial deciliation of *Paramecium* using 5% ethanol (Ogura 1977) served to separate the mechanoreceptor responses from the action potential which originated from the ciliary membranes (Ogura and Takahashi 1976; Dunlap 1977). Mechanical stimulation of a deciliated cell under voltage clamp, i.e. with the membrane potential held at the resting level by electronic feedback, revealed the existence of a reversal potential[8] of the anteriorly elicited receptor response, which was at the far positive side of the resting potential (Ogura and Machemer 1980). This level of the reversal potential is less than the inferred Ca equilibrium potential ($\leq$ 100 mV) so that two opposing receptor currents, inward as well as outward, appear to be activated by the anterior mechanical stimulus. Nernst plots of the anterior reversal potential are in fact positive with rising concentrations of external K^+ (14.8 mV/log aK_o) and Ca^{2+} (21.7 mV/ aCa_o). These results have suggested a stimulus-induced conductance ratio ($\Delta g_{Ca}/\Delta g_K$) of 3.

[8] At the reversal potential a current or voltage response of the membrane changes polarity.

5.3.2 Properties

The anteriorly elicited receptor current may be carried by a variety of divalent cations including Mg^{2+} (Deitmer 1983; Satow et al. 1983). In *Stentor*, Cl^- besides Ca^{2+} can permeate the anterior channel (Wood 1982). The current amplitude is typically around 1 nA in *Paramecium;* much larger amplitudes have been recorded in *Stylonychia* (cf. Machemer and Deitmer 1985). The function of the change in membrane conductance is to depolarize the membrane so as to induce a graded action potential. Using step-pulse stimulation under voltage clamp, the inward receptor current was shown to rise to peak within about 5 ms including rates of up to 0.2 nA ms^{-1}, and to decay with time constants of 25 ms or more. Multiple stimulation can lead to summation of the receptor potential (Naitoh and Eckert 1973; De Peyer and Machemer 1978) comparable to postsynaptic events. An electrical integration at the membrane level of various kinds of stimuli is one of the preconditions for ciliary motor coordination in *Paramecium.*

5.4 Posterior Mechanoreceptor Response

5.4.1 Bipolar Sensitivity

Touching *Paramecium* at the posterior cell end leads to an acceleration of forward swimming (Jennings 1906). Bipolarity in the sensitivity to stimulation, and a dual behavioural response, appear to apply to many free-swimming ciliates (*Paramecium:* Naitoh and Eckert 1969a; *Euplotes:* Naitoh and Eckert 1969b; *Stylonychia:* De Peyer and Machemer 1978), but is not the rule. For the gymnostome *Dileptus* (Doroszewski 1970) and *Didinium* (Hara and Asai 1980), for *Tetrahymena* (Onimaru et al. 1979) and the heterotrichous *Stentor* (Wood 1975, 1982) only depolarizing mechanoreceptor potentials or the corresponding ciliary reversal responses have been recorded.

5.4.2 Ion Mechanism

The hyperpolarizing mechanoreceptor potential is graded with the stimulus intensity comparable to the depolarizing receptor potential, but its amplitude is limited by the reversal potential which occurs typically 20 to 60 mV negative to the resting potential. Tests of the ion dependence of this posteriorly elicited receptor response revealed a slope of the reversal potential which corresponded to that of a pure potassium ion battery (Ogura and Machemer 1980). Since the reversal potential coincided with the K-equilibrium potential, it was possible to approximate the intracellular K^+ activity (34 mM). Among other alkaline ions only rubidium interfered with the reversal potential of the posterior response in *Stylonychia* (cf. Machemer and Deitmer 1985). Thus, the posterior mechanoreceptor channel is distinguished from the anterior channel in that it may be selective to cations.

5.4.3 Properties

Mechanostimulation experiments under voltage clamp reveal an outward current involving a large latency (2 ms; inward current: 1 ms), a long minimal rise time (8–10 ms; inward: 5 ms), a large maximal amplitude (17 nA; inward: 1 nA) and a short decay time constant (8 ms; inward: ≧25 ms) (cf. Machemer and Deitmer 1985). These differences in temporal properties of hyperpolarizing and depolarizing receptor responses confirm the existence of two separate mechanically sensitive conductances in *Paramecium*. The long minimal latency of the posterior K^+ receptor current and its sigmoidal activation time course suggest that an unknown mechanism intervenes between the deformation of the membrane and the eventual conductance increase. It is conceivable that the K^+ conductance rises due to a stimulus-induced Ca release from intracellular stores comparable to Ca-dependent K activation in nervous tissue (Meech 1978). This is an unlikely explanation, however, since, upon the intracellular application of a Ca-complexing buffer (EGTA), the posterior mechanoreceptor response in *Stylonychia* remained unmodified (Deitmer, unpubl.).

5.5 Role of Cilia in Mechanosensory Transduction

5.5.1 Sensing by Cilia?

Any obstacle in the way of a swimming *Paramecium* meets the cilia prior to the soma membrane, suggesting that a cilium may serve as a stimulus-sensing organelle. Deciliation experiments in *Paramecium* have indicated, however, that the mechanoreceptor potential arises in the membrane of the cell soma and not of the cilia (Ogura and Machemer 1980). Thus, the cilium and the sites of transduction appear to be separated in space. A plausible view is that the ciliated cortex is structurally organized so as to mechanically isolate the soma membrane from the beating cilium. If this were not the case, mechanical "noise" produced by thousands of cilia could impair, or even prevent, the reception of a meaningful stimulus. On the other hand, there are specialized, elongated and immobile cilia in places of pronounced hyperpolarizing mechanosensitivity in *Paramecium* and *Stylonychia*. This raises the question of the proper function of these cilia.

5.5.2 Local Stimuli

Calibrated, local stimulation (1) of the tail cilia in *Paramecium* and (2) of the soma membrane near the tail cilia showed that these cilia were less sensitive than the soma membrane, and a mechanical impact upon the cilia led to a much delayed receptor response (Machemer-Röhnisch and Machemer 1984). The response latency doubled following stimulation of the base of the tail cilia as compared to the neighbouring soma, and rose to values beyond 10 ms with the stimulating probe placed to the distal ends of the cilia. Together with an increase in response latency, the receptor current amplitude decreased to an average of 20% of the response of the somatic centre of the tail (cf. Machemer and Deitmer 1985). These data indicate that upon a localized mechanical disturbance, the tail cilia in *Paramecium* serve to attenuate and delay the receptor response in the manner of mechanical low-pass filters.

5.5.3 Non-Local Stimuli

Interpretations of the role of "sensory cilia" from local stimulation experiments have limited applications. Local impacts are rare events at the rear end of the cell, whereas omnidirectional mechanical shocks appear to be more common types of stimuli. Generation of a non-local stimulus, such as bumping the culture dish on the table, leads to an acceleration of forward swimming of the paramecia. A hydrodynamic shock wave was experimentally simulated by a mechanical pulse focused at the somatic centre of the tail so as to induce in anteriorly directed passive retreat of the cell followed by a backward swing due to elastic release of the holding needles. During this delayed back-swing of the cell, the hitherto straight tail cilia were strongly deflected. Inspection of the electrophysiological records has confirmed that even sub-threshold pulses to the cell soma may induce a long-latency receptor current following cumulative deflection of the tail cilia (Machemer and Machemer-Röhnisch 1984). Hydromechanical forces of this kind also occur during backward swimming of *Paramecium* and may have a function in limiting the period of ciliary reversal (see Chap. 14, this Vol.). For non-local, multidirectional mechanical forces the tail cilia appear to focus a mechanical disturbance to the most sensitive part of the soma membrane. In summary, the role of cilia in reception of mechanical stimuli is limited to events preceding transduction and consists in transmission of the mechanical energy to the soma membrane. Transmission always involves an additional delay and, under certain circumstances, an attenuation of the receptor response.

5.6 Responses to Temperature

5.6.1 Behavioural Parameters Affected

Heat energy can affect every molecule of the cell which is a problem for the signalling-chain specificity of this modality (Poff 1985). Early observations have established that *Paramecium* avoids extreme temperatures and aggregates in regions near the cultivation temperature (Mendelssohn 1902). The cells are sensitive to temperature throughout: paramecia escape a posteriorly applied stimulus using enhanced forward swimming, while they swim backward when the stimulus is applied anteriorly (Koehler 1935). The principle of sensing temperature in *Paramecium* appears to be "comparison in time" with the entire cell body being the sensor. Cell aggregation resembles in some respects the "tumble and run" mechanism seen in bacterial thermotaxis and chemotaxis (Maeda et al. 1976).

Raising of the environmental temperature increases the rate of ciliary beating (Machemer 1972), raises the swimming velocity (Tawada and Oosawa 1972) and the probability of avoiding reactions (Nakaoka and Oosawa 1977; Hennessey and Nelson 1979). Lowering of the temperature not only reverses the heat-dependent behaviour but, in addition, increases the duration of a depolarization-induced reversed beating of the cilia (Machemer 1974) and backward swimming in *Paramecium* (Hildebrand 1978). *Paramecium* mutants with defective Ca channels show little or no thermal avoidance (Hennessey and Nelson 1979).

5.6.2 Temperature and Membrane Response

Applying slow shifts in the bathing temperature the voltages of the resting and excited membrane are essentially insensitive to temperature (*Paramecium:* Yamaguchi 1960; Martinac and Machemer 1984; *Stylonychia:* De Peyer and Machemer 1977); in these cells the observed negative shift of 2 mV per 10° C increase largely corresponds to predicted shifts in equilibrium potentials of Ca and K. The kinetics of the membrane responses slow down with decreasing temperature (cf. Machemer and De Peyer 1977). Culturing *Paramecium* at reduced temperature raises the voltage-activated membrane Ca conductance indicating that a feedback loop exists to maintain a constant Ca entry into the cilium at different reference (= culturing) temperatures (Martinac and Machemer 1984). Rapid heating (24° C min^{-1}) has led to temperature-graded depolarizations in deciliated *Paramecium* (Hennessy et al. 1983). In ciliated paramecia high heating rates, applied within physiological limits, induce frequent ciliary reversals and can eventually lead to cell death. The evidence suggests that upon heating unspecified somatic conductances may increase with a reversal potential slightly positive to the membrane resting potential (Hennessey et al. 1983). In summary, thermoreception in *Paramecium* appears to be unrelated to particular heat-detecting molecules; sensitivities to temperature of ciliary and somatic membrane ion conductances participate in the determination of the membrane potential, the intraciliary Ca concentration and the regulation of the thermotaxis. The pathway of this regulation is not yet sufficiently understood.

5.7 Responses to Light

5.7.1 Behaviour in White and Green Cells

Early investigations have shown that protozoa, colourless or coloured, may respond to illumination by avoidance (for example, *Amoeba:* Harrington and Leaming 1900; *Stentor:* Jennings 1906). Traditionally, most heterotrophous ciliates, including the "white" paramecia, were thought to be indifferent towards light, unless extreme intensities were applied (cf. Hempelmann 1926) or the cells were otherwise sensitized (Best 1954). The first qualitative tests of the photobehaviour of *Paramecium* indicated little or no avoidance of light (Bramstedt 1935; Soest 1937). In the hypotrich ciliates *Stylonychia* and *Keronopsis* an average of 42 to 47% of cells were found to reside in the illuminated fields using light/dark chessboard patterns for mass experimentation (Machemer 1966). The same method applied to a population of *Paramecium caudatum* resulted in 51% residence in the illuminated fields (Voß and Machemer 1987).

The green *Paramecium bursaria* accumulates in illuminated areas due to avoiding reactions triggered by a decrease in the light intensity (Engelmann 1882). The action spectrum for the photoaccumulation peaks around 520 nm; it thereby differs from that for photosynthesis of the symbiotic alga *Chlorella* (440 and 680 nm). Inhibitors of photosynthesis do not interfere with the photoaccumulation (Iwatsuki and Naitoh 1981).

Chlorella-free specimens of *P. bursaria* disperse in light due to avoidance of a step-up light stimulus. The same type of response occurs in green cells of *P. bursaria* which have been previously kept in the dark. Ingestion of *Chlorella* by the *Paramecium* or exposure to light of the dark-adapted green cells leads to a recovery of the normal photoaccumulation (Iwatsuki and Naitoh 1981). These data show that the responsiveness to changes in light intensity is a property of *Paramecium,* which can be modulated by the algae. Since the symbionts are enclosed by vacuoles, the signal to the host must pass the perialgal membrane (Reisser 1980).

5.7.2 Action Spectra Indicate a Dual Transduction Pathway

Subsequent tests in *Paramecium multimicronucleatum* have confirmed the results obtained in the *Chlorella*-free *P. bursaria.* The photophobic response includes a latent period of at least 0.3 ms. Photodispersal and avoidance of the step-up stimulus have the same action spectra including peaks at 520 and 680 nm. The swimming velocity increases when the white light intensity is raised; the action spectrum of this response peaks at 440 nm. No light-direction oriented movement, or phototaxis, has been observed in these cells (Iwatsuki and Naitoh 1983).

In *Paramecium* a phobic response (= transient backward swimming) occurs due to depolarization of the membrane; a membrane hyperpolarization induces an increase in the rate of forward swimming (cf. Eckert et al. 1976). It has been suggested, therefore, that two independent photoreceptor systems exist in the membrane of *Paramecium:* hyperpolarizing transduction activated by 440 nm (blue) light, and depolarizing transduction activated by 520 nm (green) and 680 nm (dark-red) light (Iwatsuki and Naitoh 1983). The manner by which these spectral sensitivities can lead to photodispersal of cells in white light, and to accumulation in the dark, is not well understood at present.

6 Voltage-Dependent Responses

6.1 Channels for Excitation and Homeostasis

Two major types of membrane conductance in *Paramecium* are activated by shifts in potential. One is specifically permeable to calcium, the other to potassium. The depolarization-sensitive Ca conductance, often briefly addressed as the "Ca channel", is the basis for excitation of the membrane including rapid activation of the ciliary motor response. The voltage-dependent K-conductance is more complex, being based on presumably three types of K channels with different properties of activation. One of these K channels is in effect directly Ca-activated, but only depolarization commonly activates the Ca influx. K channels are antagonists of Ca channels[9] and promote electric homeostasis of the membrane . An indirect voltage dependence also exists for Na conductance which is activated following the rise in intracellular Ca. Table 1 gives an overview of the voltage-dependent channels in *Paramecium.* In the following sections some basic properties of these channels will be described. For more detailed analyses the reader is referred by Ramanathan et al. (Chap. 15, this Vol.) and to the previous review literature (Naitoh and Eckert 1974; Eckert et al. 1976; Eckert and Brehm 1979; Naitoh 1982; Kung and Saimi 1985).

6.2 Action Potential

6.2.1 Activation of Ca Conductance

Depolarizing the membrane potential by 5 mV or more leads to a "regenerative" upstroke of the potential trace as visualized on the oscilloscope screen. The digression of the membrane potential from the course of the ohmic or stimulus-induced depolarization is due to a positive feedback mechanism between depolarization and rise in conductance for the Ca ion. Comparable mechanisms of excitation involving Ca^{2+} and Na^{+} are well known in tissues of nerve and muscle (Hagiwara and Byerli 1981; Reuter 1984). The action potential in *Paramecium* has been related to a rapidly activated Ca conductance, because it is dependent upon the presence of external Ca or analogues (Sr, Ba) (Naitoh and Eckert 1968a, b). Step-pulse depolarizations under voltage clamp have isolated the transient inward Ca current (Naitoh and Eckert 1974; Satow and Kung 1979) and demonstrated that this current does not flow when the driving force for Ca has been experimentally neutralized (Brehm and Eckert 1978). In *Stylonychia* two voltage-dependent Ca currents have been characterized (Deitmer 1984), which occur in different ciliary organelles (Ivens and Deitmer 1986) and are the basis of the two-peak action potential in this cell (De Peyer and Machemer 1977).

[9] In principle, also voltage-activated K channels may be excitatory: a regenerative K-dependent hyperpolarization has been described in *Paramecium* (Satow and Kung 1977), which was restricted, however, to K-free solutions; see Sect. 6.3.6.

6.2.2 *Gradedness*

The Ca action potential is graded in amplitude depending on the rate and size of the depolarizing stimulus. The varying size of the action potential is due, among other factors, to the comparatively slow kinetics of Ca-channel activation (the time to peak of the Ca current being minimally three times that of the Na current in nerve), so that activation of the K conductance, which promotes repolarization, can effectively interfere with the regenerative process. Suppression of K-channel activation, for instance, by exchanging external Ca for Ba, transforms the graded response to an all-or-none action potential (Naitoh and Eckert 1968 b).

6.2.3 *Rates and Amplitudes*

The rate of activated current across the Ca channel rises with the depolarizing voltage, as would be expected for a regenerative process. In addition, the nature of the permeating ion affects the time to peak of the transient inward current and the current decay after termination of the depolarization. The rates of current rise and decay are maximal in Ca solution; they decrease in Sr and even more in Ba solution (Saimi and Kung 1982) corresponding to observations in other ciliates (*Stylonychia:* De Peyer and Deitmer 1980; *Didinium:* Pape and Machemer 1986) and metazoan tissue (Adams and Gage 1980; Ashcroft and Stanfield 1981; Almers and Palade 1981). These properties of Ca (Sr, Ba) permeation through the voltage-dependent channel have been recently explained assuming two Ca binding sites of the channel (Hess and Tsien 1984).

The full-size Ca action potential includes depolarizations of commonly 20 to 40 mV with maximal rates of rise ranging between 6 V s^{-1} and 8 V s^{-1} (Naitoh et al. 1972) and the peak overshooting the zero potential line. Similar amplitudes and rates of rise of the action potential have been recorded in *Stylonychia* (De Peyer and Machemer 1977) and *Didinium* (Pape and Machemer 1986). Raising the temperature increases the rate and amplitude of the action potential which may be 50 mV with maximal stimulation at 25 °C (Martinac and Machemer 1984).

6.2.4 *Inactivation*

Termination of the regenerative process is due to the intracellular accumulation of free Ca^{2+}[10]. It thus differs from the intrinsic inactivation during depolarization seen in the Na channel of excitable tissues. Application of high positive voltage to the membrane suppressing the influx of Ca failed to inactivate the

[10] Under conditions which prevent Ca-dependent Ca channel inactivation, such as injection of the Ca buffer EGTA, the Ca conductance inactivates very slowly; the rate of this slow inactivation, which occurs in the course of tens of seconds, increases with rising depolarization and is thus voltage-dependent (Hennessey and Kung 1985).

Ca channel in *Paramecium* (Brehm and Eckert 1978) and in molluscan neurons (Tillotson 1979). The Ca-dependent inactivation of Ca channels is now recognized to be widespread in the animal kingdom. Modelling of the inactivation process suggests a 1:1 binding of Ca^{2+} to a (channel?) protein at the inner face of the membrane and a saturable Ca-buffering capacity of the cytoplasm (Eckert and Chad 1984).

6.2.5 Refractoriness

A consequence of channel inactivation being dependent on the permeating ion is a large relative refractoriness of the membrane [*P. caudatum:* 250 ms (Naitoh et al. 1972); *P. calkinsi:* 300 ms (Deitmer and Machemer 1982); *Stylonychia:* $\geqq$ 100 ms (De Peyer and Machemer 1977)]. Estimates of the Ca^{2+} charge entering the ciliary space[11] during an action potential lead to final transient concentrations between 10^{-5} and 10^{-3} M (Eckert 1972; Oertel et al. 1977; Machemer, unpubl.). The pumping time required to reduce the raised Ca to its resting level (near 10^{-7} M) affects refractoriness and determines the period of reversed ciliary beating following an action potential (see Chap. 14, this Vol.).

6.2.6 Persistent Ca Current

Independent voltage-clamp experiments using methods of (1) suppression of outward K current, (2) *Paramecium* mutants with a defective Ca channel and (3) cell deciliation have revealed an inward Ca current of minor amplitude which does not inactivate during sustained depolarization (Brehm et al. 1980; Hinrichsen and Saimi 1984; Pape and Machemer 1986). This persistent Ca current, which is normally masked by the prevailing outward K current, is thought to result from incomplete inactivation of the Ca channel (Eckert and Brehm 1979); in *Didinium* its time course is also consistent with activation of a separate Ca conductance (Pernberg and Machemer, unpubl. observations). The persistent Ca current explains why *Paramecium* continues to swim backward under sustained depolarization (Machemer and Eckert 1975; see also Chap. 14, this Vol.).

6.2.7 Membrane Repolarization

Repolarization from the peak of the action potential rests, in addition to Ca-channel inactivation, on a K conductance which is activated by depolarization together with the Ca conductance ("voltage-dependent K channel"). A "fast" component of K outward current has been described in *Paramecium* (Satow

[11] Voltage-activated Ca channels in ciliates are largely restricted to the ciliary membrane (see Sects. 4.3 and 7.3).

and Kung 1980a) and *Stylonychia* (cf. Machemer and Deitmer 1987). This current rises after the early Ca current has gained its maximum; it peaks within a few milliseconds and may hasten the inward current decline. The decay is rapid as well. Similar fast K currents have been known from invertebrate neurons (Hagiwara and Saito 1959; Neher 1971). Fast K activation depends on the presence of Ca in the medium, because it is not observed in solutions of other divalent cations (Salkow 1983). A second, much slower K current develops in the course of tens or hundreds of milliseconds following depolarization. This "delayed" K conductance fully repolarizes the membrane and generates the negative after-potential of the action potential (see Sect. 6.3).

6.2.8 Role of Sodium

Ca-action potentials are a general characteristic of ciliate membrane excitability including those species living in marine environments. The action potential of *Paramecium calkinsi,* which lives in brackish coastal waters, may be generated in artificial seawater (436 mM Na^+) and in Na-free solutions, whereas the removal of external Ca^{2+} suppresses the action potential (Deitmer and Machemer 1982). Ca-action potentials were also generated by *Fabrea salina,* a ciliate tested in solutions equivalent to its extreme saline habitat [2.5 times the concentration of seawater (Kubalski 1983)]. With sodium ions present in freshwater, the rates of avoiding reactions and backward swimming periods of *Paramecium tetraurelia* increase. Following step depolarization under voltage clamp, a minor transient Na influx is seen, which activates due to the previous entrance of Ca into the cell (Saimi and Kung 1980; Saimi 1986). The Na current develops in the course of seconds and decays slowly; this conductance is thus not comparable to the early Na current in excitable metazoan cells. In wild-type paramecia and with Na^+ concentrations below 2 mM, the modulating effect of sodium on the electric membrane properties and behaviour is presumably less significant; with raised Na concentrations the cell may escape, in maintaining backward swimming, from the dangerous environment.

6.3 Membrane Rectification

6.3.1 Definitions

Membranes of living cells modify their ohmic-capacitive properties when the potential is shifted away from the resting level. Comparable to its technical correlate (the rectifier), the input resistance changes with the voltage drop across the membrane. Membrane depolarization leads to a reduction in input resistance due to activation of a potassium conductance. This property is called "outward rectification", indicating that a raised outward current of K^+ passes the membrane during depolarization. Because of the slow development of this property, the synonym "delayed rectification" is in common use. Many cell membranes including those of ciliates exhibit, in addition, reductions in input resistance upon hyperpolarization (cf. Eckert and Brehm 1979). Historically, this property was called "anomalous rectification", because it did not correspond to theoretical expectations. A more descriptive term is "inward rectification" implying the facilitation of inward current. Inward rectification of the membrane develops in a voltage- and time-dependent manner similar to outward rectification.

6.3.2 Late Current-Voltage Relationship

In *Paramecium* the first signs of rectification occur with hyperpolarizations and depolarizations of 10 mV or less, so that the late[12] (or "steady-state") current-voltage (or I−V) relationship is sigmoidal (Naitoh and Eckert 1968a; Satow 1978; Machemer and Ogura 1979; Deitmer and Machemer 1982). A similar late I−V curve applies to *Stylonychia* (De Peyer and Machemer 1977). In the hyperpolarizing direction, digression of the I−V curve from linearity starts at voltages which are clearly positive to the K equilibrium potential, indicating an outward-directed K driving force[13]. Beyond the K-equilibrium potential this driving force reverses so that K currents are inward.

6.3.3 Outward Rectification is Ca-Dependent

Outward rectification was depressed in a pawn mutant of *Paramecium tetraurelia* with non-functional Ca channels (Satow and Kung 1980a) and absent in *P. calkinsi* in Ca-free artificial seawater (Deitmer and Machemer 1982). This confirms previous findings in nerve tissue that Ca ions entering the cell can activate an outward rectifier channel (Meech 1978). An early observation in *Paramecium* that rising external Ba concentrations increasingly linearize the positive limb of the late I−V relation (Naitoh and Eckert 1968a) is understood in the light of this view, because Ba competes successfully for passage through the Ca channel at relatively low Ca concentrations (Hess and Tsien 1984). In *Stylonychia* the total replacement of Ca by Ba abolishes the delayed outward current (De Peyer and Deitmer 1980; Ballanyi and Deitmer 1984). The deciliation of *Paramecium* using 5% ethanol did not interfere with the late I−V relationship[14] (Machemer and Ogura 1979), suggesting that the outward rectifier channels are not necessarily activated by Ca from ciliary channels (see also Sect. 7).

---

[12] Late measurement (after $\geqq 100$ ms) because of the delayed K-current activation; an early I−V relationship is generated only under voltage clamp and describes the peak inward action current for different step potentials.

[13] Note that charges of a constant hyperpolarizing (= inward) current are not carried by K^+, as long as the driving force for K is outward; here, raised K efflux is compensated by an even more raised influx of positive charges, presumably of Ca, across the soma membrane.

[14] Experiments employing chloral hydrate for the deciliation led to a linearization of the I−V relation (Brehm et al. 1978); whether this was an effect of the drug or of deciliation remained untested.

6.3.4 A Two-Component Outward Rectification?

While outward rectification is clearly Ca-dependent in ciliates, an additional facilitating voltage dependence has been noted (Satow and Kung 1980a, b; Deitmer 1984) in accordance with observations in neurons (Gorman et al. 1981). A slow inactivating, Ca-independent component of the outward current has been isolated in *Paramecium* using mutant analysis (Saimi et al. 1983). In *Didinium* outward rectification includes two slow current components. The first component starts rising after 10 ms and is not abolished in solutions with Ca fully replaced by Ba. The second component superimposes on the first current after about 50 ms; it is suppressed in deciliated cells or with external Ca missing. Since the time constants of decay of these currents are unaffected by the manipulations mentioned, the two components of the outward current may pass the same K channel suggesting the possibility that Ca from external sources modulates a basically voltage-dependent outward rectifier. It is possible that similar slow components of the outward current, voltage- and Ca-dependent, exist in *Paramecium.* In squid neuron, voltage-dependent K channel activation was suppressed after the removal of Ca from the external medium (Armstrong and Lopez-Barneo 1987).

6.3.5 Inward Rectification

Injection of constant current reveals a delayed relaxation of the voltage response in various ciliates (Naitoh and Eckert 1968a; Machemer and De Peyer 1977; Pape and Machemer 1986). Hyperpolarizations showing first signs of rectification range from a few millivolts to 50 mV or more, depending on the type of cell and composition of the bath solution. Rising hyperpolarizations increase rectification, which is even more enhanced at raised external K (Oertel et al. 1978; Pape and Machemer 1986). Removal of Ca from seawater virtually eliminated inward (as well as outward) rectification in *Paramecium calkinsi* (Deitmer and Machemer 1982). In *Didinium* raised external Ca facilitated rectification (Pape and Machemer 1986). Ba has been shown to block the inward rectifier in *Stylonychia* (Ballanyi and Deitmer 1984). Tests of the reversal potential of tail currents in *Paramecium* clearly indicate that the activated conductance is specific for K (Oertel et al. 1978). The same conclusion was drawn from a frequency analysis of membrane current noise in *Paramecium* (Moolenaar et al. 1976). These properties of inward rectification in ciliates correspond to those found in other metazoan cells including vertebrate heart muscle (Noble 1984) and squid axon (Armstromg and Binstock 1965).

6.3.6 K-Spike

A peculiar property of inward rectification in *Paramecium* becomes evident upon removing K^+ from the external solution, thus shifting E_K far beyond -150 mV. Upon hyperpolarization beyond $\Delta 60$ mV the "active" membrane response (slow K conductance activated) turns "regenerative" (positive, albeit slow, feedback between increase in K conductance and hyperpolarization) reaching potentials near −120 mV (Satow and Kung 1977). Electrogenicity is evident from the reduced input resistance during this extreme response. With K ions present in the solution, the slow-spike producing power of inward rectification (voltage dependence of K conductance; E_K negative of V_m) is commonly absent (Satow and Kung 1977: Fig. 1) or much limited (Machemer and De Peyer 1977, Fig. 1).

7 Topology of Ion Channels

Uneven distributions of populations of channels in the membrane may be eminently important for proper functioning of the cell, a fact which is well recognized in the case of neurons employing segments specialized for the reception, integration and transmission of signals. That also ciliates depend on similar specializations of membrane territories, came nevertheless as a surprise, after the recording techniques had become suitable for in-depth studies. This review gives a brief systematic survey; it does not attempt to follow the historical pathway.

7.1 Channels for Cell Homeostasis Restricted to the Soma

Experiments in *Paramecium* using ethanol for deciliation have established that the resting potential, membrane input resistance and the sigmoidal late I−V relationship are virtually unmodified following the loss of the cilia, i.e. 50% of the total membrane surface (Machemer and Ogura 1979). Consequently, the conductances for the Ca and K leakage current, and for outward and inward rectification, occur in the soma membrane. Such specialization is useful, because permanent inward fluxes of Ca and outward fluxes of K pass the membrane resting conductance; passive fluxes and the following uphill transport of these ions are optimally placed in a compartment providing a small surface-to-volume ratio such the cell soma. It should be noted that the resting soma membrane is a source for steady Ca influx, whereas ciliary membranes supply, in the first place, brief and large Ca currents.

As pointed out in Section 4.3, the Ca resting conductance is slightly voltage-dependent, thus levelling the late I−V relationship near the resting potential (Fig. 3)[15]. In *Paramecium* it has not yet been crucially determined by deciliation experiments whether this minor voltage dependence of the Ca conductance comes from ciliary or somatic membranes. The identity of I−V relationships in ciliated and deciliated cells suggests, however, that this is a somatic conductance.

[15] Caution is therefore advised when using this section of the I−V relation for the determination of the resting input resistance!

7.2 Mechanoreceptor Channels Restricted to the Soma

7.2.1 Evidence of Location

Receptor potentials, depolarizing as well as hyperpolarizing, occur in the soma membrane, as indicated by deciliation experiments in *Paramecium* (Ogura and Takahashi 1976; Ogura and Machemer 1980). Testing ciliary mechanosensitivity with localized stimulation revealed longer response latencies and smaller amplitudes; this crucially excludes the possibility that some mechanoreceptor channels reside in the ciliary membrane (Machemer-Röhnisch and Machemer 1984). The somatic receptor channels may occur in distinct spatial patterns with respect to the bases of the tail cilia in *Paramecium* (see Sect. 5.5.3). The relevance of restriction of mechanosensation to the soma is considered in Section 5.5.1.

7.2.2 Pattern of Distribution

Local stimulation in *Paramecium* and *Stylonychia* using calibrated pulse stimulation has shown that depolarizing sensitivity decreases in the posterior direction, and hyperpolarizing sensitivity decreases in the anterior direction of the cell. At intermediate latitudes of stimulation, in a narrow zone near the cytosome, the receptor response is missing (Ogura and Machemer 1980; De Peyer and Machemer 1978). Voltage-clamp analysis of the receptor currents, recorded with the holding potential clamped to the reversal potential of the posterior response and to the anterior response respectively, has revealed the existence of two gradients in mechanosensitivity in the *Paramecium* soma membrane. The Ca-dependent gradient decreases posteriorly and extends beyond the middle of the cell; it is absent only at the posterior soma apex; the K-dependent gradient decreases anteriorly and is still existent at the front end of the cell. Overlapping of the two opposing sensitivity gradients implies that at some critical latitude two equal and opposite mechanoreceptor currents may be shortcircuiting so that a receptor potential does not arise (Ogura and Machemer 1980). The gradient-type distribution of mechanoreceptor channels in ciliates bears upon the reconstruction of patterns of mechanosensitivity in daughter cells following cell division.

More detailed tests of somatic mechanosensitivity have indicated that the orientation of the gradients is not strictly longitudinal. In *Paramecium* the depolarizing responses were maximal upon stimulation slightly posterior and ventral to the front end of the cell; maximal hyperpolarizing responses were elicited a small distance anterior and dorsal to the rear apex (Ogura and Machemer 1980). Here, the sensitivity decreased unidirectionally from dorsal over the left and ventral to the right side of the cell (Machemer-Röhnisch and Machemer 1984) suggesting the possibility that the suture of kineties constitutes a further starting line of the gradient system. The existence of a similar dual gradient system (longitudinal and extending between kinetal suture and cytostome) has been previously inferred from patterns of coordinated ciliary movement

(Machemer 1969). At present, it is not understood how the receptor channels can be distributed in a gradient-type manner.

7.3 Voltage-Dependent Ca Channel Restricted to Cilia

Membrane excitability of *Paramecium* is missing following removal of the cilia (Ogura and Takahashi 1976), and reappears as they grow again (Dunlap 1977; Machemer and Ogura 1979). The data do not allow the conclusion that the Ca channels are evenly distributed in the ciliary membrane because ciliary regrowth is asynchronous (Ogura 1981). The occurrence of rapid electromotor coupling and plausible reasoning suggest a distribution providing the shortest possible distance between the Ca channels and the target for Ca influx, the axoneme, which generates active sliding over its entire length. In *Paramecium* all membranes of active cilia are presumably endowed with Ca channels, perhaps including the tail cilia, which can tilt at their bases and may become active under extreme conditions. In the more specialized ciliary organelles of *Stylonychia* the classical Ca channel (having a wide range of thresholds) was found to occur in the cirri, while a second low-threshold type of Ca channel predominates in the membranelles (Deitmer 1984; Ivens and Deitmer 1986).

The present evidence is favourable to the view that the ciliary membrane is impermeable to ions except those passing the voltage-dependent Ca channel (Sect. 7.1). The cilia having an extreme surface-to-volume ratio and being narrow cables (Sect. 4.6) may be exempted from those membrane processes which are not essential for, or may even impair, their proper function including conduction of signals, excitation and motion.

8 Conclusion

The electrophysiological organization of *Paramecium,* as viewed from the present state of the data, is summarized in a diagram including relations between the membrane conductance of particular compartments, ions and potentials (Fig. 4). It is seen that the membrane potential is regulated in such a way as to swing between the equilibrium potentials of Ca and K. External stimuli are transduced in changing the membrane conductance for Ca and/or K. Signal integration and conduction occurs on the basis of potentials (= shifts in membrane potential), the species of the permanent ion being irrelevant. The Ca ion is central for regulating the Ca and K membrane conductances for homeostasis and the excitable Ca channel of the cilia. All these regulations serve one major function: the guidance of the behaviour of the organism so that it can adequately respond to the stimuli from the outside world. It is clear that the membrane functions of a cell are embedded in a multitude of other provisions of the eukaryotic cell organization, such as structure, morphogenesis, biochemistry and genetics. This book, in assembling various aspects of the biology of *Paramecium,* may serve to provide a more complete view of the relevance of bioelectric control for the daily life of the single-cell organism.

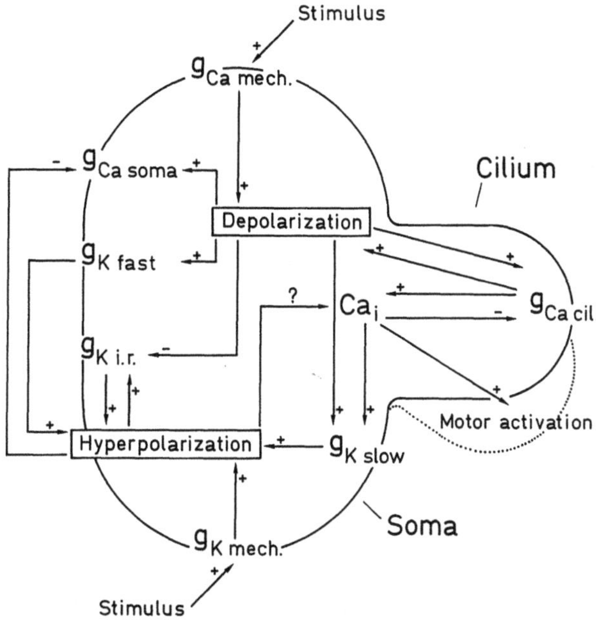

Fig. 4. Simplified diagram of relations between external stimuli, membrane conductances, potentials, intracellular Ca concentration and the ciliary motor activation. Enhancing effects are indicated by a *positive sign,* depressing effects by a *negative sign.* Note the positive feedback loops between depolarization and $g_{Ca\,cil}$, and between hyperpolarization and $g_{K\,i.r.}$. Stabilizing negative feedback loops exist between Ca_i and $g_{Ca\,cil}$, and between the membrane potential and $g_{K\,slow}$. The activation of the cilia by hyperpolarization is potentially mediated via Ca_i (see Chap. 14, this Vol.). The conductances indicated refer to those of Table 1

Acknowledgments. This work was supported by the Deutsche Forschungsgemeinschaft, SFB 114 TP A5, and Konzell-3. I would like to thank Mrs. J. Ruthmann for suggesting improvements of the manuscript.

References

Adams D, Gage P (1980) Divalent ion currents and the delayed potassium conductance in an *Aplysia* neurone. J Physiol 304:297–313

Almers W, Palade PT (1981) Slow calcium and potassium currents across frog muscle membrane: measurements with a vaseline-gap technique. J Physiol 312:159–176

Armstrong CM, Binstock L (1965) Anomalous rectification in squid giant axon injected with tetraethylammonium chloride. J Gen Physiol 48:859–872

Armstrong CM, Lopez-Barneo J (1987) External calcium ions are required for potassium channel gating in squid neurons. Science 236:712–714

Ashcroft FM, Stanfield PR (1981) Calcium dependence of the inactivation of calcium currents in skeletal muscle fibres of an insect. Science 213:224–226

Ballanyi K, Deitmer JW (1984) Concentration-dependent effects of Ba on action potentials and membrane currents in the ciliate *Stylonychia.* Comp Biochem Physiol 78A:575–581

Best JG (1954) The photosensitization of *Paramacium aurelia* by temperature shock. J Exp Zool 126:87–99

Boheim G, Hanke W, Eibl J, Schultz JE (1981) Characterization of a Ca^{2+}-dependent cationic channel from cilia of wild-type and pawn-mutant *Paramecium* by incorporation into planar bilayers. Hoppe Seyler's Z Physiol Chem 362:192

Bramstedt F (1935) Dressurversuche mit *Paramecium caudatum* und *Stylonychia mytilus.* Z Vergl Physiol 22:490–516

Brehm P, Eckert R (1978) Calcium entry leads to inactivation of calcium channel in *Paramecium.* Science 202:1203–1206

Brehm P, Dunlap K, Eckert R (1978) Ca-dependent repolarization in *Paramecium.* J Physiol 274:639–654

Brehm P, Eckert R, Tillotson D (1980) Calcium-mediated inactivation of Ca current in *Paramecium.* J Physiol 306:193–203

Deitmer JW (1983) Ca channels in the membrane of the hypotrich ciliate *Stylonychia.* In: Grinnell A, Moody WJ (eds) The physiology of excitable cells. Liss, New York, pp 51–63

Deitmer JW (1984) Evidence for two voltage-dependent calcium currents in the membrane of the ciliate *Stylonychia.* J Physiol 355:137–159

Deitmer JW, Machemer H (1982) Osmotic tolerance of Ca-dependent excitability in the marine ciliate *Paramecium calkinsi.* J Exp Biol 97:311–324

De Peyer JE, Deitmer JW (1980) Divalent cations as charge carriers during two functionally different membrane currents in the ciliate *Stylonychia.* J Exp Biol 88:73–89

De Peyer JE, Machemer H (1977) Membrane excitability in *Stylonychia:* properties of the two-peak regenerative Ca-response. J Comp Physiol 121:15–32

De Peyer JE, Machemer H (1978) Hyperpolarizing and depolarizing mechanoreceptor potentials in *Stylonychia.* J Comp Physiol 127:255–266

Doroszewski M (1970) Responses of the ciliate *Dileptus* to mechanical stimuli. Acta Protozool 7:353–362

Dunlap K (1977) Localization of calcium channels in *Paramecium caudatum.* J Physiol 271:119–133

Eckert R (1972) Bioelectric control of ciliary activity. Science 176:473–481

Eckert R, Brehm P (1979) Ionic mechanisms of excitation in *Paramecium.* Annu Rev Biophys Bioeng 8:353–383

Eckert R, Chad JE (1984) Inactivation of Ca channels. Prog Biophys Mol Biol 44:215–267

Eckert R, Naitoh Y (1970) Passive electrical properties of *Paramecium* and problems of ciliary coordination. J Gen Physiol 55:467–483

Eckert R, Naitoh Y, Machemer H (1976) Calcium in the bioelectric and motor functions of *Paramecium.* Symp Soc Exp Biol 30:233–255

Ehrlich BE, Finkelstein A, Fuorte M, Kung C (1984) Voltage-dependent calcium channels from *Paramecium* cilia incorporated into planar lipid bilayers. Science 225:427–428

Engelmann TW (1882) Über Licht- und Farbenperception niederster Organismen. Arch Ges Physiol 29:387–400

Ettisch G, Peterfi T (1925) Zur Methodik der Elektrometrie der Zelle. Pflügers Arch 208:454–466

Frankenhaeuser B, Hodgkin AL (1957) The action of calcium on the electrical properties of squid axons. J Physiol 137:218–244

Gorman ALF, Thomas MV, Hermann A (1981) Intracellular calcium and the control of neuronal pacemaker activity. Fed Proc 40:2233–2239

Hagiwara S, Byerli L (1981) Calcium channel. Annu Rev Neurosci 4:69–125

Hagiwara S, Saito N (1959) Voltage-current relations in nerve cell membrane of *Onchidium verruculatum.* J Physiol 148:161–179

Hara R, Assai H (1980) Electrophysiological responses of *Didinium nasutum* to *Paramecium* capture and mechanical stimulation. Nature (London) 283:869–870

Harrington NR, Leaming E (1900) The reaction of *Amoeba* to light of different colours. Am J Physiol 3:9–16

Hempelmann F (1926) Tierpsychologie. Akademische Verlagsanstalt, Leipzig

Hennessey TM, Kung C (1985) Slow inactivation of the calcium current of *Paramecium* is dependent on voltage and not internal calcium. J Physiol 365:165–179

Hennessey TM, Nelson DL (1979) Thermosensory behaviour in *Paramecium tetraurelia:* a quantitative assay and some factors that influence thermal avoidance. J Gen Microbiol 112:337–347

Hennessey TM, Saimi Y, Kung C (1983) A heat-induced depolarization of *Paramecium* and its relationship to thermal avoidance behavior. J Comp Physiol 153:39–46

Hess P, Tsien RW (1984) Mechanism of ion permeation through calcium channels. Nature (London) 309:453–456

Hildebrand E (1978) Ciliary reversal in *Paramecium:* temperature dependence of K^+-induced excitability decrease and of recovery. J Comp Physiol 127:39–44

Hinrichsen RD, Saimi Y (1984) A mutation that alters properties of the calcium channel in *Paramecium tetraurelia.* J Physiol 351:397–410

Ivens I, Deitmer JW (1986) Inhibition of a voltage-dependent Ca current by concanavalin A. Pflügers Arch 406:212–217

Iwatsuki K, Naitoh Y (1981) The role of symbiotic *Chlorella* in photoresponses of *Paramecium bursaria.* Proc Jpn Acad Sci Ser B 57:318–323

Iwatsuki K, Naitoh Y (1983) Behavioral responses in *Paramecium multimicronucleatum* to visible light. Photochem Photobiol 37:415–419

Jennings HS (1906) Behavior of the lower organisms. pp 1–366, Columbia Univ Press, New York

Kaissling KE (1982) Molekulares Erkennen: Biophysik der Chemorezeption. In: Hoppe W, Lohmann W, Markl H, Ziegler H (eds) Biophysik. Springer, Berlin, Heidelberg, New York, pp 722–734

Kamada T (1934) Some observations on potential differences across the ectoplasm membrane of *Paramecium.* J Exp Biol 11:94–102

Kinosita H, Dryl S, Naitoh Y (1964) Changes in the membrane potential and the responses to stimuli in *Paramecium.* J Fac Sci Univ Tokyo, Sect IV, 10:291–301

Koehler O (1935) Beiträge zum Verhalten von *Paramecium*-Teilstücken. Verh Dtsch Zool Ges 36:74–84

Kubalski A (1983) Electrical properties of the cell membrane of a marine ciliate *Fabrea salina.* Acta Protozool 22:219–228

Kung C, Saimi Y (1985) Ca^{2+} channels of *Paramecium:* a multidisciplinary study. Curr Top Membr Transport 23:45–66

Machemer H (1966) Versuche zur Frage nach der Dressierbarkeit hypotricher Ciliaten unter Einsatz hoher Individuenzahlen. Z Tierpsychol 6:641–654

Machemer H (1969) Eine 2-Gradientenhypothese für die Metachronieregulation bei Ciliaten. Arch Protistenk 111:100–128

Machemer H (1972) Temperature influences on ciliary beat and metachronal coordination in *Paramecium.* J Mechanochem Cell Motility 1:57–66

Machemer H (1974) Frequency and directional responses of cilia to membrane potential changes in *Paramecium.* J Comp Physiol 92:293–316

Machemer H (1976) Interactions of membrane potential and cations in regulation of ciliary activity in *Paramecium.* J Exp Biol 65:427–448

Machemer H, Deitmer JW (1985) Mechanoreception in ciliates. In: Hardie R et al. (eds) Progress in sensory physiology, vol 5. Springer, Berlin Heidelberg New York Tokyo, pp 81–118

Machemer H, Deitmer JW (1987) From structure to behaviour: *Stylonychia* as a model system for cellular physiology. In: Corliss JO, Patterson DJ (eds) Progress in protistology, vol 2. Biopress, Bristol, pp 213–330

Machemer H, De Peyer J (1977) Swimming sensory cells: electrical membrane parameters, receptor properties and motor control in ciliated Protozoa. Verh Dtsch Zool Ges 1977:86–110

Machemer H, Eckert R (1975) Ciliary frequency and orientational responses to clamped voltage steps in *Paramecium.* J Comp Physiol 104:247–260

Machemer H, Machemer-Röhnisch S (1984) Mechanical and electrical correlates of mechanoreceptor activation of the ciliated tail of *Paramecium.* J Comp Physiol A 154:273–278

Machemer H, Ogura A (1979) Ionic conductances of membranes in ciliated and deciliated *Paramecium.* J Physiol 296:49–60

Machemer-Röhnisch S, Machemer H (1984) Receptor current following controlled stimulation of immobile tail cilia in *Paramecium caudatum.* J Comp Physiol A 154:263–271

Maeda K, Imae Y, Oosawa F (1976) Effect of temperature on motility and chemotaxis of *Escherichia coli.* J Bacteriol 127:1039–1046

Martinac B, Machemer H (1984) Effects of varied culturing and experimental temperature on electrical membrane properties in *Paramecium.* J Exp Biol 108:179–194
Martinac B, Saimi Y, Gustin MC, Kung C (1986) Single-channel recording in *Paramecium.* Biophys J 49:167a
McLauglin SGA (1977) Electrostatic potentials at membrane-solution interfaces. Curr Top Membr Transp 9:71–144
Meech RW (1978) Calcium-dependent potassium activation in nervous tissues. Annu Rev Biophys Bioeng 7:1–18
Mendelssohn M (1902) Récherches sur la thermotaxie des organismes unicellulaires. J Physiol Pathol Gen 4:393–409
Moolenaar WH, deGoede J, Verveen AA (1976) Nature (London) 260:344–346
Naitoh Y (1982) Protozoa. In: Shelton GAB (ed) Electrical conduction and behaviour in 'simple' invertebrates. Clarendon, Oxford, pp 1–48
Naitoh Y, Eckert R (1968a) Electrical properties of *Paramecium caudatum:* modification by bound and free cations. Z Vergl Physiol 61:427–452
Naitoh Y, Eckert R (1968b) Electrical properties of *Paramecium caudatum:* All-or-none electrogenesis. Z Vergl Physiol 61:453–472
Naitoh Y, Eckert R (1969a) Ionic mechanisms controlling behavioral responses in *Paramecium* to mechanical stimulation. Science 164:963–965
Naitoh Y, Eckert R (1969b) Ciliary orientation: controlled by cell membrane or by intracellular fibrils? Science 166:1633–1635
Naitoh Y, Eckert R (1973) Sensory mechanisms in *Paramecium.* II. Ionic basis of the hyperpolarizing mechanoreceptor potential. J Exp Biol 54:53–65
Naitoh Y, Eckert R (1974) The control of ciliary activity in Protozoa. In: Sleigh MA (ed) Cilia and flagella. Academic Press, London New York, pp 305–352
Naitoh Y, Eckert R, Friedman K (1972) A regenerative calcium response in *Paramecium.* J Exp Biol 56:667–681
Nakaoka Y, Oosawa F (1977) Temperature-sensitive behavior of *Paramecium caudatum.* J Protozool 24:575–580
Neher E (1971) Two fast transient current components during voltage clamp on snail neurons. J Gen Physiol 58:36–53
Noble D (1984) The surprising heart. J Physiol 353:1–50
Oertel D, Schein SJ, Kung C (1977) Separation of membrane currents using a *Paramecium* mutant. Nature (London) 268:120–124
Oertel D, Schein SJ, Kung C (1978) A potassium conductance activated by hyperpolarization in *Paramecium.* J Membr Biol 43:169–185
Ogura A (1977) Non-lethal deciliation of *Paramecium* with ethanol. M Sci Thesis, Fac Sci, Univ Tokyo, pp 1–10
Ogura A (1981) Deciliation and reciliation in *Paramecium* after treatment with ethanol. Cell Struct Funct 6:43–50
Ogura A, Machemer H (1980) Distribution of mechanoreceptor channels in the *Paramecium* surface membrane. J Comp Physiol 135:233–242
Ogura A, Takahashi M (1976) Artificial deciliation causes loss of Ca-dependent responses in *Paramecium.* Nature (London) 264:170–172
Onimaru H, Naitoh Y, Ohki K, Nozawa Y (1979) Electrophysiological studies on the membrane of *Tetrahymena.* Dobutsugaku Zasshi (Zool Mag Tokyo) 88:529
Pape HC, Machemer H (1986) Electrical properties and membrane currents in the ciliate *Didinium.* J Comp Physiol A 158:111–124
Poff KL (1985) Temperature sensing in microorganisms. In: Colombetti G, Lenci F, Song PS (eds) Sensory perception and transduction in aneural organisms. Plenum, New York, 299–307
Reisser W (1980) The metabolic interactions between *Paramecium bursaria* Ehrbg. and *Chlorella* spec. in the *Paramecium bursaria*-symbiosis. Arch Microbiol 125:291–293
Reuter H (1984) Ion channels in cardiac cell membranes. Annu Rev Physiol 46:473–484
Roos A, Boron WF (1981) Intracellular pH. Physiol Rev 61:296–434
Saimi Y (1986) Calcium-dependent sodium currents in *Paramecium:* Mutational manipulations and effects of hyper- and depolarization. J Membr Biol 92:227–236

Saimi Y, Kung C (1980) A Ca-induced Na-current in *Paramecium.* J Exp Biol 88:305–325

Saimi Y, Kung C (1982) Are ions involved in the gating of calcium channels? Science 218:153–156

Saimi Y, Hinrichsen RD, Forte M, Kung C (1983) Mutant analysis shows that the Ca^{2+}-induced K^+ current shuts off one type of excitation in *Paramecium.* Proc Natl Acad Sci USA 80:5112–5116

Salkow L (1983) Drosophila mutants reveal two components of fast outward current. Nature (London) 302:249–251

Satow Y (1978) Internal calcium concentration and potassium permeability in *Paramecium.* J Neurobiol 9:81–91

Satow Y, Kung C (1977) A regenerative hyperpolarization in *Paramecium.* J Comp Physiol 119:99–110

Satow Y, Kung C (1979) Voltage-sensitive Ca-channels and the transient inward current in *Paramecium tetraurelia.* J Exp Biol 78:149–161

Satow Y, Kung C (1980a) Membrane currents of pawn mutants of the pwA group in *Paramecium tetraurelia.* J Exp Biol 80:57–71

Satow Y, Kung C (1980b) Ca-induced K^+-outward current in *Paramecium tetraurelia.* J Exp Biol 88:293–303

Satow Y, Murphy AD, Kung C (1983) The ionic basis of the depolarizing mechanoreceptor potential of *Paramecium tetraurelia.* J Exp Biol 103:253–264

Soest H (1937) Dressurversuche mit Ciliaten und rhabdocoelen Turbellarien. Z Vergl Physiol 24:720–748

Tawada K, Oosawa F (1972) Responses of *Paramecium* to temperature change. J Protozool 19:53–57

Tillotson D (1979) Inactivation of Ca conductance dependent on entry of Ca ions in molluscan neurons. Proc Natl Acad Sci USA 77:1497–1500

Umrath K (1930) Potentialmessungen an *Nitella mucronata* mit besonderer Berücksichtigung der Erregungserscheinungen. Protoplasma 9:576–597

Voß HJ, Machemer H (1987) Das Experiment: Können Einzeller lernen? Prüfung am klassischen Konditionierungsexperiment. BIUZ 17:122–127

Wood DC (1975) Protozoa as models of stimulus transduction. In: Eisenstein EM (ed) Aneural organisms in neurobiology. Advances in behavioral biology, vol 13. Plenum, New York, pp 5–23

Wood DC (1982) Membrane permeabilities determining resting, action and mechanoreceptor potentials in *Stentor coeruleus.* J Comp Physiol 146:537–550

Yamaguchi T (1960) Studies on the modes of ionic behavior across the ectoplasmic membrane of *Paramecium.* I. Electric potential differences measured by the intracellular microelectrode. J Fac Sci Univ Tokyo, Sect IV, 8:573–591

Chapter 14 Motor Control of Cilia

HANS MACHEMER[1]

CONTENTS

1 Introduction

Observing a *Paramecium* in its natural environment, it is easily recognized that this cell can behaviourally perform with a perfection and subtleness comparable to the behaviour of some of its metazoan companions, such as rotifers, oligochaetes or crustaceans. The similarities in movements between multicellular or-

1 Arbeitsgruppe Rezeptoren, Fakultät für Biologie, Ruhr-Universität, D-4630 Bochum, FRG

Paramecium. Ed. by H.-D. Görtz

ganisms, using specialized cells and a central nervous system, and unicells, are nevertheless superficial; regulation of the motor response in a ciliate is radically decentralized because it occurs, independently at the same time, in each of the thousands of active cilia. The reasons for synchronization of many independent motor organelles in a single cell have been outlined in Chapter 13 (this Vol.).

Ciliary activity includes two different aspects: the mechanisms of how cilia are caused to move and how they are caused to change their movements. The present chapter deals, in the first place, with the second aspect, whereas the basic machinery of the cilium is briefly treated in Sections 4 and 5. On ciliary motion, in general, the reader is referred to the recent review literature (Sleigh 1974; Machemer 1974b; Satir et al. 1981; Sleigh and Barlow 1982; Sanderson 1984; Satir 1984; Machemer 1986).

2 Galvanotaxis: A Classic Reviewed

The spectacular forced locomotion of *Paramecium* in a DC field was first described by Verworn (1889) but had been little understood until the menbrane theory of excitation became applicable to experimentation in ciliates. The linear voltage gradient extending between anode and cathode may be considered as subjecting *Paramecium* to an "extracellular voltage clamp". The voltage gradient over the medium acts to simultaneously hyperpolarize one cell end (that pointing towards the anode) and to depolarize the other cell end (being oriented cathodically; Fig. 1). Because of the graded transition of the extracellular voltage, depolarization of the cell gradually decreases in the anodal direction, and hyperpolarization decreases in the cathodic direction. The membrane resting potential is thus limited to the middle of the cell body. The cilia, which only "sense" the potential of their particular membrane, respond according to the local membrane potential. They produce depolarization-dependent ciliary activity at sites of depolarization, and hyperpolarization-dependent activity at sites of hyperpolarization.

With positive and negative membrane polarization implicated in extracellular electric stimulation, the *Paramecium* cilia automatically display, in topographical order, their complete inventory of motor responses. The mechanical consequences of this order, forced swimming of *Paramecium* with its anterior end directed towards the cathode, have been recognized early (Ludloff 1895, see Koehler 1926); the mechanism is briefly summarized in Fig. 1 (for more detailed treatment see Machemer 1988a). It should be noted that galvanotaxis is a behavioural response based on the voltage sensitivity of the ciliary membranes. The size of the electric current traversing the medium and the cell in the electric field is meaningless for the establishment of the circuit of voltage dividers (see Fig. 1). Galvanotaxis is thus the simplest exemplification of the voltage dependence of ciliary motor control in *Paramecium.*

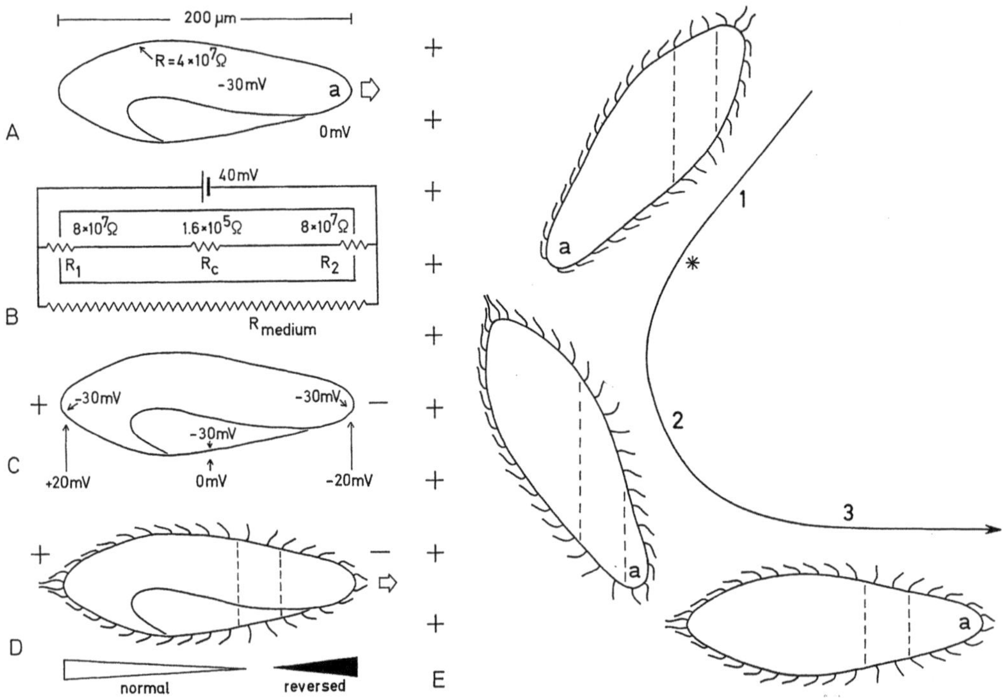

Fig. 1. Electrophysiological basis of galvanotaxis in *Paramecium*. **A** Cell at resting potential of −30 mV and resting input resistance of 40 megohms swimming forward (*arrow, a:* anterior cell end). **B** Medium and cell serving as voltage dividers after application of a 2 V cm^{-1} potential gradient to the bath. The voltage drops linearly along the resistance of the medium (0.2 mV µm^{-1}), but not so along the longitudinal resistance of the cell (R_1, $R_2 = 500 \times R_c$; R_1, R_2: input resistance of posterior, anterior halves of cell; R_c: cytoplasmic resistance). **C** The externally applied voltage (see +,−) declines across the cell membrane, whereas the drop across the cytoplasm is negligible according to Ohm's law. As a result, the anodal cell end is hyperpolarized ($V_m = -50$ mV), and the cathodic cell end depolarized ($V_m = -10$ mV); the middle of the cell remains at the resting potential. **D** Cilia of the anodal field augment the normal beating; cilia of the distant cathodic field inactivate (zone marked by *dashed lines*); cilia of the near cathodic field beat in reverse. Summation of ciliary force vectors leads to forward swimming in the cathodic direction. *Wedges:* voltage-controlled gradients of frequency of cilia beating in the normal (posterior) direction (*white wedge*) and in the reversed (anterior) direction (*black wedge*). Ciliary response symbolized by angle to cell surface. **E** Stages of forced reorientation of *Paramecium* along the potential gradient: *1* At time mark (*) a randomly swimming cell starts turning towards the cathode because the normal beating of anodal cilia predominates over reversed beating of the cathodic cilia; *2* continuation of the cathodic turning manoeuver for the same reasons, until the axis of locomotion is parallel to the voltage gradient (*3*); this orientation is stable because the cilia are symmetrically activated. (After Machemer and De Peyer 1977)

3 Current Methods

Procedures aiming at the assessment of mechanisms of ciliary control have two major targets, the cell membrane and the intraciliary space surrounding the axoneme. The electrophysiological method of choice is the voltage clamp allowing the membrane potential to be set at predetermined levels and to observe the ciliary response. Ciliary activity following stimulation and membrane excitation is more difficult to interpret in a cell with the membrane potential floating free. The responses of the cilia are best monitored using high-frequency recording techniques (cine film or video) with framing rates $\geqq 200$ Hz providing a number of ciliary images per ciliary cycle. A more convenient, albeit less direct, method is watching the swimming response following stimulation of *Paramecium*, in particular by a change in ionic composition of the medium. The ciliary motor response may be inferred from the behaviour, whereas the membrane response is deduced from the type of stimulation. Behavioural screening of *Paramecium* has been successfully applied for isolation of membrane mutants (Kung 1971; Takahashi and Naitoh 1978; see Chapter 17, this Vol.). This method is less suitable for the identification of particular ciliary responses. Mutagenizing for structural defects has also been successfully applied in *Chlamydomonas* and *P. tetraurelia*, thus opening pathways of correlation between axonemal structure and function (Brokaw et al. 1982; Huang et al. 1982; Hinrichsen and Kung 1984).

Experiments which change the intracellular environment of the axoneme, such as the concentrations of Ca and nucleotides, have been performed in *Paramecium* to monitor the ciliary motor response (Brehm and Eckert 1978a; Gustin et al. 1983; Schultz et al. 1984). Because the injected buffers or permeating drugs can interfere with the membrane conductance, results are equivocal, unless the membrane potential is kept under voltage-clamp control (Hennessey et al. 1985).[1] Cell demembranation and reactivation studies using nucleotides have clearly established the ionic and ATP requirements of ciliary activity (see Sect. 6), although discrepancies between in vivo and reactivation experiments have not been fully ruled out. Previous biochemical approaches in characterizing membrane channel functions and ciliary nucleotide metabolism are promising to elucidate steps in ciliary electromotor coupling (see Schultz and Klumpp, Chapt. 16, this Vol.).

4 The Cilium is a Rotary Sliding Machine

4.1 Microtubular Sliding and Ciliary Gyration

A quarter of a century has passed since the cilium was recognized as a motor organelle employing sliding-filament mechanisms (Afzelius 1959; Gibbons

[1] In these experiments it was shown that injection of cyclic AMP hyperpolarizes the membrane and raises the frequency of the cilia; with the membrane voltage fixed to the resting potential the ciliary frequency was not affected by cyclic AMP and analogues.

1965; Satir 1968; Brokaw 1971). A generally accepted view of this mechanism is that unidirectional active sliding occurs between neighbouring microtubular doublets of the axoneme. The force-generating sliding is based on unidirectional attachment-detachment cycles of dynein arms extending from subfibres A to subfibres B of the neighbouring doublet (Satir et al. 1981). Sliding between two microtubular doublets transforms to bending of the axoneme because the nine peripheral doublets are firmly anchored in the ciliary base. From the unidirectionality of the ciliary cycle in ciliates (counterclockwise as viewed tip to base) it has been inferred that active sliding within a particular section of the cilium rotates unidirectionally (counterclockwise) around the axoneme with the antipode microtubules of active sliders being passive backsliders for mechanical reasons (Machemer 1977). While the sliding process takes a rotational course within the axoneme, the resulting movement of the cilium is a gyration. The pathway of the cilium in space is roughly approximated by the surface of a cone, the peak of the beating cone being at the origin of the cilium. Temporal and spatial polarity of the ciliary cycle lead to deformation of the beating cone, which is further modified by external factors such as temperature and viscosity (Machemer 1972a, b). The structural enantiomorphism of the ciliary cross-section and the unidirectionality of ciliary gyration are independent aspects of the same functional design of the ciliary machine.

4.2 Rotation of the Central Complex

The central microtubules including the "sheath" of periodic projections from these microtubules end at the level of the axosome, and only one central microtubule enters the axosome (Dute and Kung 1978). The central complex, being virtually detached from the ciliary base, is strung by the radial spokes which can attach to and detach from the sheath during ciliary motion (Warner and Satir 1974). Some freedom of movement of the central sheath was indicated from functional analysis of fixed cilia showing that the common plane of the central tubules was always normal to the axonemal bending plane (Tamm and Horridge 1970). In *Paramecium* a rotation of this configuration during the ciliary cycle was established (Omoto and Kung 1980) and central-tubule rotation directly demonstrated in *Micromonas* (Omoto and Witman 1981). Central-tubule rotation does not occur during the planar performance of beating in ctenophore comb plates (Tamm and Tamm 1981). This finding is unfavourable to the view that the central core controls the ciliary beat (Satir 1984). Tests of the mechanical implications of rotation of the central core within the axoneme using a scale model cilium correspond to observations and suggest that orientation of the central core is a process of passive adjustment minimizing stress between the central tubules (Machemer 1986). In summary, the central core of the axoneme can perform rotational movements depending on the form of the ciliary cycle. The role of the central core, potentially an active regulator (Satir 1984), a passive stabilizing element (Machemer 1986) or both, remains so far undetermined.

5 Parameters of Ciliary Activity

5.1 Frequency

The most general property of cyclic ciliary activity is the rate of beating expressed in hertz (Hz, cycles per second). The nomenclature suggests steadiness of this parameter, at least over a period of 1 s. Few ciliary organelles in the ciliate organization display steady frequency. In *Paramecium* the frequency of the oral groove cilia is steadier than that of the somatic cilia, but stroboscopic determinations of the frequency, near 15 Hz at room temperature, are difficult even in these paired cilia. High-speed cinematography of cilia has shown that the rate of beating fluctuates from cycle to cycle, and these fluctuations are particularly large under free-floating membrane potential. The frequency is nevertheless strictly controlled by the membrane potential as seen after averaging the course of momentary frequencies (= reciprocal of duration of one cycle) of several cilia (Fig. 4).

5.2 Polarization of the Cycle

The performance of the cycle includes polarity in two ways: The power stroke occurs faster than the return stroke (polarization in time). The path of the cilium tends to be planar during the power stroke and more circular during the return stroke; at the same time, the ciliary shaft is largely straight during the power stroke and curved during the return stroke (polarization in space) (Fig. 2). The "direction of the power stroke" has been either derived from the ciliary orientation at the end of the power stroke or inferred from the orientation of the metachronal waves in *Paramecium* (parallel to the lines of synchrony). In an undisturbed cell the beating direction of the somatic cilia (excluding those of the oral groove) is towards the rear and right. Polarization of the ciliary cycle, although intrinsically regulated, is modified by the viscosity of the medium (Machemer 1972a) and by temperature (Machemer 1972b). Rising viscosity decreases polarization of the performance of the cycle, and the power stroke orientation is shifted in the clockwise direction. Paramecia in high viscosity media therefore spin in right-handed helices as compared to cells in normal media which spin in left-handed helices. Reduction in temperature first depresses polarization in time together with the rate of the ciliary cycle. A similar coupling between the beating rate and polarization of the cycle applies to changes in membrane potential under constant environmental conditions (Sects. 7 and 8).

5.3 Inclination, a Non-Cyclic Response

The performance of any ciliary cycle is such that the power stroke is kept "upright" (rectangular) with respect to the cell surface, while the return stroke occurs largely parallel to the cell surface (Fig. 2). The underlying beating cone

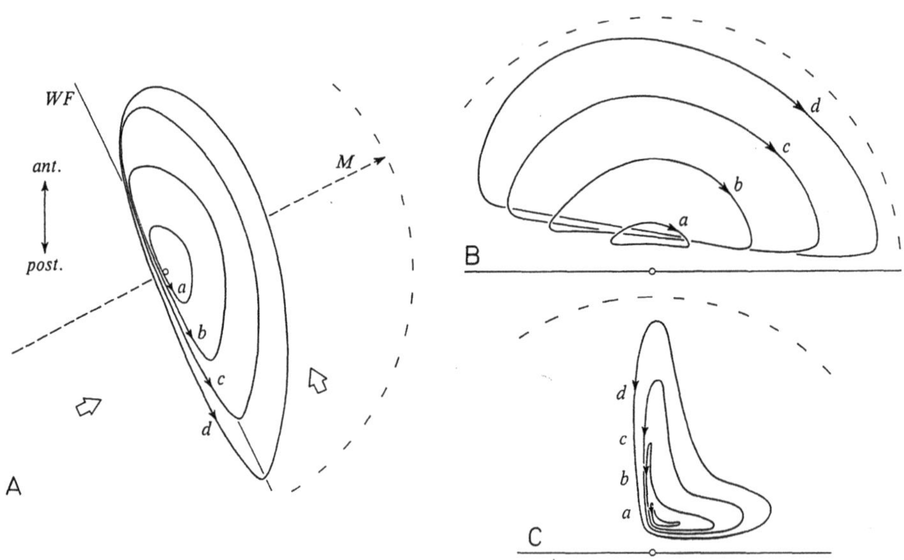

Fig. 2. Conicoidal envelope of the ciliary cycle of the unstimulated cell as represented by the projected tracts of four reference points (*a–d*) on model cilia shaped according to observed profile views of live cilia in *Paramecium* (Machemer 1972c). **A** Projection parallel to the cell surface observing the cilium tip to base. *Double-headed arrow:* antero-posterior cell axis. *WF:* line of ciliary synchronization, or wave front; *M:* direction of metachronal wave propagation. *Wide arrows:* viewing directions of ciliary cycle in **B** (*left*) and **C** (*right*). **B** Same cycle as seen parallel to the power stroke. **C** Same cycle as seen facing the power stroke direction. Note the anteriad-right inclination of the ciliary beating envelope. The entire beating program, including inclination, rotates counterclockwise upon depolarization, and clockwise upon hyperpolarization of the membrane (Machemer 1972a)

is thereby tilted. The steady angle of tilt from the vertical orientation of the beating cone axis[2] has been termed "inclination" (Machemer and Sugino 1986). Inclination of ciliary beating changes with reorientation of the ciliary power stroke. Inclination and power stroke orientation differ by an angle approximating 90°. While the latter parameter is cyclic in nature, the former is non-cyclic. In *Stylonychia* it has been possible to separate a non-cyclic orientational parameter from a cyclic one by applying voltage ramps of extremely low rate of rise under voltage clamp (De Peyer and Machemer 1983). Inclination has not yet been further explored in *Paramecium* because it is difficult to depict dense populations of cilia in an axial view.

[2] That is, parallel to the axis of the ciliary basal body.

6 Reactivation of Ciliary Axonemes

6.1 Roles of Ca and Mg

The first successful attempt to reactivate cilia in previously demembranated cell "models" employed glycerination of cells which were reactivated in solutions of ATP (Hoffmann-Berling 1955; Brokaw 1961; Gibbons 1965). *Paramecium* cilia which pointed posteriad in EDTA buffer reoriented anteriorly in a non-beating mode, when ATP, Ca and Zn were added to the medium (Naitoh 1969). Using the non-ionic detergent Triton-X 100 for demembranation, the cilia of the model could be induced to beat repetitively in the normal (posterior) direction in the presence of millimolar concentrations of ATP-Mg and EDTA as a Ca buffer. Raising the Ca concentration of the reactivation medium beyond 10^{-7} M led to ciliary beating in the "reversed" (anterior) direction. Omission of Mg in the reactivation medium suppressed the cyclic response of the cilia but left the Ca-dependent orientational response intact (Naitoh and Kaneko 1972). These data have been interpreted in terms of two mechanisms regulating the ciliary motor response: an Mg-dependent "beating mechanism" and a Ca-dependent "reversal mechanism" (Naitoh 1982).

The highly instructive reactivation experiments in *Paramecium* have been previously expanded. In employing EGTA (instead of EDTA) as Ca buffer the previous data were confirmed. It was shown, in addition, that the direction of beating of the reactivated cilia depends on the relative concentrations of Mg and Ca in the medium, so that the reversed beat orientation is produced by increases in Ca and/or reductions in Mg (Nakaoka and Toyotama 1979). Ca may also alter the frequency of the reactivated cilia using the EGTA buffer; the frequency rose with Ca dropping below and rising beyond the critical concentration of 6×10^{-7} M (Nakaoka et al. 1984).

6.2 Is Ca the Universal Regulator?

An assessment of the presently available data on ciliary reactivation in *Paramecium* suggests that Ca can interfere with the regulation of frequency and beat orientation. The reorientations of non-beating cilia in glycerinated preparations or following exposure of live cells to nickel (Naitoh 1966) may be reflecting shifts in steady inclination rather than in power stroke orientation; a genuine power stroke was never seen as separated from cyclic activity, and the observed angular reorientations of non-beating cilia are limited to angles of 90° or less (Naitoh 1969; Eckert and Naitoh 1970). At present, it appears that all ciliary motor parameters are under the control of Ca. Ca is a prime candidate for regulating the motor because of its low concentration baseline in the ciliary cytosol. Mg, on the other hand, is kept at millimolar intracellular concentrations and is virtually impermeable to the ciliary membrane (see Machemer, Chap. 13, this Vol.). A regulating function for Mg in ciliary motor control appears, therefore, to be restricted to competition or cooperativeness with intraciliary Ca (Machemer 1986).

7 Depolarization-Induced Ciliary Activity (DCA)

7.1 Inclination

The ciliary cycle of the undisturbed ("resting") *Paramecium* includes an inclination of the beating cone towards the anterior and right. At the posterior face of the cone the power stroke occurs towards the posterior and right (Fig. 2) and thus determines the lines of synchrony of the metachronal wave system (Machemer 1972c). Following a membrane depolarization, this beating program rotates in the counterclockwise direction with the inclination of the beating cone pointing anteriad-left, the power stroke and metachronal waves oriented anteriad-right. Applying slow depolarizing voltage ramps under voltage-clamp, the counterclockwise reorientation of the cycle may be visualized[3] as being graded in nature, although not fully linear to depolarization. At weak depolarizations the non-cyclic orientation persists while cyclic ciliary activity dwindles. This phenomenon has been called "ciliary inactivation" (Machemer 1974a, 1975). The depolarization-dependent reorientations of non-beating cilia in *Paramecium* (Naitoh 1966; Eckert and Naitoh 1970) are presumably part of the non-cyclic inclination response.

7.2 Cyclic Orientational and Frequency Responses

A depolarization of the membrane, such as by an action potential or a square step pulse of similar size under voltage clamp, activates a response commonly described as "ciliary reversal". The motor response of the cilia consists of an anteriad[4] reorientation of the power stroke together with a rise in ciliary frequency. This response, which commonly lasts for about 200 to 600 ms, is the basis of the transient backing up of cells which encounter a depolarizing stimulus.

Recording ciliary activity of *Paramecium* under voltage-clamp conditions and employing high-frequency cinematography has shown that the rapid ciliary reversal reveals only a fraction of the existing depolarization-dependent motor inventory. The same depolarization applied at a more reduced rate leads first to a decrease in frequency, then to a rise in frequency with depolarizations $\geqq 5$ mV (Fig. 3). While the frequency drops, the beat orientation is still posteriad; when the frequency rises again, the beat orientation has sufficiently shifted counterclockwise to be now anteriad. During rapid depolarizations such as the action potential, the cilia fail to perform a timely response[5]. Consequently, the ciliary reversal following an action potential or 20 mV depolarizing

---

[3] The reorientation of the cycle is inferred from the modification of profile views of the ciliary metachronal waves (Machemer 1974a, b).

[4] A more precise description of the new orientation of the power stroke is "towards the anterior and right side".

[5] Remember that the performance of a full cycle lasts 20 ms at 50 Hz, while the upstroke of an action potential is commonly finished within 10 ms.

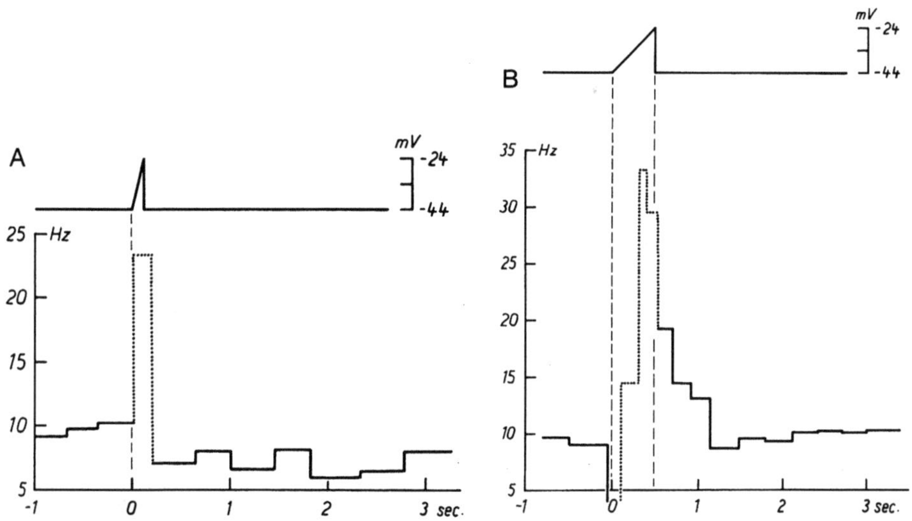

Fig. 3. Masking of the voltage-dependent ciliary response (*lower trace*) by rapid depolarization under voltage clamp (*top trace*). The time courses of the membrane potential and the momentary ciliary frequency (histogram) were recorded using high-speed cinematography. The *continuous line* of the frequency histogram indicates normal beat orientation; a *dotted line* indicates reversed beating of the cilia. **A** Direct switch from normal to reversed beating cycles applying a fast sawtooth ramp (160 mV s^{-1}). This sequence resembles the motor events following an action potential (see Fig. 5). **B** A depolarization performed at a more reduced rate (40 mV s^{-1}) allows the cilium to more adequately respond to the membrane potential; here, it is seen that reversed beating of the cilium is preceded by its transient inactivation. Compare with frequency time course shown in Fig. 4. See text for more explanations (Machemer 1975)

step is an all-or-none response. With sufficient decrease in rate of depolarization, the bimodal nature of the frequency response becomes apparent, being superimposed upon the monotonous increase in counterclockwise ciliary beat reorientation (Fig. 4).

7.3 The "Transient Inactivation" Phenomenon

A breakdown in regular cyclic beating of the cilia was first observed during the sequence of late motor events following the activation of "reversed" ciliary beating (Machemer and Eckert 1973). A similar inactivation turned up in the records of ciliary frequency evoked during low rates of depolarization (Machemer 1975). Here, inactivation intervened between the sequences of decreasing frequency and rising frequency (Figs. 3 and 4). During inactivation individual cilia rarely become immobile but do show irregular low-amplitude wriggles conveying the impression of being comparatively rigid.

Low-amplitude, sustained voltage-clamp steps at depolarizations of near 5 mV invoked only transient inactivation responses which eventually transformed into reversed beating of the cilia. Also, an evoked slow normal beating of the cilia, seen with minor sustained depolarization, was gradually fading

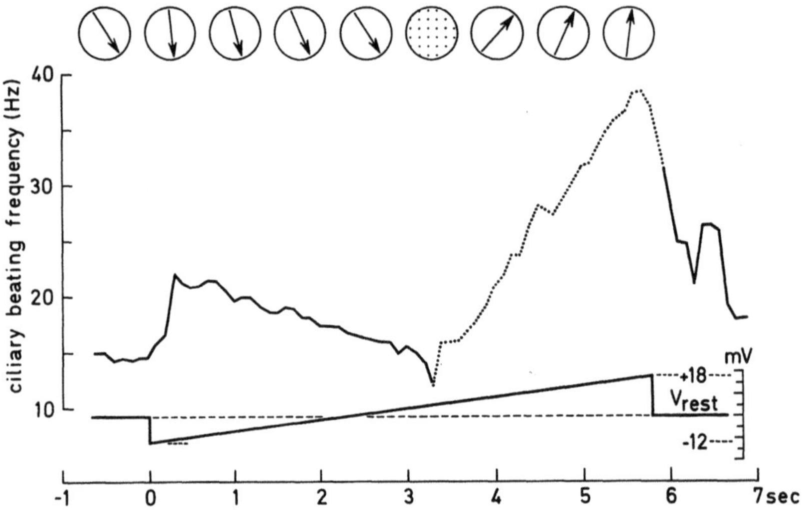

Fig. 4. Electromotor coupling of the cilia in *Paramecium.* Means of frequency of cilia from five cells under voltage clamp hyperpolarized by 12 mV and subsequently depolarized at a low rate up to 18 mV along a linear voltage time course. The resulting "frequency profile" roughly illustrates the correlations between the membrane potential and the ciliary frequency. Beating directions with posteriad vectors are indicated by a *continuous line,* and those with anteriad vectors by a *dotted line; arrows in circles* illustrate approximate orientations of the power stroke (anterior cell end up). Note the monotonous continuation of the frequency decline from the hyperpolarized to the weakly depolarized state. Minimal frequency results from ciliary inactivity slightly scattering in different cells near depolarizations of a few millivolts. The voltage-dependent frequency and directional response profile of the cilium correspond to that obtained while applying an extracellular voltage gradient (see Fig. 1 D) and in cell reactivation experiments by Nakaoka et al. (1984) with the Ca concentration varied between 10^{-7} and 2×10^{-6} M (Machemer and De Peyer 1977)

towards inactivation (Machemer and Eckert 1975). These observations have suggested the possibility that ciliary inactivation is Ca-dependent and occurs due to a depolarization-induced, slight rise in intraciliary Ca concentration above normal. The view that the voltage-dependent full "frequency profile" of the cilium in *Paramecium* (Fig. 4) is due to the gradual increase in intraciliary Ca is further supported by experiments employing very high sustained voltage-clamp steps near +100 mV; with the Ca driving force critically reduced during these steps, the cilia were seen to slowly proceed from normal beating towards inactivation, or from inactivation towards reversed beating (Machemer and Eckert 1975).

During inactivation of cyclic activity, the cilia may be in a process of reorientation (see Sect. 7.1). In *Stylonychia,* where the cirri are inactive at the membrane resting potential, minor depolarizations lead to a non-cyclic anteriad reorientation; cyclic activation of the cirri occurs with larger depolarizations (De Peyer and Machemer 1983). By analogy, it is presumed that the inactivation phenomenon in *Paramecium* applies to cyclic ciliary activity only.

7.4 Sequence in Ciliary Responses Following an Action Potential

When an action potential has been produced by injection of a square current pulse into the cell, the first post-stimulus ciliary cycle shows a peak frequency raised two- or three-fold, and the beat orientation has changed by an angle exceeding 100°. Together with the new direction of the power stroke the metachronal wave system has been restored within tens of milliseconds. The transition from "normal" to "reversed" beating of the cilia includes a rapid, near-synchronous stroke of the cilia towards the anterior and left side of the cell. Instantaneously fixed cells often show the cilia at the end of this first orientational response (see, e.g. Machemer 1986; Fig. 15) which resembles a regular power stroke. Several arguments suggest that the first powerful reorientation is not part of a cyclic ciliary process: (1) all cilia are synchronized; (2) no metachronal wave system is seen corresponding to this first orientational response; the arising wave fronts indicate that the power strokes are pointing to the anterior and right (Machemer 1974a; Machemer and Eckert 1977); (3) nickel-paralyzed paramecia can perform a similar, albeit abortive, swing of the non-beating cilia upon depolarization (Naitoh 1966; Eckert and Naitoh 1970).

Following completion of the first ciliary cycle in reverse, the rate of the subsequent cycles decays exponentially in the course of a few hundred milliseconds (Fig. 5). The cyclic beating then gives way to ciliary inactivation, but cyclic beating in the normal direction is quickly resumed at a low rate and gradually rises to the prestimulus level. Figure 5 illustrates that this sequence in motor

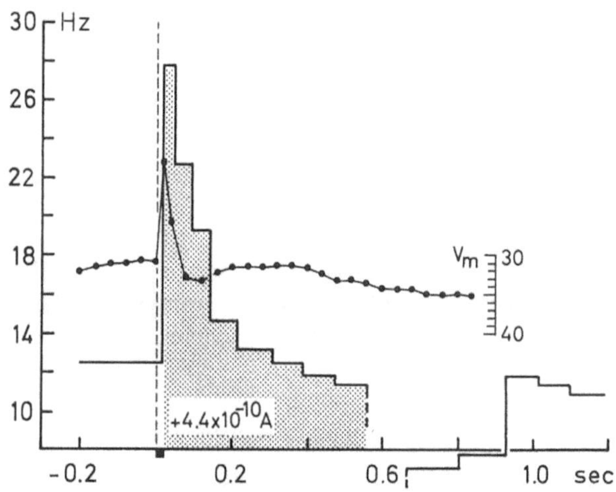

Fig. 5. Ciliary frequency and directional responses following an action potential induced by injection of a 20-ms outward current pulse. The time course of the membrane potential (*filled circles*) and the frequency (*histogram*) were recorded cinematographically. The first cycle registered after depolarization is at peak frequency and reversed in orientation (*shaded histogram*). The raised beating frequency rapidly declines along an exponential time course after completion of membrane repolarization; the cilia eventually inactivate transiently. Thereafter, the frequency of beating is resumed in the normal direction at a low level and rises to the prestimulus rate (Machemer 1974a)

events occurs largely after repolarization of the membrane potential. Such partial uncoupling of the ciliary motor events from the action potential is explained by assuming that the action potential triggers a rapid inward surge of Ca ions, raising the intraciliary Ca concentration by at least three orders of magnitude. The time course of the depolarization-induced ciliary activity is then determined by the comparatively slow removal of Ca ions from the ciliary space. It is worth noting that the sequence in ciliary events following the action potential, "reduction in frequency of reversed beating → inactivation → increase in frequency of normal beating", is inverse to the sequence observed during slow depolarizations (see Sect. 7.2). This fact supports the view that the frequency and orientational responses are under direct Ca control in either case.

8 Hyperpolarization-Induced Ciliary Activity (HCA)

Behavioural responses of *Paramecium* that follow from stimuli other than depolarizing stimuli have received little attention because the mechanism of control is less well understood. For the daily life of *Paramecium* either type of ciliary activation, speeding forward due to hyperpolarization, or reversing the direction of swimming due to depolarization, may be equally meaningful. The electrophysiological bases of these types of responses are different in principle: depolarization involves membrane excitation; hyperpolarization of the membrane is based on non-regenerative processes (see Machemer, Chap. 13, this Vol.). The result of hyperpolarization-induced ciliary activation (HCA) is, nevertheless, in some respects a "mirror image" of depolarization-induced ciliary activation (DCA): clockwise reorientation of beating and rise in frequency.

8.1 Inclination

Hyperpolarization induces the inclined beating cycle (Fig. 2) to rotate in the clockwise direction so that the orientation of the power stroke shifts from posteriad-right to posteriad (Machemer 1974a). This reorientation of the beating program has not yet been investigated in detail. The rotation has been indirectly documented by methods employing optical sections of the metachronal wave system (Machemer 1974b; Machemer and Eckert 1977).

8.2 Cyclic Orientational and Frequency Responses

Upon hyperpolarization of the membrane the power stroke reorients towards the posterior cell end and the frequency rises. This response is clearly seen under conditions of voltage clamp applying rising steps of hyperpolarization (Machemer and Eckert 1975) or with the membrane potential slowly shifting in a linear manner (Machemer 1976). Responses recorded during application of a positive-going voltage ramp reveal that the frequency and orientation of "normal" beating change monotonously between the hyperpolarized and a moder-

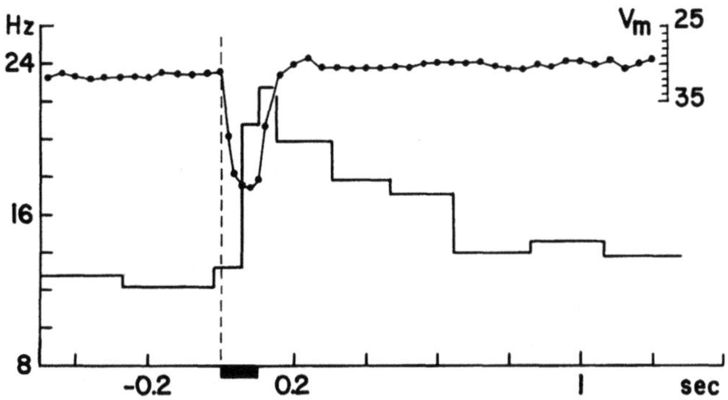

Fig. 6. Ciliary motor responses following a brief (100 ms) membrane hyperpolarization. The definitions of traces are the same as in Fig. 5. Note (1) the comparatively slow rise in frequency of normal beating which continues past the peak time of the ohmic-capacitive hyperpolarization, (2) the frequency decline extending beyond the restoration of the resting potential. See text for more explanations (Machemer and De Peyer 1977)

ately depolarized state of the membrane (Fig. 4). A similar alteration in ciliary responses applies to *Stylonychia* at voltages between the hyperpolarized and resting membrane (see Machemer 1986; Fig. 7). Regarding *Paramecium* it is therefore a straightforward assumption that ciliary motor responses following polarization to the negative and positive sides of the resting potential are subject to the same controlling mechanism.

8.3 Sequence in Ciliary Responses Following a Hyperpolarizing Stimulus Pulse

The rise in frequency and ciliary reorientation following a square hyperpolarizing current pulse of a negative voltage step under voltage clamp differs in its time course from the response induced by depolarization (Fig. 6). It takes at least 50 ms for a substantial rise in ciliary frequency and up to 1 s to rise to the peak (Machemer and Eckert 1975). The peak frequencies may be similar to depolarization peak frequencies (40 to 45 Hz), given sufficient amplitude and duration of hyperpolarization. Also, the reorientation of the beat direction towards the posterior cell end is not an instantaneous one and takes more than one ciliary cycle (Machemer 1974a; Machemer and Eckert 1977). For brief hyperpolarizations the slow rise in ciliary frequency leads to a premature abortion of the ciliary frequency response which is then restricted to a few cycles. An overall mechanical stimulus such as pipetting or otherwise agitating cells can induce enhanced forward swimming of *Paramecium* which lasts for tens of seconds or even minutes (Machemer, unpubl.). This underlines the separation of the duration of the ciliary motor response from the activating hyperpolarization (see Fig. 6). The frequency decay rate constant remains coupled, however, to the stimulus: it is reduced with rising stimulus amplitude. Corresponding observations were made with *Stylonychia* (De Peyer and Machemer 1982).

9 Adaptation

When a *Paramecium* is transferred into a depolarizing medium, such as a solution with raised concentration of K ions, it may change from forward to backward swimming. Conversely, an augmented forward swimming of *Paramecium* can be induced following transfer of the cell from high-potassium to low-potassium solutions which causes membrane hyperpolarization. A common property of these chemically induced behavioural responses is their waning with time (typically in the course of tens of seconds), so that the normal behaviour reappears after some transitions (Kamada and Kinosita 1940). The typical sequence in behaviour during chemical DCA is: reduction in the rate of backward swimming → stationary spinning → slow, later normal forward swimming (Hildebrand and Dryl 1983). The chemical HCA includes a slow decrease towards normality in the previously augmented rate of forward swimming. Although the bulk potential does not much change in these cases (Machemer 1976), the behaviour resembles the sequences in ciliary responses following transient depolarization (Fig. 5) and hyperpolarization (Fig. 6). The relaxation of reversed swimming in depolarizing solutions has been related to a decrease in membrane excitability ("adaptation") following the entrance and binding of ionic Ca to the cytoplasmic face of the Ca channel (Hildebrand and Dryl 1976). A Ca-dependent Ca channel inactivation, albeit more rapid, was substantiated in *Paramecium* (Brehm and Eckert 1978b). It has been speculated that the reduction in negative surface charges by raised cation concentrations (result: a more negative transmembrane potential) can facilitate the transfer of ions across the membrane so that hyperpolarization of the membrane is later compensated by depolarization (Eckert and Brehm 1979). This model does not explain how neutralization of changes in surface charges by subsequent slow shifts in transmembrane potential can be regulated.

A relaxation of DCA (and, in part, of HCA) in the course of 1 min was also noted in a constant environment during sustained depolarizations and hyperpolarizations under voltage clamp (Machemer and Eckert 1975). In appearance, this time-dependent shift in voltage sensitivity of the membrane resembles membrane accomodation in nerve. The residual Ca current, which persists in the depolarized state of the *Paramecium* membrane even after Ca-dependent Ca current inactivation (Brehm et al. 1980; Hinrichsen and Saimi 1984), is thought to maintain reversed ciliary beating under sustained electrical and chemical stimulation (Eckert and Brehm 1979). Previous evidence from *Paramecium* shows that even this residual current inactivates in the course of tens or hundreds of seconds (Hennessey and Kung 1985). These findings can account for the motor response relaxation, or "adaptation", of the depolarized cell; the mechanism of slow channel inactivation, however, remains to be explained (for more detailed treatment see Machemer 1988b).

10 Steps in Electromotor Coupling

Two types of ciliary activation occur in *Paramecium*, a depolarization-induced (DCA) and a hyperpolarization-induced activation (HCA). For DCA, the major steps in coupling the ciliary response to the membrane signal are comparatively well understood. For HCA a variety of hypotheses have been proposed as to how a hyperpolarization of the membrane can lead to a graded response of the intracellular axoneme. These possible pathways have been extensively discussed elsewhere (Machemer 1986). They are briefly summarized in the following sections.

10.1 DCA

Depolarization of the membrane by a few millivolts opens ciliary Ca channels. The increase in Ca conductance causes a further positive shift in membrane potential which then feeds the positive feedback circuit between Ca conductance and depolarization and generates the action potential. Ca influx into the narrow ciliary space rises to peak in the course of the action potential. Ca ions accumulate near the axoneme reaching concentrations of at least 10^{-4} M within a few milliseconds. The size of the action potential and the final Ca concentration are coupled to the size of the stimulus for two reasons: (1) the Ca-activation kinetics are comparatively slow so that depolarization-dependent activation of K channels tends to terminate the regenerative process; (2) Ca ions, after entering the cilium, inactivate the Ca channel and activate a Ca-dependent K channel (see Machemer, Chap. 13, this Vol.). Ca ions diffuse to Ca-sensitive sites along the ciliary axoneme; they modulate chemomechanical transduction at one or more types of dyneins, presumably after association with a Ca-Mg binding axonemal protein. The dyneins produce force-generating sliding of the peripheral microtubular doublets such as to generate non-cyclic (inclination) and cyclic bending responses of the cilium (power stroke oriented anteriad-right) and an increase in the rate of circular transfer of active sliding around the axoneme (Sugino and Machemer 1987), the beating frequency.

Under conditions of continued depolarization the Ca influx is not fully inactivated because a balance is established between Ca entry (channel inactivation) and Ca removal (channel de-inactivation) (Eckert and Brehm 1979). A minor sustained Ca current cancels the rate of metabolically driven Ca efflux so that a still raised intraciliary Ca concentration signals continuation of the reversed ciliary beating. This type of motor response is eventually terminated because also the sustained Ca current inactivates slowly in a voltage-dependent manner.

Reduction of the previously raised intraciliary Ca concentration leads to a graded renormalization of the ciliary response: a clockwise shift in inclination and ciliary power stroke and a relaxation of the frequency of beating. With the Ca concentration being slightly above normal, the cyclic activity of the cilium ceases. This inactivation occurs while the beating has been increasingly oriented towards the right side. Further Ca pumping causes the gradual resumption of cyclic activity and polarization of the beat which is eventually oriented to the posterior-right side of the cell.

10.2 HCA

Hyperpolarization of the membrane does not activate a particular ciliary membrane conductance. The ciliary Ca channel is thought to be in the close state at the membrane resting potential. Further conductances are restricted to the somatic membrane (see Machemer, Chap. 13, this Vol.). A hyperpolarization-dependent rise in the Ca driving force (Eckert 1972) does not raise Ca leakage sufficiently to affect the intraciliary Ca concentration; if Ca_i were to be raised, one would expect DCA to occur (e.g. a drop in the frequency of ciliary beating), rather than HCA. Loops of hyperpolarizing transmembrane current are presumably largely restricted to the soma membrane for the above mentioned reasons. Current entering the cilia is therefore unlikely to mediate the signal by electrophoretic redistribution of charges in the ciliary space (Brehm and Eckert 1978a). Possible minor acidification of the cytosol due to an influx of H^+ presumably does not cause HCA according to conflicting data from reactivation experiments in *Paramecium* (Naitoh and Kaneko 1973).

It has been proposed that membrane hyperpolarization leads to a decrease in intraciliary free Ca (Machemer 1974a). A drop in Ca_i due to a decrease in Ca conductance (Eckert et al. 1976) is not supported by data from voltage-clamp analysis of the *Paramecium* Ca channel (Naitoh 1979). A possible modulation of the Ca pump by hyperpolarization has been explored testing electrically stimulated ciliary beating at different temperatures; an evaluation of the data suggests that the Ca pump is not augmented by hyperpolarization (Martinac and Machemer, unpubl.). A direct regulating effect of free ionic Ca upon the 10-μm axoneme is less probable because the number of free Ca ions per cilium is roughly ten at 10^{-7} M ionic Ca. A modified view of how the membrane potential can control HCA suggests that Ca associated in part with membrane-integrated and axonemal Ca-binding proteins equilibrates at modified Ca-Mg axonemal binding ratios upon voltage-dependent alteration in Ca-binding of the membrane (Machemer 1986). In summary, a general acceptance of an "all-Ca hypothesis" for the regulation of DCA and HCA still awaits the establishing of crucial data, although the present, more circumstantial evidence is favourable to this hypothesis (Fig. 4; Sects. 6–8).

11 Perspectives

The understanding of the ciliary machine and its control in *Paramecium* has greatly improved after electrophysiological and cell reactivation methods have been systematically applied to this organism. It is now known that a firm link exists between the membrane potential and ciliary function and that Ca ions play a crucial mediating role. Unless the membrane of the viable cell was destroyed, attempts to uncouple the axoneme from the electric membrane signal have been unsuccessful so far (Hennessey et al. 1985). The manner of translation of the bipolar and graded membrane signal into an equally fine-tuned ciliary motor response remains largely unexplained. Assuming that a Ca concentration transmits the membrane signal to the axoneme, the question arises as to how a uniform Ca concentration at a particular time can act to induce nine pairs of microtubular doublets to perform different sliding programs in sequence, and to perform these sequences differently at different levels of the axonemal cross-section. In addition, we are asking how a change in the intraciliary Ca concentration can reprogram the sliding of the nine sets of doublets such as to maintain the polarity of the cycle, but with the power stroke redirected. These questions illustrate our vast ignorance about the physiological principles of the cilium. In the near future, it will take a lasting and interdisciplinary effort to further unveil the mysteries of this tiny, but most elegant eukaryotic sliding machine.

Acknowledgements. This work was supported by the Deutsche Forschungsgemeinschaft, SFB 114, TP A5. I would like to thank Mrs. J. Ruthmann for suggesting improvements of the manuscript.

References

Afzelius B (1959) Electron microscopy of the sperm tail. J Biophys Biochem Cytol 5:269–278

Brehm P, Eckert R (1978a) An electrophysiological study of the regulation of beat frequency in *Paramecium*. J Physiol 283:557–568

Brehm P, Eckert R (1978b) Calcium leads to inactivation of calcium channel in *Paramecium*. Science 202:1203–1206

Brehm P, Eckert R, Tillotson D (1980) Calcium-mediated inactivation of calcium current in *Paramecium*. J Physiol 306:193–203

Brokaw CJ (1961) Movement and nucleoside polyphosphatase activity of isolated flagella from *Polytoma uvella*. Exp Cell Res 22:151–162

Brokaw CJ (1971) Bend propagation by a sliding filament model for flagella. J Exp Biol 55:289–304

Brokaw CJ, Luck DJL, Huang B (1982) Analysis of the movement of *Chlamydomonas* flagella: the function of the radial-spoke system is revealed by comparison of wild-type and mutant flagella. J Cell Biol 92:722–732

De Peyer JE, Machemer H (1982) Electromechanical coupling in cilia II. Effects of hyperpolarizing voltage steps. Cell Motil 2:497–508

De Peyer JE, Machemer H (1983) Threshold activation and dynamic response range of cilia following low rates of membrane polarization under voltage-clamp. J Comp Physiol 150:223–232

Dute R, Kung C (1978) Ultrastructure of the proximal region of somatic cilia in *Paramecium tetraurelia*. J Cell Biol 78:451–464

Eckert R (1972) Bioelectric control of ciliary activity. Science 176:473–481

Eckert R, Brehm P (1979) Ionic mechanisms of excitation in *Paramecium*. Annu Rev Biophys Bioeng 8:353–383

Eckert R, Naitoh Y (1970) Passive electrical properties of *Paramecium* and problems of ciliary coordination. J Gen Physiol 55:467–483

Eckert R, Naitoh Y, Machemer H (1976) Calcium in the bioelectric and motor functions of *Paramecium*. Symp Soc Exp Biol 30:233–255

Gibbons IR (1965) Reactivation of glycerinated cilia from *Tetrahymena pyriformis*. J Cell Biol 25:400–402

Gibbons IR (1965) Reactivation of glycerinated cilia from *Tetrahymena pyriformis*. J Cell Biol 25:400–402
Gustin MC, Bonini MN, Nelson DL (1983) Membrane potential regulation of cAMP: control mechanism for the swimming behaviour in *Paramecium*. Soc Neurosci Abstr 9:167
Hennessey TM, Kung C (1985) Slow inactivation of the calcium current of *Paramecium* is dependent on voltage and not internal calcium. J Physiol 365:165–179
Hennessey TM, Machemer H, Nelson DL (1985) Injected cyclic AMP increases ciliary beat frequency in conjunction with membrane hyperpolarization. Eur J Cell Biol 36:153–156
Hildebrand E, Dryl S (1976) Significance of K^+ and Ca^{2+} ions for the excitation of the protozoan membrane. Bioelectrochem Bioenerget 3:543–544
Hildebrand E, Dryl S (1983) Dependence of ciliary reversal in *Paramecium* on extracellular Ca^{2+} concentration. J Comp Physiol 152:385–394
Hinrichsen RD, Kung C (1984) Genetic analysis of axonemal mutants in *Paramecium tetraurelia* defective in their response to calcium. Genet Res 43:11–20
Hinrichsen RD, Saimi Y (1984) A mutation that alters properties of the calcium channel in *Paramecium tetraurelia*. J Physiol 351:397–410
Hoffmann-Berling H (1955) Geißelmodelle und Adenosintriphosphat. Biochim Biophysica Acta 16:146–154
Huang B, Ramanis Z, Luck DJL (1982) Suppressor mutations in *Chlamydomonas* reveal a regulatory mechanism for flagellar function. Cell 28:115–124
Kamada T, Kinosita H (1940) Calcium-potassium factor in ciliary reversal of *Paramecium*. Proc Imp Acad Tokyo 16:125–130
Koehler O (1926) Galvanotaxis. In: Bethe A, Bergman G von, Embden G, Ellinger A (eds) Handbuch der normalen und pathologischen Physiologie, vol 11. Springer, Berlin, pp 1027–1049
Kung C (1971) Genic mutations with altered system of excitation in *Paramecium aurelia*. I. Phenotypes of the behavioural mutants. Z Vergl Physiol 71:142–164
Kung C, Saimi Y (1985) Ca^{2+} channels of *Paramecium:* a multidisciplinary study. Curr Top Membr Transp 23:45–66
Ludloff K (1895) Untersuchungen über den Galvanotropismus. Pflügers Arch Ges Physiol 59:525–554
Machemer H (1972a) Ciliary activity and the origin of metachrony in *Paramecium:* effects of increased viscosity. J Exp Biol 57:239–259
Machemer H (1972b) Temperature influences on ciliary beat and metachronal coordination in *Paramecium*. J Mechanochem Cell Motility 1:57–66
Machemer H (1972c) Properties of polarized beat in *Paramecium*. Acta Protozool 11:295–300
Machemer H (1974a) Frequency and directional responses of cilia to membrane potential changes in *Paramecium*. J Comp Physiol 92:293–316
Machemer H (1974b) Ciliary activity and metachronism in Protozoa. In: Sleigh MA (ed) Cilia and flagella. Academic Press, London, New York, pp 199–286
Machemer H (1975) Modification of ciliary activity by the rate of membrane potential changes in *Paramecium*. J Comp Physiol 101:343–356
Machemer H (1976) Interactions of membrane potential and cations in regulation of ciliary activity in *Paramecium*.J Exp Biol 65:427–448
Machemer H (1977) Motor activity and bioelectric control of cilia. Fortschr Zool 24:195–210
Machemer H (1986) Electromotor coupling in cilia. In: Lüttgau HC (ed) Membrane control of cellular activity. Fortschr Zool/Progr Zool 33:205–250
Machemer H (1988a) Galvanotaxis: Grundlagen der elektromechanischen Kopplung und Orientierung bei *Paramecium*. In: Zupanc GHK (ed) Praktische Verhaltensbiologie. Paul Paray, Berlin
Machemer H (1988b) Ions in modulating cellular behaviour: electrophysiological implications. In: Stoeckenius W (ed) Signal transduction in chemotaxis. Alan Liss, New York (in press)
Machemer H, De Peyer JE (1977) Swimming sensory cells: electrical membrane parameters, receptor properties and motor control in ciliated Protozoa. Verh Dtsch Zool Ges 1977:86–110

Machemer H, Eckert R (1973) Electrophysiological control of reversed ciliary beating in *Paramecium.* J Gen Physiol 61:572–587
Machemer H, Eckert R (1975) Ciliary frequency and orientational responses to clamped voltage steps in *Paramecium.* J Comp Physiol 104:247–260
Machemer H, Eckert R (1977) Electromechanical coupling of ciliary activity in *Paramecium.* Fortschr Zool 24:211–215
Machemer H, Sugino K (1986) Parameters of the ciliary cycle under membrane voltage control. Cell Motil Cytoskel 6:89–95
Naitoh Y (1966) Reversal response elicited in nonbeating cilia of *Paramecium* by membrane depolarization. Science 154:660–662
Naitoh Y (1969) Control of the orientation of cilia by adenosine-triphosphate, calcium and zinc in glycerol-extracted *Paramecium caudatum.* J Gen Physiol 53:517–529
Naitoh Y (1979) Membrane currents in voltage-clamped *Paramecium* and their relations to ciliary motion. Acta Protozool 18:1–6
Naitoh Y (1982) Protozoa. In: Shelton GAB (ed) Electrical conduction and behaviour in 'simple' invertebrates. Clarendon, Oxford, pp 1–48
Naitoh Y, Kaneko H (1972) Reactivated Triton-extracted models of *Paramecium:* modification of ciliary movement by calcium ions. Science 176:523–524
Naitoh Y, Kaneko H (1973) Control of ciliary activities by adenosine triphosphate and divalent cations in Triton-extracted models of *Paramecium caudatum.* J Exp Biol 58:657–676
Nakaoka Y, Toyotama H (1979) Directional change of ciliary beat effected with Mg^{2+} in *Paramecium.* J Cell Sci 40:207–214
Nakaoka Y, Tanaka H, Oosawa F (1984) Ca^{2+}-dependent regulation of beat frequency of cilia in *Paramecium.*J Cell Sci 65:223–231
Omoto CK, Kung C (1980) Rotation and twist of the central-pair microtubules in the cilia of *Paramecium.* J Cell Biol 87:37–46
Omoto CK, Witman GB (1981) Functionally significant central-pair rotation in a primitive eukaryotic flagellum. Nature (London) 290:708–710
Sanderson MJ (1984) Cilia. In: Bereiter-Hahn J, Matoldsy AG, Richards KS (eds) Biology of the integument, vol 1. Invertebrates. Springer, Berlin Heidelberg New York, pp 17–42
Satir P (1984) The generation of ciliary motion. J Protozool 31:8–12
Satir P, Wais-Steider J, Lebduska S, Nasr A, Avolio J (1981) The mechanochemical cycle of the dynein arm. Cell Motil 1:303–327
Schultz JE, Grünemund R, Hirschhausen R von, Schönefeld U (1984) Ionic regulation of cyclic AMP levels in *Paremecium tetraurelia.* FEBS Lett 167:113–116
Sleigh MA (1974) Patterns of movement of cilia and flagella. In: Sleigh MA (ed) Cilia and flagella. Academic Press, London New York, pp 79–92
Sleigh MA, Barlow DI (1982) How are different ciliary beat patterns produced? In: Amos WB, Duckett JG (eds) Prokaryotic and eukaryotic flagella. Symp Soc Exp Biol 35:139–157
Sugino K, Machemer H (1987) Axial-view recording: an approach to assess the third dimension of the ciliary cycle. J Theor Biol 125:67–82
Takahashi M, Naitoh Y (1978) Behavioral mutants of *Paramecium caudatum* with defective membrane electrogenesis. Nature (London) 271:656–658
Tamm SL, Horridge GA (1970) The relation between the orientation of the central fibrils and the direction of beat in cilia of *Opalina.* Proc R Soc Lond Ser B 175:219–233
Tamm SL, Tamm S (1981) Ciliary reversal without rotation of axonemal structures in ctenophore comb plates. J Cell Biol 89:495–509
Verworn M (1889) Die polare Erregung der Protisten durch den galvanischen Strom. Pflügers Arch Ges Physiol 45:1–36
Warner FD, Satir P (1974) The structural basis of ciliary bend formation. Radial spoke positional changes accompanying microtubule sliding. J Cell Biol 63:35–63

Chapter 15 A Genetic Dissection of Ion-Channel Functions

RAJEEV RAMANATHAN, YOSHIRO SAIMI, ROBERT HINRICHSEN, ANTHONY BURGESS-CASSLER, and CHING KUNG[1]

CONTENTS

1 Introduction

Mutations are commonly used to enumerate and to understand the components in biological phenomena, including locomotion and behavior. In *Paramecium*, mutants have been isolated and studied, having in mind a genetic dissection of the behavioral mechanisms (Kung 1971 a, b). To the genetic and behavioral analyses of the mutants was soon added an electric dimension, since the *Paramecium* membrane generates a Ca^{2+} action potential (Naitoh et al. 1972). The advent of the voltage clamp in this field greatly refined the electrical measurements (Naitoh and Eckert 1974; Oertel et al. 1977). Purification of the ciliary membrane was successful (Hansma and Kung 1975; Adoutte et al. 1980). However, the search for ion-channel proteins through the comparison of electropho-

1 Laboratory of Molecular Biology and the Department of Genetics, University of Wisconsin-Madison, Madison, WI 53706, USA

Paramecium. Ed. by H.-D. Görtz

retic patterns of wild type and mutants was frustrated by physiological variations of membrane proteins (see Adoutte et al. 1983). In 1980, Hiwatashi and co-workers (Haga et al. 1982) reported success in restoring normal behavior in a mutant by microinjecting it with the cytoplasm of wild type. This functional assay has been used in a new search for the products of the *cnrC* gene (Haga et al. 1984b) and the *pntA* gene (Hinrichsen et al. 1986). The product of *pntA* gene either is or affects calmodulin.

This chapter first reviews the ion currents to be discussed before summarizing several types of mutants whose defects in ion currents are understood through voltage-clamp analyses. We will then review how the mutants have been used in behavioral, physiological, and biochemical research. Finally, the cases wherein the gene products have been enriched for or purified using the microinjection-restoration procedure are reviewed.

2 Ion Currents

We will review briefly the normal (wild-type) currents as seen under controlled (clamped) voltages before describing the mutant defects. Due to space limitation, we will only review the four currents observed upon depolarizations from cells bathed in millimolar Ca^{2+}, K^+, and Na^+. See Machemer (Chap. 13, this Vol.) for a more systematic and detailed treatment on the electrical properties of the *Paramecium* membrane.

2.1 *The Ca^{2+} Current*

The wild-type membrane clamped at its resting voltage level of ca. −40 mV (inside negative) has little or no net current crossing it. A sudden shift of the resting level to 0 mV, an example of a step depolarization, is followed by a surge of current flowing into the cell. This inward current has been proven to be carried by Ca^{2+} and is the source of Ca^{2+} for ciliary reversal (Oertel et al. 1977). In *P. tetraurelia,* this current rises to its peak of about 7 nA at about 2 ms and relaxes in the next 10 ms to a low steady-state level of less than 1 nA. The Ca^{2+} current is larger in *P. caudatum,* approximately 12 nA at its peak (Fig. 1A). The relaxation of the Ca^{2+} current after the peak is called the inactivation of the current and has been shown to be due to the increase in intracellular concentration of Ca^{2+} (Brehm and Eckert 1978a). The steady-state inward Ca^{2+} current is usually obscured by the concomitant outward current, also induced by the depolarization.

2.2 *The Depolarization-Activated K^+ Current*

A step depolarization also induces an *outward* current carried by K^+ besides the inward Ca^{2+} current. This K^+ current reaches its peak within a few tens of milliseconds. Thus, its activation is slightly slower than that of the inward current.

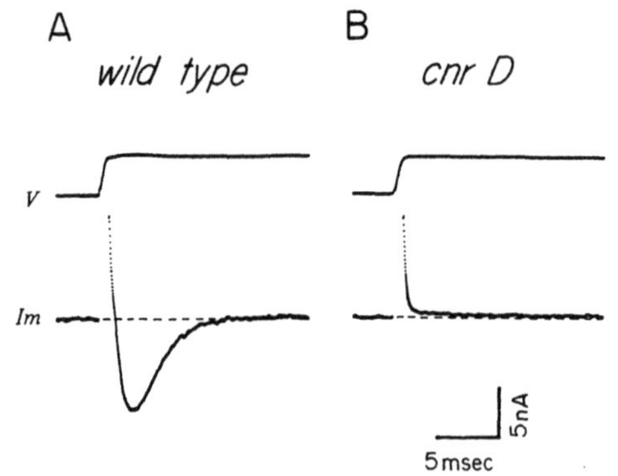

Fig. 1 A, B. The Ca^{2+} inward current in the wild type (**A**) and its absence in *cnrD* (**B**) of *P. caudatum. The membrane (V)* was held at −40 mV and then stepped up to −7 mV under the voltage-clamp conditions. The resulting change in the membrane current (I_m) showed first a capacitive surge at the onset of the voltage step and then the Ca^{2+} inward current (*below the dashed line*) in the wild type. The Ca^{2+} inward current peaked in 5 ms and relaxed to a very low level in the next 10 ms. This Ca^{2+} current was not detected in *cnrD* which is an inexcitable mutant lacking the avoiding reaction upon stimulation. The recording conditions were such that most of the outward K^+ current was suppressed by Cs^+ and tetraethylammonium. The remaining, small outward current in *cnrD* reflected leakage current. (After Takahashi et al. 1985)

Because of the temporal overlap, the approximate trajectory of its development can be better seen with mutants without the inward current (Satow and Kung 1980b) or upon the application of W-7 which blocks the Ca^{2+} inward current (Hennessey and Kung 1984).

2.3 The Ca^{2+}-Activated Na^+ Current

When the external medium contains a few tens of millimolar Na^+, depolarization indirectly activates a Na^+ inward current through the increase in internal free Ca^{2+} due to the Ca^{2+} current. This Na^+ current develops slowly to peak over hundreds of milliseconds to a magnitude of less than 1 nA (Saimi 1986) (Fig. 3A).

2.4 The Ca^{2+}-Activated K^+ Current

The same internal Ca^{2+} activates a K^+ outward current up to 3 nA (Satow and Kung, 1980a) (Fig. 4A). It is important to realize that this K^+ outward current develops more slowly and is therefore delayed compared to the Na^+ current.

The membrane potential is not clamped in cells behaving naturally, but varies as the currents flow. Upon proper stimulation, a complex, time-dependent interplay of the currents results in an active depolarization, called the action potential. The Ca^{2+} which enters the cilia during the action potential causes them to beat in the reverse direction. The resulting transient backward swimming has long been observed and termed the avoiding reaction (Jennings 1906). The earlier treatment of membrane excitation in *Paramecium* (Eckert et al. 1976) dealt with a form of action potential, resembling that of nerve and muscle in its brevity (tens of milliseconds), recorded from cells bathed in solution with Ca^{2+} but no Na^{+}. This type of action potential can be explained by the two fast, voltage-dependent currents (current 1 and 2 above), whereby the inward Ca^{2+} current is responsible for the upstroke through a Hodgkin cycle and the delayed outward K^{+} current for the downstroke. A more realistic treatment of the compound action potential lasting 0.5 to 1 s (Satow and Kung 1974), underlying the natural behavior in freshwater which contains Na^{+}, includes all four currents: the early currents 1 and 2, as well as the two late currents (3: the Ca^{2+}-activated Na^{+} inward current and 4: the Ca^{2+}-activated K^{+} outward current). The Ca^{2+} which carries the peak and the sustained current through the Ca^{2+} channel eventually activates the Na^{+} channel. The Na^{+} current, being inward, competes with the outward K^{+} currents to maintain the depolarization over hundreds of milliseconds. The full activation of the Ca^{2+}-dependent K^{+} current eventually overpowers the inward currents and thus repolarizes the membrane to its resting level. This four-current model has been presented briefly in Saimi et al. (1983) and will be detailed elsewhere.

Mutants with altered currents 1, 3, and 4 are described below. The lack of mutants of the second current in presumably unbiased screening for behavioral mutants in mutagenized populations of cells suggests that the current is either of little behavioral significance or, more likely, vital to survival.

3 Mutants with Defective Currents

The mutants described in this section are specific, i.e., the mutations affect only one ion current at a time, unless other currents depend on the current affected. The mutants have either a decreased current (the "down" mutants) or an enhanced current (the "up" mutants) (Table 1).

3.1 Down Mutants of the Ca^{2+} Current

The Ca^{2+} current is partially or completely shut down by the mutations *pwA, pwB,* and *pwC* (pawns) of *P. tetraurelia* (Kung 1971a, b; Oertel et al. 1977; Satow and Kung 1980b) or *cnrA, cnrB, cnrC,* and *cnrD* (*C*audatum *N*on-*Re*versal) of *P. caudatum* (Takahashi and Naitoh 1978; Takahashi et al. 1985). Figure 1 shows the case of the mutant *cnrD* where the inward current is absent. *Any* pawn complements *any* CNR mutant as tested by the microinjection-restoration procedure (see below, and Haga et al. 1982). The simplest interpretation of this unexpected result is that the *pw* and *cnr* loci represent different genes, i.e., there are at least seven complementation groups, mutations in each of which lead to the loss of the Ca^{2+} current. Since this Ca^{2+} current is responsible for the upstroke of the action potential, these mutants typically have no action potential. The behavioral consequence of such a loss is the failure to swim

Table 1. Summary of ion-currents affected by identified mutations in *Paramecium*

Type	Complementation group	Ion channel [a] affected	Reference
Pawn	*pwA, pwB, pwC*	$I_{Ca}\downarrow$	Oertel et al. (1977) Satow and Kung (1980 b)
CNR [b]	*cnrA, cnrB, cnrC, cnrD*	$I_{Ca}\downarrow$	Takahashi and Naitoh (1978) Takahashi et al. (1985)
Dancer	*Dn*	$I_{Ca}\uparrow$	Hinrichsen and Saimi (1984)
Fast-2	*fna*	$I^{Ca}_{Na}\downarrow$	Saimi (1986)
Paranoiac	*PaA, PaC*	$I^{Ca}_{Na}\uparrow$	Saimi and Kung (1980) Saimi (unpublished)
Pantophobiac	*pntA, pntB*	$I^{Ca}_{K}\downarrow$	Hinrichsen et al. (1985 a)
TEA-insensitive	*TeaA*	$I^{Ca}_{K}\uparrow$	Hennessey and Kung (unpubl.)

[a] ↓ "Down mutants": currents smaller or missing; ↑ "up mutants": current stronger or longer; $I_{Ca} = Ca^{2+}$ current; I^{Ca}_{K} = Ca-dependent K^+ current; I^{Ca}_{Na} = Ca-dependent Na^+ current.
[b] CNRs are mutants of *P. caudatum;* all others listed are of *P. tetraurelia.*

backward upon proper stimulation because of the failure to allow Ca^{2+} entry (Kung 1971 a, b; Takahashi and Naitoh 1978). Hence, the name "pawn", after the chess piece. An additional secondary consequence is the loss of the two Ca^{2+}-dependent currents (currents 3 and 4 above).

3.2 Up Mutants of the Ca^{2+} Current

Mutations can strengthen ion currents by various means. *Dn (Dancers)* prevent the proper inactivation of the Ca^{2+} current (Hinrichsen and Saimi, 1984). The incomplete inactivation results in a prolonged and larger influx of Ca^{2+} (Fig. 2, arrows). It is interesting that the down mutations, *pws* or *cnrs* are recessive while the up mutation, *Dn* is semidominant. The behavioral consequence of the *Dn* lesion is an overreaction to certain stimuli, consistent with a lower excitation threshold and higher internal Ca^{2+} upon excitation (Hinrichsen et al. 1984). The altered kinetics of the mutant, as opposed to simple increase or decrease in current, together with the semidominance of the mutation, suggest that *Dn* may code for the structure of the Ca^{2+} channel.

3.3 Down Mutants of the Ca^{2+}-Dependent Na^+ Current

A mutant called *fast-2* (*fna*) has almost no Na^+ current (Fig. 3). It has, however, normal Ca^{2+} current and normal Ca^{2+}-dependent K^+ currents (Saimi 1986). This mutant swims rapidly forward instead of exhibiting avoiding reactions (as wild-type cells would) when confronted with a Na^+ solution. Its behavioral response to solutions containing no Na^+ appears to be normal (Kung 1971).

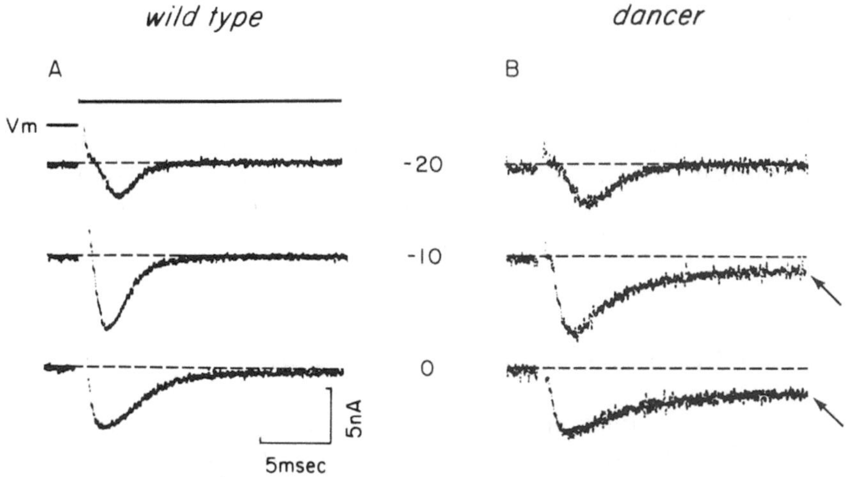

Fig. 2 A, B. The Ca^{2+} current in the wild type (**A**) and an enhanced Ca^{2+} current in the *Dancer* (*Dn*) mutant of *P. tetraurelia.* Under similar conditions to that for Fig. 1, where K^+ current was suppressed, the Ca^{2+} current was revealed with voltage steps from a holding level of −40 mV to different levels as indicated by the *traces.* In the wild type the Ca current peaked within 5 ms and relaxed to very low levels in the next 10 ms. In *dancer* (**B**), it peaked normally but its relaxation was incomplete, leaving substantially sustained Ca^{2+} current over a longer period (Hinrichsen and Saimi 1984)

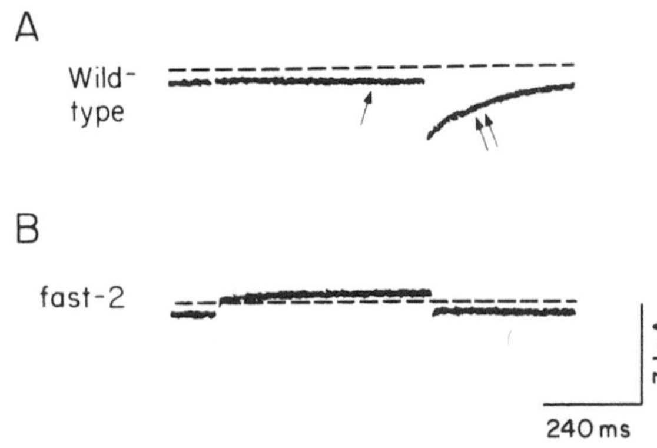

Fig. 3 A, B. The mutant *fast-2* is missing the Ca-dependent Na current. Cells were bathed in a solution of 8 mM NaCl, 1 mM $CaCl_2$, 10 mM TEA-Cl, 0.01 mM EDTA, 1 mM HEPES, pH 7.2−7.4 and were recorded with electrodes containing 4 M CsCl. A step depolarization from −40 to −5 mV induces a Na inward current (*arrow*) and a large tail current (*double arrows*) in wild-type cells. Note that the Na current and the tail are absent in the mutant *fast-2,* while the Ca current is intact (Saimi, unpubl.)

3.4 Up Mutants of the Ca^{2+}-Dependent Na^+ Current

A class of mutants called *Paranoiacs* was first noted for their spontaneous long backward swimming in culture medium which normally contains Na^+ (Kung 1971a, b). This behavioral aberration is now best explained by the enhancement of their Ca^{2+}-dependent Na^+ current (Saimi and Kung 1980). The other currents of this class of mutants are normal. Note that the *Paranoiac* mutations are dominant while *fast-2* is partially dominant (Hinrichsen, unpubl. observa tion).

3.5 Down Mutants of the Ca^{2+}-Dependent K^+ Current

Pantophobiac mutations block this K^+ current nearly completely (Fig. 4) (Saimi et al. 1983; Hinrichsen et al. 1985a). The Ca^{2+} current which activates this K^+ current is normal, however. This class of mutants overreacts to a variety of stimuli, responding with prolonged backward swimming. Curiously, these mutants also have difficulties in mating, i.e. either few or no fertile pairs can be found when mated to competent wild-type cells of the opposite mating type. Some of these mutants were found to be defective in their calmodulin (see below).

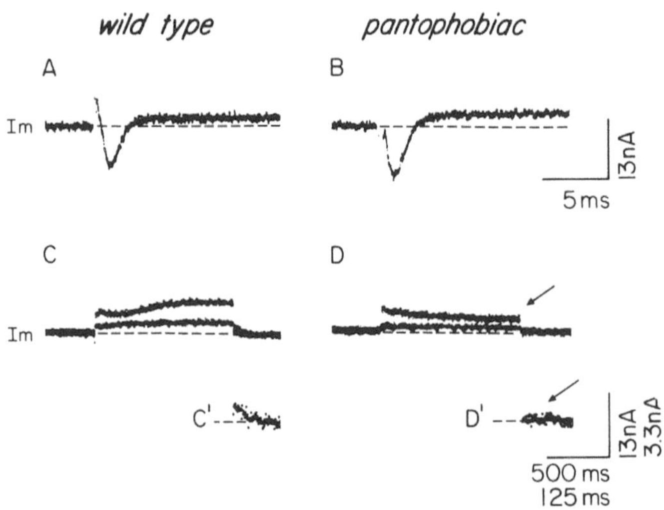

Fig. 4A–D. Membrane-current responses to step depolarizations (**A–D**) of the wild type (*left*) and the pantophobiac mutant d4-620 (*right*) bathed in a solution containing 1 mM $CaCl_2$, 1 mM KCl, 0.01 mM EDTA, and 1 mM HEPES, pH 7.3. **A, B** show the early current response (I_m) to a step depolarization from −40 to −10 mV under voltage clamp. The inward Ca current is not altered in the mutant (**B**) compared to the wild type (**A**). **C, D** show the currents induced by a 1-s step from −40 to −15 mV (*lower trace*) and by a step from −40 to −5 mV (*upper trace*). Note that the larger step clearly induces a slowly activating outward current and a tail current in the wild type (**C**), while these are reduced or missing in the mutant (**D**). The tail currents are expanded in **C′** and **D′** for the wild type and mutant, respectively (Saimi et al. 1983)

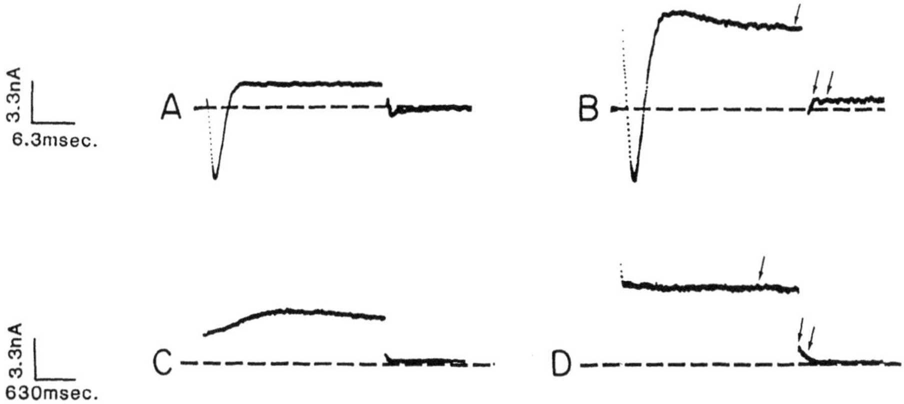

Fig. 5 A – D. The Ca-dependent K current is large in the *TeaA* mutant (**B**, **D**) than in the wild type (**A**, **C**). Cells were bathed in 4 mM KCl, 1 mM $CaCl_2$, 1 mM MOPS, 0.01 mM EDTA, pH 7.2 and were recorded with 2 M KCl electrodes. The membrane was abruptly depolarized from −40 to −4 mV which triggered the early Ca current (*upper traces*) followed by a set of late currents (*lower traces*) including the very slow Ca-dependent K current. *TeaA* (**B**, **D**) has a normal Ca transient but has a rapidly developing and larger Ca-dependent K current (*arrow*) and the corresponding tail current (*double arrows*) compared to wild type (**A**, **C**) (Hennessey and Kung, unpubl.)

3.6 Up Mutants of the Ca^{2+}-Dependent K^+ Current

The *TEA-insensitive* mutant, *TeaA*, was first isolated because of its poor behavioral reaction to the K^+ channel blocker, tetraethylammonium (Chang and Kung 1976). The mutation was found to be semidominant over the wild-type allele (Hinrichsen, unpublished result). Voltage-clamp experiments show that it has a K^+ current which activated far more efficiently than the wild type (Fig. 5, arrows). Less than 2 ms of the Ca^{2+} current instead of the normal 500 ms is needed to half activate the same K^+ current (Hennessey et al. 1985). All other currents appear normal.

4 The Use of the Mutants

Mutations can be used as blockers in biological investigations, much like pharmacological inhibitors or blockers. What the mutants are still capable of tells us what functions are independent of the steps governed by the mutations. Pawn mutants can no longer be attracted by acetate or lactate (Van Houten et al. 1975) or repelled by a heated zone (Hennessey and Nelson 1979) indicating the importance of the avoiding reaction in these behaviors. *Pawns*, however, can still perform geotaxis, galvanotaxis (Kung 1971), rheotaxis (Kung, unpubl.), and can be trapped in a Ba^{2+}-rich zone (Van Houten 1979), showing that these behaviors do not require the avoiding reactions. These latter taxes can be explained by physical effects on swimming paramecia or by a component of the control of ciliary beating frequency which is retained in the *pawns* (Brehm and Eckert 1978 b).

The depolarizations induced by various stimuli (the receptor potential or generator potential) can trigger the action potential and are often obscured by the latter. Mutations in *pawns* and *CNRs* erase the action potential but not the receptor potentials. This observation supports the notion that the receptor potential and the action potential have independent mechanisms, probably governed by different types of channel molecules. Impact at the anterior end of a *Paramecium* induces a depolarizing receptor potential (Naitoh and Eckert 1969). This first wave of depolarization has been studied, unobscured by the action potential, in *CNRs* and *pawns* (Takahashi and Naitoh 1978; Kung 1979). It was found that this depolarization is usually also because of a Ca^{2+} influx, although the pathway (channel), unlike the Ca^{2+} channel for the action potential, also conducts other divalent cations (Satow et al. 1983). This conclusion was also reached independently in the study of this receptor potential in *Stylonychia* by different means (de Peyer and Deitmer 1980). Heating from 23° to 35° C causes a depolarization which can also trigger action potentials. This first depolarization upon heating is also intact in the *pawns* (Hennessey et al. 1983).

The *Dancer* mutation, which increases the Ca^{2+} current by blocking its inactivation, has been used to inject large amounts of Ca^{2+} into the cell. It has been employed to see what would happen to other ion currents when we caused a flood of Ca^{2+} in the cell. In other words, *Dancer* mutation has been used to "biologically" inject a large amount of Ca^{2+} along the normal pathway to the relevant compartment(s), as opposed to the shotgun delivery of Ca^{2+} through ionophores or via microinjection pipettes. For example, we can test whether the loss of the Ca-dependent Na current in the *fast-2* mutant is due to a loss in that current or an increased threshold for its activation by Ca^{2+}. When the *Dancer-fast-2* double mutant was constructed and tested, the Ca^{2+} current was enhanced as expected, but the Ca-dependent Na^{+} current was not rescued, indicating that a large and sustained dose of internal Ca^{2+} is unable to activate the defective Na channel.

A guanylate cyclase has been found to be present in the ciliary membrane and is activated by Ca^{2+} (Klumpp and Schultz 1982). To test whether the Ca^{2+} entering *during* excitation, as opposed to other intracellular sources of Ca^{2+}, activates the guanylate cyclase in vivo, *Dancer* and *pawn* mutants were used to increase or decrease the Ca^{2+} concentration in the cilia. The results showed that cGMP concentration in the cell increases transiently in the wild type upon a stimulation. This response is strengthened and is prolonged in *Dancer* and is abolished in the *pawn* mutants. It was therefore concluded that Ca^{2+} ions, entering through the Ca^{2+} channel during the action potential, also initiate a cascade of signal amplification via the formation of cGMP (Schultz et al. 1986).

The mutants (Table 1) have abnormal action potentials resulting from the ion currents turned up, down, or off, and this results in the behavioral aberrations (Satow and Kung 1974). The abnormalities in the action potentials and the ion currents are part of the bases on which the four-current model of the action potential (above) is built.

Mutations also mark the gene products and therefore make their identification possible. The following section reviews the progress made toward their identification.

5 A Search for the Gene Products

Haga and Hiwatashi (1982) first showed that it is possible to restore the behavioral aberration of a mutant to normal by microinjecting cytoplasm of the wild type into the mutant. Not all mutants can be cured through the microinjection of soluble wild-type cytoplasm. Table 2 summarizes what is known concerning whether a mutant can be restored to the wild-type behavior through microinjection. Restoration, when possible, in the form of a measurable behavioral correction, can be used as a quantifiable assay of the active element injected. A search for the active element can thus be carried out by fractionating the wild-type cytoplasm and assaying the fractions for the restoration activity. The active element could be the RNA or the protein which corresponds to the gene, or an enzymatic reaction product of the protein. The following section reviews the study of two classes of restoration factors, one corresponding to the mutations affecting the Ca^{2+} channel, the other the mutations affecting the Ca^{2+}-dependent K^+ channel.

Table 2. Summary of the curability of different mutations in Paramecium

Channel defect	Mutant type	Cured?	Time[a]	Curing factor[b]	Reference
Ca channel:					
$I_{Ca}\downarrow$	Pawns: *pwA, pwB, pwC*	+	6–8	Particulate	Haga et al. (1984b)
	CNR:				
	cnrA, cnrB, cnrD	+	6–8	Particulate	Haga et al. (1984b)
	cnrC	+	6–8	Soluble	Haga et al. (1984a)
	cnrD	+	6–8	ND	
$I_{Ca}\uparrow$	Dancer: *Dn*	–	–	–	Hinrichsen and Saimi (1984)
Ca-dependent K channel:					
$I_K^{Ca}\downarrow$	Pantophobiac: *pntA, pntB*	+	2–3	Soluble	Hinrichsen et al. (1986)
	pntD	–	–	–	Wallen-Friedman and Hinrichsen (unpubl.)
$I_K^{Ca}\uparrow$	TEA-insensitive *TeaA*	–	–	–	Hinrichsen and Natarajan (unpubl.)

[a] Time required to achieve maximal curing (h).
[b] ND: not determined.

5.1 Proteins Which Restore the Ca^{2+} Current Missing in Pawns and CNRs

Among the seven mutations known to shut off the Ca^{2+} current (Table 1), *pwA*, *pwB*, and *pwC* were found to be temporarily restored by microinjection of a particulate fraction of the cell homogenate. Further characterization of the particulate proteins responsible for the curing of the *pawn* mutants showed that these activities were associated with internal membranes. Purification of these proteins, especially the *pwC* factor, has progressed to some degree, but the complexities of working with membrane proteins, and the difficulty of using the microinjection assay for such a purpose has precluded further progress (Haga et al. 1984a). Interestingly, the *pwB* factor seems to be amplified in sequential microinjections into *pwB* cells, implying an enzymatic activity or a positive feedback system for this factor.

The *cnrC* mutant in *P. caudatum*, on the other hand, was cured by a soluble fraction of the cell homogenate (Table 2). The *cnrC*-curing activity is heat-labile, N-ethylmaleimide-sensitive and RNase-resistant. Restoration occurred in the presence of G-418 which blocks protein synthesis. The return of excitability in the microinjected *cnrC* mutants is detectable as early as 2 h after microinjection, peaks by 6–8 h and lasts for 2–3 days. The *cnrC* curing factor is apparently not amplified in the cell, as is the *pwB* factor. Furthermore, deciliation of cured cells followed by behavioral testing after reciliation suggests that the curing activity can be depleted by removing the ciliary membranes which also exclusively contain the Ca channels (Haga et al. 1984b).

This protein was more amenable to characterization and purification using the multifaceted approach of microinjection followed by behavioral tests to assay the Ca-channel function. The curing activity was followed through differential centrifugation, ammonium sulfate precipitation, gel filtration, hydroxylapatite chromatography, and anion-exchange chromatography. This series of steps has resulted in over 500-fold purification of the active protein from the soluble cytoplasm (about 3,000-fold from total protein). The final samples contain 2.5% of the curing activity (Table 3). Though not yet purified to homogeneity, this highly purified protein fraction is fully capable of restor-

Table 3. Purification of the *cnrC* curing factor. (After Haga et al. 1984b)

Fraction	Specific activity (s mg^{-1} protein $\times 10^7$)	Total activity (s $\times 10^9$)	Protein (mg)	Purification factor
Cell homogenate	–	–	3700	
S_1	1.2	7.4	616.7	1
S_2	4.3	7.7	179.5	4
$>60\%$ $AmSO_4$	17.9	3.5	19.6	15
Sephadex G-75	44.5	2.7	6.0	37
HAP 100	58.7	0.53	0.9	49
DEAE 6	532.5	0.13	0.024	444
DEAE 7	686.7	0.06	0.009	572

Fig. 6. **A** SDS-PAGE of purified *cnrC* curing factor. *Lane 1:* soluble cytoplasmic fraction (S_2). *Lane 2:* curing-active fractions following chromatography on Sephadex G-75. *Lanes 3–5:* the most active fractions following elution from DEAE-Sephadex (Haga et al. 1984a).
B Restoration of the Ca current in the *cnrC* mutants. The Ca current was recorded at 6 h after an injection of a Sephadex G-75 column fraction from the wild type (*left*). A voltage step from the holding level at −40 to −5 mV induced the maximal peak Ca current of ~6 nA in this "cured" *cnrC*. The same voltage step in this series of experiments induced up to 9.5 nA current in the uninjected wild type (not shown). A similarly injected soluble cytoplasmic fraction, S_2, from *cnrC*, however, is not capable of curing their sister cells (*right*). Not more than 1.5-nA Ca current was ever detected in *cnrC*. The bath contained 10 mM TEACl, 4 mM CsCl, 1 mM $CaCl_2$, 1 mM HEPES at pH 7.3 and electrodes were filled with 4 M CsCl; all designed to better the resolution of the Ca current. (After Haga et al. 1984b)

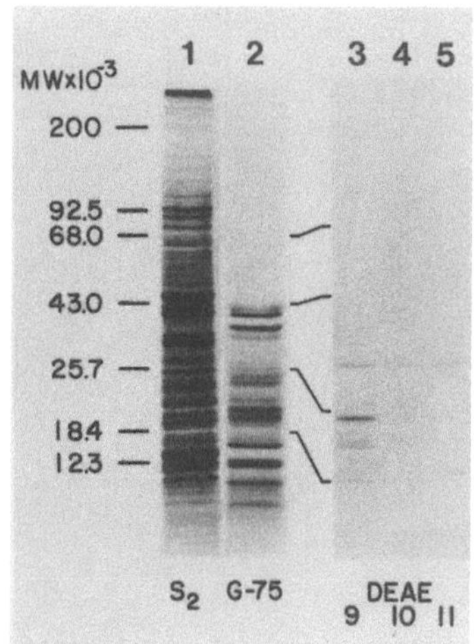

a

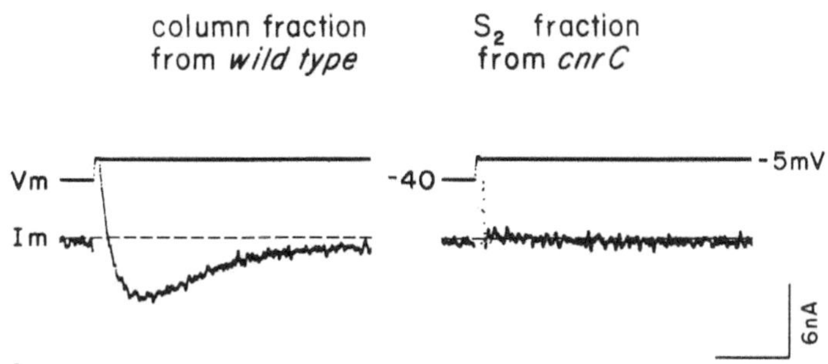

b

ing the Ca^{2+} current missing in the *cnrC* mutant (Fig. 6B) when injected into the mutant.

The behavior of this factor using DEAE chromatography and isoelectric focusing indicated an isoelectric point of 4.5 to 5.0, and the elution profile from a gel-filtration column showed that its apparent molecular weight is between 17000–30000 daltons. The overall protein composition of several fractions at different steps of the purification scheme is shown in Fig. 6A. A correlation of the curing activity with the intensity of protein gel bands in the most purified DEAE fractions indicates that proteins of molecular weights 22, 24, 19, and < 10 kD may be responsible for curing activity (Haga et al. 1984b). Although purified in a manner similar to that for calmodulin, we have determined that the *cnrC* curing factor does not have the functional characteristics of calmodulin, namely, activation of brain phosphodiesterase and Ca-dependent mobility shifts on electrophoresis (Haga et al. 1984a, b). The curing-active fractions contain a family of proteins that are heavily phosphorylated by an endogenous casein-kinase activity. The identity and density of these phosphoproteins correlates well with the biological curing activity (Ramanathan, Hin-

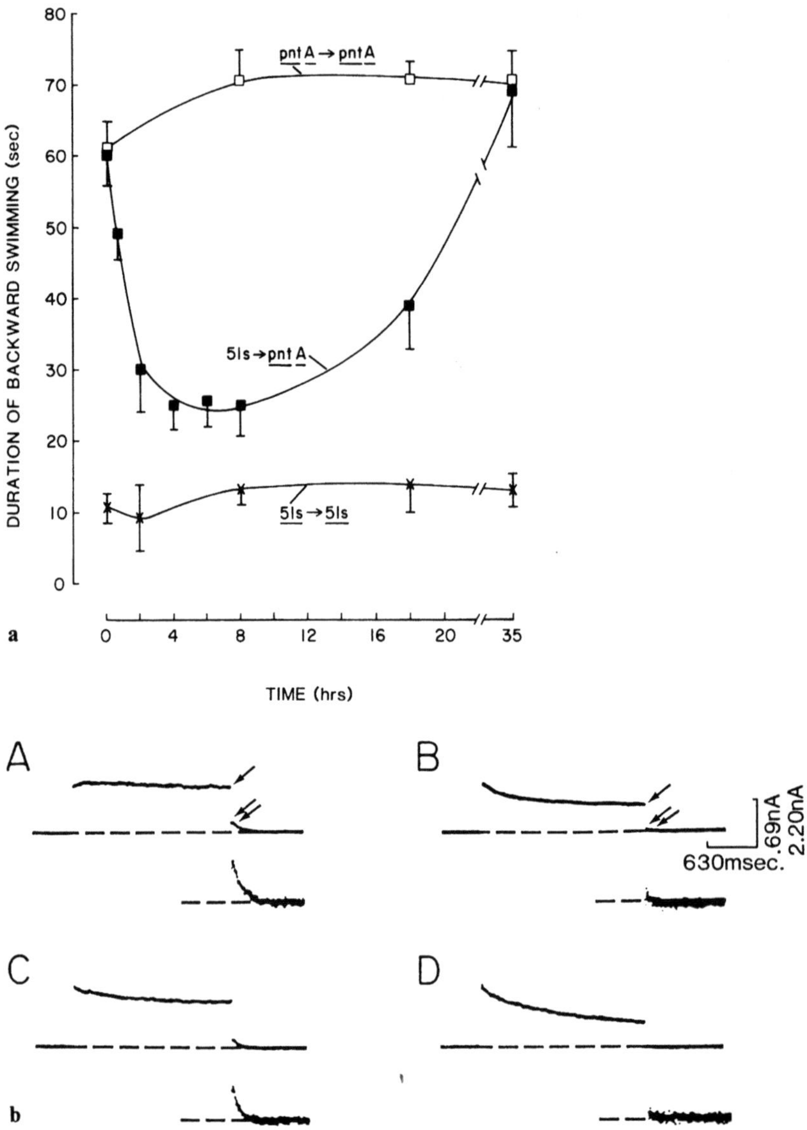

Fig. 7. a The time course for the restoration of the wild-type phenotype in *pntA* (■) after the microinjection of the wild type S_2 fraction. The cells were injected at time zero with 20 pl of the high-speed supernatant fraction (S_2) of wild type cytoplasm. The injected cells were tested at various times for their duration of backward swimming in a solution containing 10 mM NaCl and 5 mM TEA^+. *pntA* also was injected with their own S_2 (□) as a control, X, wild type cells injected with wild type S_2. Four to six cells were injected and tested at each time point (±SD) (Hinrichsen et al. 1985a).

b Voltage-clamp analysis of the Ca^{2+}-dependent K^+ current in *Paramecium* cells injected with calmodulin or buffer. The electrodes contained 3 M KCl, and the cells were bathed in 1 mM KCl, 1 mM $CaCl_2$, 1 mM MOPS (morpholinopropanesulfonic acid), and 10^{-2} mM EDTA (pH 7.2). A depolarization step from −40 to −5 mV was held for 2 s before returning to −40 mV. The cells were microinjected with either 20 pl of buffer solution [4 mM KCl, 1 mM $CaCl_2$, 1 mM Hepes (N-2-hydroxyethyl piperazine-N'-2-ethanesulfonic acid) and 10^{-2} mM EDTA at

richsen, and Kung, unpublished). Work is in progress to determine whether phosphorylation is causally related to the curing of *cnrC* cells.

5.2 Calmodulin, the Protein Which Restores the Ca^{2+}-Dependent K^+ Current Missing in the Pantophobiac A Mutant

Pantophobiacs are the mutants with much reduced Ca^{2+}-dependent K^+ current (Table 1; Fig. 4). This deficiency leads to prolonged backward swimming in a variety of solutions. Injection of the wild-type cytoplasm into one of the *pantophobiacs, pntA,* normalized its behavior. Figure 7a shows the time course of the behavioral normalization after the injection. The restoration is almost complete by 2 h and persists for about 24 h (Hinrichsen et al. 1985a).

The curing protein for *pntA* has now been identified to be calmodulin and it restores the Ca^{2+}-dependent K^+ current in the mutant (Fig. 7b) (Hinrichsen et al. 1986). The wild-type calmodulin is heat stable, and about 17000 daltons in molecular weight. It shows the characteristics Ca-dependent mobility shifts upon gel electrophoresis. Like calmodulins of other species, *Paramecium* calmodulin activates bovine-brain phosphodiesterase and myosin light-chain kinase (Rauh and Nelson 1981; Hinrichsen et al. 1986; Hinrichsen, unpubl.).

The mutant has a variant form of the calmodulin. It has the same molecular weight as determined by SDS-polyacrylamide gel electrophoresis and activates brain phosphodiesterase to a similar extent. However, on acid-urea gel systems, its relative mobility is different from that of the wild type (Fig. 8). Also, monoclonal antibodies directed against calmodulin from the wild type exhibit a lower affinity for calmodulin from the *pntA* mutant (Burgess-Cassler, Hinrichsen, Kung, submitted). How the mutant calmodulin differs from that of the wild type in terms of its structure is not yet clear at this writing. Amino acid sequencing, currently underway should reveal whether they differ in their primary structures. In that case, *pntA* would be the structural gene for calmodulin in *Paramecium.* On the other hand, they may differ in a posttranslational modification which could infer an enzymatic function for the *pntA* gene product.

There are other *pantophobiac* mutants that have behavioral aberrations similar, though not identical to, those of *pntA.* Two such mutants, *pntC* and

---

pH 7.2] or calmodulin (1 pg) 4 h before the voltage-clamp analysis was performed. In **A** to **D**, the *inset* shows the tail current amplified four times. **A** Wild-type cells injected with 20 pl of buffer solution showed the Ca^{2+}-dependent K^+ current. The *single arrow* indicates the late outward current upon depolarization, primarily through the Ca^{2+}-dependent K^+ channel. Upon repolarization, there was an obvious tail current (*double arrows*). **B** The *pntA* mutant, injected with 20 pl of buffer solution, virtually devoid of the Ca^{2+}-dependent K^+ current. There was a declining late outward current (*arrow*) upon depolarization, which indicates a nearly complete loss of the Ca^{2+}-dependent K^+ current. There is little or no tail current upon repolarization (*double arrows* and *inset*). **C** The *pntA* mutant after the injection of wild type calmodulin showed a return of both the late outward current and the tail current. **D** The *pntA* mutant injected with *pntA* calmodulin showed no restoration. There was virtually no Ca^{2+}-dependent K^+ current, as indicated by the lack of both the late outward current and the tail current (Hinrichsen et al. 1986)

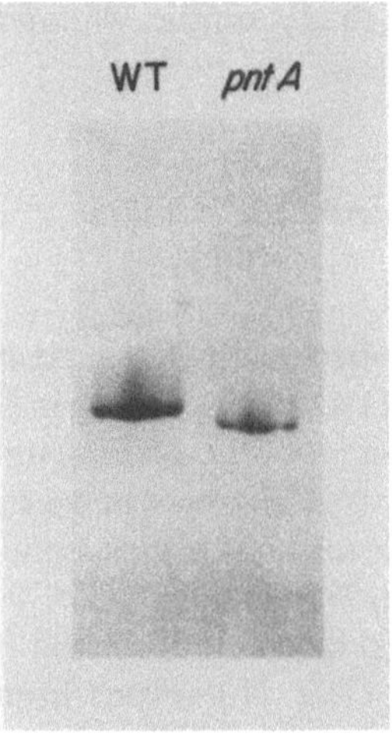

Fig. 8. Polyacrylamide gel electrophoresis showing that the *pntA* calmodulin exhibits a different relative mobility from wild type. Purified calmodulins from wild type and the *pntA* mutant were isolated and subjected to electrophoresis under acidic conditions (pH 3.0) in polyacrylamide gels containing urea. The anode is toward the bottom of the picture

pntD, also have defects in their Ca^{2+}-dependent K^+ current. They also have defects in their calmodulins, recognizable in nondenaturing gel electrophoresis or isoelectric focusing (Elbaum et al. 1986; Wallen-Friedman et al. 1986, Neuroscience Abstracts). The mutational "alterations" in each pantophobiac mutant appear to be unique based on their properties on various gel systems.

6 Conclusion

Paramecium is viewed as a swimming neuron in this study. While there are obvious differences between the ciliated protozoan and the neurons of our body, there are also fundamental similarities. The focus of our attention, ion channels, for example, must have evolved early. The molecular functional characteristics of channels such as ion selectivity, voltage activation, and time-dependent inactivation are found in the channels of *Paramecium*, as are the channel interactions in the generation of receptor potentials and action potentials. The multidisciplinary study of the *Paramecium* excitable membrane takes advantage of many factors. The cell is large enough to tolerate voltage clamp and microinjection. It has a short doubling time which allows rapid culture. Autogamy and cytogamy make possible the expression of induced recessive mutations. Methods have also been adopted in recent years to further the field of research. For example, a culture method with airlift fermenters, yielding hundreds of grams of paramecia, has been developed and made inexpensive (Schoenfeld et al. 1986). Polyclonal and monoclonal antibodies have been generated against membrane proteins (Eisenbach et al. 1983; Ramanathan et al. 1983) and monoclonal antibodies against calmodulin (Burgess-Cassler et al. 1987). The patch-clamp method has been applied to the *Paramecium* membrane to study the activities of individual ion channels (Martinac et al. 1986). The application of the recombinant DNA technology to this field of research is in its infancy. The identification of a deviant genetic code as the problem in heterologous translation in vitro (Caron and Meyer 1985; Preer et al. 1985) and the successful transformation in *Tetrahymena* (Tondravi and Yao 1986) brighten the prospect of such an application.

The ciliate body, in general, and the cilia, in particular, carry out a very active cyclic-nucleotide metabolism and much is known of the phosphorylation machinery in the cell (e.g., Schultz and Jantzen 1980; Lewis and Nelson 1981; Klumpp and Schultz 1982; Bonini et al. 1986; Hennessey et al. 1985a; see also Schultz and Klumpp, Chap. 16, this Vol.). The connection between the Ca^{2+}-controlled events and the cyclic-nucleotide controlled events are now being sought.

The restoration factors point out a connection between the soluble cytoplasm and the membrane-bound ion channels which has been previously underappreciated. How the soluble calmodulin affects the Ca^{2+}-dependent K^+ channel is of great interest. Whether calmodulin is posttranslationally modified before it is used to control the Ca^{2+}-dependent channels is of immediate concern and the answer may provide clues to the unsuspected chemistry of this molecule. An application of the recombinant DNA technology to clone the *Paramecium* calmodulin gene, now in progress (Forte, Hinrichsen, Ramanathan, unpublished), should further our understanding of this molecule and how it figures in the regulation of ion-channel activities.

Acknowledgments. Supported by grants NIH GM 22714-12 and NIH GM 36386-01.

References

Adoutte A, Ramanathan R, Lewis RM, Dute RR, Ling K-Y, Kung C, Nelson DL (1980) Biochemical studies of the excitable membrane of *Paramecium tetraurelia.* III. Proteins of cilia and ciliary membranes. J Cell Biol 84:717–738

Adoutte A, Ling K-Y, Chang S, Huang F, Kung C (1983) Physiological and mutational protein variations in the ciliary membrane of *Paramecium.* Exp Cell Res 148:387–404

Bonini NM, Gustin MC, Nelson DL (1986) Regulation of ciliary motility by membrane potential in *Paramecium:* a role for cyclic AMP. Cell Motil Cytoskel 6:256–272

Brehm P, Eckert R (1978a) Calcium entry leads to inactivation of calcium channel in *Paramecium.* Science 202:1203–1206

Brehm P, Eckert R (1978b) An electrophysiological study of the regulation of ciliary beating frequency in *Paramecium.* J Physiol 283:557–568

Burgess-Cassler A, Hinrichsen RD, Maley ME, Kung C (1987) Biochemical characterization of a genetically altered calmodulin in *Paramecium.* Biochim Biophys Acta (in press)

Caron F, Meyer E (1985) Does *Paramecium primaurelia* use a different genetic code in its macronucleus? Nature (London) 314:185–188

Chang S-Y, Kung C (1976) Selection and analysis of a mutant *Paramecium tetraurelia* lacking behavioral response to tetraethylammonium. Genet Res 27:97–107

de Peyer JE, Deitmer JW (1980) Divalent cations as charge carriers during two functionally different membrane currents in the ciliate *Stylonychia.* J Exp Biol 88:73–89

Eckert R, Naitoh Y, Machemer H (1976) Calcium in the bioelectric and motor functions on *Paramecium.* In: Calcium in biological systems. Symp Soc Experimental biology 30. Univ Press Cambridge, pp 233–255

Eisenbach L, Ramanathan R, Nelson DL (1983) Biochemical studies of the excitable membrane of *Paramecium tetraurelia.* IX. Antibodies against ciliary membranes. J Cell Biol 97:1412–1420

Elbaum L, Wallen-Friedman M, Hennessey T, Hinrichsen R, Burgess-Cassler A, Saimi Y, Kung C (1986) The calcium-dependent potassium current is defective in several mutants of *Paramecium tetraurelia.* Neurosci 12:558 (Abstr. 154.2)

Haga N, Hiwatashi K (1982) A soluble gene product controlling membrane excitability in *Paramecium.* Cell Biol Int Rep 6:295–300

Haga N, Forte M, Saimi Y, Kung C (1982) Microinjection of cytoplasm as a test of complementation in *Paramecium.* J Cell Biol 82:559–564

Haga N, Forte M, Saimi Y, Kung C (1984a) Characterization of factors which complement Ca^{2+}-channel mutations in *Paramecium tetraurelia.* J Neurogenet 1:259–294

Haga N, Forte M, Ramanathan R, Hennessey T, Takahashi M, Kung C (1984b) Characterization and purification of a soluble protein controlling Ca-channel activity in *Paramecium.* Cell 39:71–78

Haga N, Forte M, Ramanathan R, Saimi Y, Kung C (1984c) Purification of a soluble protein controlling Ca^{2+} channel activity in *Paramecium.* Biophys J 45:130–132

Hansma HG, Kung C (1975) Studies of the cell surface of *Paramecium.* Biochem J 152:523–528

Hennessey T, Nelson DL (1979) Thermosensory behavior in *Paramecium tetraurelia:* a quantitative assay and some factors that influence thermal avoidance. J Gen Microbiol 112:337–347

Hennessey TM, Kung C (1984) An anticalmodulin drug, W-7, inhibits the voltage dependent calcium current in *Paramecium caudatum.* J Exp Biol 110:169–181

Hennessey TM, Saimi Y, Kung C (1983) A heat-induced depolarization of *Paramecium* and its relationship to thermal avoidance behavior. J Comp Physiol 153:39–46

Hennessey T, Machemer H, Nelson DL (1985a) Injected cyclic AMP increases ciliary beat frequency in conjunction with membrane hyperpolarization. Eur J Cell Biol 36:153–156

Hennessey TM, Saimi Y, Kung C (1985b) A Ca^{2+} dependent K^+ current is increased in the *Paramecium* mutant, *TeaA.* Neurosci (Abstr) 11:787

Hinrichsen RD, Saimi Y (1984) A mutation that alters properties of the calcium channel in *Paramecium tetraurelia.* J Physiol 351:397–410

Hinrichsen RD, Saimi Y, Kung C (1984) Mutants with altered Ca^{2+}-channel properties in *Paramecium tetraurelia:* isolation, characterisation and genetic analysis. Genetics 108:545–558

Hinrichsen RD, Amberger E, Saimi Y, Burgess-Cassler A, Kung C (1985a) Genetic analysis of mutants with a reduced Ca^{2+}-dependent K^+ current in *Paramecium tetraurelia.* Genetics 111:433–445

Hinrichsen RD, Saimi Y, Ramanathan R, Burgess-Cassler A, Kung C (1985b) A genetic and biochemical analysis of behavior in *Paramecium.* In: Eisenbach M, Balaban M (eds) Sensing and response in microorganisms. Elsevier, Amsterdam, pp 145–157

Hinrichsen RD, Burgess-Cassler A, Soltvedt BC, Hennessey T, Kung C (1986) Restoration by calmodulin of a Ca^{2+}-dependent K^+ current missing in a mutant of *Paramecium.* Science 232:503–506

Jennings HS (1906) Behavior of lower organisms. Indiana Univ Press, Bloomington, Ind

Klumpp S, Schultz JE (1982) Characterization of a Ca^{2+}-dependent guanylate cyclase in the excitable ciliary membrane from *Paramecium.* Eur J Biochem 124:317–324

Kung C (1971a) Genic mutations with altered system of excitation in *Paramecium aurelia.* II. Mutagenesis, screening and genetic analysis of the mutants. Genetics 69:29–45

Kung C (1971b) Genic mutations with altered system of excitation in *Paramecium aurelia.* I. Phenotypes of the behavioural mutants. Z Vergl Physiol 71:142–164

Kung C (1979) The biology and genetics of *Paramecium* behavior. In: Breakefield XO (ed) Topics in neurogenetics. Elsevier, North-Holland. Biomedical Press, New York Amsterdam, pp 1–26

Lewis RM, Nelson DL (1980) Biochemical studies of the excitable membrane of *Paramecium tetraurelia.* IV. Protein kinase activities of cilia and ciliary membrane. Biochim Biophys Acta 615:341–345

Lewis RM, Nelson DL (1981) Biochemical studies of the excitable membrane of *Paramecium tetraurelia.* V. Endogenous protein substrates for in vitro and in vivo phosphorylation in cilia and ciliary membranes. J Cell Biol 91:167–170

Martinac B, Saimi Y, Gustin M, Kung C (1986) Single channel recording in *Paramecium.* Biophys J 49:167 a (Abstr)
Naitoh Y, Eckert R (1969) Ionic mechanisms controlling behavioral responses of *Paramecium* to mechanical stimulation. Science 164:963–965
Naitoh Y, Eckert R (1974) The control of ciliary activity in Protozoa. In: Sleigh MA (ed) Cilia and flagella. Academic Press, London New York, pp 305–352
Naitoh Y, Eckert R, Friedman K (1972) A regenerative calcium response in *Paramecium.* J Exp Biol 56:669–681
Oertel D, Schein SJ, Kung C (1977) Separation of membrane currents using a *Paramecium* mutant. Nature (London) 268:120–124
Preer JR, Preer LB, Rudman BM, Barnett AJ (1985) Deviation from the universal code shown by the gene for surface protein 51 A in *Paramecium.* Nature (London) 314:188–190
Ramanathan R, Saimi Y, Peterson JB, Nelson DL, Kung C (1983) Antibodies to the ciliary membrane of *Paramecium tetraurelia* alter membrane excitability. J Cell Biol 97:1421–1428
Rauh JJ, Nelson DL (1981) Calmodulin is a component of extruded trichocysts from *Paramecium tetraurelia.* J Cell Biol 91:860–865
Saimi Y (1986) Calcium-dependent sodium currents in *Paramecium:* mutational manipulations and effects of hyper- and depolarization. J Membr Biol 92:227–236
Saimi Y, Kung C (1980) A Ca-induced Na^+ current in *Paramecium.* J Exp Biol 88:305–325
Saimi Y, Hinrichsen RD, Forte M, Kung C (1983) Mutant analysis shows that the Ca^{2+}-induced K^+ current shuts off one type of excitation in *Paramecium.* Proc Natl Acad Sci USA 80:5112–5116
Satow Y, Kung C (1974) Genetic dissection of active electrogenesis in *Paramecium aurelia.* Nature (London) 247:69–71
Satow Y, Kung C (1980a) Ca-induced K^+-outward current in *Paramecium tetraurelia.* J Exp Biol 88:293–303
Satow Y, Kung C (1980b) Membrane currents of pawn mutants of the *pwA* group in *Paramecium tetraurelia.* J Exp Biol 84:57–71
Satow Y, Kung C (1981) Possible reduction of surface charge by a mutation. J Mem Biol 59:179–190
Satow Y, Murphy AD, Kung C (1983) The ionic basis of the depolarizing mechanoreceptor potential of *Paramecium tetraurelia.* J Exp Biol 103:253–264
Schoenfeld U, Alfemann AF, Schultz JE (1986) J Protozool 33:222–226
Schultz JE, Jantzen H-M (1980) Cyclic nucleotide-dependent protein kinases from cilia of *Paramecium tetraurelia.* Partial purification and characterization. FEBS Lett 116:75
Schultz JE, Pohl T, Klumpp S (1986) Voltage-gated Ca^{2+} entry into *Paramecium* linked to intraciliary increase in cyclic GMP. Nature (London) 322:271–273
Takahashi M, Naitoh Y (1978) Behavioral mutants of *Paramecium caudatum* with defective membrane electrogenesis. Nature (London) 271:656–658
Takahashi M, Haga N, Hennessey T, Hinrichsen RD, Hara R (1985) A gamma ray-induced non-excitable membrane mutant of *Paramecium caudatum:* A behavioral and genetic analysis. Genet Res 46:1–10
Tondravi MM, Yao M-C (1986) Transformation of *Tetrahymena thermophila* by microinjection of ribosomal RNA genes. Proc Natl Acad Sci USA 83:4369–4373
Van Houten J (1979) Membrane potential changes during chemokinesis in *Paramecium.* Science 204:1100–1103
Van Houten J, Hansma HG, Kung C (1975) Two quantitative assays for chemotaxis in *Paramecium.* J Comp Physiol 104:211–223
Wallen-Friedman M, Elbaum L, Saimi Y, Hennessey T, Hinrichsen R, Burgess-Cássler T, Kung C (1986) Calmodulin defects are correlated with defects in the calcium-dependent potassium current in several mutants of *Paramecium tetraurelia.* Neursci 12:558 (Abstr. 154.1)

Chapter 16 Biochemistry of Cilia

JOACHIM E. SCHULTZ and SUSANNE KLUMPP[1]

CONTENTS

1 Introduction

In the 1960s and 1970s, biochemical investigations of *Paramecium* concentrated mainly on isozyme patterns of esterases, phosphatases, succinate-, isocitrate-, glutamate-, and β-hydroxybutyrate dehydrogenases, and fumarase (Tait 1970a, b; Esteve 1970; Allen and Gibson 1971; Allen and Nerad 1978a; Khadem and Gibson 1985). The isozyme patterns were preferably used as taxonomic tools. Earlier, the surface antigens of *Paramecium* were also a major area of biochemical research efforts (Beale and Kacser 1957; Hansma and Kung 1975; Steers and Davis 1977). These large immobilization proteins (i-antigens) are dealt with by Schmidt in Chapter 11 (this Vol.). Starting around 1970, *Paramecium* was recognized as a possible model cell for a genetic dissection of behavior (Kung 1971a, b; Kung and Eckert 1972). The advantage of using *Paramecium* was that both the genetics and electrophysiology of this genus were reasonably well understood at that time (Kung et al. 1975). Biochemistry was added to this

1 Pharmazeutisches Institut der Universität Tübingen, Morgenstelle 8, D-7400 Tübingen, FRG

Paramecium. Ed. by H.-D. Görtz

model system as a new dimension in 1976 (Browning and Nelson 1976). In 1977, it became clear that the membrane covering the motile machinery of *Paramecium*, about 5000 cilia, carries the voltage-dependent Ca channels which are responsible for the depolarizing Ca-inward current, and possibly for other ion channels as well (Ogura and Takahashi 1976; Dunlap 1977). Since then, the cilia and in particular their excitable membrane, which comprises about two-thirds of the total surface membrane of *Paramecium* (Andrews and Nelson 1979), have been the topic of numerous biochemical studies aimed at forming an idea of how ion channel regulation and translation of an ionic signal into a mechanical response, i.e., ciliary reversal, and how desensitization and adaptation are accomplished in molecular terms.

2 Axenic Mass Culture of *Paramecium*

"Biochemistry starts at the slaughterhouse": this old proverb surely has lost its literal sense, especially concerning *Paramecium*, yet, a reasonable supply of biological starting material is indispensable for a biochemical approach. Bacterized cultures of *Paramecium*, although often unavoidable, always suffer from poor yields. Due to the work of van Wagtendonk and Soldo (1970) on nutritional requirements, axenic media were developed for several strains of *Paramecium* which promised better yields. Various procedures have been proposed to transfer *Paramecium* from bacterized hay infusion to sterile cultures (van Wagtendonk and Soldo 1970; Allen and Nerad 1978b; Fok and Allen 1979). These methods are not equally successful for all strains. Currently, *Paramecium tetraurelia* and several mutant strains from this genus and *Paramecium caudatum* are firmly established axenically. The medium mostly used has originally been devised by Soldo et al. (1966). It is now used in modified versions (Table 1).

Cultures in 0.1 to 1 liter Erlenmeyer flasks (30 to 300 ml operating volume) grow fastest (doubling time around 14 h) and to a high density (up to 50000 cells ml^{-1}) with gentle shaking (Thiele et al. 1980). A further scale up was possible by using airlift bioreactors (Fig. 1; Thiele et al. 1980; Schönefeld et al. 1986). Using skim milk powder as the major nutritional protein source instead of expensive proteose peptone and tryptic agar base, economical production of *P. tetraurelia* and a few mutants up to a 170-l scale is possible with yields of 750 g wet wt. *Paramecium* or about 10 g dry wt. purified cilia within 8 – 10 days (Schönefeld et al. 1986). Interestingly, *Paramecium* apparently secretes proteases into the medium hydrolyzing proteinaceous nutrients since skim milk powder-based medium clears during cell growth. Carbohydrates are not utilized significantly as an energy source (Schönefeld et al. 1986).

The advances in axenic mass cultivation of *Paramecium* constitute a solid basis for detailed biochemical studies of the cilia as an excitable organelle with a built-in motor function. We will now survey the biochemistry of the individual components of which the cilia are constructed as far as they seem related to ciliary functions on an item by item basis going from the outside, the surface antigens, across the membrane to the axonemal apparatus inside.

Table 1. Components of the culture medium for axenic growth of *Paramecium tetraurelia* (all quantities in mg l^{-1}). (After Soldo et al 1966; Soldo and van Wagtendonk 1969; van Wagtendonk 1974; Thiele et al. 1980). Proteose peptone and tryptic agar base can be replaced by 8.5 g skim milk powder (Schönefeld et al. 1986)

Proteose peptone	10000
Tryptic agar base	5000
Ribonucleic acid	1000
$MgSO_4 \times 7\ H_2O$	500
Phosphatidylcholine	250
Stigmasterol	5
Calcium pantothenate	5
Nicotinamide	5
Riboflavin	5
Pyridoxamin HCl	5.8
Folic acid	2.5
Thiamine HCl	15
Biotin	0.00625
Lipoic acid	0.05
Palmitic acid	7.5
Stearic acid	5
Oleic acid	1
Linoleic acid	0.5
Linolenic acid	0.15

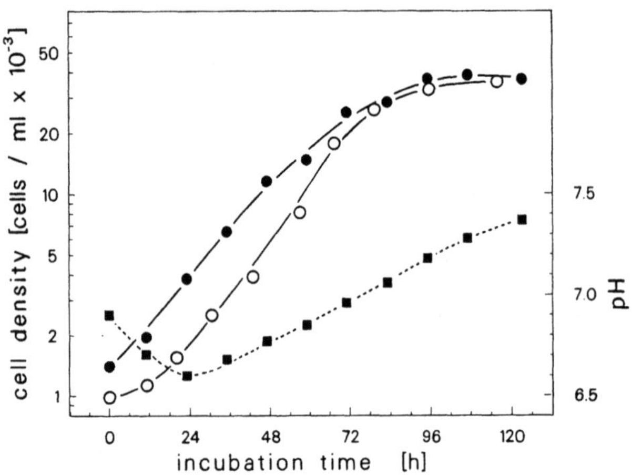

Fig. 1. Growth curves of *Paramecium tetraurelia* stock 51s in a 20-liter airlift bioreactor using medium with proteose peptone and tryptic agar base (Table 1; ○—○) or skim milk powder (8.5 g l^{-1}; ●—●) as major nutrient source. The pH during the culture is indicated by the *dotted line* (■). (After Thiele et al. 1980; Schönefeld et al. 1986)

3 Ciliary Surface Proteins

Paramecium is completely covered by a protein coat which constitutes a 20–30-nm-thick fuzzy layer. Beale and Kacser (1957) have shown that tagged antibodies against these large (250 kD) proteins react specifically with the entire surface of *Paramecium* including the ciliary membrane. These polyclonal antibodies also immobilize the cells, hence the name i-antigen. The biological function of this protein family is not understood, e.g., i-antigens can be removed by pronase digestion without distinct differences in the swimming behavior of the treated cells (Ramanathan et al. 1981).

Surface antibody binding reduces the current through the ciliary Ca^{2+} channels as monitored by a voltage clamp (Ramanathan et al. 1983). This reduction in excitability is unrelated to simultaneous immobilization since monovalent F_{ab} fragments also diminish the current, but do not immobilize the cells (Eisenbach et al. 1983; Ramanathan et al. 1983). Further, loss of the inward Ca current is apparent using antibodies against another abundant family of surface proteins of 42 to 45 kD (Ramanathan et al. 1983). Other electrical membrane properties are not altered by the antibody treatment (Ramanathan et al. 1983). It does not appear that the antibodies directly bind to a part of the Ca channel itself. The possibility exists that in the low-ionic environment of a ditch, the major surface proteins of the ciliary membrane somehow define and eventually buffer the external charges in the neighborhood of the Ca channel which operates the inwardly directed current.

4 Ciliary Lipid Composition

An essential factor which affects membrane properties, in particular membrane fluidity, is the lipid composition. Changes in growth temperature are known to alter the membrane lipid composition of cells from very different origins. Therefore, quite a few studies have been concerned with the nature of lipids from *Paramecium* membranes, the positional distribution of fatty acids in these lipids, and their biosynthesis (Andrews and Nelson 1979; Kaneshiro et al. 1979, 1984; Rhoads and Kaneshiro 1979a, 1984; Kaneshiro 1980; Hennessey and Nelson 1983; Henessey et al. 1983; Matesic and Kaneshiro 1984). The lipid composition of the ciliary membrane was of special interest in view of the excitable properties associated with this organelle (Dunlap 1977). In Table 2A, the major classes of lipids identified in the ciliary membrane and their approximate share of the total are compiled from the references cited above. The fraction of a particular lipid can vary considerably depending on the age of the culture and the medium used (bacterized or axenic). Cilia do not contain triglycerides.

Paramecium requires sterols for growth (Hennessey et al. 1983). Usually, in monoxenic medium the sterol requirement is satisfied by the stigmasterol or sitosterol originating from the plant powder used for medium preparation. Additional sterol supplementation enhances growth. In axenic medium a considerable amount of cholesterol is introduced as an impurity of lecithin (Hennessey et al. 1983). This cholesterol is readily taken up and incorporated as such or as dehydrocholesterol into cell and cilia membranes. However, cholesterol as sole sterol will not sustain growth. This strongly indicates that stigmasterol very specifically satisfies a sort of vitamin requirement of *Paramecium*. Possibly, it is involved as cofactor in a vital biochemical reaction in addition to its role as membrane constituent. The total sterol content depends largely on the supply by the medium (Hennessey et al. 1983). In general, phospholipid to sterol ratios are 5:1, i.e., up to 20% (by mass) of the ciliary lipids are unesterified sterols. A further remarkable feature of the lipid composition is the high proportion of

Table 2. (A) Phospholipid composition of *Paramecium* cilia (% of total) [a]

Glyceryl-aminoethylphosphonate	30 (24–42)
Phosphatidyl ethanolamine	37 (9–37)
1-Alkyl-phosphatidylcholine	11 (8–18)
Ceramide aminoethylphosphonate	15 (11–20)
Phosphatidylinositol	2 (1–5)
Lysophosphatidylethanolamine	2 (2–5)
Minor lipids (mainly sphingolipids)	2 (2–6)

(B) Fatty acid composition of ciliary phospholipids (% of total)

Palmitic acid	9
Stearic acid	1
Oleic acid	4.5
Linoleic acid	8
Linolenic acid	1.5
Arachidonic acid	30
Eicosapentaenoic acid	6

[a] The percentage ranges found under different culture conditions are indicated in parentheses.

phosphonolipids (up to 40%) (Andrews and Nelson 1979; Rhoads and Kaneshiro 1979a).

The nature of the principal fatty acids (listed in Table 2B), which make up the lipids, has been studied and a possible role in thermosensitivity was examined (Kaneshiro 1980; Hennessey and Nelson 1983; Kaneshiro et al. 1984). Again, culture age and conditions, such as temperature and medium, have a profound effect. In general, amide-linked fatty acids as in sphingolipids and ether-linked fatty alcohols are preferentially saturated chains of 16 and 18 carbons (Hennessey and Nelson 1983). Since *Paramecium* lacks the capacity to synthesize fatty acids from scratch, it requires oleic acid for growth (Rhoads and Kaneshiro 1984), and thus directed alterations of membrane lipid composition may become possible in the future.

Inositol phosphatides have been related to Ca^{2+} metabolism and Ca-mediated regulation of vital cellular functions (Hawthorne 1983). Phosphoinositides were consistently found as constituents of the ciliary membrane (Table 2a). In an abstract (Rhoads and Kaneshiro 1979b) a possible role of phosphatidylinositol in cilia from *Paramecium* was suggested. Certainly, a more thorough study is needed to firmly establish an eventual role for phosphoinositides in ciliary operation.

Earlier, several behavioral mutants with defects in normal electrogenesis were investigated with respect to their lipid composition (Andrews and Nelson 1979). Although some minor differences were apparent, for most mutants this was not investigated vigorously enough to exclude a link between the mutational defect and an altered lipid pattern. Later, Forte et al. (1981) reported a marked alteration of the sphingolipid and phosphonolipid composition of the ciliary membrane in a family of Ba^{2+} shy mutants. This defect was present in independently isolated alleles and segregated in crosses with the behavioral phenotype.

Electrophysiologically, the mutation reduced several ion conductances. Interestingly, sterol supplementation of the medium restored wild-type lipid patterns, behavior, and electrophysiology (Forte et al. 1981). This is further evidence for a cofactorlike activity of stigmasterol in *Paramecium*. The Ba^{2+} shy mutant also provides strong evidence that membrane lipids have a profound effect on functions of membranous proteins.

5 Ciliary Membrane Proteins

The protein patterns of cilia, its membranes, and axonemes were investigated by one- and two-dimensional electrophoresis by several groups (Adoutte et al. 1980, 1983; Brugerolle et al. 1980; Merkel et al. 1981; Thiele et al. 1982). This survey indicated the presence of about 175 proteins in the ciliary membrane, ranging in molecular weight from the huge dyneins and i-antigens (around 300 kD) to very small peptides. The hope to find distinct and reproducible differences between membranes from wild-type and mutant cells remained unfulfilled. Either the mutational alterations are very minor as far as the primary structures are concerned or the sheer amount of the mutated gene products is too small to be detected by these methods. To date, only a few bands on SDS gels have been tentatively assigned to particular enzymatic functions (Eistetter et al. 1983). Current efforts are concentrated, therefore, on the identification, isolation, and characterization of such ciliary proteins, which are thought of by analogy with other systems, to be involved in the regulation of vital functions in the cilia, such as excitability of the membrane or mechanical properties.

5.1 Guanylate Cyclase

A guanylate cyclase is localized primarily in the ciliary membrane (Schultz and Klumpp 1980). The exceptionally high activity measured in the presence of physiological Mg^{2+} concentrations is comparable to guanylate cyclase activities reported for sea urchin sperm and rod outer segments (Gray and Drummond 1976; Fleischman and Denisevich 1979). Incidently, the latter systems are phylogenetically rudimentary cilia. The *Paramecium* guanylate cyclase is unaffected by redox mechanisms or atrial natriuretic peptide. Instead, it is regulated most specifically by Ca ions (Klumpp and Schultz 1982). Its activity is nil at 100 nM Ca^{2+}, the usual intraciliary Ca^{2+} concentration. At 10 μM Ca^{2+} the guanylate cyclase is half-maximally activated. Note that intraciliary Ca^{2+} can rise to 40 μM during a depolarizing Ca^{2+} inward current (Kung and Saimi 1985). Ca^{2+} ions increase V_{max}, while K_m is unaltered (Klumpp and Schultz 1982). Ba^{2+} ions are completely inactive and do not affect the action of Ca^{2+} (Schultz et al. 1986; Fig. 2). Antibodies against calmodulin reduce membrane-bound guanylate cyclase activity to a large extent (Schultz and Klumpp 1984). Therefore, it seems that similarly to the guanylate cyclase from *Tetrahymena* the Ca^{2+} sensitivity of guanylate cyclase from *Paramecium* is mediated by calmodulin itself or a related Ca-binding protein (Schultz and Klumpp 1982, 1984;

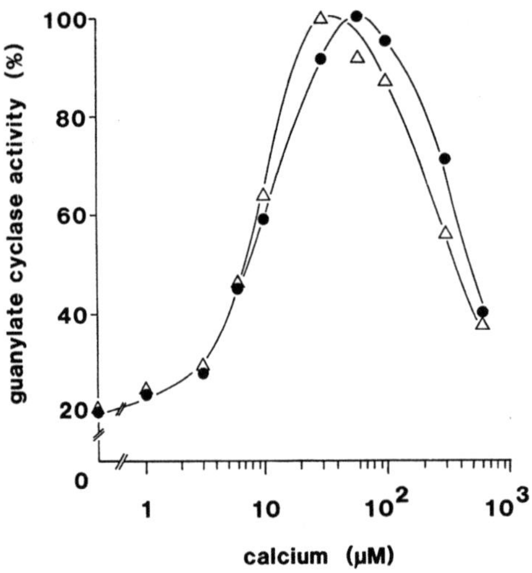

Fig. 2. Dose-response curve for Ca^{2+}-stimulated guanylate cyclase in ciliary membranes from *Paramecium*. Membranes were washed with 60 μM EGTA followed by two washes with buffer to remove the chelator. All reagents were purified by treatment with Chelex. The remaining Ca^{2+} concentration was <1 μM. Activity was determined for various concentrations of Ca^{2+} in the absence (●—●) or presence △—△) of 1 mM Ba^{2+} (Schultz et al. 1986)

Klumpp et al. 1983a). The presence of calmodulin in the cilia has been proven by biochemical isolation and by immunohistochemical techniques (Walter and Schultz 1981; Klumpp et al. 1983b). Recently, the amino acid sequence of *Paramecium* calmodulin was established (Schaefer et al. 1987). In addition to trimethyllysine at position 115, which is present in most calmodulins, it contains a dimethyllysine at position 13. Furthermore, amino acids 79 and 146 are replaced by glutamine and valine, respectively. This is unusual with respect to calmodulin sequences reported from other organisms. A functional significance of these unique substitutions is suggested by the peculiar specificity of guanylate cyclase for activation by calmodulins from different sources (Schultz and Klumpp 1984) and the reported curing capacity of an ion channel mutant of *Paramecium* by *Paramecium* wild-type calmodulin (Hinrichsen et al. 1986). Guanylate cyclase can be solubilized by detergents and may in the future be accessible for a more detailed investigation of the mechanism of its regulation.

5.1.1 Regulation of cGMP Biosynthesis in Vivo

Due to its activation by Ca^{2+} the guanylate cyclase is a likely candidate as an intraciliary Ca sensor which can translate the incoming Ca signal during an action potential into a biochemical reaction sequence via cGMP-stimulated protein phosphorylation. Indeed, it was possible to demonstrate a direct link between the Ca^{2+} influx through voltage-gated Ca channels and intraciliary cGMP production measured by radioimmunoassay (Schultz et al. 1986). A sudden depolarizing increase of K^+ from 5 to 25 mM in the incubation resulted in only a slight, yet reproducible increase in cGMP levels, probably because the influx of Ca^{2+} leads to a rapid inactivation of the Ca^{2+} channels (Brehm and Eckert 1978; Eckert and Brehm 1979). Therefore, Ba^{2+} was used to increase the open time of the Ca^{2+} channels by decreasing the rate of inactivation, thus pro-

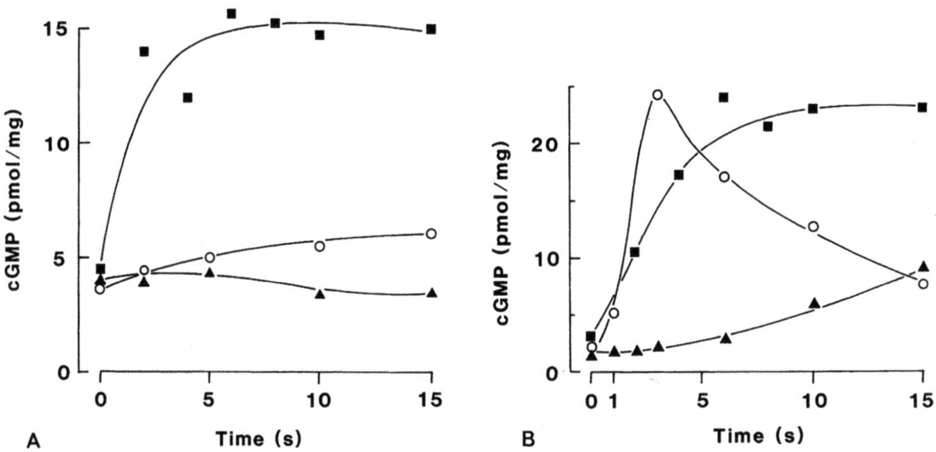

Fig. 3A, B. Time-dependent increase in cGMP concentrations in *Paramecium* wild type 51s (○–○), double-mutant pawn AB (▲–▲), and mutant dancer (■–■) after a sudden increase in the concentration of K^+ from 5 to 25 mM (**A**) or after addition of 3 mM Ba^{2+} (**B**) (Schultz et al. 1986)

ducing a more sustained (if somewhat slower) Ca^{2+} entry. Stimulation of *Paramecium* with Ba^{2+} also provokes a more violent and prolonged behavioral response than K^+. Addition of 3 mM Ba leads to a rapid, large, and transient increase in cGMP, culminating after 5–10 s. cGMP levels thus increase from about 5 to 40 pmol cGMP mg^{-1} protein. The duration of backward swimming outlasts the cGMP response indicating that the level of cGMP does not control swimming direction. The effect of Ba^{2+} is specific, as 3 mM Mg^{2+}, Ca^{2+}, or Sr^{2+} does not concomitantly change cGMP levels and behavior.

Further evidence for a direct link between the Ca^{2+} inward current and intraciliary cGMP biosynthesis comes from studies of a pair of mutants with opposite defects. First, pawns lack voltage-gated Ca^{2+} channels and cannot swim backward (Kung and Saimi 1985). Addition of 3 mM Ba^{2+} to two cell lines of the pawn family does not stimulate cGMP formation as in wild-type cells (Fig. 3). Second, dancer cells possess Ca channels which inactivate more slowly and incompletely resulting in enhanced Ca^{2+} inward currents with respective behavioral consequences (Hinrichsen and Saimi 1984). In this mutant, addition of a depolarizing concentration of K^+ evokes a fourfold increase in cGMP levels (Fig. 3). The increase in cGMP can also be visualized immunocytochemically to occur mainly in the cilia which are wrapped by the Ca-channel carrying membrane (Klumpp et al. 1983b).

This is the first example in *Paramecium* in which a proven relationship between an ion current and a biochemical event was established by a combination of genetic, behavioral, electrophysiological, biochemical, and immunocytochemical methods. The behavioral mutants generated by Kung and associates (Kung et al. 1975; see also Ramanathan et al., Chap. 15, this Vol.) which carry defects in electrogenesis or mechanics of the ciliary beat regulation, will certainly enable further similarly designed experimental approaches.

5.2 Adenylate Cyclase

The ciliary membrane from *Paramecium* contains a considerable fraction of the total cellular adenylate cyclase activity (Schultz and Klumpp 1983). The adenylate cyclase is a separate enzyme entity, not just a guanylate cyclase which happens to use ATP as substrate. Adenylate cyclase and guanylate cyclase are differentially localized in the ciliary membrane (Schultz and Klumpp 1983).

Fractionation of ciliary homogenates by an isoosmotic Percoll gradient yields about 50% of ciliary adenylate cyclase in membrane vesicles of low buoyant density (Fig. 4). The other half of adenylate cyclase is in membrane fragments of higher density (Schultz and Klumpp 1983). This second peak reproducibly precedes the main peak of guanylate cyclase which is found mainly in the heavier fraction (Schultz and Klumpp 1983). The functional significance of this differential localization of the cyclases in the ciliary membrane is unclear at present. Guanylate cyclase and adenylate cyclase can also be physically separated after solubilization by molecular sieve chromatography (Klumpp et al. 1984). Temperature dependence, stability, pH optima, and divalent ion requirements differ between the cyclases (Klumpp et al. 1984). *Paramecium* adenylate cyclase does not resemble that found in most higher organisms, i.e., its activity is unaffected by GTP, forskolin, cholera toxin, and pertussis toxin. The protozoan adenylate cyclase is allosterically activated by Mg^{2+} via a single metal binding site separate from the substrate site with a Ks of 0.27 mM (Schultz et al. 1987). Ca^{2+} inhibits in an uncompetitive manner, the extent of inhibition is dependent on the free Mg^{2+} concentration. Around the physiologically relevant and likely free Mg^{2+} levels of 1 mM, half-maximal inhibition is accomplished by 10 µM Ca^{2+}, a meaningful Ca^{2+} concentration in the normal cellular setting (Schultz et al. 1987). With MgATP as substrate, K^+ very specifically activates adenylate cyclase at low millimolar concentrations (Klumpp et al. 1984). Other alkaline ions are inactive and do not interfere with the potas-

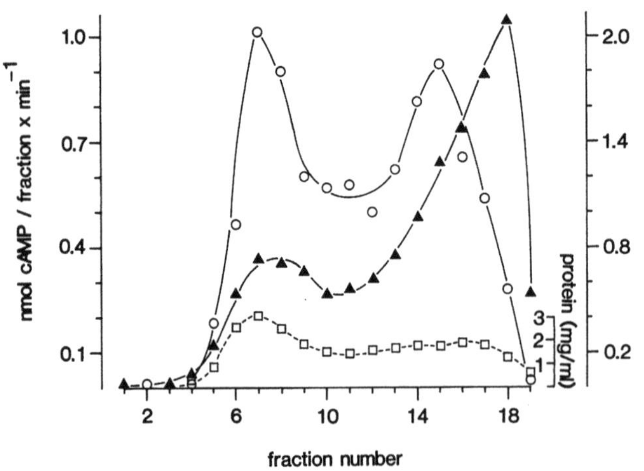

Fig. 4. Distribution of adenylate cyclase (○–○) and guanylate cyclase (▲–▲) activities in a Percoll gradient (protein □–□) (Schultz and Klumpp 1983)

sium activity. Again, the magnitude of the K^+ effect is related to the free Mg^{2+} concentration (Schultz et al. 1987). At 1 mM free Mg^{2+}, 20 mM K^+ doubles adenylate cyclase activity. The inhibitory Ca^{2+} and stimulatory K^+ inputs are independent of each other (Schultz et al. 1987). One can envisage a physiological regulation of the ciliary adenylate cyclase by dynamic changes in membrane potential established in *Paramecium* mainly by K^+ and Ca^{2+}.

5.2.1 Regulation of Cyclic AMP Biosynthesis in Vivo

Although K ions stimulate adenylate cyclase "in vitro", a sudden increase of K concentration in the medium of free swimming *Paramecium* does not elevate intracellular cAMP levels (Schultz et al. 1984). However, when Ca ions are abruptly increased from 50 μM to 2 mM, a dramatic rise in cAMP levels occurs (from 3 to 20 pmol mg^{-1} protein; Schultz et al. 1984). At 160 ms cAMP concentrations are already more than doubled, they peak after 3–10 s before declining rapidly to a new steady-state level of about 175% of the original basal value (Schultz et al. 1984). The behavioral response to Ca^{2+} is an immediate acceleration of forward swimming speed lasting for up to 20 min. Such a behavior is usually considered as indicator of hyperpolarization.

Depending on the buffering system used (e.g., TRIS or MOPS) cAMP increases by Ca^{2+} addition can also be elicited without a speeding response of the cells. Therefore, cAMP most likely does not take part in enhanced frequency of ciliary beating (Gustin et al. 1983; Hennessey et al. 1985) and a clear behavioral correlate for it has yet to be identified. In Triton X-100 extracted *Paramecium* ciliary reversal induced by Ca^{2+} seems to be affected by cAMP (Nakaoka and Ooi 1985).

Currently, there is no sufficient experimental data to prove that the stimulatory input of K^+ and the inhibitory input of Ca^{2+} upon adenylate cyclase occur "in vivo". Since the intracellular K^+ concentration is estimated to be 20 to 30 mM (see Machemer, Chap. 14, Sect. 5, this Vol.), adenylate cyclase ought to be effectively shielded from it under resting conditions. Because of the peculiar localization of adenylate cyclase in the ciliary membrane this does not seem inconceivable. A hyperpolarizing event may alter the internal surface membrane charges such that K^+ can then activate adenylate cyclase.

5.3 Cyclic Nucleotide-Dependent Protein Kinases and Phosphorylation

What are the intraciliary functions of cyclic nucleotides? This problem seems particularly relevant for cGMP in view of a recent report of a direct action of cGMP on Ca conductances in rod outer segments not involving phosphorylation processes (Fesenko et al. 1985; Koch and Kaupp 1985). Whether cGMP in *Paramecium* cilia will have such an action cannot be answered at present. Two papers concerned with phosphorylation of ciliary proteins by endogenous protein kinases have appeared (Lewis and Nelson 1981; Eistetter et al. 1983). In the absence of cyclic nucleotides phosphorylation of endogenous proteins is

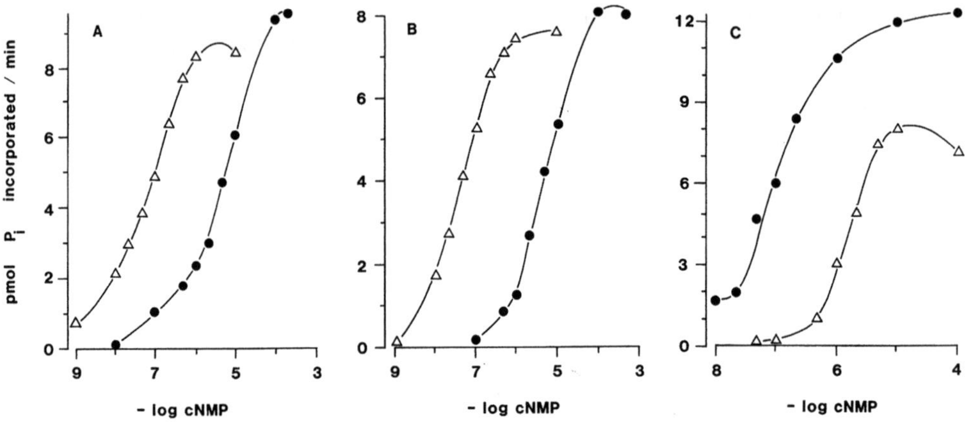

Fig. 5A–C. Effect of cyclic nucleotides on protein kinase activities. Combined fractions of DEAE chromatography peaks [cAMP-dependent protein kinase I (**A**) and II (**B**) and of the DEAE- and CM-cellulose chromatography peak (cGMP-dependent protein kinase) (**C**)] were tested with cAMP (●–●) and cGMP (△–△) using histones as substrate. Basal activity was subtracted

rather low, while in the presence of either cAMP or cGMP many proteins are phosphorylated. Phosphorylation is rapid (within 10 s) and only slowly reversible (within minutes) (Eistetter et al. 1983).

The cilia of *Paramecium* contain at least three different cyclic nucleotide-dependent protein kinases, two are stimulated by cAMP and correlate somewhat to cAMP protein kinases type I and II isolated from many organisms (Lewis and Nelson 1980; Schultz and Jantzen 1980). In addition, there is one protein kinase activated with a 3:1 preference by cGMP over cAMP (Fig. 5). Affinity labeling with 8-azido-[^{32}P]-cAMP indicates a band of 85 kD to represent the cGMP-stimulated protein kinase since labeling was specifically inhibited by concomitant incubation with unlabeled cGMP, while proteins with apparent MW of 43, 39, 37, 31, and 30 kD are probably related to the regulatory subunits of cAMP-dependent protein kinases as evidenced by inhibition of affinity labeling by low concentrations of cAMP (Eistetter et al. 1983). Further, a Ca^{2+}-stimulated protein kinase unrelated to protein kinase C has been identified and partially purified in *Paramecium* (Gundersen and Nelson 1987).

Currently, it is difficult to assign with any degree of certainty particular substrate proteins of the ciliary membrane or axoneme to a specific cAMP- or cGMP-stimulated phosphorylation since there is a broad cross-reactivity of the cyclic nucleotide-dependent protein kinases. A great effort was made to identify those proteins which are phosphorylated "in vivo" during stimulatory episodes induced by various ionic stimuli. For that purpose 3×10^5 cells in 0.5 mL 5 mM HEPES, 1.5 mM K^+, pH 7, were labeled with 100 µCi $H_3{}^{32}PO_4$. Phosphate uptake is increased by the presence of 5–10 mM mono- or divalent cations. It is a time-dependent process: labeled bands appear in the order of 38/65/14/28/41–52/22 kD. The phosphorylation intensity of most of these proteins reaches a maximum after 6 h and remains constant for at least another 6 h. Only the activity of the 22-kD protein continues to increase in labeling beyond that period. It finally constitutes the major labeled moiety. This protein is part of the cell body, and is not present in the cilia.

In vivo stimulations were carried out under conditions where known changes in cyclic nucleotide levels occur [additions of Ca^{2+} (cAMP increase) and Ba^{2+} (cGMP increase)]. After 10 s to 10 min incubations were stopped by heating in a gel sample buffer. The ion concentrations of unstimulated controls were adjusted to those of the stimulated ones prior to SDS-PAGE. No change in patterns of phosphorylated proteins is seen compared to control cells. Similarly discouraging results were obtained independently by ionic stimulation of Triton X-100 permeabilized cells (Bonini and Nelson, pers. commun.). Furthermore, labeling was about the same comparing the mutant pawns and restless to wild-type cells.

Phosphorylation/dephosphorylation of a 65-kD protein seems to be involved in trichocyst discharge (Gilligan and Satir 1982, 1983; Zieseniss and Plattner 1985; see Adoutte, Chap. 20, this Vol.).

Another possibility to identify a phosphorylation/dephosphorylation cycle possibly linked to excitation is the study of phosphoprotein phosphatases and their substrates.

5.4 Phosphatases

The Ca/calmodulin-dependent protein phosphatase 2B (also termed calcineurin) was tentatively identified immunocytochemically in the cilia of *Paramecium* (Klumpp et al. 1983b). However, so far it was impossible to detect such an enzyme activity in cilia. The protein subunits of 61 and 15 kD usually representing phosphoprotein phosphatase 2B (Klee et al. 1979) were also not found on SDS gels. Instead, several other calmodulin-binding proteins were isolated that await further characterization with respect to their possible physiological function.

Starting with homogenates from either cell bodies or purified cilia, a new type of phosphatase activity was discovered in the breakthrough fractions of a calmodulin affinity chromatography column. Using purified cilia from 340 liters *Paramecium* culture a 6000-fold purification was achieved by using subsequently ammonium sulfate precipitation, DEAE ion exchange-, blue sepharose affinity-, and TSK G-2000 SW gel filtration chromatography. Final specific activity is 60 μmol min^{-1} mg^{-1} using p-nitrophenylphosphate as substrate. The apparent MW is estimated by SDS-PAGE to be 35 kD. Enzyme activity is inhibited by Ca^{2+}. The purified enzyme does not use phosphorylated forms of histone, casein, and phosphorylase a as substrates. Furthermore, addition of phosphatase to (1) ciliary proteins phosphorylated by cyclic nucleotide-stimulated protein kinases, (2) to autophosphorylated ciliary protein kinases, or (3) to in vivo labeled ciliary proteins from *Paramecium* failed to reveal any protozoan protein substrate. Therefore, a possible involvement in lipid metabolism may be considered for this enzyme.

In addition to the soluble phosphatase described above, another type of phosphatase has been identified which is associated exclusively with the ciliary membrane fraction. Enzyme activity is maximal at pH 5–6 and is unaffected by earth alkaline ions. This phosphatase can be solubilized with detergents. It is an acid phosphoprotein phosphatase which dephosphorylates ^{32}P-histone at a rate of 3 nmol min^{-1} mg^{-1}.

5.5 Phosphodiesterases

Two cyclic nucleotide phosphodiesterases have been partially purified from *Paramecium*, one for cAMP and another for cGMP degradation (Manneck 1985). The specificity of these enzymes for their respective cyclic nucleotide substrates is almost exclusive. So far, we failed to detect any particular regulation of the phosphodiesterases. Enzyme purification from cilia has yet to be accomplished together with more detailed kinetic analysis.

5.6 ATPases

During normal electrical activity in *Paramecium*, Ca^{2+} enters the cell as a depolarizing inward current through the ciliary membrane. It is responsible for the momentary ciliary reversal. Ca^{2+} must then be sequestered and finally transported out of the cell. *Paramecium* clearly has a system for extruding Ca^{2+} into the medium, since such a system has been shown to be inhibited at low temperature (Browning and Nelson 1976). On deciliation of *Paramecium* by a Ca^{2+} shock, about 1% of the total cellular protein is released. Among the set of proteins released is a high Ca-dependent ATPase activity (Riddle et al. 1982). This release is rather specific, since marker enzymes for cytoplasm, endoplasmic reticulum, peroxisomes, and lysosomes are not significantly lost during deciliation, i.e., unspecific protein release due to partial cell damage and lysis is highly unlikely (Riddle et al. 1982). It is speculated that this ATPase originates from the ciliary lumen (Riddle et al. 1982). This has been made even more likely by the work from Andrivon et al. (1983). They reported that the amount of a Ca-dependent ATPase released depends on the deciliation method. In these studies most of the Ca-ATPase was loosely associated with the ciliary membrane. By a single cycle of freeze-thawing more than 50% are solubilized (Andrivon et al. 1983). Thus, this ATPase is not an integral membrane protein. Further, Doughty (1978) partially purified a Ca-dependent ATPase by affinity chromatography from Triton extracts of whole cilia. The molecular weights of these Ca-dependent ATPases from different groups were always around 68 kD. It is likely that they are related or identical. They are clearly different from the axonemal dynein ATPases (see below). The ciliary ATPases characterized by Andrivon et al. (1983), Riddle et al. (1982), and Doughty and Kaneshiro (1983, 1985) are specifically activated by Ca^{2+} and to a lesser extent by Sr^{2+} and Ba^{2+}, Mg^{2+} is not needed. The ATPases are not at all specific with regard to the nucleotide triphosphate required: ATP, GTP, ITP, CTP, and UTP are used. For optimal activity Ca^{2+} is required at millimolar concentrations. Therefore, it is rather unlikely that this type of ATPase is involved in processes which are directly linked to a lowering of the intraciliary Ca^{2+} level to 100 nM after an excitation. Such an ATPase, if it resides in the cilia, has as yet to be detected. Obviously, the possibility exists that removal of Ca^{2+} from the cilia can involve sequestering mechanisms directed in the first instance toward the cell body at the ciliary base.

6 Axonemal Dynein ATPases

The ciliary motility is generated by sliding between doublet microtubules (A and B tubules). The arms projecting from the A tubules contain an ATPase activity, dynein, which couples in an ATP-dependent fashion the A tubules to the adjacent B neighbors to produce mechanical force. These ATPases have been isolated and studied quite extensively from several cell types, among them *Paramecium* (Gibbons 1965; Hayashi and Takahashi 1979). The dynein ATPases are bound to axonemal structures and can easily be detached by EDTA treatment (Gibbons 1965). In sucrose density gradients, ATPase activities comigrate with 14S and 30S fractions. On SDS-PAGE, these dynein ATPases exhibit usually large molecular masses between 280 and 388 kD. The biochemical details of how the mechanical force is generated and how the frequency and direction of the ciliary beat are modulated must await further studies.

References

Adoutte A, Ramanathan R, Lewis RM, Dute RR, Ling KY, Kung C, Nelson DL (1980) Biochemical studies of the excitable membrane of *Paramecium tetraurelia*. III. Proteins of cilia and ciliary membranes. J Cell Biol 84:717–738

Adoutte A, Ling KY, Chang S, Huang F, Kung C (1983) Physiological and mutational protein variations in the ciliary membrane of *Paramecium*. Exp Cell Res 148:387–404

Allen SL, Gibson I (1971) Intersyngenic variations in esterases of axenic stocks of *Paramecium aurelia*. Biochem Genet 5:161–181

Allen SL, Nerad TA (1978a) Effect of acetate on esterase C activity during the growth cycle of *Paramecium*. J Protozool 25:273–279

Allen SL, Nerad TA (1978b) Method for the simultaneous establishment of many axenic cultures of *Paramecium*. J Protozool 25:134–139

Andrews D, Nelson DL (1979) Biochemical studies of the excitable membrane of *Paramecium tetraurelia*. II. Phospholipids of ciliary and other membranes. Biochim Biophys Acta 550:174–187

Andrivon C, Brugerolle G, Delachambre D (1983) A specific Ca^{2+}-ATPase in the ciliary membrane of *Paramecium tetraurelia*. Biol Cell 47:351–364

Beale GH, Kacser HJ (1957) Studies on antigens in *Paramecium aurelia* with the aid of fluorescent antibodies. J Gen Microbiol 17:68–75

Brehm P, Eckert R (1978) Calcium entry leads to inactivation of calcium channel in *Paramecium*. Science 202:1203–1206

Browning JL, Nelson DL (1976) Biochemical studies of the excitable membrane of *Paramecium aurelia*. I. ^{45}Ca-fluxes across resting and excited membrane. Biochim Biophys Acta 448:338–351

Brugerolle G, Andrivon C, Bohatier J (1980) Isolation, protein pattern and enzymatic characterization of the ciliary membrane of *Paramecium tetraurelia*. Biol Cell 37:251–260

Doughty MJ (1978) Ciliary Ca^{2+}-ATPase from the excitable membrane of *Paramecium.* Some properties and purification by affinity chromatography. Comp Biochem Physiol 60B:339–345
Doughty MJ, Kaneshiro ES (1983) Divalent cation-dependent ATPase activities associated with cilia and other subcellular fractions of *Paramecium:* an electrophoretic characterization on Triton-polyacrylamide gels. J Protozool 30:565–573
Doughty MJ, Kaneshiro ES (1985) Divalent cation-dependent ATPase activities in ciliary membranes and other surface structures in *Paramecium tetraurelia:* comparative in vitro studies. Arch Biochem Biophys 238:118–128
Dunlap K (1977) Localization of Ca channels in *Paramecium caudatum.* J Physiol 271: 119–133
Eckert R, Brehm P (1979) Ionic mechanisms of excitation in *Paramecium.* Annu Rev Biophys Bioeng 8:353–383
Eisenbach L, Ramanathan R, Nelson DL (1983) Biochemical studies of the excitable membrane of *Paramecium tetraurelia.* IX. Antibodies against ciliary membrane proteins. J Cell Biol 97:1412–1420
Eistetter H, Seckler B, Bryniok D, Schultz JE (1983) Phosphorylation of endogenous proteins of cilia from *Paramecium tetraurelia* in vitro. Eur J Cell Biol 31:220–226
Esteve JC (1970) Distribution of acid phosphatase in *Paramecium caudatum:* its relation with the process of digestion. J Protozool 17:24–35
Fesenko EE, Kolesnikow SS, Lyubarsky AL (1985) Induction by cyclic GMP of cationic conductance in plasma membrane of retinal rod outer segment. Nature (London) 313:310–313
Fleischman D, Denisevich M (1979) Guanylate cyclase of isolated retinal rod axonemes. Biochemistry 18:5060–5066
Fok AK, Allen RD (1979) Axenic *Paramecium caudatum.* I. Mass culture and structure. J Protozool 26:463–470
Forte M, Satow Y, Nelson DL, Kung C (1981) Mutational alteration of membrane phospholipid composition and voltage-sensitive ion channel function in *Paramecium.* Proc Natl Acad Sci USA 78:7195–7199
Gibbons IR (1965) Chemical dissection of cilia. Arch Biol (Liège) 76:317–352
Gilligan DM, Satir BH (1982) Protein phosphorylation/dephosphorylation and stimulus-secretion coupling in wild type and mutant *Paramecium.* J Biol Chem 257:13903–13906
Gilligan DM, Satir BH (1983) Stimulation and inhibition of secretion in *Paramecium:* role of divalent cations. J Cell Biol 97:224–234
Gray JP, Drummond GI (1976) Guanylate cyclase of sea urchin sperm: subcellular localization. Arch Biochem Biophys 172:31–38
Gunderson RE, Nelson DL (1987) A novel Ca^{2+}-dependent protein kinase from *Paramecium tetraurelia.* J Biol Chem 262:4602–4609
Gustin MC, Bonini NM, Nelson DL (1983) Membrane potential regulation of cAMP: control mechanisms for swimming behavior in the ciliate *Paramecium.* Neurosci (Abstr) 9:167
Hansma HG, Kung C (1975) Studies of the cell surface of *Paramecium.* Ciliary membrane proteins and immobilization antigens. Biochem J 152:523–528
Hawthorne JN (1983) Polyphosphoinositide metabolism in excitable membranes. Biosci Rep 3:887–904
Hayashi M, Takahashi M (1979) Ciliary adenosinetriphosphatase from a slow swimming mutant of *Paramecium caudatum.* J Biol Chem 254:11561–11565
Hennessey T, Nelson DL (1983) Biochemical studies of the excitable membrane from *Paramecium tetraurelia.* VIII. Temperature-induced changes in lipid composition and in thermal avoidance behavior. Biochim Biophys Acta 728:145–158
Hennessey TM, Andrews D, Nelson DL (1983) Biochemical studies of the excitable membrane of *Paramecium tetraurelia.* VII. Sterols and other neutral lipids of cells and cilia. J Lipid Res 24:575–587
Hennessey T, Machemer H, Nelson DL (1985) Injected cyclic AMP increases ciliary beat frequency in conjunction with membrane hyperpolarization. Eur J Cell Biol 36:153–156
Hinrichsen RD, Saimi Y (1984) A mutation that alters properties of the calcium channel in *Paramecium tetraurelia.* J Physiol 351:397–410
Hinrichsen RD, Burgess-Cassler A, Soltvedt BC, Hennessey T, Kung C (1986) Restoration by calmodulin of a Ca^{2+}-dependent K^+ current missing in a mutant of *Paramecium.* Science 232:503–506

Kaneshiro ES (1980) Positional distribution of fatty acids in the major glycerophospholipids of *Paramecium tetraurelia.* J Lipid Res 21:559–570

Kaneshiro ES, Beischel LS, Merkel SJ, Rhoads DE (1979) The fatty acid composition of *Paramecium aurelia* cells and cilia: changes with culture age. J Protozool 26:147–158

Kaneshiro ES, Matesic DF, Jayasimhulu K (1984) Characterizations of six ethanolamine sphingophospholipids from *Paramecium* cells and cilia. J Lipid Res 25:369–377

Khadem N, Gibson I (1985) Enzyme variation in *Paramecium caudatum.* J Protozool 32:622–626

Klee CB, Crouch TH, Krinks MH (1979) Calcineurin: a calcium- and calmodulin-binding protein of the nervous system. Proc Natl Acad Sci USA 76:6270–6273

Klumpp S, Schultz JE (1982) Characterization of a Ca^{2+}-dependent guanylate cyclase in the excitable ciliary membrane from *Paramecium.* Eur J Biochem 124:317–324

Klumpp S, Kleefeld G, Schultz JE (1983a) Calcium/calmodulin-regulated guanylate cyclase of the excitable ciliary membrane from *Paramecium:* dissociation of calmodulin by La^{3+}: calmodulin specificity and properties of the reconstituted guanylate cyclase. J Biol Chem 258:12455–12459

Klumpp S, Steiner AL, Schultz JE (1983b) Immunocytochemical localization of cyclic GMP, cGMP-dependent protein kinase, calmodulin and calcineurin in *Paramecium tetraurelia.* Eur J Cell Biol 32:164–170

Klumpp S, Gierlich D, Schultz JE (1984) Adenylate cyclase and guanylate cyclase in the excitable ciliary membrane from *Paramecium:* separation and regulation. FEBS Lett 171:95–99

Koch KW, Kaupp UB (1985) Cyclic GMP directly regulates a cation conductance in membranes of bovine rods by a cooperative mechanism. J Biol Chem 260:6788–6800

Kung C (1971a) Genetic mutations with altered system of excitation in *Paramecium aurelia.* I. Phenotypes of the behavioral mutants. Z Vergl Physiol 71:142–164

Kung C (1971b) Genetic mutations with altered system of excitation in *Paramecium aurelia.* II. Mutagenesis, screening and genetic analysis of the mutants. Genetics 69:29–45

Kung C, Eckert R (1972) Genetic control of action potential in *Paramecium aurelia.* Proc Natl Acad Sci USA 69:93–97

Kung C, Saimi Y (1985) Ca^{2+} channels of *Paramecium:* a multidisciplinary study. Curr Top Membr Transp 23:45–66

Kung C, Chang SY, Satow Y, van Houten J, Hansma H (1975) Genetic dissection of behavior in *Paramecium.* Science 188:898–904

Lewis RM, Nelson DL (1980) Biochemical studies of the excitable membrane of *Paramecium.* IV. Protein kinase activities of cilia and ciliary membrane. Biochim Biophys Acta 615:341–353

Lewis RM, Nelson DL (1981) Biochemical studies of the excitable membrane from *Paramecium tetraurelia.* VI. Endogenous protein substrates for in vitro and in vivo phosphorylation in cilia and ciliary membranes. J Cell Biol 91:167–174

Manneck HE (1985) Anreicherung und Trennung der cAMP und cGMP Phosphodiesterasen aus *Paramecium tetraurelia.* Diplomarbeit, Univ Tübingen

Matesic DF, Kaneshiro ES (1984) Incorporation of serine into *Paramecium* ethanolamine phospholipid and phosphonolipid head groups. Biochem J 222:229–233

Merkel SJ, Kaneshiro ES, Gruenstein EI (1981) Characterization of the cilia and ciliary membrane proteins of wild-type *Paramecium tetraurelia* and a pawn mutant. J Cell Biol 89:206–215

Nakaoka Y, Ooi H (1985) Regulation of ciliary reversal in Triton-extracted *Paramecium* by calcium and cyclic adenosine monophosphate. J Cell Sci 77:185–195

Ogura A, Takahashi K (1976) Artificial deciliation causes loss of calcium-dependent responses in *Paramecium.* Nature (London) 264:170–172

Ramanathan R, Adoutte A, Dute RR (1981) Biochemical studies of the excitable membrane of *Paramecium tetraurelia.* V. Effects of proteases on the ciliary membrane. Biochim Biophys Acta 641:349–365

Ramanathan R, Saimi Y, Peterson JB, Nelson DL, Kung C (1983) Antibodies to the ciliary membrane of *Paramecium tetraurelia* alter membrane excitability. J Cell Biol 97:1421–1428

Rhoads DE, Kaneshiro ES (1979a) Characterizations of phospholipids from *Paramecium tetraurelia* cells and cilia. J Protozool 26:329–338

Rhoads DE, Kaneshiro ES (1979b) Phosphoinositide metabolism and backward swimming in *Paramecium*. J Cell Biol 83:265a

Rhoads DE, Kaneshiro ES (1984) Fatty acid metabolism in *Paramecium*. Biochim Biophys Acta 795:20–29

Riddle LM, Rauh JJ, Nelson DL (1982) A Ca^{2+}-activated ATPase specifically released by Ca^{2+} shock from *Paramecium tetraurelia*. Biochim Biophys Acta 688:525–540

Schaefer WH, Lukas TJ, Blair JA, Schultz JE, Watterson DM (1987) Amino acid sequence of a novel calmodulin from *Paramecium tetraurelia* that contains dimethyllysine in the first domain. J Biol Chem 262: 1025–1029

Schönefeld U, Alfermann AW, Schultz JE (1986) Economic mass cultivation of *Paramecium tetraurelia* on a 200 liter scale. J Protozool 33:222–225

Schultz JE, Jantzen HM (1980) Cyclic nucleotide-dependent protein kinases from cilia of *Paramecium tetraurelia:* partial purification and characterization. FEBS Lett 116:75–78

Schultz JE, Klumpp S (1980) Guanylate cyclase in the excitable ciliary membrane of *Paramecium*. FEBS Lett 122:64–66

Schultz JE, Klumpp S (1982) Lanthanum dissociates calmodulin from the guanylate cyclase in the excitable ciliary membrane from *Paramecium*. FEMS Lett 13:303–306

Schultz JE, Klumpp S (1983) Adenylate cyclase in cilia from *Paramecium*. FEBS Lett 154:347–350

Schultz JE, Klumpp S (1984) Calcium/calmodulin-regulated guanylate cyclases in the ciliary membranes from *Paramecium* and *Tetrahymena*. Adv Cycl Nucl Res 17:275–284

Schultz JE, Grünemund R, Hirschhausen R von, Schönefeld U (1984) Ionic regulation of cyclic AMP levels in *Paramecium tetraurelia* in vivo. FEBS Lett 167:113–116

Schultz JE, Uhl DG, Klumpp S (1987) Ionic regulation of an adenylate cyclase from the excitable ciliary membrane of *Paramecium*. Biochem J 246:187–192

Schultz JE, Pohl T, Klumpp S (1986) Voltage-gated calcium entry into *Paramecium* linked to intraciliary cyclic GMP formation. Nature (London) 322:271–273

Soldo AT, van Wagtendonk WJ (1969) The nutrition of *Paramecium tetraurelia*, stock 299s. J Protozool 16:500–506

Soldo AT, Godoy GA, van Wagtendonk WJ (1966) Growth of particle-bearing and particle-free *Paramecium aurelia* in axenic culture. J Protozool 13:492–497

Steers E, Davis RH (1977) A reexamination of the structures of the immobilization antigen from *Paramecium aurelia*. Comp Biochem Physiol 56B:195–199

Tait A (1970a) Enzyme variation between syngens in *Paramecium aurelia*. Biochem Genet 4:461–470

Tait A (1970b) Genetics of NADP isocitrate dehydrogenase in *Paramecium aurelia*. Nature (London) 225:181–182

Thiele J, Honer-Schmid O, Wahl J, Kleefeld G, Schultz JE (1980) A new method for axenic mass cultivation of *Paramecium tetraurelia*. J Protozool 27:118–121

Thiele J, Klumpp S, Schultz JE, Bardele CF (1982) Differential distribution of voltage-dependent calcium channels and guanylate cyclase in the excitable ciliary membrane from *Paramecium*. Eur J Cell Biol 28:3–11

Wagtendonk WJ van (1974) Nutrition of *Paramecium*. In: Wagtendonk WJ van (ed) *Paramecium* – a current survey. Elsevier, New York, pp 339–376

Wagtendonk WJ van, Soldo AT (1970) Methods used in the axenic cultivation of *Paramecium aurelia*. Meth Cell Physiol 4:117–130

Walter MF, Schultz JE (1981) Calcium receptor protein calmodulin isolated from cilia and cells from *Paramecium*. Eur J Cell Biol 24:97–100

Zieseniss E, Plattner H (1985) Synchronous exocytosis in *Paramecium* cells involves very rapid (1 s), reversible dephosphorylation of a 65-kD phosphoprotein in exocytosis-competent strains. J Cell Biol 101:2028–2035

Chapter 17 Behavioral Genetics in *P. caudatum*

MIHOKO TAKAHASHI[1]

CONTENTS

1 Introduction

Since the extensive and classic work by Jennings (1906), *Paramecium* has been one of the favorite materials for the study of behavior in unicellular organisms and, among a few species of *Paramecium, P. caudatum* has been used in many studies that are important for understanding the physiological mechanisms of behavior. The electrophysiological methods including voltage clamping (Kamada 1934; Naitoh and Eckert 1968a, b, 1969, 1974; Eckert et al. 1976), demembranated modeling (Naitoh and Kaneko 1972), and deciliation and reciliation procedures (Dunlap 1976; Ogura and Takahashi 1976) were first developed in the study of *P. caudatum.* A new approach to understanding the mechanisms of behavior in *Paramecium* using *P. tetraurelia* was initiated by Kung (1971a, b) and has been adding more extensive information relating to the mechanism of behavior by using mutants (see Ramanathan et al., Chap. 15, this Vol.). *P. tetraurelia* is a very useful organism for genetic studies because it has autogamy. This makes the genetic work of *P. tetraurelia* easy as in haploid organisms. *P. caudatum* lacks autogamy, but this disadvantage has been overcome by developing techniques of artificial induction of autogamy (Tsukii and Hiwatashi 1979) or cytogamy (Takahashi and Shono 1980), and thus many be-

1 Institute of Biological Sciences, University of Tsukuba, Ibaraki, 305, Japan

Paramecium. Ed. by H.-D. Görtz

havioral mutants have been obtained (Takahashi 1979; Takahashi et al. 1985). These mutants have been used for genetic studies in *P. caudatum* as convenient genetic markers (Karino and Hiwatashi 1981; Tsukii and Hiwatashi 1983; Tsukii and Hiwatashi 1985).

2 Outline of *P. caudatum* Genetics and Isolation of Recessive Mutants

2.1 *Nuclear Reorganization in P. caudatum*

When cells of complementary mating types of *P. caudatum* are brought together under appropriate conditions, they stick together to form large agglutinates. This reaction is called the mating reaction. In about an hour, the agglutinates disintegrate and release pairs uniting at the anterior part of the cell. These pairs are called the holdfast unions. The third step of cell union is called the paroral unions in which unions of cells extend to the oral regions.

Immediately after the formation of the paroral unions, the micronucleus swells as it enters the premeiotic S-phase. Two consecutive meiotic divisions of the micronucleus produce four haploid nuclei. Only one of the four survives, and the others degenerate. The surviving nucleus undergoes one mitotic division and thus produces two genomically identical daughter nuclei. One of them migrates into the partner and fuses with the remaining nucleus in the partner to form the synkaryon. The synkaryon divides three times. Four of the eight products become macronuclear anlagen, one becomes a micronucleus, and three degenerate in each mate. Therefore, eight caryonidal clones are produced from one pair. All of them are identical genetically (see Wichterman 1986; see also Mikami, Chap. 8, this Vol.).

2.2 *Induction of Homozygosity*

Since most mutations will be recessive, it is hardly expected that we can obtain many mutants from ordinary conjugation. One of the methods to be used for obtaining recessive mutants is selfing conjugation which occurs in a single clone. In this case, one-fourth of the progeny would express the recessive mutant phenotype. For the use of selfing conjugation, however, isolation of single cells and then establishment of clones are essential after mutagenesis. Takahashi (1979) used this laborious method to obtain many behavioral mutants.

Autogamy known well in *P. tetraurelia* is a type of fertilization in single cells. At autogamy, fertilization occurring at the fusion of two sister gametic nuclei brings about homozygosity for all genes simultaneously. This type of fertilization has never been found to occur spontaneously in *P. caudatum.* Several attempts at artificial induction of autogamy were made in this species. Mikami and Koizumi (1979) used separation of conjugating pairs by the treatment with trypsin for inducing autogamy. Tsukii and Hiwatashi (1979) also succeeded in inducing autogamy by inhibiting chemically induced conjugating pairs by papain. However, recessive mutants have not been obtained by these methods.

Another method of the induction of homozygosity in *P. caudatum* is artificial induction of cytogamy (Takahashi and Shono 1980). The method is based on the same principle as the artificial induction of cytogamy in *Tetrahymena* (Orias et al. 1979). At the period of nuclear transfer conjugating pairs are treated with 1% methylcellulose. By this treatment nuclear transfer is prevented and then genomically identical sister nuclei form the synkaryon resulting in homozygosity for all genes. This method was successfully used to obtain recessive mutants (Takahashi et al. 1985).

3 Behavioral Mutants

All behavioral mutants so far isolated in *P. caudatum* are listed in Table 1.

Table 1. Behavioral mutants in *P. caudatum*

Mutant	Gene locus	Characteristics
CNR (Caudatum Non-Reversal)	*cnrA, cnrB, cnrC, cnrD*	No avoiding reaction to any kinds of stimuli
Kag (K^+-agitated)	$cnrB^{Kg}$	Respond to K^+ solution with long backward swimming
Kr (K^+-resistant)	$cnrC^{kr}$	Can grow in culture medium containing a high concentration of K^+
Tsb (Temperature shock Behavioral)	*tsb*	Frequent repeated backward swimming when temperature is raised rapidly
Slow swimmer	*sl*	Slow swimming
Ba^{2+}-insensitive	*bai*	Barely respond to Ba^{2+} solution

3.1 CNR Mutants

The behavioral mutant showing no ciliary reversal upon stimuli is called CNR which denotes *c*audatum *n*on-*r*eversal. The numbers of strains belonging to CNR are the largest among all behavioral mutants in *P. caudatum,* probably because screening of this kind of mutant is rather easy. Nine original strains belonging to CNR were obtained (Takahashi 1979; Takahashi et al. 1985). Their characteristics, in general, are very much like the "pawn" in *P. tetraurelia* reported by Kung (1971 a, see also Ramanathan et al., Chap. 15, this Vol.). No strains of CNR show a spontaneous avoiding reaction in the culture medium. When wild-type cells are transferred into a high K^+ solution, they respond with continuous backward swimming. When they are transferred into a Ba^{2+} solution, they respond with the so-called barium dance (Dryl 1961). However, when

CNR cells are transferred into the K^+ solution from culture medium, they do not show an avoiding response and disperse in the solution with slow forward swimming. When the cells of CNR are transferred into the Ba^{2+} solution, they whirl a few seconds, but show no avoiding response.

When demembranated with Triton X-100 by the method of Naitoh and Kaneko (1972), all CNR mutants can swim backwards in the presence of $>10^{-6}$ M Ca^{2+} after reactivation with ATP and Mg^{2+}. Therefore, the lesion in the CNRs resides not in the axonemal components but in the membranes.

Electrophysiological evidence showed that all strains of CNR examined lack the function of voltage-dependent Ca^{2+} channels (Takahashi and Naitoh 1978; Haga et al. 1983; Takahashi et al. 1985). Malfunction of CNR seems to be restricted to the voltage-dependent Ca^{2+} channels and does not involve other channels. For example, the depolarizing mechanoreceptor potential carried by Ca^{2+} is found to be quite normal (Takahashi and Naitoh 1978; Takahashi et al. 1985).

Genetically, nine strains of CNR were grouped into four different loci by the genetic complementation test. They are designated as *cnrA, cnrB, cnrC*, and *cnrD*. Each locus has its own characteristics. Some alleles of *cnrB* are leaky, but no strains of *cnrA* are leaky. The *cnrC* strain, the only strain isolated originally, changes its phenotype to wild type probably depending on its clonal age (Takahashi 1979). The underlying mechanism of this phenotypic change in *cnrC* is not well understood. The *cnrD* is the tightest CNR and shows not only a lack of voltage-dependent Ca^{2+} current, but also of the sustained inward Ca^{2+} current (Takahashi et al. 1985). Further locus-dependent characteristics of each CNR are as follows (see also Table 2).

Ion Resistance. In a high Ba^{2+} solution (15 mM $BaCl_2$, 1 mM KCl, 0.5 mM $CaCl_2$, and 1 mM Tris-HCl, pH 7.2–7.4), wild-type cells are killed within 30 min, while most CNR survive. Some difference in sensitivity to Ba^{2+} is ob-

Table 2. Pleiotropic phenotypes in CNR mutants

	Resistance to Ni^+ paralysis	Growth in high K^+	Phenotypic expression during conjugation with wild-type cells	Chemical induction of conjugation	
				in 0.06 mM Ca^{2+}	in 0.6 mM Ca^{2+}
Mutant					
cnrA	S[a]	–[b]	Stable[c]	Very high	Low
cnrB	S	–	Very stable	High	Low
cnrC	R	+	Very unstable	Very high	No
cnrD	S	–	Unstable	Very high	Low
Wild type	S	–		High	Very low

[a] S denotes immobilizing within 50 s in 1 mM NiCl, 1 mM $CaCl_2$, 1 mM Tris-Cl (pH 7.2–7.4); R denotes no immobilization over 100 s in the same solution.
[b] – denotes no growth in culture medium containing 40 mM KCl; + denotes growth in the same medium.
[c] See text.

served among loci. Clones of *cnrB* appear more sensitive and those of *cnrC* more resistant. The strains belonging to *cnrC* are also resistant to paralysis by Ni^+.

Chemical Induction of Conjugation. Conjugation of *Paramecium* can be induced by chemical agents without mixing complementary mating types. This technique was discovered by Miyake (1968) and the conditions for the induction are similar to those of ionic stimulation to induce ciliary reversal, i.e., high in K^+ and low in Ca^{2+}. In the induction of conjugation by chemical agents, locus-dependent differences were observed among different CNRs. Locus-dependent characteristics of the inducibility of conjugation are well reflected by tests using two concentrations of $CaCl_2$, 0.06 and 0.6 mM. The mutant *cnrB* was more inducible than wild type in the high concentration of Ca^{2+} (0.6 mM). Cells with *cnrC* were induced to conjugate very well in 0.06 mM, but no induction occurred in 0.6 mM $CaCl_2$. Conjugations of *cnrA* and *cnrD* were induced well in both Ca^{2+} concentrations.

Phenotypic Instability During Conjugation with Wild Type. When cells of CNR are conjugated with wild type, a change of phenotype from CNR to wild type sometimes occurs, and this phenomenon is thought to be due to exchange of a small amount of cytoplasm between mates (Berger 1976; Hiwatashi et al. 1980). In this instability of phenotype during conjugation with wild type, locus dependency is also observed. Phenotypic expression of *cnrB* during conjugation is the most stable among the four CNR mutant loci. Even immediately after separation of pairs, a *cnrB* exconjugant shows no response to K^+ stimulation, while in other CNRs some responses are observed. A slight effect of contact with wild type on *cnrA* was observed. This expression of wild-type phenotype by contact with wild type, however, is very unstable and disappeared within 30 min after pair separation. The stability of the *cnrD* phenotype during conjugation is intermediate. Phenotypic transformation of *cnrD* to wild type occurs gradually after the formation of pairs with wild type and the degree of excitability of the *cnrD* mate is about half of wild type immediately after separation of the pairs. Expression of the mutant *cnrC* phenotype during conjugation is the most unstable, and 1 h after formation of tight pairs, the phenotype of *cnrC* cells in indistinguishable from wild type. The gene product of *cnrC* is soluble and is known to be an acidic protein of MW 15000–30000 (Haga et al. 1984). These different stabilities of phenotype during conjugation observed among the CNR loci may reflect the nature, mostly of diffusibility, of each gene product controlling the voltage-dependent Ca^{2+} channel function.

3.2 Other Behavioral Mutants

Slow Swimmer. The slow swimmer swims slowly at about 62% of the wild-type speed in the standard physiological solution and is controlled by the recessive gene, *sl* (Takahashi and Naitoh 1978). Triton-extracted models of this mutant retain *sl* characteristics after reactivation with ATP and Mg^{2+}, the speed being about 73% of the wild type. This suggests that the mutant has some defect in the motile system of the cilia. The Tris/EDTA extract from wild type shows two ATPase activities, one in the 14S and one in the 30S region, whereby that from the slow swimmer gives very low ATPase activity in the 14S region but normal in the 30S region compared with the wild type (Fig. 1) (Hayashi and Takahashi 1979). Thus, the decrease in ATPase activity of the 14S dynein causes slow swimming speed in the slow swimmer mutant.

Temperature Shock Behavioral (Tsb) Mutant. This mutant responds to a rapid elevation of environmental temperature of 10 °C or more, with long backward swimming interrupted by short rounds of forward swimming. This response also appears when the mutant is agitated mechanically. Tsb expresses another pleiotropic phenotype. More than 90% of Tsb cells are killed by K^+ stimulation (20 mM KCl, 0.25 mM $CaCl_2$, and 1 mM Tris-HCl), while no wild-type cells are killed under these same conditions. Thus, the Tsb has a K-sensitive character.

K^+ Agitated Mutant (Formally Called K^+ Sensitive). The K^+ agitated mutant (Kag) responds to K^+ stimulation with extraordinarily long backward swimming. In the culture medium this mutant frequently shows a spontaneous avoiding reaction. Behavioral characteristics are very much like "dancer" in *P. tetraurelia* (Hinrichsen and Saimi 1984).

K^+ Resistant Mutant. This mutant can grow in medium containing high KCl (Endo and Hiwatashi 1981) and is also resistant to paralysis by Ni^+. The cells can respond to Ba^{2+} stimulation with the so-called Ba dance even after incubation for 3 days in high K^+ solution, while wild type looses this ability gradually under the same conditions. Electrophysiological evidence shows that they

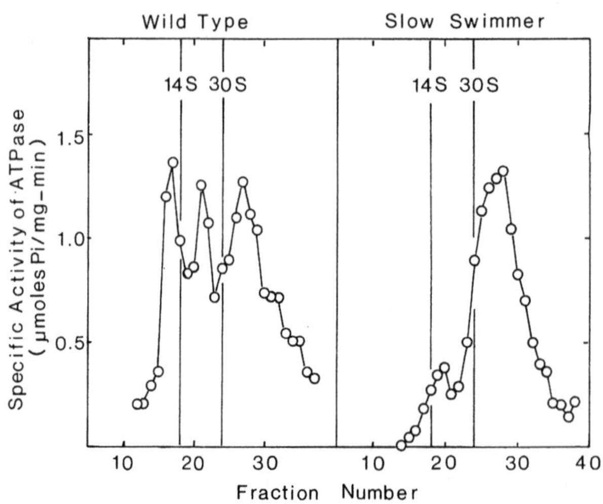

Fig. 1. Comparison of dynein ATPase activity between wild type and the mutant slow swimmer. *Left:* wild type; *right:* the slow swimmer. (After Hayashi and Takahashi 1979)

maintain deep resting potentials at the high outer concentration of K^+ (Oka, unpublished). The mutation appears to be an alteration in some mechanism of "adaptation" and is very much like the K-resistant mutant in *P. tetraurelia* (Shusterman et al. 1978).

Ba^{2+}-Insensitive Mutant. A reduced response to Ba^{2+} stimulation appears in this mutant. No response is observed in 1 mM $BaCl_2$ and the latent time for the Ba dance in 2 mM $BaCl_2$ is about 60 s, while no latent time is observed in wild type. The response to K^+ stimulation is not different from wild type (Takahashi and Shono 1980).

4 Relation of CNRs to Pawns in *P. tetraurelia*

The pawns in *P. tetraurelia* have been mapped at three different loci and CNRs in *P. caudatum* are mapped at four loci. All of them are nonexcitable mutants with defects in voltage-dependent Ca^{2+} channels. It was of interest to see if any CNRs are equivalent genetically to pawns. Since the usual complementation test by cross-breeding analysis is impossible between *P. caudatum* and *P. tetraurelia*, cytoplasmic transfers of all combinations among the mutants of seven different loci of the two species were made. Cytoplasmic transfers among pawns have shown that cytoplasmic complementation strictly parallels genetic complementation (Haga et al. 1982). Malfunctions of Ca^{2+} channels in all CNRs and pawns could be cured by injection not only of wild-type cytoplasm, but also of cytoplasm of mutants controlled by different loci. The cytoplasmic transfer experiments showed that all CNRs complemented all pawns and vice versa (Haga et al. 1983; Takahashi et al. 1985). Thus, seven genetic loci among the two species are independent and each has a different role in Ca^{2+} channel function. There were different degrees of restoration of wild-type phenotype by cytoplasmic transfer among recipient loci, but no significant difference was observed among donors of different loci.

Cronkite (1976) reported that pawn mutants cannot be induced to conjugate by chemical agents. It is not well understood why conjugation of pawns cannot be induced chemically while that of CNRs can be (see Sect. 3.1). One possibility, however, might be that pawns (or cells of *P. tetraurelia* in general) are extremely sensitive to the Ca^2 inhibition of the chemical induction of conjugation.

5 Genic and Allelic Interactions of CNR Loci Observed in *P. caudatum*

5.1 Alleles Expressing Completely Different Phenotypes

The mutants CNRs and Kag have opposite characteristics, that is, CNRs show no response to the K^+ stimulation, but Kag shows an over-response to the same stimulation with a long duration of backward swimming. To determine the genetic relationship between the over-reactor and the nonreactor, a cross-breeding analysis between CNRs and Kag was performed. The results revealed that Kag was controlled by a dominant allele of *cnrB*, designated as $cnrB^{Kg}$. It is very interesting that different mutations at the same *cnrB* locus cause opposite phenotypes. Though Kag is controlled by the dominant allele, a small difference of phenotype was observed between the heterozygote of $cnrB^{Kg}/+^{cnrB}$ and the homozygote of $cnrB^{Kg}/cnrB^{Kg}$ in behavioral response to K^+ stimulation. When cells of the homozygote ($cnrB^{Kg}/cnrB^{Kg}$) were transferred into the stimulation solution, the posterior half of the cell shrank. In this phenotype, $cnrB/cnrB^{Kg}$ was indistinguishable from $cnrB^{Kg}/cnrB^{Kg}$. This suggests that *cnrB* might be a null allele.

The K^+ resistant (Kr) was found to be controlled by an allele of *cnrC*, designated as $cnrC^{kr}$ (Sato and Hiwatashi 1982). The heterozygote of $cnrC/cnrC^{kr}$ can grow in culture medium containing 40 mM KCl and responds to K^+ stimulation with long backward swimming.

In the chemical induction of conjugation, Kag showed the *cnrB* character, that is, it was highly inducible at a high concentration of Ca^{2+}. Conjugation of Kr was induced well in low Ca^{2+} but not in high Ca^{2+}, just as in *cnrC*. When Kag was conjugated with wild type, phenotypic transformation from Kag to wild type never occurred in the Kag cell. On the other hand, this transformation occurred in Kr cells when mated with wild type and the Kr cell in the conjugant lost resistance to Ni^+ paralysis. Thus, Kag and Kr show different phenotypes from CNR in their behavior, but show characteristics of CNR alleles of their loci in their pleiotropic phenotypes, i.e., chemical inducibility of conjugation and the stability of phenotype during conjugation with wild type.

5.2 Interactions Between Different Loci

Double mutants of different loci often give us important clues to the functions of the mutant loci. All double mutants of CNRs and Kag or Kr, such as *cnrA* and Kag, *cnrA* and Kr, *cnrB* and Kr or *cnrD* and Kag expressed CNR phenotype in their behavior, that is, they showed no avoiding response to any stimulus. The double mutants of *cnrA* and Kr ($cnrA/cnrA$, $cnrC^{kr}/cnrC^{kr}$) or *cnrB* and Kr ($cnrB/cnrB$, $cnrC^{kr}/cnrC^{kr}$) are Ni^+ resistant and also K^+ resistant. So both mutant genes, *cnrA* (or *cnrB*) and $cnrC^{kr}$, express their phenotype independently, showing the CNR character in behavior, but the Kr character in ion resistance. The double mutant of *cnrA* and Kag, however, shows Ni^+ resistance but not K^+ resistance. Ni^+ resistance is a characteristic of *cnrC*, but not of *cnrA* nor of the *cnrB* locus to which Kag belongs. Thus, in this double mutant, a new phenotype different from either of the two mutant loci appeared. This kind of result also appeared in another pleiotropic character of the behavioral loci, the inducibility of conjugation by chemical agents. Conjugation in the double mutant of *cnrA* and Kag was induced very well in low Ca^{2+}, but not in high Ca^{2+}. This pattern of chemical induction of conjugation is the character of *cnrC*, but of neither the *cnrA* nor the *cnrB* locus. Why does the double mutant of *cnrA* and Kag express a character of the third locus? One possible explanation is shown in Fig. 2. In this model, the wild-type product of the *cnrC* locus controls the Ca-channel function through an interaction with molecules pro-

Fig. 2. Model showing the interrelation of gene products of the three *cnr* loci

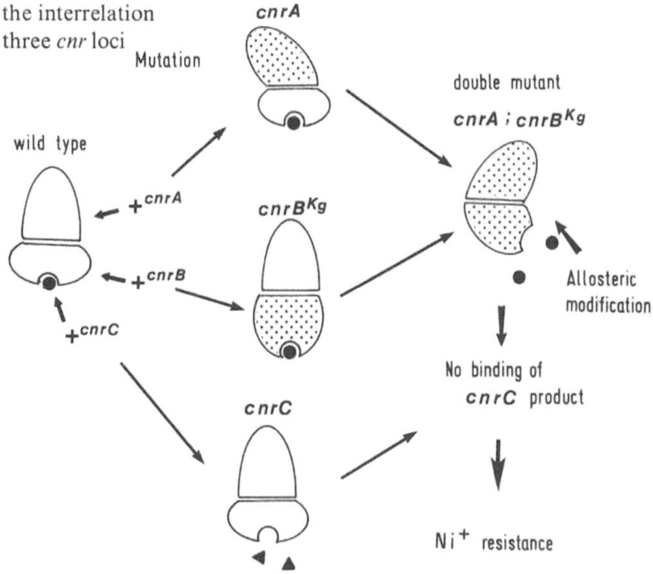

duced by wild-type alleles of *cnrA* and *cnrB*. A single defect in either the molecule of $+^{cnrA}$ or $+^{cnrB}$does not change the binding site of the $+^{cnrC}$ molecule to the $+^{cnrB}$ product. Additional defects of *cnrA* and Kag mutants result in an allosteric modification of the binding site of the $+^{cnrB}$ product. This causes no binding with $+^{cnrC}$ products and produces the *cnrC*-like character in the double mutant of *cnrA* and Kag (Fig. 2).

6 Conclusion

Studies of mutants concerning the voltage-dependent Ca^{2+}-channel function are progressing in *P. tetraurelia* and in *P. caudatum.* Mutants in these two species are interrelated and thus give us integrated information about genic control of the Ca channels. The gene products controlling Ca channels are functional in both species. Thus, for the isolation of a curing factor of *cnrC* in *P. caudatum,* cells of *P. tetraurelia* were used because the latter can grow faster.

In *P. caudatum,* interesting behavioral mutants, such as Tsb- and Ba-insensitive, have been isolated which have not been obtained in *P. tetraurelia.* Full physiological analysis and analysis at the molecular level of these mutants have not yet been performed and thus remain for future study. Moreover, isolation of additional interesting and important mutants for the understanding of the behavior of *Paramecium* should be possible. Mutants are a rich source of information for the study of behavior in *Paramecium.*

Acknowledgments. I thank Drs. D.L. Cronkite and K. Hiwatashi for critical reading of the manuscript. The work was supported in part by grants 58540403, 59127035, and 60115008 from the Ministry of Education, Science, and Culture of Japan.

References

Berger JD (1976) Gene expression and phenotypic change in *Paramecium tetraurelia* exconjugants. Genet Res Cambr 27:123–134

Cronkite DL (1976) A role of calcium ions in chemical induction of mating in *Paramecium tetraurelia*. J Protozool 23:431–433

Dryl S (1961) The ciliary reversal in *Paramecium caudatum* induced by simultaneous action of barium and calcium ions. J Protozool 8:s55

Dunlap K (1976) Ca channels in *Paramecium* confined to ciliary membrane. Am Zool 16:185

Eckert R, Naitoh Y, Machemer H (1976) Calcium in the bioelectric and motor functions of *Parmaecium*. In: Duncan CJ (ed) Calcium in biological systems. Cambridge Univ Press, London, pp 233–255

Endo H, Hiwatashi K (1981) Chemical induction of conjugation in K^+-resistant mutants of *Paramecium caudatum*. Jpn J Genet 56:s591

Haga N, Forte M, Saimi Y, Kung C (1982) Microinjection of cytoplasm as a test of complementation in *Paramecium*. J Cell Biol 82:559–564

Haga N, Saimi Y, Takahashi M, Kung C (1983) Intra- and interspecific complementation of membrane-inexcitable mutants of *Paramecium*. J Cell Biol 97:378–382

Haga N, Forte M, Ramanathan R, Hennessey T, Takahashi M, Kung C (1984) Characterization and purification of a soluble protein controlling Ca-channel activity in *Paramecium*. Cell 39:71–78

Hayashi M, Takahashi M (1979) Ciliary adenosine triphosphatase from a slow swimming mutant of *Paramecium caudatum*. J Biol Chem 254:11 561–11 565

Hinrichsen RD, Saimi Y (1984) A mutation that alters properties of the calcium channel in *Paramecium tetraurelia*. J Physiol 351:397–410

Hiwatashi K, Haga N, Takahashi M (1980) Restoration of membrane excitability in a behavioral mutant of *Paramecium caudatum*. J Cell Biol 84:476–480

Jennings HS (1906) Behavior of the lower organisms. Columbia Univ Press, New York

Kamada T (1934) Some observations on potential differences across the ectoplasm membrane of *Paramecium*. J Exp Biol 11:94–102

Karino S, Hiwatashi K (1981) Analysis of germinal aging in *Paramecium caudatum* by micronuclear transplantation. Exp Cell Res 136:407–415

Kung C (1971a) Genic mutants with altered system of excitation in *Paramecium aurelia*. I. Phenotypes of the behavioral mutants. Z Vergl Physiol 71:142–164

Kung C (1971b) Genic mutants with altered system of excitation in *Paramecium aurelia*. II. Mutagenesis, screening and genetic analysis of the mutants. Genetics 69:29–45

Kung C, Eckert R (1972) Genetic modification of electric properties in an excitable membrane. Proc Natl Acad Sci USA 69:93–97

Mikami K, Koizumi S (1979) Induction of autogamy by treatment with trypsin in *Paramecium caudatum*. J Cell Sci 35:177–184

Miyake A (1968) Induction of conjugation by chemical agents in *Paramecium*. J Exp Zool 167:359–380

Naitoh Y, Eckert R (1968a) Electrical properties of *Paramecium caudatum:* modification by bound and free cations. Z Vergl Physiol 61:427

Naitoh Y, Eckert R (1968b) Electrical properties of *Paramecium caudatum:* all-or-none electrogenesis. Z Vergl Physiol 61:453–471

Naitoh Y, Eckert R (1969) Ionic mechanisms controlling behavioral responses in *Paramecium* to mechanical stimulation. Science 164:963–965

Naitoh Y, Eckert R (1974) The control of ciliary activity in Protozoa. In: Sleigh MA (ed) Cilia and flagella. Academic Press. London New York, pp 305–352

Naitoh Y, Kaneko H (1972) Reactivated Triton-extracted models of *Paramecium:* modification of ciliary movement by calcium ions. Science 176:523–524

Ogura A, Takahashi K (1976) Artificial deciliation causes loss of calcium-dependent responses in *Paramecium*. Nature (London) 264:170–172

Orias E, Hamilton HP, Flacks M (1979) Osmotic shock prevents nuclear exchange and produces whole-genome homozygotes in conjugating *Tetrahymena*. Science 203:660–663

Sato M, Hiwatashi K (1982) Genetic relationship between two mutants, cnrC and K^+ resistant in *Paramecium caudatum.* Jpn J Genet 57:s702

Shusterman CL, Thiede RW, Kung C (1978) K^+-resistant mutants and "adaptation" in *Paramecium.* Proc Natl Acad Sci USA 75:5645–5649

Takahashi M (1979) Behavioral mutants in *Paramecium caudatum.* Genetics 91:393–408

Takahashi M, Naitoh Y (1978) Behavioral mutants of *Paramecium caudatum* with the defective membrane electrogenesis. Nature (London) 271:656–659

Takahashi M, Shono M (1980) A new behavioural mutant in *Paramecium caudatum:* isolation by artificial induction of cytogamy. Zool Mag (Tokyo) 89:s566

Takahashi M, Haga N, Hennessey T, Hinrichsen RD, Hara R (1985) A gamma ray-induced non-excitable membrane mutant in *Paramecium caudatum;* a behavioral and genetic analysis. Genet Res Cambr 46:1–10

Tsukii Y, Hiwatashi K (1979) Artificial induction of autogamy in *Paramecium caudatum.* Genet Res Cambr 34:163–172

Tsukii Y, Hiwatashi K (1983) Genes controlling mating-type specificity in *Paramecium caudatum:* three loci revealed by intersyngenic crosses. Genetics 104:41–62

Tsukii Y, Hiwatashi K (1985) Meiotic nondisjunction and aneuploids in intersyngenic hybrids of *Paramecium caudatum.* Genetics 111:779–794

Wichterman R (1986) The biology of *Paramecium.* Plenum, New York

Chapter 18 Chemokinesis

JUDITH VAN HOUTEN and ROBIN R. PRESTON[1]

CONTENTS

1 Introduction

Paramecia are sensitive to many kinds of changes in their environment. The stimuli can take the forms of heat, mechanical jostling, or chemicals. The response to each is generally a change in swimming that causes cells to approach or move away from the stimulus source. This chapter will trace the progress of the study of organic chemical stimuli on *Paramecium* behavior. The tracing will lead from some fairly remarkable observations using simple equipment to modern membrane biochemistry. From this will emerge a picture, albeit incomplete, of the chemosensory transduction pathway in *Paramecium* from receptor to motile apparatus, perception to response. At the end we will offer some models to fill the gaps in our picture of how detection of a chemical cue ultimately results in altered swimming behavior.

This chapter will deal primarily with organic stimuli as opposed to inorganic cations, which have well-documented effects on *Paramecium* behavior (Kung and Saimi 1982). The reason for the dichotomy is that the cells are likely to have evolved receptors on the cell surface to specifically detect the presence of these organic stimuli. Inorganic ions have been studied extensively and the effects of inorganic ions on cell physiology and behavior will be examined in other chapters of this volume.

1 Department of Zoology, University of Vermont, Burlington, VT 05405, USA

Paramecium. Ed. by H.-D. Görtz

2 Assays of Chemoresponse

For a unicellular organism such as *Paramecium*, life in a freshwater environment presents a number of problems. Aside from the unremitting need to regulate and expel water from the cell interior, the organism is subject to rapid changes in external temperature and ion composition that are characteristic of the small bodies of stagnant water in which *Paramecium* abounds. In addition, most species of *Paramecium* are wholly dependent on external sources of nutrition, the availability of which may show considerable seasonal variation.

Clearly, it would be of advantage for the survival of the organism if it were to possess the capacity to both sense these environmental changes and then respond by moving toward or away from the stimulus. As previously discussed, the surface of *Paramecium* is covered with several thousand motile cilia. The metachronal beating of these organelles propels the cell at up to 2 mm s^{-1} through its surrounds; the cell is thus readily capable of a locomotory response to environmental change. That *Paramecium* is able to sense such change has been recognized since the turn of the century, when H. S. Jennings concluded a series of observations on the behavior of a variety of protists. These studies culminated in the publication of his treatise *Behavior of the Lower Organisms* (1906), a work still relevant to modern research on protist behavior. Many of Jennings' studies focused on the behavior of *Paramecium*, providing precise details of its responses to touch, gravity, electric and water currents, acidity, and chemicals. His experimental approach to the study of chemosensation was simple but effective. A suspension of paramecia was layered on a glass slide beneath a cover glass, supported by thin, glass rods. The cells could be seen randomly swimming in all directions until a drop of a test solution was pipetted into the center of the suspension. The drop was initially free of cells, and if the test solution were repellent to the cells, would remain this way. However, if the introduced solution were a positive stimulus, the drop would become filled with cells and the density of cells would be higher in the test solution than in the surrounds. Jennings noted that a strongly attractant solution would "trap" the cells within the confines of the drop, the cells turning upon leaving the area of stimulation and hence behaviorally defining its perimeter (Fig. 1). Jennings called this stimulus-induced turning an "avoiding reaction", a term that persists to this day in describing either stimulus-induced or spontaneous changes in swimming direction.

Jennings' studies were important in many respects. Firstly, he recognized the existence of a concentration threshold that had to be crossed for a cell to

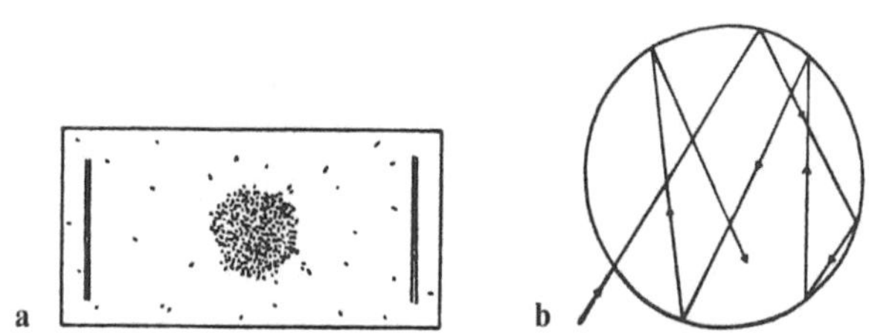

Fig. 1 a, b. Jennings' studies of chemoresponse. **a** Collection of paramecia in a drop of 0.02% acetic acid; **b** path followed by a single *Paramecium* in a drop of acid (Jennings 1906)

show a response to a stimulus. Secondly, he noted that responses to environmental stimuli are mediated through generation of an avoidance reaction; the avoidance response brings about a change in the direction of a cell's swimming path that causes eventual accumulation in attractants and prevents a cell from entering a repellent area. Thirdly, he emphasized that cells tend to collect in areas of optimal stimulation, and that it is movement away from this optimum that triggers a behavioral response. The importance of the latter is that it recognizes the fact that the terms attractant and repellent are relative. Thus, a weak repellent is attractive relative to a stronger repellent. Conversely, when presented with two attractant stimuli, cells will tend to collect in the solution that elicits a stronger positive response. The second solution is thus perceived by an observer as being repellent, despite the fact that the same solution may elicit a positive response when tested against a neutral or noxious solution. In concluding his studies, Jennings noted that paramecia are generally attracted to areas of weak acidity and aeration, and are repelled by a variety of monovalent and divalent cation salts.

Despite such eloquent beginnings, the study of *Paramecium's* responses to chemical stimuli languished until 1957, when Ferguson described a dark field photographic technique that could be used to record the swimming paths of single cells over the course of several seconds exposure. This technique was subsequently refined by Dryl (1958) in first attempts to quantify the behavioral responses to chemicals. Paramecia trace a helix as they swim, so that a two-dimensional photographic image of their swimming path appears as a now-familiar undulating white line against a black background. Since the time of exposure was known, the length of this line was a direct measure of the cells' swimming speed, and by widening the lens aperture during the final second of exposure, the last cell image (and hence direction) of the path is recorded as a brighter spot (Fig. 2). The simplicity of the recording system belies its ability to faithfully and clearly reproduce the individual swimming paths of many cells and was later used in a complete description of the characteristic behaviors exhibited by paramecia in response to stimuli (Dryl and Grebecki 1966).

Dryl (1959b) extended Jennings' observations of *Paramecium's* responses to chemicals by pipetting drops of test solution to each of the four corners of a glass plate supporting a thin suspension of cells. The number of cells collecting within these areas was counted from a time exposure photograph, and these data expressed as a percentage of the number of cells collecting within a central, control drop of distilled water. Inevitably a degree of mixing occurred when the test and control solutions were introduced to the cell suspension, but Dryl (1959b) was able to demonstrate reproducible repulsion of *P. caudatum* from ten lower alcohols. The concentration of alcohols required to both repel and kill paramecia was proportional to their molecular weight, perhaps reflecting a direct interaction of the chemicals with the lipid portion of the plasma membrane. In later studies, Dryl (1961a) determined the optimum pH for the motility of *P. caudatum*, showed quinine to be a strong repellent (1961b), and described an "adaptation" phenomenon, whereby a prolonged (24 h) incubation in high concentrations of $MgCl_2$, $CaCl_2$, NaCl, or KCl caused a slowly reversible insensitivity of the cells to subsequent cation stimulation (Dryl 1959a).

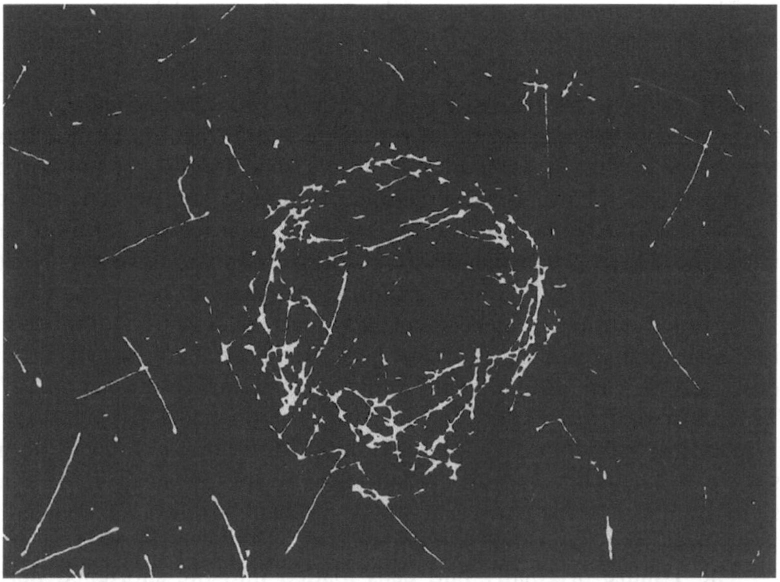

Fig. 2. Photomacrographic assay of attraction of paramecia to slightly acidic medium (Dryl and Grebecki 1966)

Recognizing the inherent diffusion and dilution problems associated with adding drops of a test solution to a layer of cells, Nakatani (1968) developed a capillary tube assay for testing the relative attractiveness of starch, milk, latex particles, acids, ethanol, and ions to *P. caudatum.* Capillary tubes are routinely used to assay the responses of bacteria to chemicals (Adler 1966), but have since been deemed less reliable for use with the larger, faster-swimming ciliates (Levandowsky and Hauser 1978). Nakatani's (1968) assay involved filling 0.6 mm diameter capillary tubes with test solutions, and then placing one end in a trough containing paramecia. Cells attracted to the test solution swam into the capillary and then became trapped as they migrated upwards under a negative geotactic influence. At 5-min intervals, the upper ends of the capillaries were photographed and the number of accumulated cells estimated by densitometric analysis of the resultant negative. Of all the various compounds tested, only $BaCl_2$ elicited a strong negative response, while the effects of acetic acid and ammonium chloride were concentration-dependent, being attractant at concentrations below the millimolar range, but repellent above these concentrations. Such diphasic behavior is also observed for bacteria capillary tests and may be due to the geometry of the assay rather than due to the cells making a discrimination between high and low concentrations (Adler 1969; Adler et al. 1973). Salts of acetic acid at neutral pH have subsequently been shown to be attractant to *P. tetraurelia* (Van Houten 1978), so it is likely that Nakatani's observations were of responses to changing pH rather than to acetic acid as a specific ligand. These studies emphasize the need to transfer paramecia to a buffered salts solution for experimentation, while previous authors had used culture fluid or tap water.

The chemoresponse assays of Jennings, Dryl, and Nakatani each provided some improvements over previous assays and significantly contributed to the progress of understanding *Paramecium* chemoreception. The macrophotographic method of Dryl made it clear that the avoiding reaction must figure into the behavioral mechanism of repulsion and attraction, but the draw back of this assay is that it is semiquantitative. Nakatani's capillary method later proved useful for the rapid screening of mutants and progeny of crosses (Van Houten 1977; DiNallo et al. 1982). However, in order to infer more information about the chemoresponse pathway, it was necessary to have a more quantitative assay. In response to this need, Hansma designed an assay (Van Houten et al. 1975) based on the countercurrent flow method used by Dusenbury (1973) for isolating chemoresponse mutants of *Cenorhabditis elegans*. Another method, described by Van Houten in the same article (Van Houten et al. 1975), used a T-maze made from a modified three-way stop cock. Of the two methods, the T-maze has been routinely used and adapted to other organisms because it was simple, fast, and uncomplicated conceptually and physically. The T-maze assay satisfied the need to carry out many replicate studies for concentration and structure-activity studies, for example.

The T-maze originally had cells loaded into the shortened perpendicular arm of a three-way stop cock. When the stop cock was opened, cells entered the plug and distributed between the two longer side arms. An index of chemokinesis was calculated based on the fraction of cells that distributed into the arm with the test compound. B. Diehn suggested a modification of using a two-way plug in the three-way stop cock. This change made the T-maze even more convenient (Fig. 3). Kinetic studies showed that cells distribute rapidly between the two arms of the T-maze, and that the distribution of cells in the arm with attractant, acetate, for example, was not due to a subpopulation of cells that could detect acetate with the cells in the control arm representing another subset of cells that were unable to detect the attractant (Van Houten et al. 1982). Instead, all the cells in the population seemed able to detect the attractants tested and distribution depended upon the average amounts of time spent by the cells in test and control solutions.

Van Houten used the T-maze to systematically search for organic attractants of *P. tetraurelia*. In this search, the test and control solutions of the T-maze were identical in pH and ionic strength and only one compound being tested was

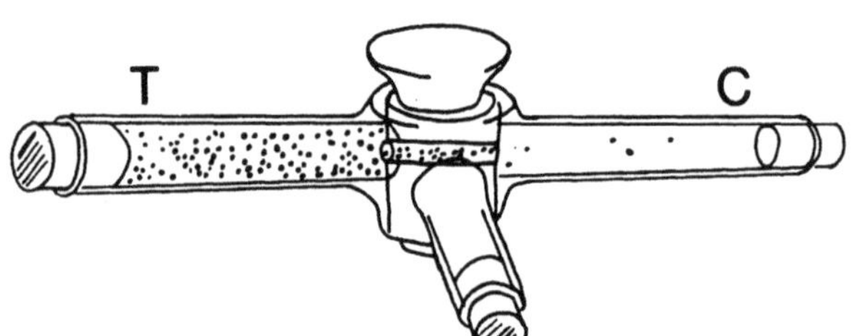

Fig. 3. T-maze apparatus for assaying chemoresponse. This T-maze is modified from the original (Van Houten et al. 1975) by the use of a two-way plug in the three-way stop cock

varied, e.g., the acetate anion was tested against Cl^- (Van Houten et al. 1975). This approach was essential because inorganic cations varying in ionic strength or pH would affect the *P. tetraurelia* swimming behavior and hence would complicate the search for a response to an organic compound for which *P. tetraurelia* could have evolved a receptor-response pathway. The T-maze assays demonstrated that attraction behavior was saturable and specific (Van Houten 1976, 1978). For example, acetate was attractive relative to Cl^- even when saturating amounts of another attractant, lactate, were included in all solutions as a competitor (Van Houten 1976, 1978). Saturability and specificity could be accounted for by specific surface receptors for stimuli. Doughty later used a modification of the T-maze in similar "jamming" experiments to argue for specific receptor sites for nicotinic cholinergic ligands (Doughty 1979).

Once a few organic attractants, such as acetate, lactate, and folate, were identified, the stage was set to select mutants defective in chemokinesis and to use these in comparison with normal cells to determine the behavioral mechanism of attraction and repulsion. Clones of mutants were screened for among mutagenized populations using capillaries filled with K-acetate and KCl (Van Houten 1977). The mutant that proved to be most useful in behavioral analysis was d4-530, a line of cells that were repelled from and not attracted to acetate relative to Cl^-.

3 Swimming Behavior

The previous work of Jennings and Dryl firmly established the avoiding reaction or turn as an important behavioral component in keeping *P. caudatum* cells accumulated or dispersed. Nakatani (1970) recognized that chemoeffectors also change the swimming speed of *P. caudatum.* In this study, cells were transferred to a test solution where they remained for 10–20 min, a time required for the "chemotactic response of *Paramecium* to reach a maximum" (Nakatani 1970). A few specimens in test solution were then placed under the objective of a microscope and their motion photographed under stroboscopic illumination. Detailed analyses of these records revealed that compounds that had previously been noted to repel paramecia (Nakatani 1968) also decreased their swimming speed, while chemicals that were attractant to *P. caudatum* increased swimming speed. Nakatani (1968) had shown that acetic acid and NH_4Cl become repellent at high concentrations; paradoxically these same concentrations caused even greater increases in cell swimming speed. Thus, there is a point at which an increase in swimming speed acts to disperse paramecia rather than attract them to the area of stimulation. This relationship is shown in Fig. 4.

Nakatani's observation of increased speed of swimming did not intuitively fit into the long accepted framework of the mechanism of attraction and repulsion by turning frequency. Dryl even considered speed changes a "contamination" of a pure "chemotactic" response (Dryl 1973). However, speed had to be incorporated somehow into a comprehensive mechanism of chemoresponse. Using both long exposure macrophotographs to measure speed and ob-

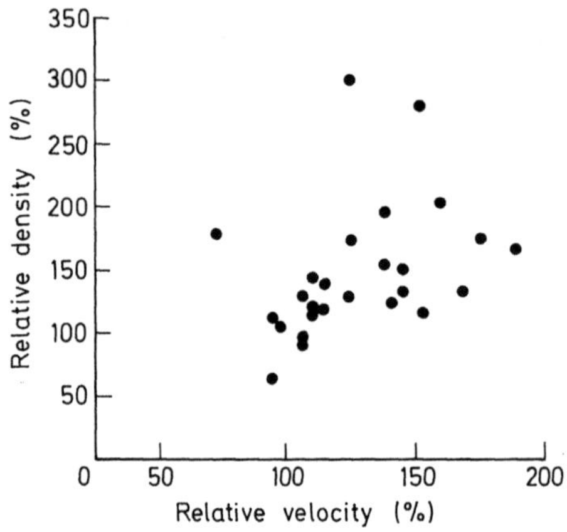

Fig. 4. Nakatani's measurements of speed correlated with density of cells accumulating in capillaries of chemical stimuli (Nakatani 1970)

servations of turns over time, Van Houten was able to build a coherent picture of both speed and turning in chemoresponse behavior for *P. tetraurelia* that had not been done for *P. caudatum*. *P. tetraurelia* both decreased frequency of turning and increased speed in attractants like acetate and increased frequency of turning and decreased speed in repellents like quinidine. However, these two behavioral changes were at odds when fit into the two mechanisms of accumulation and dispersal described for small organisms (Fraenkel and Gunn 1961). There was no evidence of oriented movement to chemical stimuli in *P. caudatum* or any ciliate and, therefore, the cells must have accumulated or dispersed by a kinesis mechanism: klinokinesis or orthokinesis. Klinokinesis requires the modulation of turning frequency, the equivalent of the biased random walk so nicely described in bacteria (Berg 1986). Orthokinesis requires the modulation of speed with accumulation occurring where cells move slowly. Somehow in *Paramecium* these two kineses must combine because both frequency of turning and speed were changed with chemical stimuli, but a complication arose for *Paramecium* in that the change in turning frequency and speed seemed to be at cross purposes: In attractants such as acetate, cells slightly decreased frequency of turning, which would cause attraction by klinokinesis, but also increased speed, which by orthokinesis would work against attraction. Conversely, repellents increased frequency of turning, consistent with repulsion by klinokinesis, but also decreased speed inconsistent with repulsion by orthokinesis. There were at least two possible explanations: (1) There were conditions under which the speed change was nonconsequential and the klinokinesis mechanism was all important and other conditions under which the speed change was all important; or (2) the combining of turning frequency and speed changes produced yet a different mechanism of attraction and repulsion not described by Fraenkel and Gunn.

In order to explore these possibilities, pawn, fast-2 (Kung et al. 1975) and mutant d4-530 (Van Houten 1976, 1977) were tested for attraction and repulsion (Van Houten 1978). Pawn mutants could not turn for lack of ciliary Ca influx, which would transiently reverse ciliary beating and make normal cells turn. However, pawns could modulate speed with external stimuli. Pawns were not attracted to acetate or similar attractants, implicating turning as essential for these attractants. However, pawns could be repelled from high pH in which the cells swim extremely fast. Fast-2 mutant showed smooth, fast swimming in Na solutions and relatively normal behavior in K solutions. Fast-2 mutants could not be attracted to Na salts of organic anion attractants such as acetate, but were normally attracted to the K salts, in which the cells showed normal turning control and speed. Mutant d4-530 was repelled from Na-acetate and, therefore, it was expected that this mutant would show the converse of the wild-type behavioral responses in acetate, that is, it would turn more frequently and swim more slowly in acetate than in chloride. Instead, mutant d4-530 moved faster and more smoothly in Na-acetate than the normal cells. This appeared to be a repulsion by orthokinesis and was reminiscent of the unusual observation by Nakatani of repulsion from high concentrations of attractants that induced fast swimming in *P. caudatum*.

From these mutant studies, it was clear that turning was essential in some responses but that speed control was sufficient for others. Therefore, there must be the potential for two mechanisms of response, and both must be compatible with *Paramecium* physiology. When the behavior of cells in attractants and repellents was plotted as frequency of turns vs speed, normal *Paramecium* behavior fell into two quadrants: a small increase in speed and decrease in frequency of turning was associated with attractants such as acetate; a large increase in speed and a large decrease in frequency of turning was associated with repellent action of high pH; a small decrease in speed and increase in frequency of turning was associated with repellents such as low pH or quinidine (Fig. 5).

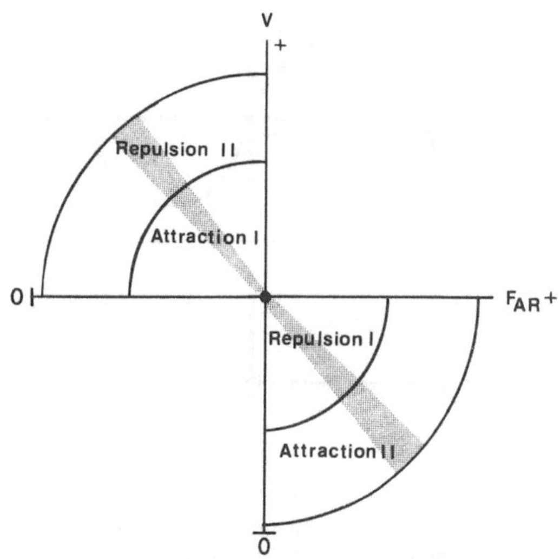

Fig. 5. Graphical description of *Paramecium* behavior. Behavior of cells in control solution is at the origin. An increase or decrease in velocity (V) from control is plotted on the y-axis. An increase or decrease in the frequency of avoiding reactions (F_{AR}) that cause turns is plotted on the x-axis. Behavior of normal animals falls in the *upper left* or *lower right* quadrants. Behavior is restricted to an area represented by a *shaded sector* that is determined by the membrane potential (ΔVm) (Van Houten 1978)

It was clear from this plot that the key to the behavior in attractants and repellents was the membrane potential of the cell. A change in membrane potential of *Paramecium* controls both the frequency of Ca action potentials, hence turns, by setting the potential closer to or farther from the threshold for action potential; and a change in membrane potential from rest changes swimming speed by altering the frequency and angle of ciliary beating (Machemer 1976; Machemer and de Peyer 1982).

From Fig. 5 it could be predicted that cells in attractant acetate would experience a small hyperpolarization relative to control and this ΔVm would cause the behavioral changes that lead to attraction; a larger hyperpolarization as in KOH (high pH) would cause repulsion. Conversely, a small depolarization of cells in repellent quinidine relative to control would lead to repulsion; a larger depolarization as in $BaCl_2$ relative to KCl would cause attraction. These predictions of change in membrane potential (ΔVm) were replotted as in Fig. 6a. Here, it is more evident that there should be a continuum of change in membrane potential relative to control and that depending on the magnitude of the ΔVm, a cell would be attracted or repelled. Direct testing using standard electrophysiological techniques gave credence to this hypothesis of ΔVm control of chemoresponse (Fig. 6b) (Van Houten 1979, 1980; Van Houten et al. 1984a). Therefore, even a relatively complex population behavior like chemoresponse could be explained on the basis of the known physiology of individual cells.

Frequency of turning and speed were the primary parameters that were necessary to describe *P. tetraurelia* behavior using computer simulations (J. L.

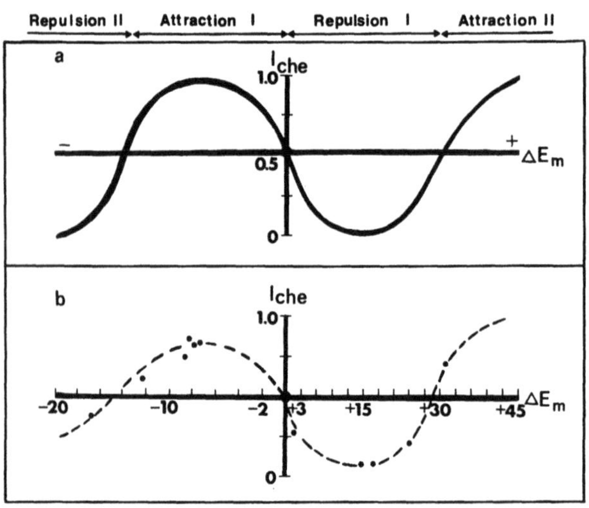

Fig. 6a, b. Membrane potential control of chemokinesis. **a** Graphical description of membrane potential control of chemokinesis. Change of membrane potential (ΔE_m or ΔVm) from control at origin is plotted against the index of chemokinesis: $I_{che} > 0.5$ indicates attraction; < 0.5 indicates repulsion. As chemical stimuli change E_m relative to control, animals will be attracted or repelled, depending on the magnitude and direction of the E_m change. **b** Data from direct recordings of ΔE_m from individual cells (Van Houten 1979) are plotted as ΔE_m produced by the attractant or repellent versus I_{che}. Scale of ΔE_m is different for depolarizing and hyperpolarizing stimuli (Van Houten 1979)

Van Houten and J. Van Houten 1982). Empirically determined speeds and turning frequencies were used to simulate accumulation in T-mazes. To properly mimic *Paramecium* behavior, however, adaptation had to be built into the simulation because the cells' behavior adapted and changed toward a basal level despite the continued presence of a stimulus. Adaptation seemed to prepare the cells for an immediate response at the boundary between solutions (Van Houten 1976, 1978; Machemer and de Peyer 1979; Van Houten et al. 1981; J. L. Van Houten and J. Van Houten 1982). The exact kinetics of this adaptation were not crucial to the outcome of computer simulation of attraction and repulsion for mutants and normal cells. However, the immediate response of the cells upon encountering the change in solution was very influential in the behavioral result.

4 Stimuli

For the most part, attractants of *P. tetraurelia* were found among fermentation or other products of bacteria, *Paramecium*'s food. At neutral pH, acetate and lactate anions, folic acid (a vitamin for the ciliate), cAMP, NH_4^+, and amino acids were among the compounds found to be attractive (Van Houten 1976, 1978; Schulz et al. 1982; Preston 1983). Repellents of *P. tetraurelia* include extremes of pH and salt and organic compounds that were "membrane active" and elicit action potentials (Van Houten 1976, 1978). Toxicity per se did not disperse the cells (Van Houten 1976). It appeared that *P. tetraurelia* accumulate in areas of food and optimal conditions for pH and osmotic balance and did not use organic stimuli as repellent cues. Compounds like the repellents quinine and quinidine (Dryl 1961b; Van Houten 1978) or attractant cholinergic ligands (Doughty 1979) may not normally occur in *P. tetraurelia*'s environment, but were useful in perturbing the sensory transduction pathway for study. *P. caudatum* had a slightly different array of attractants and repellents. Alcohols were repellent to *P. caudatum* as were areas of high salt and extremes of pH (Dryl 1959b). However, *P. caudatum*'s pH optimum was more acidic than that for *P. tetraurelia.*

Generally, the stimuli need to be present in high, near millimolar concentrations to elicit responses from *Paramecium,* which may be advantageous to the ciliates as they swim through a noisy aqueous environment. Exquisite sensitivity would lead the cells to many nonproductive areas for food gathering. Fermenting *E. coli* were found to be capable of producing lactate in millimolar concentrations around the soft agar into which they were inoculated. It was also nearest to this plug of agar that *P. tetraurelia* accumulated (Van Houten, unpubl. results).

Mating-reactive paramecia do not appear to release any diffusible mating substances for the complementary mating type (Van Houten et al. 1981) nor do paramecia release diffusible products detected by its predator *Didinium.* In T-mazes, *Didinium* was attracted to bacterial products of low molecular weight and, therefore, sought out its prey's food and not the prey itself (Antipa et al. 1983).

5 Chemoreceptors

A general model for a sensory transduction pathway begins with binding of ligand to receptor and this binding is subsequently transduced into a second messenger and response. Receptors can account for at least some of the specificity and saturability of the chemosensory systems and would be expected to be affected by surface-acting agents such as glutaraldehyde (Doughty and Dodd 1978). However, the task of identifying receptors is difficult and to date there has been only one external chemoreceptor protein identified, the receptor for cAMP in *Dictyostelium* (C. Klein et al. 1986; P. Klein et al. 1985). In *Paramecium*, the chemosensory behavior and hyperpolarization are saturable and specific, which argues for a receptor-mediated system. The attractant stimuli were effective in a charged form (Schulz et al. 1984; Preston et al. 1987) and, therefore, they cannot act internally without first binding to a site on the surface for transport into the cell. Therefore, regardless of whether the stimuli act exclusively externally or must enter the cell to be active, there must be a specific binding site, which we will refer to here as a receptor.

The initial description of receptors generally comes from structure-activity studies of the behavior that determine the important site on the stimulus that is detected by the receptor and from binding studies that provide information on the affinity, number, and location of receptor sites. The attractant molecules studied for most of the preliminary work on *P. tetraurelia* and *P. caudatum* were too small for or not amenable to manipulation for the structure-activity studies that would be necessary to make inferences about the nature of the receptor from behavioral studies. Therefore, the discovery of folate as an attractant molecule for *P. tetraurelia* provided a useful stimulus for receptor studies (Van Houten 1978). The folate molecule has three distinct moieties, making possible competition experiments (Van Houten 1976, 1978) to determine that the pterin group was necessary and sufficient for attraction (Schulz et al. 1984). Folate was found to bind specifically to whole cells and this binding was diminished to nonspecific, low levels in mutants (Schulz et al. 1984) and in the presence of a weak competitor for the behavioral response, cAMP (Schulz et al. 1984; Preston et al. 1987). The half-maximal values for behavior and ΔVm were approximately 100 μM, while the binding Kd was measured at approximately 30 μM. Discrepancies between binding and response data are not uncommon, but part of this discrepancy could be due to the need to measure whole cell binding as instantaneous binding (Schulz et al. 1984). Folate was found to be transported into cells and cold temperatures would not eliminate this transport to make equilibrium binding measurements possible. Therefore, instantaneous binding was used and though it should not have underestimated the Kd, it might have underestimated the binding capacity of the cell. Nonetheless, the instantaneous binding measured by centrifugation methods using [^{3}H]-folate had strong correlations with behavior and hyperpolarization in stimulus and probably represented binding involved in chemoreception: mutants did not show this binding; cAMP affected attraction, and binding and characteristic hyperpolarization (Schulz et al. 1984; Preston et al. 1987); binding was primarily to the cell body and not cilia (Schulz et al. 1984). This latter point was significant because electrical recording from deciliated cells indicated that the cell body membrane had all the components necessary for the characteristic hyperpolarization in folate and other attractants (Preston and Van Houten 1987b).

The folate receptor should reside among the folate binding proteins of the cell body membrane. Affinity chromatography has been used to identify approximately 20 proteins of the cell body membrane that bind folate specifically. Only five were accessible to [^{125}I] labeling and, therefore, externally facing and among these only two were much more abundant in the cell body membrane compared to the ciliary membrane (Schulz et al. 1985a, b; J.L. Van Houten et al. 1986; Van Houten and Preston 1987c). However, low affinity posed a problem for the use of traditional affinity chromatography in identifying chemoreceptors. It was likely that the folate binding protein of interest would bind too weakly to be specifically eluted and thereby display differences in affinity chromatography between normal and mutant. Therefore, antiligand antibody was developed to circumvent the problems of low affinity. The rationale for its use was as follows: when folate was cross-linked onto whole cells, they specifically lost their attraction response to folate, but not to acetate (Schulz et al. 1985a, b; Sasner 1986; Van Houten and Preston 1987a, c). Therefore, the receptors should have been among the surface proteins that were cross-linked with folate. To detect the proteins that were cross-linked, a polyclonal antifolate antibody was used to develop electroblots (Sasner et al. 1986) and to precipitate folate cross-linked proteins from [^{35}S]-labeled cells. This alternative approach will be potentially useful for other low affinity receptor systems.

A parallel picture has emerged for cAMP, which was found to be an attractant to *P. tetraurelia* (Schulz et al. 1982; Smith et al. 1987). Cyclic AMP and not a hydrolysis product acted externally as an attractant (Smith et al. 1987). Cyclic AMP attraction and characteristic hyperpolarization could be blocked by 5′-AMP but not by 5′-GMP or cGMP (Smith et al. 1987). Likewise, binding of cAMP to whole cells was saturable, specific, and blocked only by 5′-AMP. Binding once again was primarily to the cell body membrane with a Kd of approximately 250 μM, which was approximately one-fourth of the concentrations for hyperpolarization and behavioral half-maximal responses.

There was primarily one protein of approximately 48000 daltons that specifically eluted from a cAMP affinity column with cAMP or 5′-AMP, but not with cGMP or 5′-GMP (Van Houten and Preston 1987a, c). This protein appeared to be missing from affinity column fractions of a mutant, cyc^{-1}, that was not attracted to cAMP (Smith et al. 1987). The 48000 dalton protein was close in molecular weight to the immobilization antigen-related proteins in the 40000 molecular weight range (Eisenbach et al. 1983). However, immunoblots of whole cell body membranes and cAMP affinity column fractions have demonstrated that they are not the same proteins (Sasner and Van Houten, unpubl. results).

Like folate, cAMP can be covalently cross-linked to its receptor. 8-azido-cAMP will covalently bond to nearby proteins in the presence of *UV* light (Haley 1975). Treatment of cells with *UV* light in the presence of 8-azido-cAMP eliminated attraction to cAMP in the T-maze assay, while irradiation in the presence of cAMP did not. This inhibition was specific for cAMP; acetate

Table 1. Effects of 8-N_3-cAMP on chemoresponse

Pretreatment with UV irradiation in:	Test for chemoresponse		I_{che} [a]	n
	Test solution	Control solution		
2.5 mM NacAMP	2.5 mM NacAMP	2.5 mM NaCl	0.66 ± 0.05	25
2.5 mM N_3cAMP	2.5 mM NacAMP	2.5 mM NaCl	0.50 ± 0.06	14
	2.5 mM Na_2folate	5 mM NaCl	0.62 ± 0.07	11
	5 mM NaOAc	5 mM NaCl	0.66 ± 0.06	12

[a] Data are averages of n T-maze tests $\pm$ one standard deviation. $I_{che} > 0.5$ indicates attraction; $= 0.5$ indicates no response (Baez, unpubl. results).

and folate attraction were not affected (Table 1). This development makes possible the use of [^{32}P]-azido-cAMP to specifically label the chemoreceptors.

Chemoreceptors probably are not evenly distributed over the cell surface. There is a partitioning of components for the hyperpolarization between the ciliary membrane and the contiguous cell body membrane; generally, cells could be deciliated without loss of the attractant-induced hyperpolarization (see Preston and Van Houten 1987b for discussion and an exception). Additionally, there is partitioning between the anterior and posterior ends of the cell for folate reception. Pressure perfusion of folate onto deciliated cells elicited the largest hyperpolarization on the anterior end with decreased responses as the perfusion pipette was moved toward the posterior of the cell (Fig. 7) (Preston and Van Houten 1987b). Gradients of chemoresponsiveness anterior to posterior could not be distinguished as a gradient of fluorescence from normal cells stained with fluorescein-folate. Fluorescein-folate bound specifically to normal cells and normal cells could be distinguished in blind tests from folate chemore-

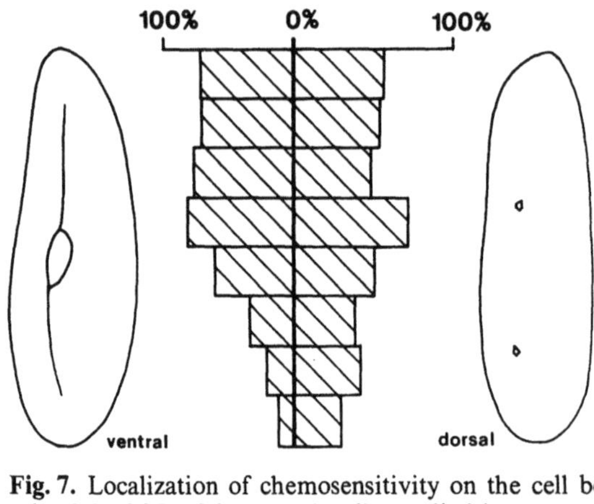

Fig. 7. Localization of chemosensitivity on the cell body membrane of deciliated *P. tetraurelia.* 2.5 mM K_2 folate was locally applied by pressure perfusion to 8 sites along the ventral and dorsal surfaces of cells. The resultant membrane hyperpolarizations are expressed as a percentage of the maximum response obtained from each cell. The data represent mean responses of 20 deciliated cells (Preston and Van Houten 1987b)

ception mutants and autofluorescence (Van Houten et al. 1975). However, this method of assaying chemoreceptor binding probably was not sensitive enough to detect gradients along the cell.

Once the identities of chemoreceptors for folate, cAMP, and others are verified, it should be possible to clone the genes for these receptors and thus obtain extremely useful reagents for their study. The isolation and use of mutants will assist in this effort. Mutants induced with nitrosoguanidine have been selected and characterized (DiNallo et al. 1982) and from these mutants there have been revertants that by genetic analysis appeared to be second-site suppressors (Van Houten, unpubl. results). However, the reversion of phenotype toward the wild type and hence instability has posed a problem with existing mutants. To address this, a different mutagenesis designed to produce null-phenotype mutants by a deletion or break in a structural gene, has produced mutants that are stable, perhaps due to lack of a functional gene product. Additionally, a newer screening procedure was developed. Previously, cells had been selected for their inability to remain in attractant (DiNallo et al. 1982). The alternative method selects for cells that do not bind folate, by scoring clones grown and stained with fluorescein-folate in microtiter (96-well) plates (Van Houten et al. 1985).

6 Characteristic Membrane Potential Changes in Chemoresponse: Models for Testing

The study of the ionic basis of the membrane potential change (ΔVm) has gone on concurrently with the receptor studies, with acetate and folate as the most intensely studied stimuli. The simplest working models for the hyperpolarization in attractants were a K or Na efflux triggered either directly by the ligand-receptor interaction or by ligand-dependent increase in Ca^{+}_{i}. These models were tested (Preston and Van Houten 1987a) by determining the effects of external ion substitution and resting ΔVm on the size of the attractant-induced hyperpolarizations. There was found to be a general dependence on ionic strength; increasing any divalent or monovalent cation to high levels decreased the size of the hyperpolarization. However, K_0 or Na_0 could be present, absent, and replaced by $Tris^{+}$, or present in amounts that should eliminate the driving forces for their net movement in or out of the cell with no effect on ΔVm. Therefore, ion substitution experiments indicated that simple efflux of K or Na could not account for the ΔVm, nor could influx of Cl^{-}. Ca_0 could not be reduced greatly without affecting cell viability and motility. However, the use of mutants with specific defects in Ca currents and W-7, a calmodulin inhibitor that has been shown to eliminate many Ca currents in *Paramecium* (Hennessey and Kung 1984), have ruled out most known Ca conductances in *Paramecium* that could trigger a hyperpolarizing ion efflux. Additionally, there appeared to be no voltage dependence and reversal potential for the hyperpolarizations in folate or acetate as measured either by current clamp or voltage clamp (Preston and Van Houten 1987a).

The lack of reversal potential and lack of effects of ion substitutions had at least two possible explanations. The ΔVm was due to surface charge changes or to electrogenic pumps. Ruthenium red, a polycation that perturbs surface charge, did not affect the ΔVm; nor did a surface charge mutant (Satow and Kung 1981) display an altered ΔVm for folate or acetate (Preston and Van Houten 1987a). Therefore, efforts are being made to investigate Ca^{2+} and H^+ pumping and exchange during chemoresponse using voltage clamping, radiotracer, and Ca-sensitive fluorescent dyes. All of these methods can be applied to *Paramecium* and should clarify some aspects of the role of Ca in chemoreception.

In addition to attractants like folate, acetate, and cAMP, amino acids were found to hyperpolarize the cells (Preston 1983). This hyperpolarization of 5–6 mV is transient, lasting approximately 9 s. All amino acids tested, except ornithine and tyrosine at low concentrations elicited a variable attraction response from the cells, with thresholds in the nM range and each amino acid caused increased swimming speed (Preston 1983). The binding of [^{3}H]-amino acids to isolated cilia indicated the presence of binding sites with Kds in approximately the 10^{-7} M range (Preston 1983, 1984). The relation of the transient hyperpolarization in amino acids to chemoresponse remains to be established.

7 Second Messengers

Stimulation of a cell with attractant was transduced into a hyperpolarization. It was expected that a second messenger system was part of the transduction of an external chemical signal into an electrical one, to which the motor apparatus responded. Cyclic nucleotides were the most obvious second messengers, since internal levels of cAMP and cGMP have been linked with control of ciliary beating (Bonini et al. 1986; Majima et al. 1986; Schultz et al. 1986). Hyperpolarization from decreased ionic strength was associated with increased $cAMP_i$ and swimming speed (Bonini et al. 1986), and therefore, it was possible that the hyperpolarization associated with attractants could also be associated with $cAMP_i$. However, isobutyl methylxanthine (IBMX), a phosphodiesterase inhibitor, did not perturb chemoresponse when it was included in both test and control arms of the T-maze, even though it was present in concentrations high enough to elevate internal cAMP levels and to increase swimming speed (Van Houten et al. 1987). Also, measurements of $cAMP_i$ during chemostimulation indicated no significant change in internal levels other than an increase as would have been predicted from the size of the hyperpolarization alone (Van Houten et al. 1987). There remains the possibility that cyclic nucleotides are messengers in chemoreception because the final motor response of the pathway involves cilia. However, it does not seem likely that cAMP is a second messenger involved in the generation of the ΔVm. Other second messengers, such as $Ca^{2+}{}_i$, H^+, IP_3, and diacyl glycerol remain possibilities. There was no effect on chemoresponse by phorbol esters that should mimic diacyl glycerol as second messenger (Wright, unpubl. results), and no strong effect of external pH, which might perturb H^+ movement if it were involved in transduction. While none of these alternative second messengers has been eliminated, we favor an efflux pump or exchange for Ca^{2+} to account for the hyperpolarization and the 0.2 nA current that can be measured by voltage clamp.

Application of amiloride, a diuretic that is known to perturb Na^+/H^+ and Na^+/Ca^{2+} exchange (Zhuang et al. 1984; Kaczorowski et al. 1985) has been shown to inhibit chemoresponse to sodium salts of acetate and folate (Van Houten and Preston 1985). Likewise, lithium has been shown to inhibit chemoresponse to acetate and folate. The study of these two effects is ongoing and may hold key information about the mechanism of transducing stimulus binding into ΔVm or into the slower process of adaptation. Covalent modification may also be involved in adaptation, as is methylation in bacteria (Boyd and Simon 1982; Hazelbauer and Harayama 1983; Ordal 1985). There is a striking effect of S-adenosyl-methionine on chemoresponse that is inhibited by a battery of methyl transfer inhibitors in *Paramecium* chemoresponse (Van Houten et al. 1984b). Likewise, phosphorylation of membrane components could potentially be involved in the mechanism of adaptation, as in *Dictyostelium* chemoreceptors (C. Klein et al. 1985; P. Klein et al. 1985).

8 Summary

Paramecium is a bona fide receptor cell. It shows clear responses to organic stimuli, for which it seems to have evolved receptors and sensory transduction pathways. The mechanisms by which cells accumulate or disperse as a population depend upon the magnitude of the ΔVm the stimulus elicits. Cells can be attracted or repelled either by primarily modulating the frequency of turning or primarily modulating speed. Both are consistent with the known physiology of individual cells. The identity of receptors is being determined and will eventually lead to the cloning of the receptor genes. The ionic mechanism of the ΔVm appears to be due to a pump perhaps specific for Ca^{2+} or H^+. Other internal messengers such as cAMP are undoubtedly involved in the control of ciliary beating, but perhaps as only third or fourth messengers for chemoresponse. The roles for methylation, phosphorylation, and amiloride-sensitive ion fluxes are being explored.

References

Adler J (1966) Effect of amino acids and oxygen on chemotaxis in *Escherichia coli.* J Bacteriol 92:121–129

Adler J (1969) Chemoreceptors in bacteria. Science 166:1588–1597

Adler J, Hazelbauer G, Dahl M (1973) Chemotaxis toward sugars in *Escherichia coli.* J Bacteriol 115:824–847

Antipa G, Martin K, Rintz M (1983) A note on the possible ecological significance of chemotaxis in certain ciliated protozoa. J Protozool 30:55–58

Berg H (1986) Random walks in biology. Princeton, NJ: Princeton University Press

Bonini NM, Gustin MC, Nelson DL (1986) Regulation of ciliary motility by membrane potential in *Paramecium:* a role for cyclic AMP. Cell Motil Cytoskel 6:256–272

Boyd A, Simon M (1982) Bacterial chemotaxis. Annu Rev Physiol 44:501–517

DiNallo M, Wohlford M, Van Houten J (1982) Mutants of *Paramecium* defective in chemoreception of folate. Genetics 102:149–158

Doughty M (1979) Control of ciliary activity in *Paramecium.* III. Evidence for specific membrane binding sites for ions and cholinergic ligands. Comp Biochem Physiol 63:183–197

Doughty M, Dodd GH (1978) Chemical modification of the excitable membrane in *Paramecium aurelia:* effect of a crosslinking reagent. Comp Biochem Physiol 59:21–31
Dryl S (1958) Photographic registration of movement of protozoa. Bull Acad Polonaise Sci 6:429–431
Dryl S (1959a) Effects of adaptation to environment on chemotaxis of *Paramecium caudatum.* Acta Biol Exp 19:83–93
Dryl S (1959b) Chemotactic and toxic effects of lower alcohols on *Paramecium caudatum.* Acta Biol Exp 19:95–104
Dryl S (1961a) The velocity of forward movement of *Paramecium caudatum* in relation to pH of medium. Bull Acad Pol Sci Ser Sci Biol 6:71–74
Dryl S (1961b) Chemotaxis in *Paramecium caudatum* as adaptive response of organism to its environment. Acta Biol Exp 21:75–83
Dryl S (1973) Chemotaxis in ciliated protozoa. In: Perez-Miravete A (ed) Behaviour of micro-organisms. Plenum, New York, pp 16–30
Dryl S, Grebecki A (1966) Progress in the study of excitation and response in ciliates. Protoplasma 62:255–284
Dusenbury D (1973) Countercurrent separation: a new method for studying behavior of small aquatic organisms. Proc Natl Acad Sci USA 70:1349–1352
Eisenbach L, Ramanathan R, Nelson DL (1983) Biochemical studies of the excitable membrane of *Paramecium tetraurelia.* IX. Antibodies against ciliary membrane proteins. J Cell Biol 97:1412–1420
Ferguson ML (1957) Photographic technique for quantitative physiological studies of *Paramecium* and other motile cells. Physiol Zool 30:208–215
Fraenkel GS, Gunn DL (1961) Orientation of animals. Dover, New York
Haley SE (1975) Photoaffinity labeling of adenosine 3′,5′-cyclic monophosphate binding sites on human red cell membranes. Biochem 14:3852–3857
Hazelbauer GL, Harayama S (1983) Sensory transduction in bacterial chemotaxis. Int Rev Cytol 81:33–70
Hennessey T, Kung C (1984) An anticalmodulin drug, W-7, inhibits the voltage-dependent calcium current in *Paramecium caudatum.* J Exp Biol 110:169–181
Jennings HS (1906) Behavior of the lower organisms. Columbia Univ Press, New York
Kaczorowski G, Barros F, Dethmers J, Trimble M, Cragoe E (1985) Inhibition of Na^+/Ca^{2+} exchange in pituitary plasma membrane vesicles by analogues of amiloride. Biochemistry 24:1394–1403
Klein C, Lubs-Haukeness J, Simmons S (1985) Cyclic AMP induced rapid and reversible modification of the chemotactic receptor in *D. discoideum.* J Biol Chem 260:715–720
Klein C, Sadeghi H, Simmons S (1986) Immunological analysis of the chemotactic receptor of *Dictyostelium discoideum.* Identification of cDNA clones. J Biol Chem 261:15192–15196
Klein P, Theibert A, Fontana D, Devreotes P (1985) Identification and cAMP induced modification of the cAMP receptor in *Dictyostelium discoideum.* J Biol Chem 260:1757–1764
Kung C, Saimi Y (1982) The physiological basis of taxes in *Paramecium.* Annu Rev Physiol 44:519–534
Kung C, Chang S-Y, Satow Y, Van Houten J, Hansma H (1975) Genetic dissection of behavior in *Paramecium.* Science 188:898–904
Levandowsky M, Hauser DCR (1978) Chemosensory responses of swimming algae and protozoa. Int Rev Cytol 53:145–210
Machemer H (1976) Interactions of membrane potential and cations in regulation of ciliary activity in Paramecium. J Exp Biol 65:427–448
Machemer H, de Peyer JE (1977) Swimming sensory cells: electrical membrane parameters, receptor properties and motor control in ciliated protozoa. Verh Dtsch Zool Ges Erlangen 1977:86–110
Machemer H, de Peyer JE (1982) Analysis of ciliary beating frequency under voltage clamp control of the membrane. Cell Motil Suppl 1:205–210
Majima T, Hamasaki T, Arai T (1986) Increase in cellular cyclic GMP level by potassium stimulation and its relation to ciliary orientation in *Paramecium.* Experientia 42:62–64
Nakatani I (1968) Chemotactic response of *Paramecium caudatum.* J Fac Sci Hokkaido Univ Ser 6 Zool 16:553–563

Nakatani I (1970) Effects of various chemicals on the behaviour of *Paramecium caudatum.* J Fac Sci Hokkaido Univ Ser 6 Zool 17:401–410
Ordal GW (1985) Bacterial chemotaxis: biochemistry of behavior in a single cell. CRC Crit Rev 64:938–983
Preston RR (1983) Studies on the responses of *Paramecium tetraurelia* to amino acids. PhD Thesis, Univ Nottingham
Preston RR (1984) A specific L-^{3}H-glutamic acid binding site on cilia isolated from *Paramecium tetraurelia.* J Cell Biol 99:187a
Preston RR, Van Houten J (1987a) Chemoreception in Paramecium: Acetate and folate-induced membrane hyperpolarization. J Comp Physiol 160:525–536
Preston RR, Van Houten J (1987b) Localization of chemoreceptive properties of the surface of *Paramecium.* J Comp Physiol 160:537–541
Preston RR, Schulz S, Van Houten J (1987) Pharmacological studies of the folate chemoresponse of *Paramecium.* J Comp Physiol (submitted)
Sasner JM, Schulz S, Van Houten J (1986) Anti-ligand antibody as a probe for a chemoreceptor in *Paramecium.* J Cell Biol 103:212a
Satow Y, Kung C (1981) A possible reduction of surface charge by a mutation in *Paramecium tetraurelia.* J Membr Biol 59:179–190
Schultz J, Pohl T, Klumpp S (1986) Voltage-gated Ca^{2+} entry into *Paramecium* linked to intraciliary increase in cyclic GMP. Nature (London) 322:271–273
Schulz S, Denaro M, Van Houten J (1982) Chemoattraction to folic acid and cAMP in *P. tetraurelia.* J Cell Biol 95:249a
Schulz S, Denaro M, Xypolyta-Bullock A, Van Houten J (1984) Relationship of folate binding and uptake to chemoreception in *Paramecium.* J Comp Physiol 155:113–119
Schulz S, Sasner JM, Van Houten J (1985a) In search of the folate chemoreceptor. J Cell Biol 10:301a
Schulz S, Preston R, Van Houten J (1985b) Characterization of putative *Paramecium* chemoreceptors. Chem Senses 10:466
Schulz S, Sasner JM, Van Houten J (1987) Progress on the identification of the folate chemoreceptor. In: Atema J (ed) From reception to perception. Ann NY Acad Sci (in press)
Smith R, Preston RR, Schulz S, Van Houten J (1987) Correlation of cyclic adenosine monophosphate binding and chemoreception in *Paramecium.* Biochim Biophys Acta 928:171–178
Van Houten J (1976) Chemoreception in *Paramecium:* A genetic approach. Ph D Diss, Univ Cal, Santa Barbara
Van Houten J (1977) A mutant of *Paramecium* defective in chemotaxis. Science 198:746–749
Van Houten J (1978) Two mechanisms of chemotaxis in *Paramecium.* J Comp Physiol A 127:167–174
Van Houten J (1979) Membrane potential changes during chemokinesis in *Paramecium.* Science 204:1100–1103
Van Houten J (1980) Chemoreception in *Paramecium:* the role of membrane potential. In: Starre H van der (ed) Olfaction and taste VII. IRL, London, pp 53–56
Van Houten J, Preston RR (1985) Effect of amiloride on *Paramecium* chemoreception. Chem Senses 10:466
Van Houten J, Preston RR (1987a) Chemoreception: *Paramecium* as a receptor cell. In: Ehrlich Y (ed) Advances in experimental medicine and biology. Plenum, New York, pp 375–384
Van Houten J, Preston R (1987b) Chemoreception in unicellular organisms. In: Silver W, Finger T (eds) Neurobiology of taste and smell. John Wiley & Sons, New York, pp 11–38
Van Houten J, Preston R (1987c) Eukaryotic unicells: how useful in studying chemoreception? In: Atema J (ed) From reception to perception. Ann NY Acad Sci (in press)
Van Houten J, Hansma H, Kung C (1975) Two quantitative assays for chemotaxis in *Paramecium.* J Comp Physiol A 104:211–223
Van Houten J, Hauser DCR, Levandowsky M (1981) Chemosensory behavior in Protozoa. In: Levandowsky M, Hutner SH (eds) Biochemistry and physiology of protozoa, vol 4. Academic Press, London New York, pp 67–124

Van Houten J, Martel E, Kasch T (1982) Kinetic analysis of chemokinesis of *Paramecium.* J Protozool 29:226–230

Van Houten J, Preston RR, Schulz S (1984a) Electrophysiological studies of chemoreception in *Paramecium.* Soc Neurosci Abstr 10:861a

Van Houten J, Wymer J, Cushman M, Preston RR (1984b) Effects of S-adenosyl-L-methionine on chemoreception in *P. tetraurelia.* J Cell Biol 99:242a

Van Houten J, Smith R, Wymer J, Palmer B, Denaro M (1985) Flurescein conjugated folate as an indicator of specific folate binding to *Paramecium.* J Protozool 32:613–616

Van Houten J, Preston RR, Gagnon M, Wright M (1987) Cyclic AMP and chemoreception in *Paramecium.* Chem Senses 12:(in press)

Van Houten JL, Van Houten J (1982) Computer analysis of *Paramecium* chemokinesis behavior. J Theor Biol 98:453–468

Van Houten JL, Preston RR, Schulz S, Sasner JM, Smith R (1986) Chemoreceptors of *Paramecium.* Soc Neurosci Abstr 12:1352

Zhuang Y-X, Cragoe E, Shaikewitz T, Glaser L, Cassel D (1984) Characterization of potent Na/H exchange inhibitors. Biochemistry 23:4481–4488

Chapter 19 The Lysosome System

AGNES K. FOK and RICHARD D. ALLEN[1]

CONTENTS

1 Introduction

By definition the term lysosome includes those organelles and vesicles in a cell which contain hydrolytic enzymes for digesting exogenous macromolecules (heterophagy) and endogenous macromolecules (autophagy). These organelles are detectable both at the light- and electron-microscope levels as they give a positive cytochemical reaction when the cells are incubated in medium containing a phosphate substrate and lead salt (Gomori 1952) or a medium containing hexazonium pararosaniline and α-naphthyl phosphate (Barka and Anderson 1962). Using these techniques Rosenbaum and Wittner (1962), Müller and Törö (1962), Esteve (1970), Karakashian and Karakashian (1973), and Fok et al. (1984b) have demonstrated the extreme heterogeneity of the lysosomes in *Paramecium.* Bodies ranging in size from tiny 70 nm diameter vesicles to vacuoles 14 μm or more in diameter contain acid phosphatase (AcPase) activity. De Duve (1964) classified lysosomes according to the content of the AcPase-containing compartments. Those which contain only hydrolytic enzymes without any substrate to act on are called primary lysosomes, while those containing both enzymes and substrates are called secondary lysosomes.

1 Pacific Biomedical Research Center and Department of Microbiology, University of Hawaii, Honolulu, HI, USA

Paramecium. Ed. by H.-D. Görtz

Paramecium is a heterotrophic organism and as such takes exogenous food particles into a phagosome (also called food or digestive vacuole). Phagosomes initially do not contain AcPase, so by definition they are not lysosomes. However, phagosomes eventually fuse with lysosomes and become phagolysosomes. But before this can happen, the phagosome must undergo internal changes, such as acidification, as well as surface changes which allow lysosomes to recognize and bind to the acidified digestive vacuole (DV).

This chapter deals with the subject of lysosomes in its broadest context. Besides lysosomes per se, phagosomes, acidosomes, the various stages of the DVs, DV formation, and defecation will be considered. Thus, sections will cover DV classification, the four processes of a digestive cycle, and the existence of a processing period. Finally, since membrane dynamics play a central role in the phagosomal-phagolysosomal cycle, a discussion of membrane replacement, flow, and recycling will complete our coverage of the lysosome system.

2 Phagosome Formation

2.1 *Membrane Morphology of the Cytopharynx*

In *Paramecium* a funnel-shaped laterally flattened invagination or opening called the vestibulum is located at the cell's mid-ventral position. In continuity with and set perpendicular to the axis of the channel through the vestibulum is the buccal cavity (gullet) where cilia and their basal bodies are arranged into three tightly packed membranelles known as the ventral peniculus, the dorsal peniculus, and the quadrulus. Covering the dorsal surface of the buccal cavity over its posterior half is the single membrane limited region comprised of the cytopharynx and the nascent phagosome or pouch (Allen 1974). (For a detailed morphological description of the oral region, see Chap. 2, this Vol.). Arrayed normal to the cytopharyngeal membrane are approximately 40 ribbons each consisting of 10 – 12 microtubules known as the cytopharyngeal ribbons. In addition, a similar number of microtubular pairs are found next to the cytopharyngeal membrane to the right of each ribbon. These ribbons and microtubular pairs appear to be held in position next to the cytopharynx by specialized microfilaments and bridges. Aligned along these cytopharyngeal ribbons are numerous flattened membrane-bound disks (0.2 – 0.5 μm in diameter and 60 nm in thickness) called the discoidal vesicles. These vesicles con-

▶

Fig. 1 A – F. Thin sections of membranes of the vesicles and vacuoles comprising the lysosomal system of *Paramecium*. The thickness of the gylcocalyx lining the vacuoles changes as the vesicle pools are added to the vacuole. See text for details. **A** Discoidal vesicles; **B** fusion of a discoidal vesicle (*DCV*) with the cytopharyngeal membrane (*arrow*); **C** acidosome (*AC*) fused with the phagosomal (*DV-I*) membrane; **D** secondary lysosomes bound to the membrane of acidified phagosomes(*DV-II*); **E** phagolysosomal membrane (*DV-III*) with thick glycocalyx and membrane of closely apposed endoplasmic reticulum (*er*); **F** membrane of late phagolysosome (*DV-IV*) showing reduction of glycocalyx and tubular formations (Allen and Fok 1984 a). All × 95000

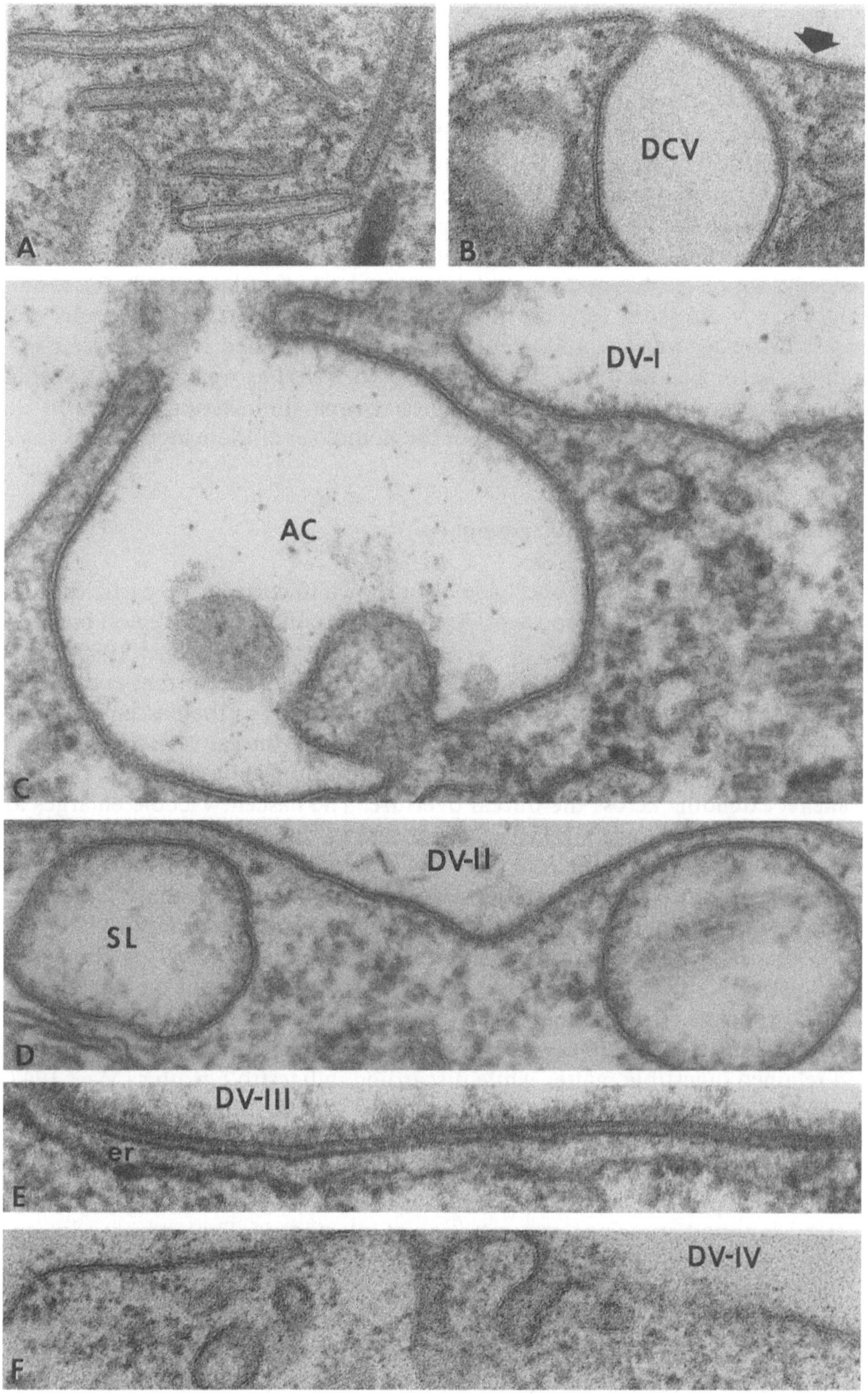
A
B
DCV
C
AC
DV-I
D
SL
DV-II
E
er
DV-III
F
DV-IV

tain a relatively prominent glycocalyx, 10 nm thick, lining their luminal surface (Fig. 1A). During phagosome formation these disks fuse with the cytopharyngeal membrane (Fig. 1B), to provide membrane for the growth of the nascent phagosome. This finding has been supported by morphological observations of vesicle fusion (Allen 1974; Schneider 1964; Yusa and Shigenaka 1966) and by freeze-fracture evidence. The latter shows similar numbers and distributions of intramembrane particles (IMPs) on the two fracture faces of the discoidal vesicles (Fig. 2A), the nascent phagosome, and the cytopharynx (Fig. 2B) (Allen and Staehelin 1981) which strongly suggests a homology of these membranes. Recently, we have obtained a monoclonal antibody that labels only the membranes of the discoidal vesicles, the cytopharynx, the nascent phagosome, and one or two early phagosomes. Labeling was accomplished by an indirect immunofluorescence assay (Fig. 3A,B) (Fok et al. 1986) as well as colloidal gold on LR Gold-embedded (Fig. 3C) and on frozen thin sections (Allen et al. 1986). This gives additional support for the homology of these membranes.

2.2 *The Role of Particles in DV Formation*

Paramecia are filter feeders and have been shown to concentrate particles from the medium. It has been proposed that the cilia around the oral region propel a current of water into the buccal cavity (e.g., Mast 1947; Fenchel 1980). Suspended particles are brought into the buccal cavity with the current and moved parallel to the long axis of the peniculi and quadrulus to the nascent vacuole. Particles circling around and moving forward in the buccal cavity are filtered and retained in the buccal cavity by a curtain of beating cilia, the endoral membranelle, extending across the buccal overture. The nascent vacuole enlarges as particles are concentrated in the posterior end of the buccal cavity and as the vacuole membrane grows by the fusion of discoidal vesicles with the cytopharyngeal membrane. When the nascent vacuole becomes sufficiently large, it is released as a DV which rapidly moves away from the cytopharynx.

A filtering or sieving mechanism according to Fenchel (1980) is characterized by a constant feeding rate and a lack of selection (except on the basis of physical properties) from among the various types of particles available in the medium. Earlier literature on paramecia suggested that these ciliates preferred and retained digestible matter while discarding indigestible matter (Mast 1947; Rosenbaum and Wittner 1962). Müller et al. (1965) could not confirm this observation, but found that both *Paramecium* and *Tetrahymena* take in both digestible and indigestible materials indiscriminately. Regarding the other criterion, in our work (Sison et al. 1985) the rate of DV formation was constant over a 2-h period at a given concentration of latex beads. Thus, paramecia conform to Fenchel's criteria for being filter feeders.

The question of whether solid particles are necessary for digestive vacuole formation has been debated. Studies on *Tetrahymena* (Müller et al. 1965; Nilsson 1979; Rasmussen and Modeweg-Hansen 1973) seem to indicate that particles are needed for DV formation. In *Paramecium* the rate of DV formation was proportional to the particle concentration when fluorescent latex beads,

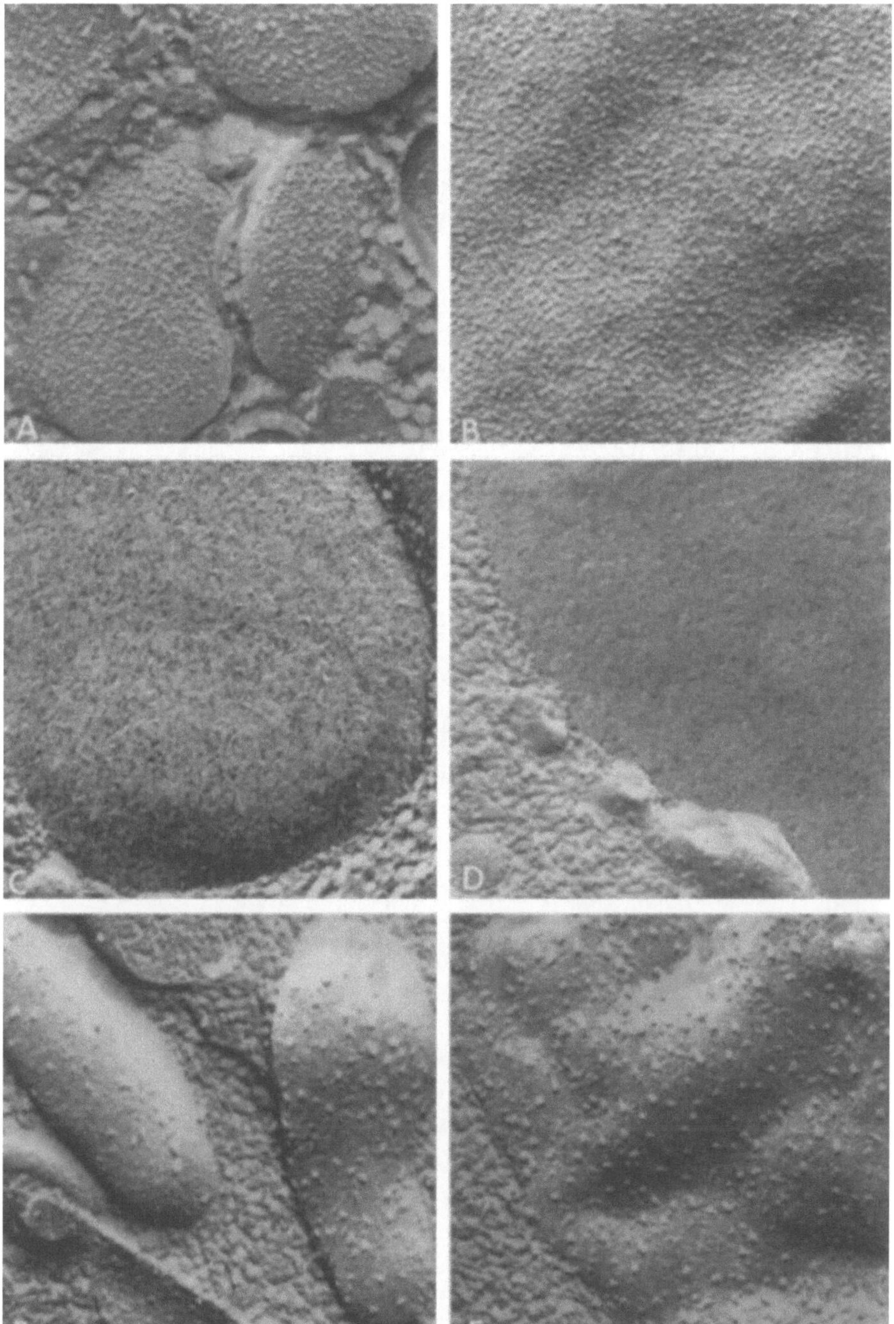

Fig. 2A–F. Freeze-fracture views of E-fracture faces of vesicles and vacuoles of the lysosomal system. Changes in vacuole membrane intramembrane particle numbers are due to the sequential fusion of different vesicle pools. **A** Discoidal vesicles; **B** cytopharynx membrane; **C** acidosome; **D** acidified phagosome (DV-II); **E** lysosomes; **F** phagolysosome (DV-III) membrane (Allen 1984). All× 94000

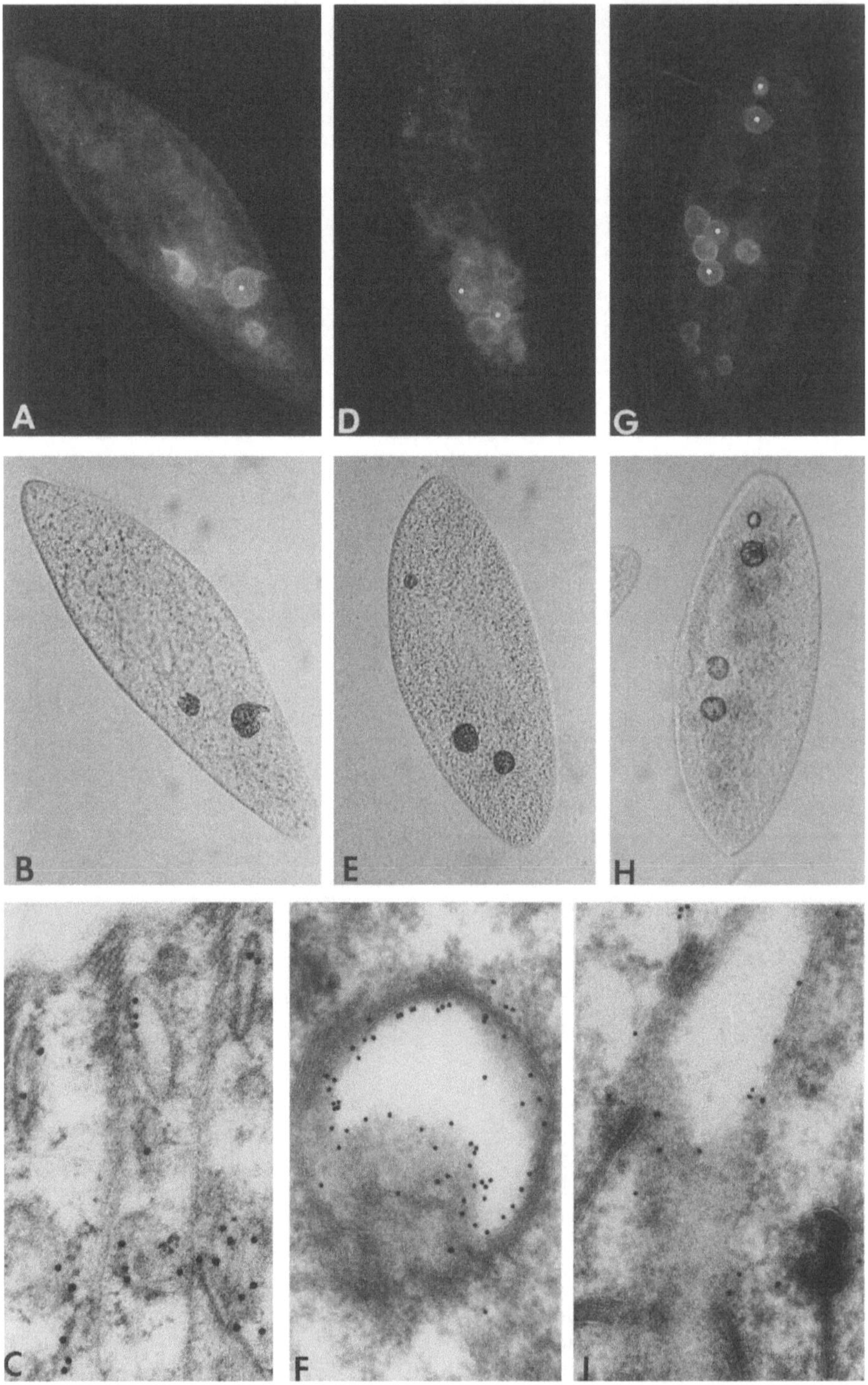
A
D
G
B
E
H
C
F
I

◀ **Fig. 3A–I.** Monoclonal antibodies (MAbs) to three stages of the phagosomal-phagolysosomal membrane system. These confirm the membrane changes seen in thin sections and freeze-fracture studies. Epitopes to MAb 132F4G6 are found on the discoidal vesicles, cytopharynx, and on membranes of vacuoles that are 1-min-old. **A** FITC; **B** same cell as **A** showing latex-labeled vacuole 1-min-old in bright field view. **C** Immunogold-labeled discoidal vesicles of LR Gold-embedded cell. Epitopes to MAbs 88B10E7 and 316C9F9 are found on the membrane of 5-min-old vacuoles and on acidosomes. **D** FITC, MAb 316C9F9; **E** same cell as **D** showing latex-labeled vacuoles 3- to 5-min-old; **F** immunogold-labeled acidosome, MAb 88B10E7; cryosectioned. Epitopes to MAbs 69D7 and 189D1G12 are found on the membrane of 20-min-old vacuoles. **G** FITC, MAb 189D1G12; **H** same cell as **G** showing latex-labeled vacuoles 20- to 22-min-old; **I** immunogold-labeled DV-III membrane and retrieved lysosomal membrane, MAb 69D7, cryosectioned. *White dots* indicate latex-labeled vacuoles that react with MAbs (Fok et al. 1986). Light and fluorescent micrographs all × 500; electron micrographs all × 57000 and previously unpublished

0.26 μm in diameter, were present at concentrations above 10^7 beads ml^{-1} (Fig. 4; Sison et al. 1985). Below this concentration some DVs labeled with one or two beads could be seen in cells continuously pulsed for 2 h. However, accurate counts were very difficult, since the vacuole margins could not be detected. Wasik (1983) and Wasik et al. (1984) also showed that in *P. bursaria* the rates of DV formation were also proportional to particle number above 10^5 particles ml^{-1} carmine or 10^6 barium chromate. Below these concentrations the rate of DV formation was very low or undetectable.

The preceding demonstrates that vacuole formation is affected by solid particles in the medium, but does not answer the question of whether particles are required for vacuole formation. As suggested above one of the most sensitive tests for studying vacuole formation is the use of fluorescent markers. Such markers can be fixed to particles or can be present in the solution. To determine

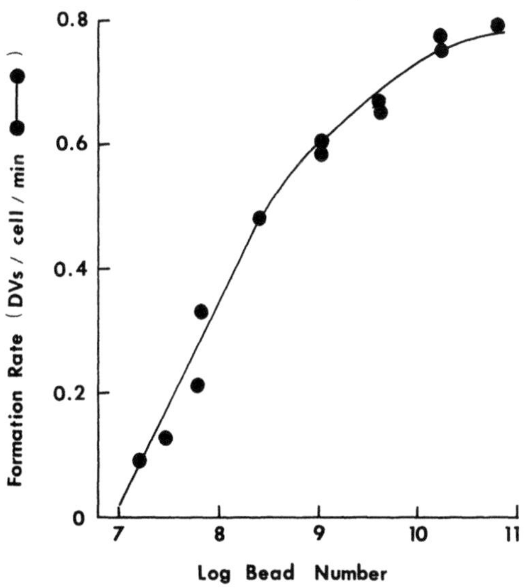

Fig. 4. The rates of DV formation were proportional to bead number above a concentration of 10^7 ml^{-1}. These rates were obtained from cells pulsed continuously with fluorescent latex beads (0.26 μm in diameter) for 30 min. (Data from two experiments)

if particles are required we added soluble markers such as bovine serum albumin labeled with fluorescein isothiocyanate (FITC) and FITC-immunoglobulins which were filtered through a 0.22 μm Millipore filter to ensure the absence of particles above this value. These markers were added to filtered axenic medium lacking phosphatidylethanolamine which forms an emulsion. Cells added to the markers were also washed thoroughly with the above filtered medium. When these precautions were taken, DVs were still formed but the rates of formation were much lower than when latex beads were present (unpublished observations). When lucifer yellow, a soluble low molecular weight fluorescent dye, was added to the filtered medium labeled DVs were formed in both *P. multimicronucleatum* (see Footnote in Chap. 2, this Vol.) and *P. tetraurelia*, but the formation rates were very low and erratic (unpublished observations). Thus, we can conclude that although it is possible for vacuoles to form in the absence of soluble markers or solid particles, as previously reported by Mast (1947), the presence of solid particles greatly enhances DV formation.

How particles stimulate DV formation has yet to be determined. It is reasonable to suspect that mechanical stimulation by the particles at the level of the beating cilia or as the particles make contact with the membranes in the cytopharynx-nascent vacuole region may stimulate DV formation. A concentration of particles in the posterior end of the buccal cavity may provide additional stimulation for vacuole growth. Localized stimulation of the oral region may open ion channels in the cytopharyngeal membrane which then sets in motion events leading to movement of discoidal vesicles toward and fusion of these vesicles with the cytopharyngeal membrane.

2.3 Phagosome Release

So far we have discussed the first of the two steps in vacuole formation, the growth of the vacuole membrane and the role of particles in bringing about the formation and enlargement of the nascent DV. The second step brings about the release of the nascent DV from the cytopharynx. Available evidence suggests that microfilaments may be involved in this step (Allen and Fok 1985; Fok et al. 1985a). Cytochalasin B (CB), a drug that prevents actin polymerization, inhibits vacuole release. Normally, DVs can form and be released as early as 15 s following a pulse in latex beads. However, in the presence of CB, a DV is not released for at least 10 to 15 min. During part of this time the vacuole continues to grow so discoidal vesicle fusion (step one) is not immediately inhibited by the CB treatment. An abnormally large nascent DV is formed and remains attached to the cytopharynx when cells are exposed to a combination of CB and particulate markers (Tolloczko 1977; Cohen et al. 1984b; Allen and Fok 1985; Fok et al. 1985a). A similar effect has been reported for *Tetrahymena* (Nilsson 1977).

Actin microfilaments have been localized in thin sections around the nascent vacuole by decorating the microfilaments with heavy meromyosin (HMM) in Triton X-100 extracted cells and by fluorescein-labeled HMM at the light microscope level (Cohen et al. 1984a, b). Kersken et al. (1986) used rhodamine-labeled phalloidin to show actin around the nascent vacuole. In our study in CB-treated cells large clumps of microfilamentlike material can be seen near the DV-I. These clumps are surrounded by discoidal vesicles (Allen and Fok 1985). Exactly how this actin is involved in pinching off the large nascent vacuole has yet to be determined. No special contactile ring around the opening to the nascent vacuole has been observed in electron micrographs and the role of the cytostomal cord which is found adjacent to the cytopharynx is still a mystery. The cytostomal cord does not appear to be made up of actin filaments (Kersken et al. 1986).

3 Phagosome Classification

After a DV is released from the cytopharyngeal membrane, it undergoes a period of maturation before it becomes defecation-competent and is defecated. During this period, known as the processing period, DVs can be grouped into stages (Fig. 5) based on their age, size, thin-section and freeze-fracture morphology, phagosomal pH, and the presence or absence of AcPase reaction product (Müller 1962; Müller and Törö 1962; Müller et al. 1963; Allen and Staehelin 1981; Fok et al. 1982, 1984b; Allen and Fok 1984a). These stages were also recognized using a wheat germ agglutinin-gold preparation on Lowicryl K4M sec-

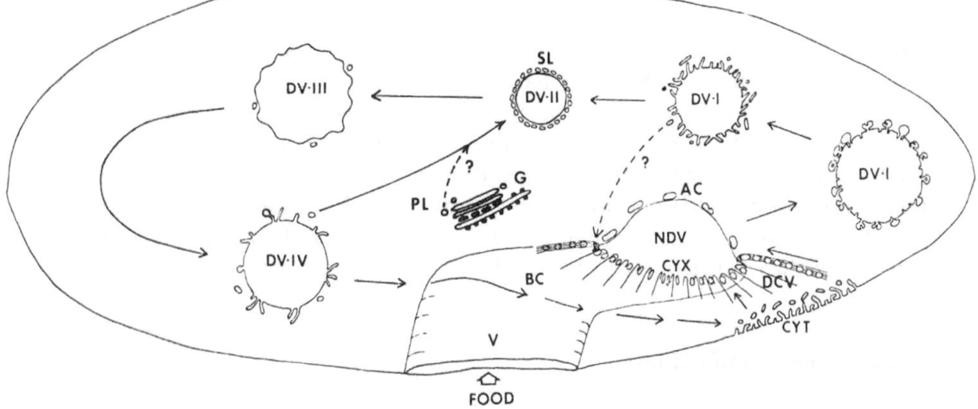

Fig. 5. The movements of vacuoles in *Paramecium* and the flow of membrane in the lysosomal system. Discoidal vesicles (*DCV*) fuse with the cytopharynx (*CYX*) supplying membrane for phagosome (*NDV*) growth. Acidosomes (*AC*) accumulate around the *NDV* and fuse with the separated phagosome (*DV-I*). Membrane of the *DV-I* is removed and possibly recycled to the cytopharynx while being replaced by acidosome membrane. Secondary lysosomes (*SL*) bind to the condensed and acidified phagosome (*DV-II*) and then fuse to form the phagolysosome (*DV-III*). Following a period of digestion lysosome membrane is retrieved and recycled to another *DV-II;* on the way the retrieved lysosome may be recharged with hydrolases by fusing with primary lysosomes (*PL*). Defecation-competent vacuoles (*DV-IV*) fuse with the plasma membrane of the cytoproct (*CYT*) where the spent vacuole membrane is retrieved and recycled to the cytopharynx. *BC* buccal cavity; *G* Golgi stack; *V* vestibulum (After Allen and Staehelin 1981)

tions of *P. tetraurelia* growing in bacterized cultures based on the degree of bacterial digestion (Lüthe and Plattner 1986).

Digestive vacuoles of stage I (DV-I) have no AcPase activity, are condensing, becoming acidic, and are 0 to ≦6-min-old. They are located in the posterior end of the cell. At first they have membranes resembling discoidal vesicles but this changes as they become DV-II. The DV-II are the most condensed, have a pH of ~3, and have no AcPase as measured cytochemically. They are from 4 to 10 min in age and their membrane ultrastructure is similar to those of the acidosomes (cf. Fig. 1C with 1D and Fig. 2C with 2D). DV-II are located near the oral region and around the macronucleus (Allen and Fok 1984a). The DV-III are expanded DVs 8- to 20-min-old whose membrane resembles that of the lysosomes (cf. Fig. 1D with 1E and Fig. 2E with 2F) and whose phagosomal pH increases from 3 to 7. The DV-III are the only DVs that contain active AcPase and in which digestion takes place (Fok et al. 1984b). They are generally located in the anterior half of the cell. Active retrieval of lysosomal membrane along with AcPase begins in late DV-III (Allen and Fok 1984b). The DV-IV stage, the last stage before defecation, is marked by the lack of AcPase. Active retrieval of lysosomal membrane may continue into this stage (Fig. 1F), but active AcPase is not present. These DVs, which are ≧20-min-old, are defecation-competent and are generally located on the ventral half of the cell as they move toward the cytoproct. Finally, because of their characteristic morphologies these four stages can be recognized in thin sections and freeze-fractured preparations even when the ages of DVs are not marked (Allen and Fok 1984a).

Cohen et al. (1984b) have divided the phagocytic sequence into eight steps based on observable events in the light microscope. These events included (1) the formation and growth of the nascent vacuole, (2) closing of the cytopharyngeal pouch, (3) rapid movement of the vacuole along the postoral fibers, (4) short random movements of the vacuole leading to vacuole release from the postoral fibers, (5) formation of spikes during membrane replacement, (6) fusion with lysosomes, (7) digestion and transit through the cell, and (8) egestion. Both classification systems are valid since one system classifies stages of the phagosomes, while the other classifies the dynamic phagocytic events.

4 Phagosome Acidification

As a nascent DV grows in size, a population of large vesicles binds to its surface. These vesicles are irregularly-shaped electron translucent vesicles ranging from a small diameter to over 1 μm in diameter. In thin sections their membrane is similar in thickness and general appearance to the plasma membrane, but their luminal surface does not bear the organized glycocalyx seen in the discoidal vesicles (Allen and Fok 1983a) (cf. Fig. 1A, C). These vesicles are devoid of AcPase activity, but since they accumulate weak bases such as neutral red and acridine orange, they must have an acidic content. They are therefore called acidosomes (Allen and Fok 1983c). Acidosomes have a distinctive pattern of IMP number and distribution. They have a smooth E-fracture face with only 50–100 IMP μm^{-2} (Fig. 2C) and a particulate P-fracture face of about 2500 IMP μm^{-2} (Allen and Fok 1983b). Microfilamentous networks can be readily seen in the regions between the acidosomes and the nascent phagosome and around acidosomes and very young DV-I. These microfilaments play an important part in the fusion of the acidosomes with the DV-I. When they are disrupted by CB, acidosome-DV fusion fails to take place and DVs fail to become acidified (Allen and Fok 1983c, 1985).

Immediately after a DV separates from the cytopharynx it rapidly moves in a straight line to the cell's posterior pole. This directed and rapid movement occurs along the postoral microtubular bundles which extend posteriorly from the ribbed wall. Actin is also temporarily associated with this pathway at the time of movement (Cohen et al. 1984b; Kersken et al. 1986). Arriving at the posterior pole of the cell the young DV-I soon fuses with its accompanying acidosomes. The pH of the DV decreases (Mast 1947; Fok et al. 1982, 1985b), this is probably responsible for killing the food organisms and denaturing proteins. The DV then begins to move anteriorly in a slower and more random fashion borne along by the process of cytoplasmic streaming (Kuznicki and Sikora 1972) which is not yet well defined mechanistically.

When cells were pulsed briefly with heat-killed yeast stained with bromcresol purple, bromcresol green, or bromphenol blue, the phagosomal pH of live cells declined from pH 7 to 3 within 5 min of formation (Fig. 6) (Fok et al. 1982). In the presence of CB, less than 10% of the labeled DVs become acidic and these become only mildly acidic (pH about 6), whereas over 70% of DVs in untreated cells become very acidic and reach a pH as low as ~3 during this 5-min period (Allen and Fok 1983c). This inhibition in DV acidification is related to a block by CB of the fusion of the acidosomes with the early DV-I. These results strongly suggest that the acidosomes are involved in phagosome acidification and that they contain the acidification mechanism which is transferred to the vacuole upon fusion.

To study DV acidification in *Paramecium*, rapid and sensitive cytochemical and biochemical assays were needed. In our lab both biochemical and cytochemical assays were developed. These assays are based on the acid pH-dependent and time-dependent inactivation of horseradish peroxidase (HRP) in which this enzyme is dissociated into an inactive protein and ferriprotoporphyrin (Fok et al. 1985b). Using these assays it was shown that the *P. aurelia* complex had a more rapid rate of acidification than did *P. multimicronucleatum*, while *Tetrahymena* had the lowest rate (Table 1). The biochemical assay could be used to estimate the phagosomal pH by comparing the residual HRP activity at the end of acidification, but before the start of degradation by lysosomal enzymes, with a standard HRP inactivation curve (Fig. 7). Due to the ease of applying these assays it became possible to explore the mechanism of phagosome acidification.

As a DV-I is being acidified, it undergoes a reduction in size as membrane tubules form from the vacuole membrane and separate into the cytosol (Allen and Staehelin 1981). The rate and the time course for DV condensation parallel those for acidification, so that DVs reach their smallest size and most acid pH by an age of about 5 min. These DVs then remain in this state for the next 3 to

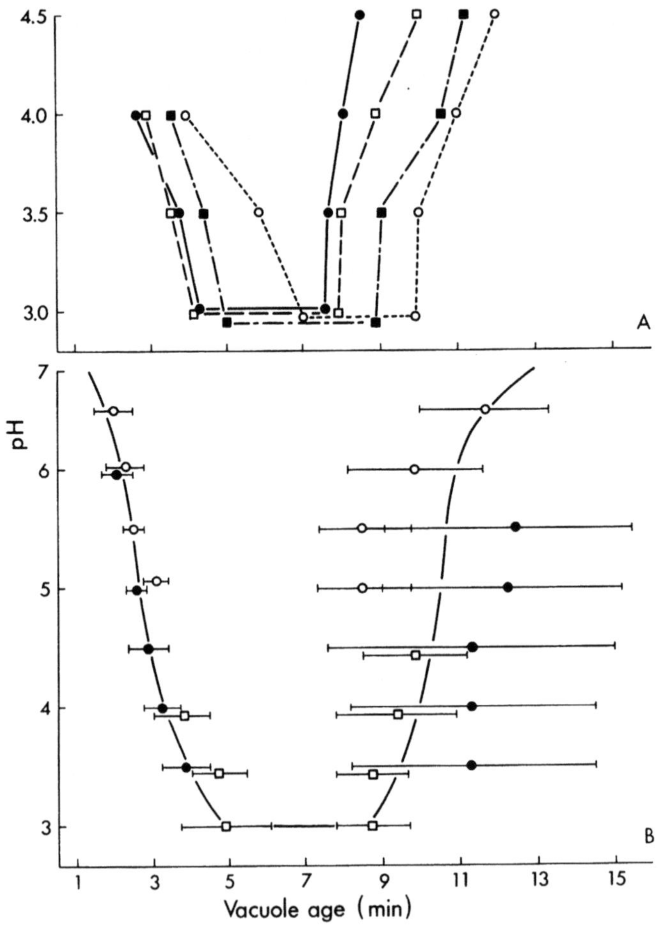

Fig. 6A,B. Change in pH of digestive vacuoles. **A** Change in pH of vacuoles marked with bromphenol blue-stained yeast. Each *symbol* represents a single vacuole whose change in pH was monitored continuously. **B** A composite of pH changes in vacuoles containing bromcresol purple (○—○), bromcresol green (●—●), and bromphenol blue (□—□). Each *point* represents the mean ± SD of 13–44 vacuoles (Fok et al. 1982)

5 min and are termed DV-II. The IMP particle number and distribution of these DV-II membranes are almost identical to those of the acidosomes (Fig. 2C,D), suggesting that at this stage the membrane of the acidosomes has completely replaced the membrane derived from the discoidal vesicles, the precursor membrane of the forming DV. We refer to this process as membrane replacement (Allen and Fok 1983b).

It must be pointed out that DVs up to now are still devoid of AcPase. However, at about this time a prominent layer of lysosomes can be seen lining the membrane of a DV-II (Allen and Fok 1984a).

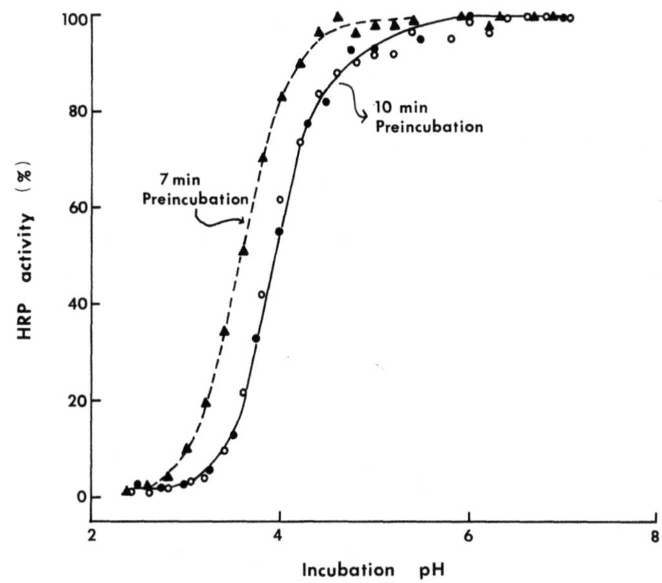

Fig. 7. Inactivation of HRP in citrate-phosphate buffer of varying pH. Aliquots of HRP (4 μg ml^{-1}) were preincubated in this buffer at varying pH for 7 min (▲) and 10 min (○, ●) at room temperature and were assayed at pH 3.5 using DAB as a hydrogen donor (●), or at pH 5.0 using o-dianisidine as a hydrogen donor (○, ▲). The residual HRP activities were expressed as % of the highest activity at pH 7.0. (After Fok et al. 1985 b)

Table 1. Cytochemical assay for acidification[a] (Fok et al. 1985 b)

Cell type	K_{ac} (min^{-1})	Processing period (min)	K_{def} (min^{-1})
P. multimicronucleatum	0.712 (0.992)	28.5	0.113 (0.998)
P. octaurelia	1.275 (0.998)	16.5	0.278 (0.998)
P. primaurelia	1.158 (0.998)	16.5	0.316 (0.995)
P. tetraurelia	1.159 (0.991)	16.5	0.274 (0.998)
T. thermophila	0.034 (0.978)	45.0	0.026 (0.993)

[a] Cells were chased with fresh culture medium after a 2.25-min pulse with fluorescent beads and HRP and fixed in 4% formalin in phosphate buffer at pH 7.0 at various time points. Cells were then incubated in the DAB-H_2O_2 medium to develop the phagosomal HRP activity. HRP-positive DVs from 100 to 300 cells were enumerated and expressed as an average DVs/cell. The K_{ac} (rate constants for acidification), K_{def} (rate constants for defecation), and correlation coefficients (enclosed in parentheses) were determined using regression analyses as described by Fok and Valin (1983). The processing period is the period when DVs are being processed and are not defecation-competent (Fok et al. 1982; Fok and Valin 1983).

5 Lysosomal Fusion and Degradation

Both primary and secondary lysosomes are present in *P. multimicronucleatum*. The primary lysosomes have an average diameter of about 75 nm (Fig. 8 A) and are scattered in the vicinity of the Golgi stacks and are often completely filled with reaction product when incubated in the Gomori medium. The secondary lysosomes are about ten times larger than the primary lysosomes and contain only scattered clumps of reaction product (Fig. 8 B) (Fok et al. 1984 b). These secondary lysosomes also have a distinct thin section (Fig. 1 D) (Fok and Allen 1979, 1981) and freeze-fracture morphology (Fig. 2 E) (Allen and Staehelin 1981). They support a 30-nm-thick fibrous glycocalyx lining their luminal membrane surface and contain paracrystalline inclusions in their lumens. In freeze fracture the E-fracture face contains very prominent IMPs. With the availability of an axenic medium for paramecia (Soldo et al. 1966) lysosomal enzymes in these cells can be studied. We have determined that both *P. multimicronucleatum* and *P. tetraurelia* have a complement of acid hydrolases (Fok and Paeste 1982) (Table 2) whose acid pH and temperature optima are quite similar to those of their mammalian counterparts. The arylsulfatase activities

Table 2. Hydrolytic activities of *Paramecium*[a] (Fok and Paeste 1982)

Enzyme (EC)	*P. multimicronucleatum*	*P. tetraurelia*
Nucleotidyl transferases		
DNase (3.1.4.5)	0.0093 (3)	0.00092 (1)
RNase (3.1.4.22)	0.127 (3)	0.026 (4)
Phosphoric monoester hydrolase		
Acid phosphatase (3.1.3.2)		
βGPase	11.0 (3)	4.40 (4)
pNPPase	6.12 (3)	2.49 (3)
Glycoside hydrolases		
β-Glucosidase (3.2.1.21)	3.890 (5)	0.890 (3)
N-acetyl-β-D-glucosaminidase (3.2.1.30)	0.750 (2)	0.267 (1)
β-Galactosidase (3.2.1.23)	0.460 (1)	0.126 (1)
α-Mannosidase (3.2.1.24)	0.194 (1)	0.050 (1)
α-Glucosidase (3.2.1.20)	0.157 (1)	0.049 (1)
β-D-fucosidase (3.2.1.38)	0.105 (1)	0.034 (1)
β-Xylosidase (3.2.1.37)	0.087 (1)	0.037 (1)
α-L-fucosidase (3.2.1.51)	0.113 (1)	0.028 (1)
β-Glucuronidase (3.2.1.31)	0.043 (2)	0.011 (1)
Peptidyl amino acid hydrolases		
Cathepsin B (3.4.22.1)	0.37 (2)	0.123 (2)
Cathepsin C (3.4.14.1)	0.42 (3)	0.06 (3)
Cathepsin D (3.4.23.5)	1.34 (1)	0.33 (1)
Cathepsin E (3.4.23.5)	0.27 (1)	–
Sulfuric ester hydrolases		
Aryl sulfatase A (3.1.6.1)	0 (2)	0 (2)
Aryl sulfatase B (3.1.6.1)	0 (2)	0 (2)

[a] Maximum activity during log phase of growth, expressed as pmoles of substrate hydrolyzed, of nucleotide released (for DNase and RNase), or of tyrosine released (cathepsin D)/min/cell. Numbers in parentheses represent numbers of cell cultures assayed.

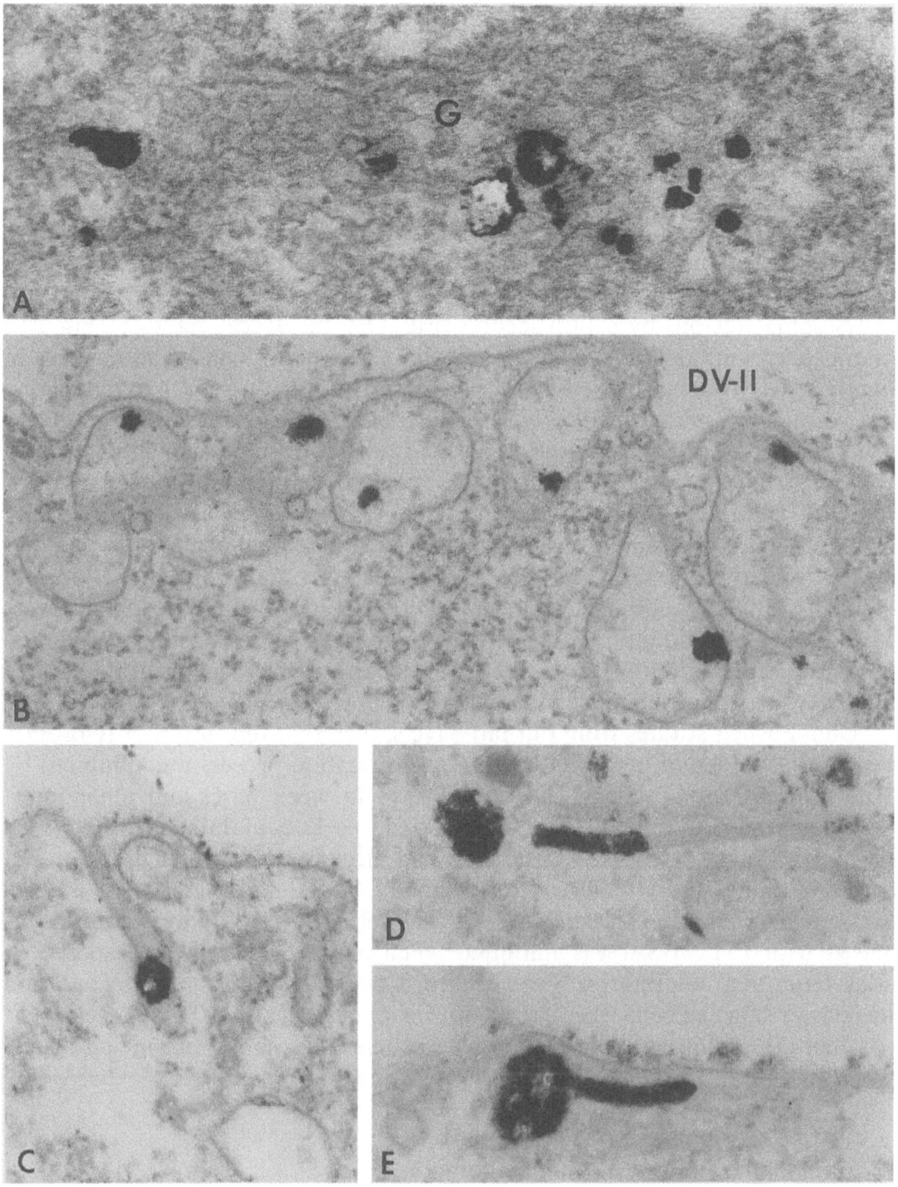

Fig. 8 A – E. Acid phosphatase (AcPase)-positive bodies. **A** Small primary lysosomes, about 70 nm diameter, lie near the trans-face of the Golgi stack (*G*) and are filled with AcPase-reaction product. **B** Larger secondary lysosomes with foci of AcPase-reaction product line the cytosolic side of a *DV-II* (Fok et al. 1984 b). **C – E** Lysosomal membrane is retrieved as tubules that expand at their distal ends. These tubules and distal expansions frequently contain AcPase-reaction product (Allen and Fok 1984 b). All pictures × 60 000

are very low, i.e., at an undetectable level. By far the most active glycosidase is β-glucosidase. Furthermore, it appears that there is only one AcPase which is capable of hydrolyzing a number of phosphate-containing compounds (Fok 1983).

The secondary lysosomes appear to preferentially bind to the DV-II (Fig. 1D) as very few lysosomes are associated with the DV-I and DV-III. This binding and the subsequent fusion could be inhibited by CB (unpublished observations). Under an unknown stimulus, these lysosomes fuse with the DV-II, thus advancing these DVs to the DV-III stage. The DV-III are generally between 8 to 20 min old. They are characterized by an increase in size as a result of lysosomal fusion, by a thin-section (Fig. 1E) and freeze-fracture (Fig. 2F) morphology similar to those of the secondary lysosomes and by an increase in vacuolar pH. The cause for the latter increase is not known but may be related to lysosomal digestion. In 1962 when Müller and his colleagues first reported a rise in vacuolar pH during digestion, this finding was received with considerable skepticism (Müller et al. 1963) and even today it remains an enigma. For lysosomal hydrolases to function at maximal efficiency, pH neutralization may be required after lysosomal fusion. This is because several of the lysosomal enzyme activities of paramecia drop precipitously at a pH below 4 (Fok and Paeste 1982). The continual increase in vacuolar pH during lysosomal degradation could presumably be due to the neutralizing effects of the degradative products. At this time we do not know how and if these neutralizations take place nor what the usual intralysosomal pH is in paramecia.

Unpublished studies from our lab have shown that the fusion of lysosomes with the DV-II could be inhibited if the acidification process was inhibited by exposing cells containing newly formed DVs to weak bases and ionophores. However, when these perturbants were added to cells containing acidified DVs, i.e., DVs 3-min-old whose acidification had already been initiated, fusion between lysosomes and DV-II occurred normally. Lysosomal degradation as measured by the increase in trichloroacetic acid-soluble FITC fluorescence in cells pulsed with FITC-BSA was inhibited when acidified DVs were exposed to weak bases and ionophores. Furthermore, a twofold inhibition of degradation was observed when cells were exposed prior to acidification. These results suggest that DV acidification is required for efficient lysosomal fusion and degradation.

6 Processing or Maturation Period

One of our recent findings on the DV cycle is the existence of a definite maturation or processing period in *Paramecium* (Fig. 9) (Fok et al. 1982, 1984a, 1985a; Fok and Valin 1983). This is also found in *Tetrahymena* (Table 1) (Ricketts 1971, 1983; Fok and Shockley 1985; Fok et al. 1985b). This period begins with the formation of the DV and ends when the DV becomes defecation-competent. Thus, a DV cycle can be divided into a processing and a defecation period. In view of the relatively close proximity of the cytopharynx to the cytoproct, it was suggested that newly formed DVs could migrate rapidly to the

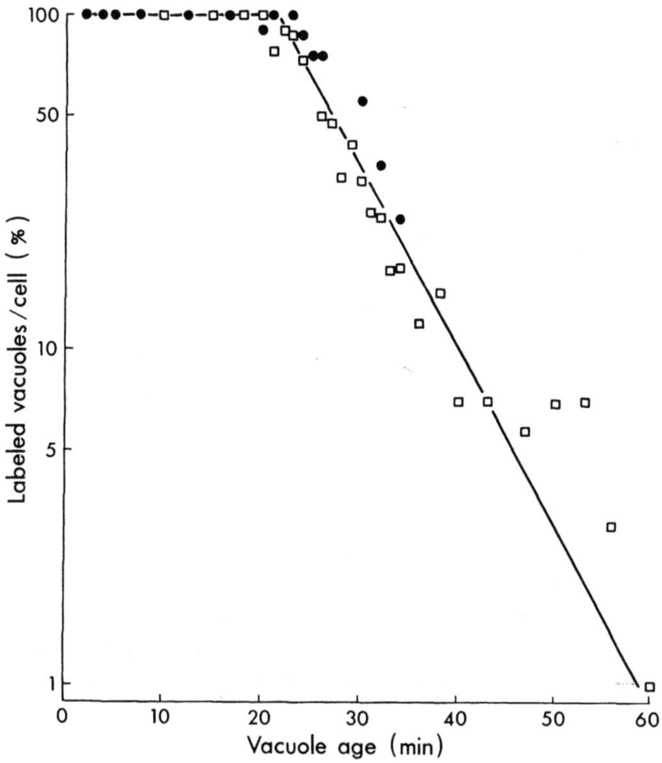

Fig. 9. The processing period of a digestive vacuole cycle in *Paramecium* is represented by the *horizontal portion* of the graph where the values for DVs/cell remain unchanged. Following the processing period is the defecation period when DVs are lost from the cells at an exponential rate. Number of DVs/cell was expressed as percent of the average number of DVs formed during a 1-min pulse with latex beads (Fok et al. 1982)

cytoproct and be defecated (Berger and Pollock 1981). However, the existence of a definite processing period suggests that this cannot occur under normal conditions. On the contrary, it suggests that the initiation of the defecation process is not random and that the DVs must first mature by passing through the steps of acidification, lysosomal fusion, and digestion before they become defecation-competent. Also, when either of these two steps are inhibited, the duration of this processing period will be greatly extended (Fok et al. 1985a).

The duration of the processing period appears to depend in part on the vacuolar content and can be as short as 15 min in several species of paramecia if polystyrene latex beads are used as a phagocytic stimulus. The use of fluorescent latex beads tends to extend the processing period to about 30 min in *P. multimicronucleatum*. When digestible markers such as *Escherichia coli*, yeast, or albumin were used, the processing period was extended to more than 40 min (unpublished observations). Furthermore, whether some special recognition factors appear on the membrane to indicate that the DVs are defecation-competent or whether defecation occurs as a result of cyclosis bringing only older DVs into close proximity to the cytoproct is not known at this time. However, since the processing period can be lengthened by changes in temperature (Fok et al. 1984a) or by adding weak bases and ionophores (unpublished observations), a more intricate control mechanism than previously suspected seems to be indicated.

7 Phagosome Defecation

Prior to the end of the processing period, as early as 15 min in age, small membrane tubules can be found radiating from the surface of the DVs. These tubules, which often contain AcPase (Fig. 8C–E), round out distally. They pinch off to give rise to retrieved secondary lysosomes which morphologically resemble the lysosomes around the DV-II, except that they are somewhat smaller in size and frequently retain a tubular tail (Allen and Fok 1984b). Following AcPase retrieval, the DVs are termed the DV-IV and are devoid of AcPase. The DV-IV have become defecation-competent, that is, they are capable of approaching and fusing with the plasma membrane at the apex of the cytoproct ridge. Following digestion and lysosome membrane retrieval, DVs migrate along the ventral half of the cell from the cell's anterior end to its posterior end where they approach the cytoproct. Microtubules extending out singly from the cytoproct ridge or in bundles from basal bodies adjacent to the cytoproct come

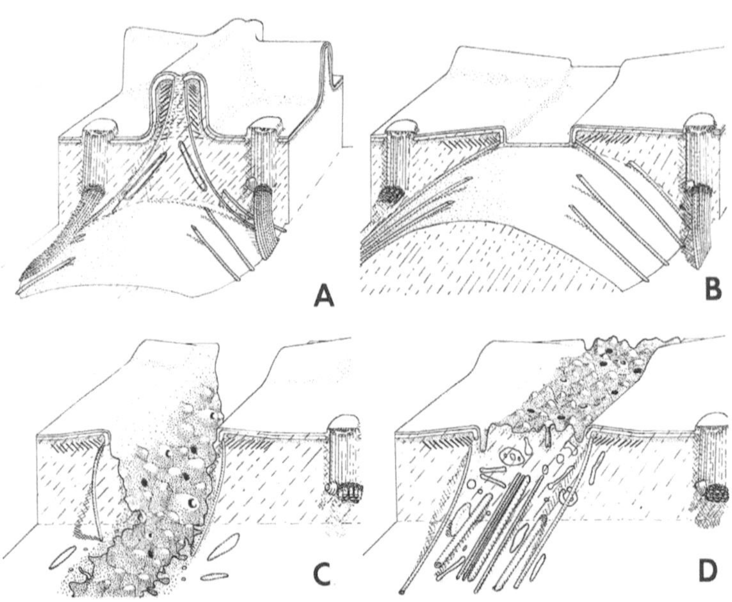

Fig. 10A–D. The opening and closing of the cytoproct. **A** Defecation-competent vacuoles are guided to the cytoproct along microtubules. **B** Cytoproct lips are pulled down against vacuole membrane. **C** Fusion between the vacuole and plasma membrane is followed by retrieval of spent vacuole membrane. **D** Cytoproct lips are brought back together as spent vacuole membrane is completely removed (Allen and Wolf 1974)

in contact with these defecation-competent DVs and direct them to the cytoproct (Fig. 10) (Allen and Wolf 1974). The two lips of the cytoproct ridge are pulled apart as a DV is pulled up to the cell surface. Close contact between the plasma membrane and DV membrane results in membrane fusion and the rapid opening of the cytoproct and release of the vacuole's residual contents.

Immediately upon opening of the cytoproct, spent DV membrane undergoes a profound remodeling into pits and tubules which are pinched off into the cytoplasm (Fig. 11 A). The contents of the DV seem to be passively expelled as the membrane is progressively retrieved. Complete membrane retrieval pulls the two lips of the cytoproct back together and the ridge is restored to its inactive configuration (Fig. 10) (Allen and Wolf 1974).

Passing near the cytoproct are long cytopharyngeal microtubular ribbons. The membrane vesicles and tubules retrieved from the spent DV at the open cytoproct bind to these ribbons (Allen 1975; Allen and Fok 1980). These tubules and vesicles subsequently undergo a modification into a discoidal shape. When defecation takes place in culture medium containing HRP, the tubules retrieved from the spent vacuole membrane will contain HRP-reaction product upon incubation in a medium containing diaminobenzidine and H_2O_2 (Fig. 11 B). HRP-labeled vesicles can then be found along the microtubular ribbons in the cytopharynx-cytoproct corridor where they are moved back to the cytopharynx to be reused for a new round of phagosome formation (Fig. 11 C) (Allen and Fok 1980).

By using DVs pulse labeled with latex beads (Fok et al. 1982, 1984a, 1985a,b, 1986) or with watercolor (Berger and Pollock 1981), the rate of defecation for paramecia (Fok et al. 1985b) and *Tetrahymena* (Fok and Shockley 1985) has been shown to follow an exponential pattern. The rate constant for defecation for the *P. aurelia* complex was almost three fold faster than that for *P. multimicronucleatum* (Table 1).

8 Membrane Recycling and Replacement

Freeze-fracture work (Allen and Staehelin 1981) supports the above interpretation of membrane flow, membrane modifications (outlined in the above sections), and membrane recycling (Fig. 2). In ultrastructure discoidal vesicles are similar to the cytopharyngeal membrane and to the membranes of the nascent and newly formed phagosomes. In striking contrast to these membranes which bear a very large number of IMPs on their E-fracture faces, the most condensed and most acid DV-II as well as the acidosomes contain almost no IMPs on their E-fracture faces (Allen and Staehelin 1981; Allen and Fok 1983b). For these phagosomes this transition has occurred in less than 3 min. Finally, the phagolysosome (DV-III) membrane resembles the lysosomal membranes which bear a moderate number of prominent IMPs on their E-fracture faces. These observations and transitions are understandable if the initial DV is formed from discoidal vesicle membrane which is subsequently removed from the phagosome at the same time that the acidosome membrane is being added to the DV-I. This removal of the discoidal-derived membrane concomitant with

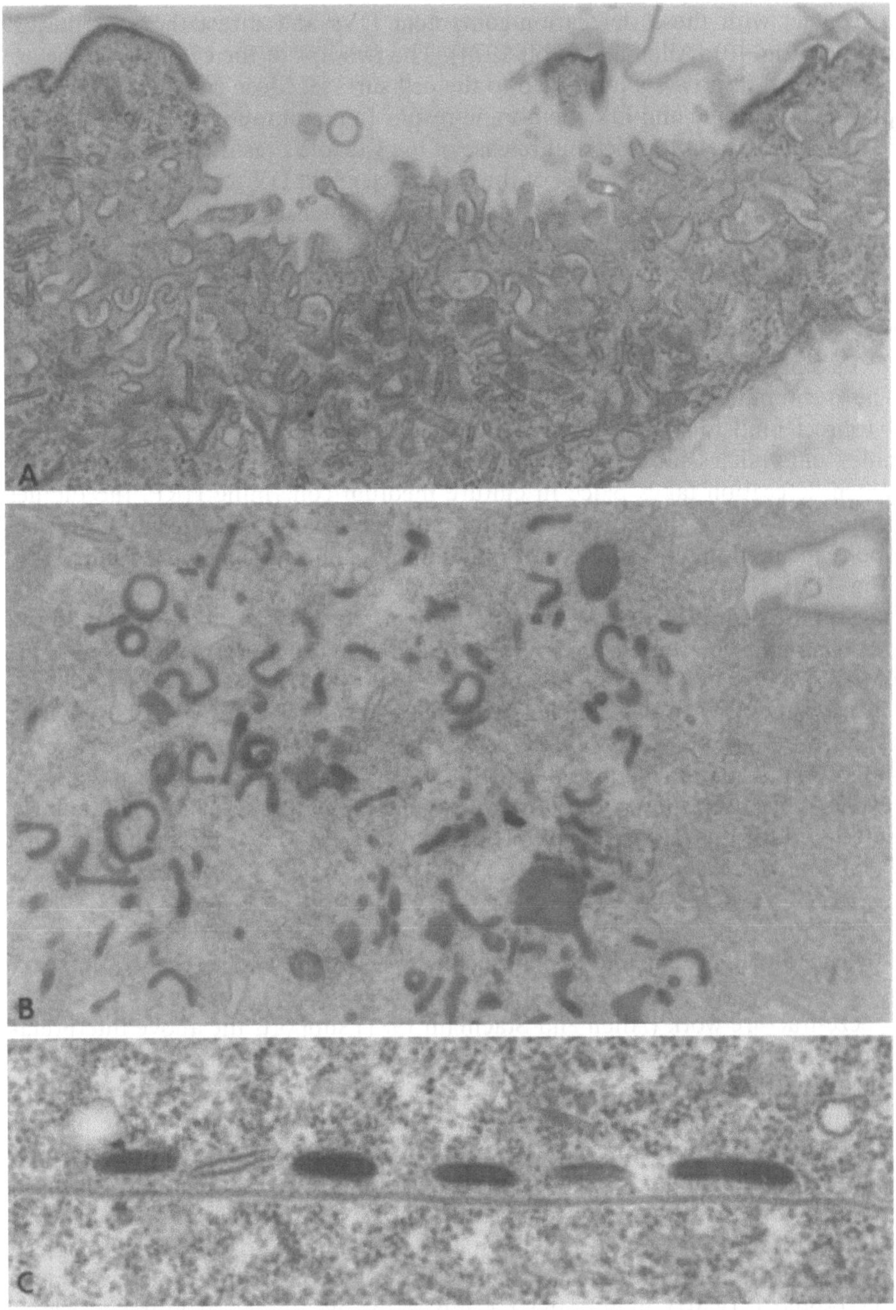

Fig. 11 A–C. Retrieval of spent vacuole membrane from cytoproct. **A** Membrane is retrieved by endocytosis at the open cytoproct. This is mediated by a network of microfilaments; ×19000. **B** Retrieved vesicles contain horseradish peroxidase (HRP)-reaction product when HRP is present in the medium at the time of defecation; × 28500. **C** Retrieved vesicles containing HRP-reaction product become discoidal vesicles and bind to cytopharyngeal microtubular ribbons for transport back to the cytopharynx; × 47500

the addition of acidosome membrane is termed membrane replacement. Lysosomes then fuse with the condensed DV-II, thereby contributing their prominent IMPs to the phagolysosomal membrane.

Not only do the freeze-fracture pictures support this interpretation but so do the thin-section pictures (Fig. 1) of the glycocalyxes lining the various membranes (Allen and Fok 1984a). Discoidal vesicles and nascent phagosomes have a smooth, 15-nm-thick glycocalyx. Acidosomes and DV-II bear an almost imperceptible glycocalyx, while lysosomes and phagolysosomes bear a thick 30-nm filamentous glycocalyx. This thick glycocalyx becomes depleted in defecation competent DV-IV and membrane at the open cytoproct again bears a reduced glycocalyx not much different from that of the discoidal vesicles.

Finally, monoclonal antibody studies also support a dynamic phagosomal-phagolysosomal membrane system in *Paramecium* (Fig. 3). Monoclonal antibodies have been raised to three stages of the digestive vacuoles (Fok et al. 1986; Allen et al. 1986). One labels only the cytopharynx, discoidal vesicles, nascent DV, and one or two DVs in the posterior end of the cell (Fig 3A–C). The second labels several DVs in the cell's posterior end and the acidosomes (Fig. 3D–F). The third labels DVs which are in the DV-III and DV-IV stages (Fig. 3G–I). A fourth labels most or all DVs (Fok et al. 1986). This shows that the membrane epitopes are added or are exposed as the vacuole membranes are growing or being modified by the addition of discoidal vesicles, acidosomes, and presumable lysosomes, although a monoclonal antibody to lysosomes per se has not yet been developed.

These findings support and are consistent with at least three forms of membrane recycling and flow: (1) rapid or short-range recycling of discoidal vesicle membrane removed from the condensing phagosome and shuttled back to the cytopharynx, (2) retrieval of lysosome membrane from the DV-III and DV-IV and recycling back as secondary lysosomes to the acidified phagosome, and (3) a slower or long-range flow of spent vacuole membrane in the form of vesicles retrieved from the cytoproct which are transported back to the cytopharynx (Fig. 5). Membrane replacement apparently satisfies the need to rapidly change the young DV membrane at the same time when it is condensing in size and becoming acid. This may also permit the incorporation into or expression of lysosomal recognition factors on the DV-II membranes. Vacuoles become larger as they become hydrolytically active by incorporating the membranes of the lysosomes into their own limiting membrane. A final tendency toward smaller vacuoles before defecation results from lysosomal membrane retrieval.

Problems relating to membranes remaining to be solved include: (1) the origin and ultimate fate of acidosome membrane, (2) the site of entry of new membrane into this system, and (3) the extent of mixing of the membranes from different vesicle pools, i.e., do they remain separate to be retrieved intact or do they intermix once they fuse with DVs so that only selected components are retrieved?

9 Concluding Remarks

Since the comprehensive reviews on the digestive lysosomal system in *Paramecium* by Mast (1947), Kitching (1956), and Müller (1967), we have in the last decade gained a much better understanding of the structures and functions of the phaogosomal-phagolysosomal system in this cell. By uncovering the many secrets of this system we have learned more about the dynamic nature of cell membranes and the cell's capacity to alter membranes and to control their physiological processes by utilizing a preprogrammed plan of selective membrane fusions and fissions. Life, as we know it, would certainly be impossible without these remarkable membrane boundaries both at the cell's surface and bordering its many internal compartments. However, unanswered questions remain both for the study of *Paramecium* and for other cells. For example, what is the nature of the factors which allow specific membranes to recognize, bind to, and subsequently fuse with other membranes and what is the mechanism and importance of DV acidification? Research in these areas will be possible once the acidosomes, lysosomes and the various stages of the DVs can be purified.

Acknowledgments. This work was in part supported by the National Science Foundation, grants PCM 84-02881 and DCB 85-02212, and by the National Institute of Health MARC grant GM 07684, MBRS grant RR 08125, and RCMI grant RR 03061.

References

Allen RD (1974) Food vacuole membrane growth with microtubule-associated membrane transport in *Paramecium.* J Cell Biol 63:904–922

Allen RD (1975) Evidence for firm linkages between microtubules and membrane-bounded vesicles. J Cell Biol 64:497–503

Allen RD (1984) *Paramecium* phagosome membrane: from oral region to cytoproct and back again. J Protozool 31:1–6

Allen RD, Fok AK (1980) Membrane recycling and endocytosis in *Paramecium* confirmed by horseradish peroxidase pulse-chase studies. J Cell Sci 45:131–145

Allen RD, Fok AK (1983a) Phagosome fusion vesicles of *Paramecium.* I. Thin-section morphology. Eur J Cell Biol 29:150–158

Allen RD, Fok AK (1983b) Phagosome fusion vesicles of *Paramecium.* II. Freeze-fracture evidence for membrane replacement. Eur J Cell Biol 29:159–165

Allen RD, Fok AK (1983c) Nonlysosomal vesicles (acidosomes) are involved in phagosome acidification in *Paramecium.* J Cell Biol 97:566–570

Allen RD, Fok AK (1984a) Stages of digestive vacuoles in *Paramecium:* membrane surface differences and location. Eur J Cell Biol 35:149–155

Allen RD, Fok AK (1984b) Retrieval of lysosomal membrane and acid phosphatase from phagolysosomes of *Paramecium caudatum.* J Cell Biol 99:1955–1959

Allen RD, Fok AK (1985) Modulation of the digestive lysosomal system in *Paramecium caudatum.* III. Morphological effects of cytochalasin B. Eur J Cell Biol 37:35–43

Allen RD, Staehelin LA (1981) Digestive system membranes: freeze-fracture evidence for differentiation and flow in *Paramecium.* J Cell Biol 89:9–20

Allen RD, Wolf RW (1974) The cytoproct of *Paramecium caudatum:* structure and function, microtubules, and fate of food vacuole membranes. J Cell Sci 14:611–631

Allen RD, Ueno MS, Fok AK (1986) Monoclonal antibodies differentiate between various membrane pools of the phagosomal-phagolysosomal system in *Paramecium.* J Cell Biol 102:356a

Barka T, Anderson PJ (1962) Histochemical methods for acid phosphatase using hexazonium pararosanilin as coupler. J Histochem Cytochem 10:741–753
Berger JD, Pollock C (1981) Kinetics of food vacuole accumulation and loss in *Paramecium tetraurelia.* Trans Am Microsc Soc 100:120–133
Cohen J, Carreau de Loubresse N, Beisson J (1984a) Actin mobilization during phagocytosis in *Paramecium.* J Submicrosc Cytol 16:103–104
Cohen J, Carreau de Loubresse N, Beisson J (1984b) Actin microfilaments in *Paramecium:* localization and role in intracellular movements. Cell Motil 4:443–468
De Duve C (1964) From cytases to lysosomes. Fed Proc 23:1045–1049
Esteve JC (1970) Distribution of acid phosphatase in *Paramecium caudatum:* its relations with the process of digestion. J Protozool 17:24–35
Fenchel T (1980) Suspension feeding in ciliated protozoa: structure and function of feeding organelles. Arch Protistenk 123:239–260
Fok AK (1983) An inhibition and kinetic study of acid phosphatase in *Paramecium caudatum* and *Paramecium tetraurelia.* J Protozool 30:14–20
Fok AK, Allen RD (1979) Axenic *Paramecium caudatum.* I. Mass culture and structure. J Protozool 26:463–470
Fok AK, Allen RD (1981) Axenic *Paramecium caudatum.* II. Changes in fine structure with culture age. Eur J Cell Biol 25:182–192
Fok AK, Paeste RM (1982) Lysosomal enzymes of *Paramecium caudatum* and *Paramecium tetraurelia.* Exp Cell Res 139:159–169
Fok AK, Shockley BU (1985) Processing of digestive vacuoles in *Tetrahymena* and effects of dichloroisoproterenol. J Protozool 32:6–9
Fok AK, Valin EL (1983) Effects of dimethylsulfoxide (DMSO) on the digestive lysosomal system in *Paramecium caudatum.* Eur J Cell Biol 32:41–51
Fok AK, Lee Y, Allen RD (1982) The correlation of digestive vacuole pH and size with the digestive cycle in *Paramecium caudatum.* J Protozool 29:409–414
Fok AK, Leung SS-K, Allen RD (1984a) Modulation of the digestive-lysosomal system in *Paramecium caudatum.* I. Effects of temperature. Eur J Cell Biol 34:265–270
Fok AK, Muraoka JH, Allen RD (1984b) Acid phosphatase in the digestive vacuoles and lysosomes of *Paramecium caudatum:* a timed study. J Protozool 31:216–220
Fok AK, Leung SS-K, Chun DP, Allen RD (1985a) Modulation of the digestive lysosomal system in *Paramecium caudatum.* II. Physiological effects of cytochalasin B, colchicine, trifluoperazine and compound 48/80. Eur J Cell Biol 37:27–34
Fok AK, Ueno MS, Azada EA (1985b) Rapid and sensitive assays for phagosomal acidification in *Paramecium* and *Tetrahymena.* Eur J Cell Biol 38:306–311
Fok AK, Ueno MS, Allen RD (1986) Differentiation of *Paramecium* phagosome membrane and stages using monoclonal antibodies. Eur J Cell Biol 40:1–8
Gomori G (1952) Microscopic histochemistry. Principles and practice. Univ Press, Chicago
Karakashian MW, Karakashian SJ (1973) Intracellular digestion and symbiosis in *Paramecium bursaria.* Exp Cell Res 81:111–119
Kersken H, Vilmart-Seuwen J, Momayezi M, Plattner H (1986) Filamentous actin in *Paramecium* cells: mapping by phalloidin affinity labeling in vivo and in vitro. J Histochem Cytochem 34:443–454
Kitching JA (1956) Food vacuoles. Protoplasmatologia III D 3:1–54
Kuznicki L, Sikora J (1972) The hypothesis of inverse relation between ciliary activity and cyclosis in *Paramecium.* Acta Protozool 11:234–250
Lüthe N, Plattner H (1986) Lectin binding sites in *Paramecium tetraurelia* cells. II. Labeling analysis predominantly of non-secretory components. Histochemistry 85:377–388
Mast SO (1947) The food-vacuole in *Paramecium.* Biol Bull 92:31–72
Müller M (1962) Studies on feeding and digestion in protozoa. V. Demonstration of some phosphatases and carboxylic esterases in *Paramecium multimicronucleatum* by histochemical methods. Acta Biol Acad Sci Hung 13:283–297
Müller M (1967) Digestion. In: Kidder GW (ed) Chem Zool 1:351–380
Müller M, Törö I (1962) Studies on feeding and digestion in protozoa. III. Acid phosphatase activity in food vacuoles of *Paramecium multimicronucleatum.* J Protozool 9:98–102
Müller M, Rohlich P, Toth J, Törö I (1963) Fine structure and enzymic activity of protozoan food vacuoles. Ciba Found Symp Lysosomes. Churchill, London, pp 201–216

Müller M, Rohlich P, Törö I (1965) Studies on feeding and digestion in protozoa. VII. Ingestion of polystyrene latex particles and its early effects on acid phosphatase in *Paramecium multimicronucleatum* and *Tetrahymena pyriformis*. J Protozool 12:27–34

Nilsson NR (1977) Fine structure and RNA synthesis of *Tetrahymena* during cytochalasin B inhibition of phagocytosis. J Cell Sci 27:115–126

Nilsson JR (1979) Phagotrophy in *Tetrahymena*. In: Levandowsky M, Hutner SH (ed) Biochemistry and physiology of protozoa, vol 2. Academic Press, London New York, pp 339–379

Rasmussen L, Modeweg-Hansen L (1973) Cell multiplication in *Tetrahymena* cultures after addition of particulate material. J Cell Sci 12:275–286

Ricketts TR (1971) Periodicity of endocytosis in *Tetrahymena pyriformis*. Protoplasma 73:387–396

Ricketts TR (1983) Dichloroisoproterenol and digestive vacuole formation, movement and egestion in *Tetrahymena pyriformis* GL-9. Protoplasma 115:25–33

Rosenbaum RM, Wittner M (1962) The activity of intracytoplasmic enzymes associated with feeding and digestion in *Paramecium caudatum*. The possible relationship to neutral red granules. Arch Protistenk 106:223–240

Schneider L (1964) Elektronmikroskopische Untersuchungen an den Ernährungsorganellen von *Paramecium*. I. Der Cytopharynx. Z Zellforsch Mikrosk Anat 62:198–224

Sison BC, Ueno MS, Fok AK, Allen RD (1985) Factors affecting phagosome formation in *Paramecium caudatum*. J Cell Biol 101:419a

Soldo AT, Godoy GA, van Wagtendok WJ (1966) Growth of particle-bearing and particle-free *Paramecium aurelia* in axenic culture. J Protozool 13:492–497

Tolloczko B (1977) Endocytosis in *Paramecium*. III. Effect of cytochalasin B and colchicine. Acta Protozool 16:185–193

Wasik A (1983) Effect of external agents on cytoplasmic streaming in *Paramecium*. I. Influence of carmine suspension. Acta Protozool 22:183–189

Wasik A, Sikora J, Kuznicki L (1984) Effects of different suspensions on phagocytosis and cytoplasmic streaming in *Paramecium bursaria*. Postepy Biol Kom 11:529–532

Yusa A, Shigenaka A (1966) The food vacuole formation and the process of digestion and absorption in *Paramecium caudatum*. 6th Int Congr for Electron microscopy, vol 20. Maruzen, Tokyo, pp 237–238

Chapter 20 Exocytosis: Biogenesis, Transport and Secretion of Trichocysts

André Adoutte[1]

Contents

1 Introduction

Trichocysts (from the Greek *thrix*, "hair" and *kystis*, "bladder") have been appropriately named by the 19th century observers of paramecia. The organelle is indeed a membranous bag whose contents are fired in the form of a thin hair-like needle. It therefore corresponds to an exocytotic vesicle and, as such, it shares a large number of characteristics common to all eukaryotic secretory vesicles (Palade 1975) as we shall emphasize in this review: synthesis of the

1 Laboratoire de Biologie Cellulaire 4 de l'Université Paris-Sud et Unité Associée 1134 du C.N.R.S., Bâtiment 444, 91405 Orsay Cedex, France

Paramecium. Ed. by H.-D. Görtz

secretory proteins on the rough endoplasmic reticulum, packaging and maturation of these proteins in the Golgi apparatus and during post-Golgi stages, formation of secretory vesicles displaying saltatory movements in the cytoplasm before reaching the plasma membrane, fusion of the vesicle's membrane with the plasma membrane to allow discharge of the vesicle's content out of the cell, involvement of calcium during the final steps of exocytosis, etc.

In addition to these "universal" attributes, the trichocysts of *Paramecium* (along with many other "extrusomes" of protists) dispaly the following specific, interesting features:

1. Trichocysts are stably docked at the plasma membrane, awaiting only the secretory stimulus to accomplish the final steps of the exocytotic pathway (membrane fusion and discharge of the secretory proteins). Exocytotic responses are therefore fast, homogeneous, and interpretations are not complicated by having to take into account earlier steps in the pathway.

2. Trichocyst docking occurs at specialized sites of the ciliate's cortex. The morphology of these sites and their transformation during exocytosis can therefore be analyzed conveniently.

3. There are several thousands of these fairly large organelles per cell and massive discharge, easily monitored by light microscopy, can be induced. This facilitates the development of simple tests to monitor exocytosis and to analyze the effects of various treatments, the detection of mutants that fail to discharge, the recovery of large amounts of the secreted material and the analysis of the intracellular events which follow massive exocytosis.

4. Trichocysts display an elaborate and reproducible architecture. The contents of the secretory vesicles are highly organized both in the intracellular and in the discharged states. They can therefore be used to understand the morphogenesis of an intracellular organelle.

5. Finally, and most importantly, trichocyst exocytosis appears to be dispensable for the cell, at least under laboratory conditions, and mutants affecting many steps of trichocyst biogenesis and exocytosis have been isolated. The power of genetic dissection has therefore been used to establish structure-function relationships throughout the trichocyst cycle by observing the consequences of the mutational impairment of specific components. The order of events in trichocyst morphogenesis and the interactions between the various components of the trichocyst have also been revealed by mutant analysis. In addition, as the electrophysiology and genetics of excitability is well advanced in *Paramecium*,the interactions between the exocytotic and excitability systems within the same cell have become amenable to analysis.

In spite of the vexing fact that trichocyst function is still not known, the properties just summarized have made the trichocyst an attractive model system for the analysis of a set of basic problems in cell biology, especially those concerned with the various steps of the exocytotic pathway. This chapter will review the trichocyst system of *Paramecium* in this perspective. Studies on *Paramecium* trichocysts have been carried out almost exclusively on *P. caudatum* and *P. tetraurelia.* The trichocysts of these two species being identical in all respects as far as we can see, we have borrowed the data indifferently from either species. Two broad reviews on the extrusomes of protists are those of Hovasse and Mignot (1975) and of Hausmann (1978).

2 Trichocyst Morphology

2.1 Overview

Trichocysts are 3–4-μm-long fusiform organelles implanted at right angles as a single layer beneath the plasma membrane. They are easily observed by phase-contrast microscopy (Fig. 1) which shows them to be composed of a carrot-shaped dense body surmounted by a tip. The trichocysts are positioned with remarkable regularity throughout the whole cell surface (except the oral region): they occupy a specific docking site located halfway within the transverse ridges of the numerous adjacent cortical units that make up the ciliate's cortex (Ehret and Powers 1959; Pitelka 1963; Jurand and Selman 1969). Within a longitudinal

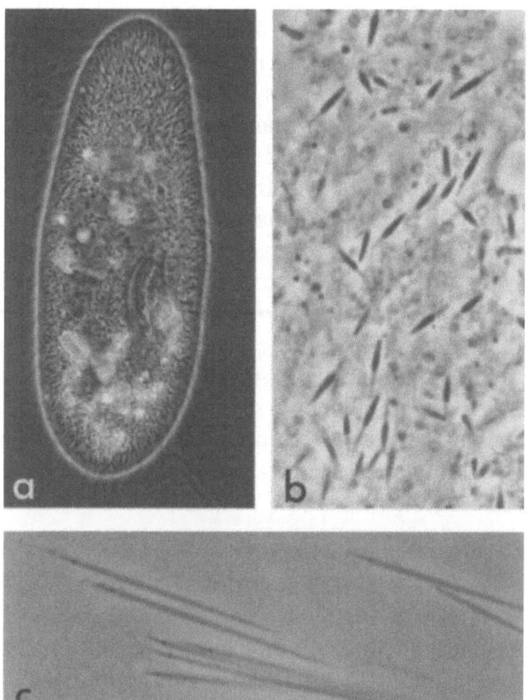

Fig. 1 a–c. Trichocysts in phase-contrast microscopy. **a** Phase-contrast view of a living *Paramecium* with its load of trichocysts lined up beneath the cell surface. Magnification: 800. **b** Condensed trichocysts with their dark carrot-shaped body and their thinner tip within the cytoplasm of a crushed cell. Magnification: 1600. **c** Excreted trichocysts in their extended configuration showing the expansion of the body and the persistence of the tip. Magnification: 1600. A few trichocysts just excreted are visible (although out of focus) at the *right-hand lower side* of the cell in **a** (Adoutte et al. 1984)

strip of cortical units (kinety) one therefore observes a perfectly regular alternation of cilia (emerging to the outside of the cell) and trichocysts (below the cell surface).

In *Paramecium tetraurelia* there are 3500–4000 cortical units (Sonneborn 1970; Iftode, unpubl.) and the number of trichocysts in stationary phase cells (in which all the docking sites are occupied) should be only slightly inferior. However, Plattner and co-workers (Plattner et al. 1984, 1985a) have evaluated the total number of discharged trichocysts to be only about half this number under conditions in which about 90% of trichocysts were discharged and only 50–60 undocked ones remained in the cytoplasm. The reason for this discrepancy is not known. One possibility is that all potential docking sites are not as fully accupied as generally assumed. The number of trichocysts in *P. caudatum* (which is a larger cell than *P. tetraurelia*) has been estimated to be 6000–8000 (Hausmann and Allen 1976).

A variety of stimuli (discussed below) induce more or less massive trichocyst discharge. The contents of the trichocyst expand and are fired in the form of a 25–35-μm needle (Fig. 1). Only the trichocyst body expands, the tip retains its intracellular size. The actual transition from the condensed to the fired state is extremely rapid. Using electrical triggering (Wohlfarth-Bottermann 1953) by a strong current pulse it has been calculated by high-speed cinematography to last only about half a millisecond from the time of stimulus application to that of full discharge (Miller et al. 1968; Matt et al. 1978; Cohen and Vaucquier, unpubl.).

The detailed studies carried out using thin sectioning and negative staining electron microscopy mainly by Bannister (1972) and Hausmann et al. (1972a, b) have revealed considerable order in trichocyst ultrastructure. We shall summarize their data and include more recent information not only because they are useful to follow the assembly steps, but also because plausible functions can now be suggested for several trichocyst components.

As expected of an exocytotic vesicle, the trichocyst is surrounded by a continuous membrane. *Within* this membrane two parts are easily recognized, the tip and the body. On the *outside* of the membrane, in the upper third portion of the tip, an additional component (resembling a gauntlet), the "collar" is tightly apposed to the trichocyst membrane. Finally, a number of additional external elements appear to establish connections between the trichocyst membrane and the surrounding membranes, at the docking site.

2.2 The Trichocyst Body

As shown in Fig. 2 the carrot-shaped body displays a remarkable crystalline pattern with several apparent periodicities, the most conspicuous in negatively stained or sectioned preparations being the 7-nm one (Fig. 2). Discrepancies concerning this value occur in the literature, however, a 12 nm periodicity having been documented by Kersken et al. (1984). Based on the comparative analysis of condensed and expanded trichocysts Hausmann et al. (1972a, b) and Bannister (1972) have proposed specific three-dimensional models of the organization of the trichocyst matrix fibrils. Recently, the problem of trichocyst matrix architecture has been actively reinvestigated using deep etching techniques (Peterson et al. 1987) as well as freeze fracture and X-ray diffraction

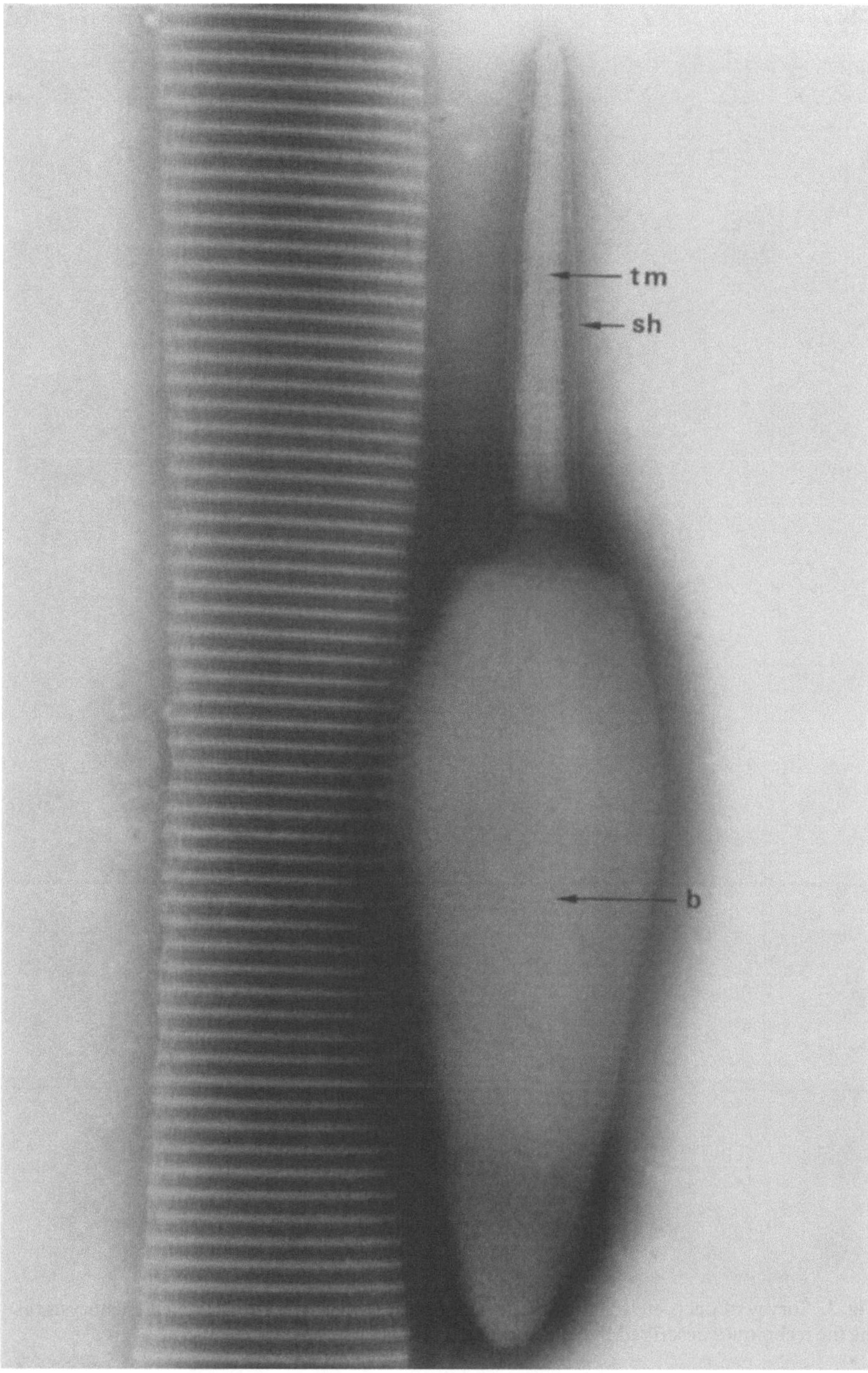

Fig. 2. Trichocysts in negative staining (1% phosphotungstic acid). A typical condensed trichocyst lies to the *right* of a portion of the decondensed body of a discharged trichocyst. Note the crystalline aspect of both the body (*b*) and the tip matrix (*tm*). The details of the sheath (*sh*) structure are best seen on sections (Peterson et al. 1987)

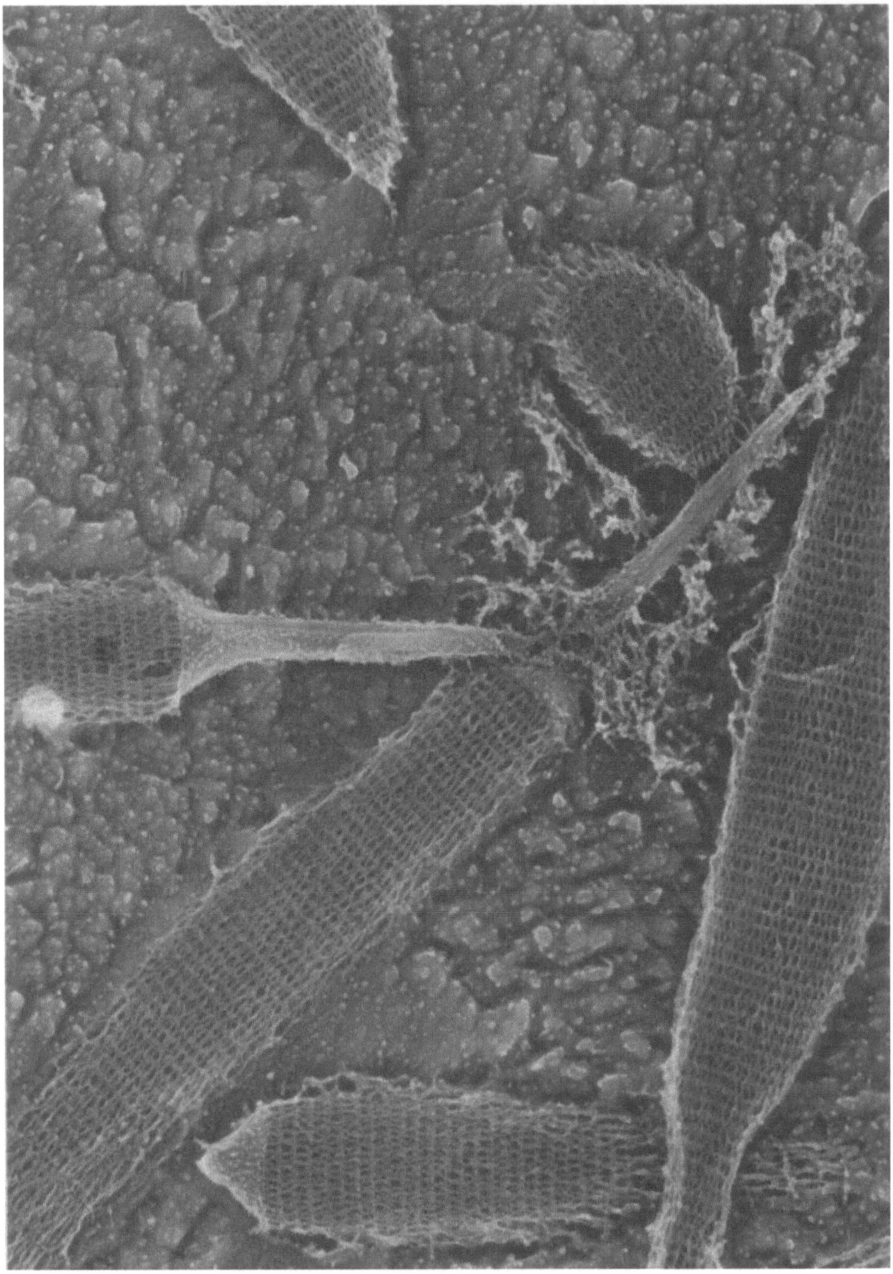

Fig. 3. Survey of deep-etch replica of a quick-frozen suspension of discharged trichocysts, using the techniques described by Peterson et al. (1987)

(Sperling et al. 1987). Peterson et al.'s study also took advantage of the fact that trichocyst matrices can be progressively solubilized by heat treatment (Pollack and Steers 1973). Examples of the images obtained by freeze etching are given in Figs. 3 and 4. The salient feature is the presence of interwoven filaments, which appear to be thickened at regular intervals. A good agreement was found by Sperling et al. between X-ray diffraction data and freeze-fracture images permitting the determination of the dimension of the "unit cell" of condensed and extended trichocysts. These approaches, which avoid the problem of image overlap and reduce the possible occurrence of artefacts due to fixatives, can be expected to yield a relatively detailed knowledge of the molecular packing of the matrix proteins both in the condensed and in the expanded forms. The crystalline matrix is surrounded by a thin, meshlike layer.

Some confusion has arisen from the fact that in standard fixations for transmission electron microscopy a large proportion of trichocysts display swollen, amorphous, lightly stained bodies. Selman and Jurand (1970; see also Jurand and Selman 1969) developed the idea that such trichocysts correspond to the final, fully differentiated stage of the organelle ("mature" trichocysts) which followed the compact, crystalline stage ("juvenile" trichocysts). Bannister (1972) clearly showed, however, that the swollen forms arise most probably from a mild fixation artefact, the fixative acting very briefly as a secretagogue. The simplest and most convincing evidence in this respect is that the size and overall shape of trichocysts observed in *living cells* by phase contrast microscopy are those of the compact crystalline organelle seen in electron microscopy and not those of the swollen forms. In addition, Garofalo et al. (1983) have shown that Ca^{2+} chelation and/or the use of anti-calmodulin drugs in vivo maintain the trichocysts in the condensed form. Furthermore, routine permeabilization of paramecia in EGTA-containing medium for immunocytochemistry shows all the trichocysts underlying the cortex in the condensed form (see for example Fig. 4 in Adoutte et al. 1984). Whether these amorphous forms correspond to a real intermediate stage in in vivo matrix expansion as suggested by Garofalo et al. (1983) or to a mere disorganization induced by the fixative remains to be established.

2.3 *The Trichocyst Tip*

The conical trichocyst tip is made up within the trichocyst membrane of several successive layers represented schematically in Fig. 5. From centre to periphery one can identify.

- a crystalline matrix which is in continuity with that of the trichocyst body and which displays identical periodicities, including a prominent 16-nm one (Bannister 1972).
- "an inner sheath made up of four spiralling envelopes with a square net substructure".
- "an outer sheath made up of a dense amorphous matrix containing longitudinal tubules and scattered fine filaments" (Bannister 1972).

2.4 *The Collar*

The collar, which is tightly apposed, externally to the membrane, at the upper third of the trichocyst tip, is composed of a "crenulated dense material whose

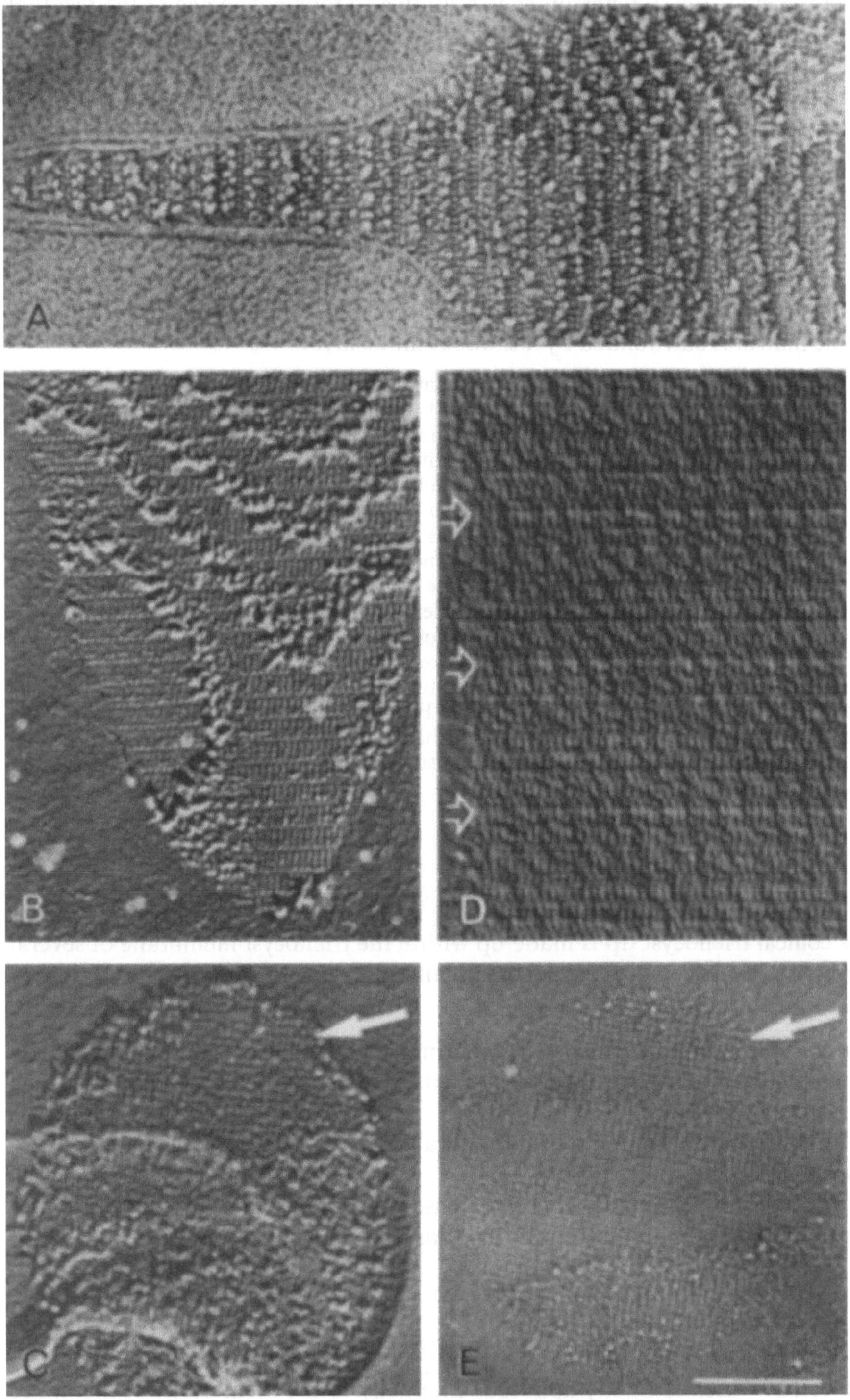
A
B
D
C
E

lacunae are filled by tubules" [Bannister's "outer microtubular ring", now known throughout immunocytochemistry (Cohen et al. 1982)] and high resolution transmission electron microscopy using tannic acid (Plattner et al. 1982) not to be made of tubulin. Pouphile et al. (1986) have shown that some mutants with unattached trichocysts display only the ribbed structure, while others possess only the dense filling. The two elements can therefore be assembled independently and have been named "collar fibers" and "collar matrix" respectively. Freeze-fracture images of the trichocyst membrane in the region covered by the collar show impressions of regular longitudinal striations most probably reflecting modelling due to the tight apposition of the tubular collar elements.

2.5 *The Insertion Site*

The trichocysts insert at predetermined sites in the cortex through openings that occur in the layers underlying the plasma membrane i.e. the epiplasm and the membranous alveoli (Pitelka 1965; Allen 1971). Their insertion induces a set of remarkable reorganizations both within the plasma membrane, the trichocyst membrane and around the trichocyst tip. The membrane reorganizations were first demonstrated using freeze-fracture electron microscopy by taking advantage of mutants that fail to dock or to discharge their trichocysts (Beisson et al. 1976). The remodelling of the site has been recently recapitulated by completing the analysis of a large number of distinct mutant strains (Pouphile et al. 1986) and is summarized in Fig. 6.

In thin sections, the unoccupied site displays a "plug" of amorphous material below the plasma membrane between the alveoli (Beisson et al. 1976; Westphal and Plattner 1981 b). In freeze fracture the plasma membrane at this site displays a specific array of particles organized in the shape of parentheses. Trichocyst insertion induces the circularization of the parenthesis, which takes the form of a double ring of $\sim 2 \times 40$ particles, and the appearance of a central rosette of ~ 8 intramembranous particles (Beisson et al. 1976), leading to the configuration most often seen in the P-face of freeze-fractured wild-type cells (Janish 1972; Satir et al. 1972; Plattner et al. 1973; Allen and Hausmann 1976).

Fig. 4A–E. Freeze-fracture electron micrographs of condensed (**A–C**) and extended (**D, E**) isolated, untreated trichocysts. **A** and **E** are rotary shadowed and **B, C, D** unidirectionally shadowed with W-Ta. Magnification for all images is 90000 (*bar* = 200 nm). **A** Condensed trichocyst. Note the continuity of the internal structure between body and tip and the presence of the inner sheath around the tip. **B** One of the longitudinal fracture views of condensed trichocysts. Note the presence of complex, regular two-dimensional arrays of what looks like short filaments. **C** Cross-fracture view of a condensed trichocyst. Note the presence of a two-dimensional rectangular array (*arrow*). **D** One of the numerous different longitudinal fracture views of the extended trichocyst. Note the presence of regular arrays of filamentlike elements and the fact that morphological units repeat every four striations (2300 Å) (*arrows*). **E** Cross-fractured extended trichocyst. Note the presence of a regular square lattice (*arrow*) and more or less wide stripes which correspond to the regularly spaced striations seen on the longitudinal fractures (**B**). (Courtesy of T. Gulik-Krzywicki and L. Sperling)

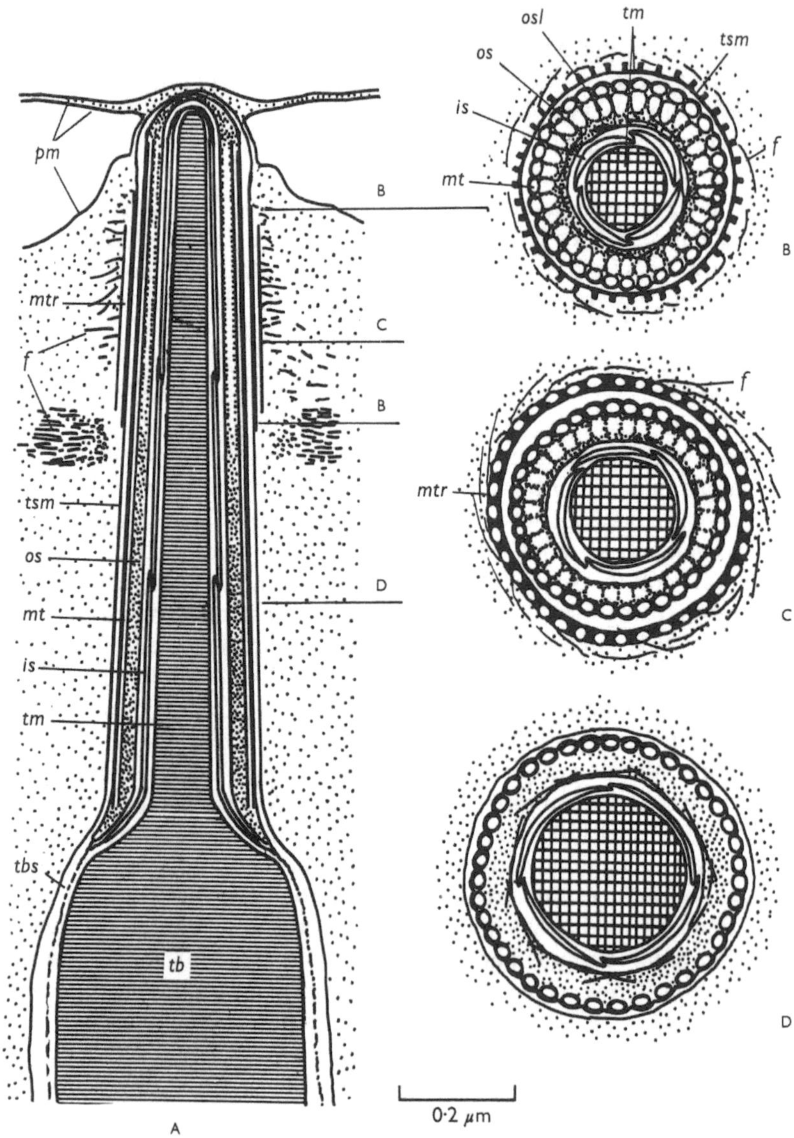

Fig. 5A–D. Diagrams showing the morphology of the apical regions of the undischarged trichocyst. **A** Longitudinal section showing the various sheathing structures surrounding the crystalline matrix of the trichocyst. **B–D** Transverse sections of the trichocyst tip taken at the levels indicated on **A**. Note that the form of the transverse section **B** is found at two levels in the trichocyst, on either side of the section at **C**. *f* Cytoplasmic filaments; *is* inner sheath of trichocyst tip; *mt* microtubule; *mtr* outer ring of microtubule (= part of the "collar"); neither *mt* nor *mtr* are made of tubulin (see text); *os* outer ring of trichocyst tip; *osl* lacuna in outer sheath of tip; *pm* pellicular membrane (composed of the continuous plasma membrane and interrupted alveolar membranes tightly interacting with the trichocyst membrane); *tb* trichocyst body; *tbs* sheath of trichocyst body; *tm* crystalline matrix of trichocyst; *tsm* trichocyst sac membrane (Bannister 1972)

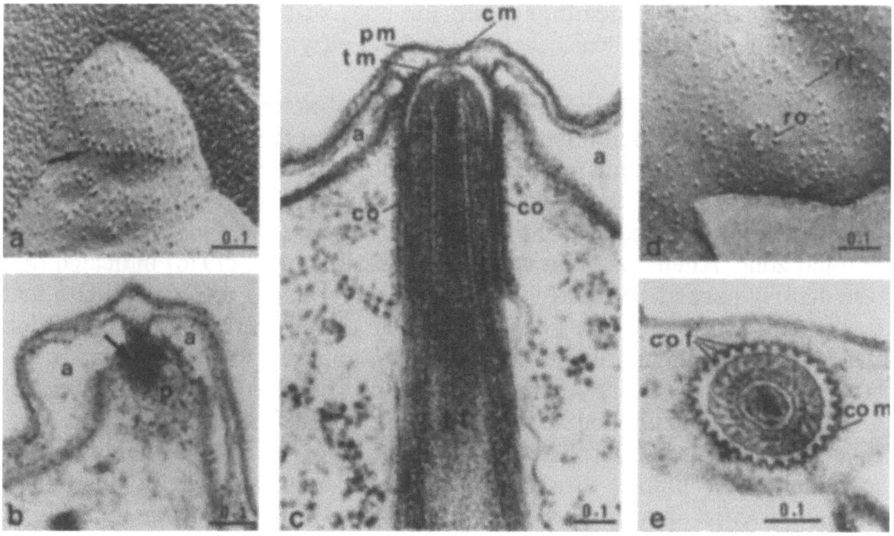

Fig. 6a–e. Recapitulation of the normal membrane and submembrane differentiations at the docking site of trichocysts *before* (**a, b**) and *after* (**c, d, e**) its attachment to the cell surface. **a** The docking site is seen by cryofracture of plasma membrane on the P-face as a specific array of particles organized in a "parenthesis" (*arrow*). **b** Thin section perpendicular to the surface: a dense fibrous material called the "plug" (*p*) with a denser upper part (*arrow*) is always observed between the subpellicular alveoli (*a*). **c** Thin section perpendicular to the cell surface through an occupied site showing the tip (*tt*) of a docked trichocyst. A dense "connecting material" (*cm*) links the membrane of the trichocyst tip and the plasma membrane. The subapical zone of the trichocyst is surrounded by a composite dense sheath, the collar (*co*). *pm* Plasma membrane; *tm* trichocyst membrane; *a* alveoli. **d** In cryofractured plasma membrane (PF face), at an occupied site, the parenthesis has been circularized into a double outer ring (*ri*) of particles, and a central rosette of large particles (*ro*) has been assembled. **e** Cross-section through an attached trichocyst tip showing the organization of the collar composed of "collar fibers" (*cof*) and "collar matrix" (*com*) as defined in the text (*Bars* = 0.1 μm) (Pouphile et al. 1986)

Upon insertion, the plug material appears to flow over the trichocyst tip (Westphal and Plattner 1981 b) giving rise to one of the collar elements: the "collar matrix", defined above. The membrane covering the very tip of the trichocyst is held at a distance of 15–30 nm below the plasma membrane, being linked to it by a set of thin, apparently amorphous or filamentous elements (Plattner et al. 1973), the "connecting material" (Plattner et al. 1980).

The "connecting material" and the "collar fibers", may not originate from the plug material at least in mutants, since they are observed on some mutant trichocysts before the organelle has reached its insertion point or, in other mutants, can be absent while the collar matrix is present (Pouphile et al. 1986). These three structures therefore assemble independently. Cytochemical analysis indicates that collar, connecting material and epiplasm have distinct proteinaceous compositions (Westphal and Plattner 1981 a).

Plattner et al. (1973) established that the outer ring of particles corresponds to points of contact between the plasma membrane and the underlying alveolar membrane while the central rosette lies over the "connecting material" which, itself, connects plasma membrane with the apex of the trichocyst membrane. The rosette particles are sensitive to proteolytic enzymes (Vilmart and Plattner 1983).

Other specializations, also best seen on the P-fracture face, occur in the region of the trichocyst membrane which is encased by the collar: this whole area is free of intramembranous particles while the narrow zone located just above it, towards the apex of the trichocyst tip, displays an aggregation of particles sometimes forming regular linear arrays as if they had been pushed away from the collar zone (Allen and Hausmann 1976). Beisson et al. (1976) indicated that these densely packed particles form at least three annuli differing in size and arrangement. One (or several) of these particle annuli correspond to the region of the trichocyst membrane which is tightly apposed to the alveolar membranes. It therefore clearly corresponds to the differentiation of an area involved in membrane to membrane attachment (Plattner et al. 1973).

On the whole, this morphological description of trichocysts should convey the notion of a highly structured, membrane-bound organelle, implanted (through specific apertures in the cortical layers) right beneath the plasma membrane and connected to the surrounding membranes both directly and through fibrillar elements, the connections being marked by conspicuous intramembranous particle arrays.

3 Trichocyst Biochemistry

3.1 The Proteins of the Crystalline Matrix

Since trichocyst matrices can be massively discharged as small rigid needles without impairment of cellular integrity, it is relatively easy to isolate the extruded material (as a fluffy layer lying on top of cell pellets after low-speed centrifugation) then submit it to various purification steps. Analysis of trichocyst matrix proteins was initiated by Steers et al. (1969) who identified two major components by SDS gel electrophoresis with molecular weights of 17000 and 36000 respectively. Reduction of the samples with 2-mercaptoethanol prior to electrophoresis resulted in a single component of 17000 molecular weight. This suggested that the elaborate crystalline matrix might be constructed from an association of as few as two polypeptide chains of similar molecular weight (ca. 17–20000 daltons) linked to each other by disulphide bridges. The name "trichynin" was given to the trichocyst structural protein. This picture has remained essentially correct except that improvements in electrophoretic resolution have now shown the 17000 and 34–36000 dalton areas to consist in a *family* of closely migrating polypeptides which should therefore be labelled trichynins. Two-dimensional gel electrophoresis revealed still more complexity since about 40 spots could be identified in reduced samples of highly purified trichocysts. These spots displayed molecular weights centered at 17000 daltons and slightly acidic isoelectric points (Adoutte et al. 1980). Recently, high resolution 2-D gels have disclosed up to 100 spots of which about 50 are quantitatively dominant (Tindall 1986).

In addition to the 15 – 20 kD bands which are by far the major components of reduced trichocysts in SDS gels, a faint band at 68 kD is sometimes observed as well as a set of very high molecular weight bands (above dyneins, i.e. > 300 kD).

Although the tips, including the inner and outer sheath, are often present in trichocyst preparations, no corresponding protein constituent has been attributed to them yet.

Amino acid composition of the bulk excreted trichocysts (Steers et al. 1969; Rauh and Nelson 1981) shows them to be especially rich in acidic residues, and to have about 1.4 half-cystine per 17000 average molecular weight molecule. The overall amino acid composition displays many similarities to that of calmodulin (Rauh and Nelson 1981). The presence of 1 – 10% authentic calmodulin within excreted trichocysts was reported by Rauh and Nelson (1981). However, this calmodulin most probably becomes associated with trichocysts after secretion and is not an intrinsic component of the trichocyst in situ (Kersken et al. 1984; Tindall et al., pers. commun.).

3.2 Possible Molecular Bases of Protein Multiplicity

The trivial explanation of artefactual generation of heterogeneity of trichocyst matrix proteins by proteolysis or charge modification during sample preparation appears to be ruled out on the basis of the very high reproducibility of the electrophoretic patterns, the use of a variety of protease inhibitors and recent careful studies of Tindall et al. which excluded in vitro generation of charge heterogeneity (pers. commun.). There are therefore two extreme hypotheses to account for the origin of the numerous protein spots obtained on 2-D gels of trichocyst matrices. One is that they correspond to as many distinct gene products, the other is that they correspond to a very small number of distinct gene products, heterogeneity being generated by post-translational modifications. Various combinations of these two hypotheses can, of course, also be considered.

At the moment, no post-translational modification of trichocyst proteins has been definitely identified except for the proteolytic cleavage of high molecular weight precursors during the course of trichocyst biogenesis to yield the 15 – 20 kD proteins, described below (Adoutte et al. 1984). Trichynins do not appear to be phosphorylated or methylated (Nelson, pers. commun.). Several lines of evidence indicate, however, that they may undergo limited glycosylation (Kersken et al. 1984). Estève (1974) has reported labelling of the trichocyst body sheath (and not the body proper) by two cytochemical techniques specific for carbohydrates.

On the whole then, although all possible post-translational modifications have not been tested, the available evidence does not favour the idea that trichynin diversity arises through such modifications. One is therefore left with the suggestion that trichynins are coded by a fairly high number of structural genes, possibly belonging to a family of evolutionarily related redundant sequences. Precedents for such a situation exist for the storage proteins of plants

(Geraghty et al. 1982). The fact that trichynin precursors also display a complex electrophoretic pattern (see below), while they can be assumed to have undergone little if any post-translational modification, is compatible with the multiple genes hypothesis. Within this framework, the immunological cross-reactions observed with antibodies raised against one set of trichynin bands towards all the other bands (Adoutte et al. 1984) would be attributed to the sharing of at least some immunological determinants by all the trichynins as a result of their evolutionary kinship. Direct molecular data at the protein and DNA levels are being sought by different laboratories to definitely solve this problem.

4 Biogenesis of the Trichocyst Matrix

4.1 The Morphological Steps

The various steps leading to a morphologically mature trichocyst have been reconstructed with reasonable certainty by electron microscope analysis of dividing cells which must at least double their trichocyst population (Ehret and de Haller 1963; Selman and Jurand 1970) or of cells regenerating their trichocysts after massive electrically induced discharge (Yusa 1963). Many of the early steps were later seen in mutants and in monensin-treated cells (Pollack 1974; Adoutte et al. 1984; Garreau de Loubresse, unpubl.).

Basically, the earliest recognizable step consists in an endoplasmic membrane-limited vesicle filled with a homogeneous amorphous material and in which a minute crystallization point can often be discerned. The subsequent steps correspond to a progressive increase in size as well as a shaping of the crystalline core from which the future crystalline axis of the tip soon emerges, in continuity with the body. When formation of the crystalline matrix and tip is nearly complete and the vesicle has enlarged and elongated, the outer sheath (and probably the inner one as well), start differentiating. All these steps occur within the endoplasm and it is only when the organelle has reached its final, mature shape that it attaches to the cortex.

This sequence has been strongly confirmed through the analysis of mutants blocked at various steps of the morphogenetic pathway (Jurand and Saxena 1974; Pollack 1974; Cohen and Beisson 1980; Pouphile et al. 1986). The phenotypic classes available (see Table 1) correspond to (a) total absence of mature or immature trichocysts ("trichless"); (b) grossly abnormal, incomplete trichocysts usually displaying the shape of an American football, occasionally with a small misplaced tip (the "football" series). The trichocyst matrix is often only partially crystalline in these mutants; (c) incompletely differentiated plump trichocysts with highly variable morphology, more advanced, however, than that of the football type ("stubby"); (d) trichocysts with a normal body ultrastructure but with a misplaced tip lying within the membrane, usually on a lateral side of the body ("pointless"); (e) elongated cigar-shaped trichocysts, otherwise normal for attachment and exocytosis capacity (screwy-cigar); (f) trichocysts displaying a normal morphology but failing to reach their cortical at-

Table 1. A summary of the relationships established between the cytological phenotype of trichocyst mutants (columns) and the site of action of the corresponding mutations, as established by microinjection (lines). Mutants not tested by microinjection have also been indicated for the record. The table is based essentially on data by Aufderheide (1978c), Lefort-Tran et al. (1981) and Cohen and Beisson (1980). (From Cohen 1984)

Phenotype of mutants / Point of impact of mutations	Mutants deficient in trichocyst biogenesis						
	Trichocyst morphology NORMAL			Trichocyst morphology ABNORMAL			Trichocysts ABSENT
				Short (stubby)	Pointless	Abortive (football)	
	Attached to the cortex	Not attached to the cortex					
	Motile		Non-motile	Motile			
Trichocyst	*nd B* *nd 7*	*tam 10*	*nd A* *tam 8* *tam 11*	*tam G*	*pt A*^{2}	*ft A*^{3} *tam 38* *ft 33*	
Cortex	*nd 6*	*tam 6*^{a} *tam 6*^{b}					
Cytoplasmic factor	*nd 9*^{a} *nd 9*^{c}						
Not tested by microinjection	*nd 3*^{a} *nd 3*^{b} *nd 3*^{c} *nd 9*^{b} *nd 12* *nd 16* *nd 17*	*tam 9*		*st A* *st B*	*pt A*^{1} *pt 2*	*ft A*^{1} *ft A*^{2} *ft B* *tam A*	*tl*

tachment site either because of a defect in trichocyst intracellular motility or because of a defect at the cortex (see below; *tam 8, nda*); (g) trichocysts with normal morphology, attached to the cortex, only defective in the very final exocytotic step (such as *nd9*). The (f) and (g) classes are not directly relevant to the problem of matrix biogenesis. Classes (a) to (d) display the epistatic interactions expected from their phenotype, that is, the phenotype of the double mutants is always that of the mutant which appears to correspond to the "earliest" block in the morphogenetic pathway (as deduced from cytological studies). Thus, the mutations can be arranged in the following sequence (Pollack 1974; Beisson and Rossignol, pers. commun.):

tl → *ftb* → *stub* → *ptl* → *ci* → *tam 8* → *nd*.

However, the expression of the "pointless" phenotype is partially independent from that of the body suggesting that positioning of the trichocyst tip occurs on a separate pathway. Similarly, acquisition of motility by the trichocyst (see below) is largely independent of the state of body maturation. The genetic data therefore suggest that body morphogenesis proceeds along an essentially linear series of interdependent steps, while acquisition of additional components and properties by the trichocyst may occur on partially independent morphogenetic pathways.

4.2 Intracellular Transit and Maturation of the Secretory Proteins

The generation of antibodies to whole purified trichocyst matrices or to trichocyst matrix protein bands cut out from polyacrylamide gels provided a tool that led to the discovery of an extensive post-translational proteolytic processing of these proteins: when extracts of whole *Paramecium* cells were electrophoresed on SDS gels then transferred to nitrocellulose filters and challenged with antibodies raised against purified trichocyst bands, positive reactions were not only observed (as expected) in the 15–20 kD region (i.e. that of "mature" trichocyst proteins) but also in an area of 40–45 kD (Adoutte et al. 1984; see Fig. 7). That these 40–45 kD bands correspond to trichocyst precursors was suggested by several observations: (1) In mutants displaying a strong decrease of the 15–20 kD bands (such as *tam 38*) the bands at 40–45 kD are intensified; (2) treatment of wild-type cells with monensin, an inhibitor of intracellular protein transit at the Golgi level, strongly increases the amount of 40–45 kD proteins. These two observations therefore reveal an inverse quantitative correlation between the amount of 40–45 kD proteins and that of the 15–20 kD ones, strongly suggestive of a precursor:product relationship; (3) finally, the analysis of immunoprecipitates from post-microsomal supernatants showed that the 40–45 kD reactive area corresponds to several distinct bands. The "precursors" therefore display a complexity that matches that of the "products".

During the course of these studies a systematic correlation was observed between the amount of 15–20 kD bands and the amount of crystallized material, as observed at the electron microscope level within secretory vesicles. For instance, monensin-treated cells which show an abundance of "immature" trichocysts in the form of large vesicles containing a crystalline centre and an amorphous periphery contain a much increased proportion of 40–45 kD proteins with respect to the 15–20 kD ones. It was therefore inferred that crystallization was coupled to proteolytic cleavage of the precursors.

On the whole, the assembly of the trichocyst matrix therefore displays many similarities with that of bacteriophages: it corresponds to a cascade of interdependent morphogenetic steps apparently involving self-assembly of protein components but also requiring some essential post-translational "triggers" in the form of proteolytic cleavage. Whether true "morphogenetic proteins" (i.e. proteins that are necessary at a given step to obtain the normal shape, but that do not belong to the final assembled structure), also play a role in trichocyst matrix biogenesis, remains to be established.

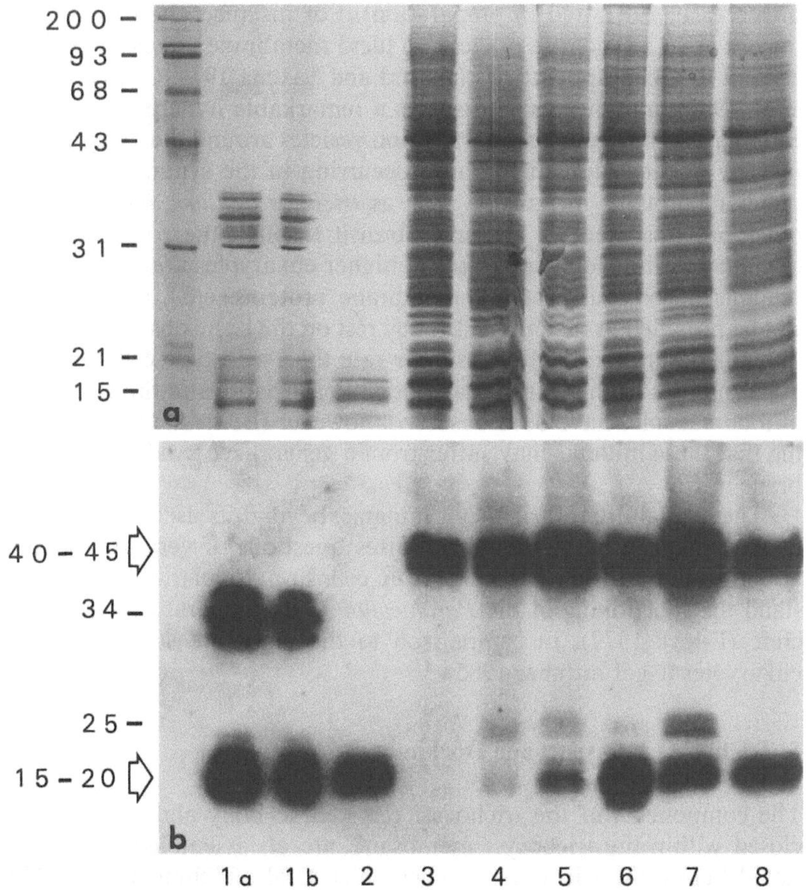

Fig. 7 a, b. Immunochemical detection of secretory products and immunologically related polypeptides in whole cell extracts. **a** Coomassive blue-stained SDS/polyarcylamide gel (15% w/v acrylamide). **b** Autoradiogram of the immunoblot (after treatment with an antibody specific for "mature" trichocyst proteins, then iodinated protein-A). *Lanes 1 a* and *1 b,* purified excreted trichocysts, unreduced samples; *lane 2,* purified excreted trichocysts, reduced sample. Note the disappearance of the 34-kD bands after β-mercaptoethanol treatment, as described in the text. *Lane 3, tl* whole cell extract; *lane 4, tam 38* whole cell extract; *lane 5,* monensin-treated *tam 8* cells; *lane 6,* control *tam 8* cells; *lane 7,* monensin-treated wild-type cells; *lane 8,* control wild-type cells. Note the higher intensity with respect to wild type of the reaction at 40–45 kD in monensin-treated and mutant *tam 38* cells (Adoutte et al. 1984)

4.3 General Implications

The studies reported above also imply that trichocyst matrix proteins transit through the Golgi apparatus in *Paramecium* as all secretory proteins do in higher eukaryotes. This was established both by the monensin effect and by the gross abnormalities in the rough endoplasmic reticulum, Golgi and post-Golgi ultrastructure, evident in several of the trichocyst mutants (Garreau de Loubresse et al. 1983, and in preparation). For instance, the *trichless* mutant displays a great abundance of electron lucid membrane vesicles, apparently located next to Golgi apparatuses (Jurand and Saxena 1974; Garreau de Loubresse et al. 1983); mutant *tam 38* displays a remarkable hypertrophy of areas of the endoplasmic reticulum or of transition vesicles around the Golgi apparatus, etc. In view of the major disturbance occurring in the synthesis and/or transit of secretory proteins in mutants such as *trichless* or *tams,* one wonders whether other proteins that are known to transit through the rough endoplasmic reticulum and the Golgi apparatus in higher eukaryotes are also affected in these mutants (lysosomal proteins, membrane proteins, etc.). Only partial answers can be provided at the moment. They rest on the basic observation that all these mutants do grow (albeit at a slower rate than wild-type cells) in spite of their abnormalities in the transit machinery. Perturbations in the size and dynamics of food vacuoles have been observed, however (Beisson, pers. comm.), suggesting that these mutants may suffer from a generalized syndrome affecting membrane traffic.

The availability of trichocyst mutants, beyond its usefulness for the analysis of the organelle itself, therefore, raises questions of very general significance about protein traffic in an eukaryotic cell. In particular, they may help understand the functioning of the *Paramecium* Golgi apparatus which is rather discrete (Estève 1972), in comparison to the elaborate stack system of higher eukaryotes (e.g. Farquhar 1985).

5 Trichocyst Transport and Docking

The components of the trichocyst (crystalline body and tip, tip sheaths), enclosed within the trichocyst membrane, are all assembled and fully differentiated before the trichocyst reaches its cortical attachment point. How is this relatively large organelle transported through the cytoplasm and how does it become firmly docked in its predetermined position? Here again, a combination of genetics and cell biological methods have provided answers of broad significance.

5.1 Motility Events

Fine cytological observations on living cells, later combined with an ingenious microinjection protocol between genetically different cells, have enabled Aufderheide (1977, 1978 a, b, c) to describe the motility events involved in transport and docking of trichocysts as well as to identify a number of mutational blocks within these processes [see Lefort-Tran et al. (1981) for a diagram of the experimental protocol]. Trichocyst insertion into the cortex is best observed in cells that have first been triggered to release the organelles massively by electrostimulation. In such cells, the 40–50 mature trichocysts that were available in the cytoplasm, undergoing cyclosis, are seen first to become immobilized out of the cyclotic flow. The trichocyst then makes a series of saltatory motions (also seen in non-stimulated cells) which apparently transport it to the cortex, tip first. If "capture" occurs, the trichocyst then undergoes a series of pivoting movements around its stationary tip until it is definitely and stably inserted.

Three mutants, *nd A*, *tam 8* and *tam 11* (Lefort-Tran et al. 1981) were specifically shown to lack saltatory motility. This totally prevents their insertion into the cortex and occurs in spite of the fact that the morphology appears to be normal in these mutants and that other mutants with marked morphological abnormalities (*pt A*, *ft A*) display normal saltatory movements. These observations point to the existence of discrete, genetically controlled components, responsible for the ability of trichocysts to undergo saltatory motion. In these mutants, the motility defect lies in the trichocyst itself since, for example, *tam 8* trichocysts microinjected in *ft A* cannot undergo exocytosis although the host's cytoplasm is known to sustain the saltatory motion of its own *ft A* trichocysts and is capable of excreting wild-type microinjected trichocysts. Other mutants (such as *tam 6* and *nd 6*) fail to dock their trichocysts properly because of a defect in the cortical membrane. These mutants display gross abnormalities in the configuration of their plasma membrane particle arrays and can be "cured" by cellular contact with a wild-type partner during conjugation (Lefort-Tran et al. 1981). A final class, that of *nd 9*, displays motile and docked trichocysts but specifically lacks the rosette and connecting material. This mutant can be cured by microinjection of wild-type cytoplasm (Garofalo et al. 1978; Beisson et al. 1980) and of partially purified post-microsomal supernatant protein extracts (Cohen, pers. commun.). The results have been summarized in Table 1.

Irrespective of the morphogenesis of the trichocyst body and tip, motility and efficient docking of trichocysts therefore require at least three types of genetically controlled elements located respectively, in the trichocyst membrane, in the cytosol and in the plasma membrane. All the results can be rationalized by suggesting the existence of "receptors" in the trichocyst membrane that act as anchoring points for specifically oriented cytoplasmic "pulling elements". These receptors would be defective in mutants such as *tam 8* and *nd A* but would remain functional in other mutants in spite of the grossly abnormal morphology of their trichocysts. In addition, the cortex would contain appropriate "binding sites" for capturing and docking the trichocysts. These elements would be defective in *tam 6* and *nd 6*.

There are a few preliminary hints as to what the cytoplasmic guiding elements might be. Firstly, Beisson and Rossignol (1975) and Aufderheide (unpubl.) showed that trichocyst motility and insertion is cytochalasin B-sensitive, suggesting that actin filaments might be involved. Secondly, Plattner and coworkers (1982) suggested the guidance of trichocysts to unoccupied docking sites by long microtubules emanating from basal bodies, pointing towards the centre of the cell and sometimes seen in contact with the sides or tips of trichocysts (see Cohen and Beisson, Chap. 21, this Vol.).

5.2 Mechanism of the Rearrangements Induced by Docking

As indicated above, trichocyst insertion induces a set of molecular reorganizations at the docking site. Pouphile et al. (1986) have recently reviewed the evidence indicating that the circularization of the parenthesis and the rearrangement of the plug around the trichocyst tip apparently correspond to "passive" events, induced by steric constraints. As for the rosette, the origin of its particles and the mechanism of their gathering at the centre of the ring are not known. However, it is clear that the presence of the rosette is strictly correlated with that of the underlying connecting material since both are simultaneously absent in four mutants (*nd 6, nd 7, nd 9, pt A2*) or simultaneously abnormal in a fifth one (*nd 12*). The rosette particle-connecting material system may therefore correspond to a situation of membrane-cytoskeleton interaction but which of the two partners guides the organization of the other remains unknown at the moment.

Following transport, docking and these molecular reorganizations, the trichocyst is now ready for discharge.

6 Trichocyst Discharge, Membrane Retrieval and Trichocyst Renewal

6.1 The Steps in Trichocyst Firing

6.1.1 Questions About the Natural Stimulus

The "natural" inducer of exocytosis unfortunately remains unknown at the moment (see Sect. 7) and experimenters have had to rely on a variety of chemical or physical triggering agents, most of which are probably unrelated to the physiological stimulus. The most common triggering agents, used since the very early work on *Paramecium*, have been solutions of weak acids (tannic acid, picric acid, etc.) or strong electrical stimuli. These agents appear nevertheless to converge at some point on a common pathway involving Ca^{2+} mobilization (see below). It also appears that the steps in trichocyst firing are, to a large extent, independent of the agent used.

More recently, Plattner and co-workers have introduced a very useful triggering agent which, to some extent, may be closer to a physiological inducer. Among a variety of polyamino compounds, they found aminoethyldextran (AED) to be the most effective (Plattner et al. 1985a). Both the spacing of the amino group (~ 1.0 nm) and the association to dextran are important. This agent has several features of great interest:

1. It can induce massive exocytosis (over 95% of trichocysts extruded) at relatively low concentrations (of the order of 10^{-6} M).
2. This exocytosis occurs during an interval of only a few seconds.
3. Cell viability shows virtually no impairment even after massive discharge which allows both repeated cycles of triggering to be carried out and a study of adaptation phenomena during trichocyst reinsertion.
4. The efficiency, rapidity and non-toxicity of AED renders all the recent studies on Ca^{2+} dependency of exocytosis, correlated phosphorylations, etc. more meaningful than those carried out with lethal secretagogues.

The precise mechanism of AED action remains speculative (Ca^{2+} desorption from the glycocalyx by polyamines to make it available for exocytosis, cross-linking of membrane components, etc.). It is nevertheless tempting to speculate that AED mimics the natural stimuli of exocytosis by undergoing molecular interactions with the surface of the *Paramecium* analogous to those possibly provided by presumptive predators or prey (see below).

6.1.2 Membrane Fusion and Matrix Expansion

The release of the secretory material out of the cell obviously requires fusion of the secretory vesicle's membrane with the plasma membrane, placing the contents of the vesicle in topological continuity with the extracellular medium. At the molecular level, mechanisms must ensure that lipid bilayers that were initially distinct become physically continuous. These mechanisms are actively studied on cellular and in vitro systems and there is as yet no uniform detailed model for secretory systems. The fact that, in *Paramecium,* sites of membrane fusion are predetermined, well identified and characterized by structural landmarks both within the membrane and in contact with it, has allowed a detailed ultrastructural description of the fusion events. The most original aspect of this system, however, resides in the discovery of mutants specifically affected in the final steps of the exocytotic process and which lack some of the structural landmarks.

In terms of the chronological succession of events, it should first be recalled that in addition to membrane fusion, trichocyst exocytosis is accompanied by a marked expansion of the fired matrix material which is already evident when the organelle is squeezed through the narrow cortical opening (see Fig. 8). Because of the very high speed of the whole process it is not easy to ascertain whether matrix expansion precedes or follows membrane fusion in vivo. Gilligan and Satir (1983) have documented cases of membrane fusion preceding matrix expansion in cells stimulated by picric acid and calcium. In contrast in our laboratory we have repeatedly observed that glutaraldehyde fixation induces swelling of the trichocyst matrix often not associated with membrane fusion or reorganization of the connecting material. Identical observations were reported by Garofalo et al. (1983) using osmium-glutaraldehyde fixation. Whether this swelling corresponds to a stage in matrix expansion also occurring during normal exocytosis (Garofalo et al. 1983) or, as seems more likely, to a disorganization caused by the fixative, is not established yet. Finally, Matt and Plattner (1983) have shown that cells first deprived of extracellular calcium and then immediately stimulated by dibucaine display fused membranes and only partially swollen trichocysts.

At the moment it may therefore be provisionally concluded that matrix expansion and membrane fusion are partially independent events, although the precise chronology of the two events and their causal relationship as they occur in the in vivo situation remain unsettled. The point is disturbing since its clarification would have had important implications in understanding the details of membrane fusion and the routes of Ca^{2+} transit (see below).

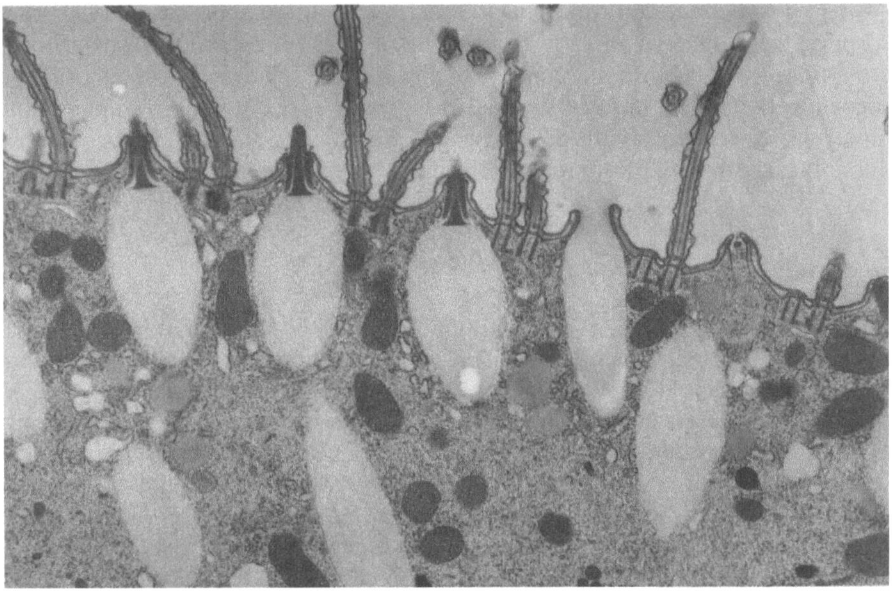

Fig. 8. A glutaraldehyde fixed cell displaying various stages of trichocyst exocytosis. All the matrices are swollen and amorphous. From *left to right:* regularly alternating with cilia, are three trichocysts with their tip emerging out of the cell (membrane fusion has already taken place). The fourth trichocyst has undergone substantial decondensation and is seen to be squeezed through the membrane openings. To the right of this trichocyst, between two cilia, is an empty trichocyst site, immediately after discharge, still displaying the collar material. (Courtesy of F. Iftode)

Irrespective of this problem, two important facts have emerged from the combined genetic/ultrastructural approach to membrane fusion in *Paramecium.*

First, the detailed analysis by Beisson and co-workers (Beisson et al. 1976, 1980; Plattner et al. 1980; Lefort-Tran et al. 1981; Pouphile et al. 1986) of a number of mutants affected in the very late stages of the secretory pathway (i.e. with docked, ultrastructurally normal trichocysts) has disclosed a number of discrete, tightly correlated structural abnormalities. The reference situation is that illustrated by the mutant *nd 9.* This conditional mutant (normal secretion at 18° C, lack of secretion at 27° C) displays the normal rosette of intramembranous particles and underlying connecting material at the permissive temperature, while it lacks both elements at the restrictive temperature. This correlation was extended to several other mutant strains as indicated earlier (sect. 5.2). These observations first demonstrate an interdependence of the rosette and connecting material assembly and secondly provide one of the strongest pieces of evidence currently available in favour of the participation of specific proteinaceous elements in membrane fusion: although in these mutants the two membrane systems are quite close to each other, the absence of rosette and connecting material totally prevents proper exocytosis. This is not simply due to lack of Ca^{2+} penetration since, contrary to an initial report (Satir and Oberg

1978), a Ca^{2+} ionophore does not restore normal exocytosis in *nd 9* (27 °C) cells, but leads instead to a "pseudo-exocytosis" process whereby the trichocyst matrix expands and the trichocyst protrudes out of the cell body by pressing on the overlying membranes. True membrane fusion fails to occur, however (Matt et al. 1980; Gilligan and Satir 1983). Some of these mutants can be transiently "cured", that is, induced to discharge trichocysts either by microinjection of wild-type cytoplasm (*nd 9;* Aufderheide 1978c; Beisson et al. 1980) or purified cytosolic proteins (Garofalo et al. 1978, *nd 9;* Cohen, pers. commun.) or by tight contact with wild-type cells during conjugation (*nd 6*), which probably allows lateral diffusion of the missing protein (Lefort-Tran et al. 1981). Using the *nd 9* mutant it was further shown that protein-lipid interactions were involved in the establishment of a functional connecting material (Beisson et al. 1980). Clearly, then, a number of cytosolic and membranar proteins and possibly membrane lipids are necessary for rosette-connecting material assembly which in turn provides an indispensable machinery for membrane fusion. In several other secretory systems there are cytological and biochemical indications for the involvement of specific proteins in membrane fusion although the absolute requirement for proteins is still debated (see Kelly et al. 1979; Plattner 1981; De Lisle and Williams 1986; Hutton 1986, for reviews). On the whole, the *Paramecium* system may be of quite general significance, its particularity along with a few other fast responding systems simply residing in the fact that the exocytotic sites are predefined. This would facilitate the visualization of the plasma membrane specializations and their underlying network which are more difficult to seize in cells displaying "random" fusions.

The second aspect of interest concerns the ultrastructural analysis of the transformations occurring during and after fusion. Following up his initial observations, Plattner has recently expanded the freeze-fracture and transmission electron microscope analysis of exocytotic sites using AED-induced discharge and fast-freezing (Plattner 1974, 1976; Olbricht et al. 1984; Plattner et al. 1985b; also see Hausmann and Allen 1976). The main observations can be summarized using Olbricht et al.'s (1984) terms: "(a) Membrane fusion starts as focal events, the smallest recognizable openings measuring 20–30 nm in diameter. (b) The exocytotic opening always forms in the center of the rosette. (c) Rosette MIPs may stay very close to the exocytotic opening, or they may partly be dispersed as the exocytotic opening is formed. (d) No diaphragm is formed during exocytotic membrane fusion. (e) The exocytotic opening is increasing to a size where it fills the total fusogenic zone contained within a ring, but not any further. (f) Rosette MIPs become further dispersed through the rings. (g) Resealing involves the transformation of rings into a collapsed form ('parenthesis'). (h) A resealed exocytosis site contains no conspicuous MIP aggregates, such as rosettes or 'annulus' structures from the trichocyst membrane, indicating a clear separation of both components."

In terms of the mechanism of membrane fusion two points are worth stressing. First, no diaphragm structure was seen at any stage, suggesting that these are not obligatory intermediary elements during membrane fusion. Second, no large-scale clearing of the rosette particles occurred *prior* to fusion. Both observations are in agreement with a "focal fusion event" model in which the apposition of the two lipid bilayers and their breakdown is confined to very small areas (Ornberg and Reese 1981; Plattner 1981; Lucy and Ahkong 1986). However, the origin of the strong driving forces necessary for intimate bilayer apposition and fusion remains elusive in *Paramecium* as in all other biological systems (Finkelstein et al. 1986; Holz 1986; Hutton 1986; Rand and Pargesian 1986). In particular, the possible role of the connecting material as a tension-generating system as opposed to (or complementary to) osmotic forces generated within the vesicle (possibly manifested as matrix expansion) is not yet clarified. In view of the demonstrated importance of the rosette particles and connecting material for membrane fusion in *Paramecium* it is nevertheless tempting to suggest that specific proteins are involved both for apposition and fusion in this biological system. In this case very interesting parallels with the "fusion proteins" of enveloped animal viruses are immediately suggested (see White et al. 1983, for review).

Whether preceding, concomitant with or posterior to membrane fusion, trichocyst matrix expansion must occur for the secretory material to be discharged out of the cell. Manipulation of the ionic concentrations both in vivo and on isolated condensed trichocysts has provided information on the mechanism of decondensation. The possibility of isolating pure trichocyst matrices in their intracellular condensed configuration (Anderer and Hausmann 1977; Matt et al. 1978; Garofalo and Satir 1984) has greatly simplified the analysis and has also enabled in vitro X-ray microanalysis of isolated organelles (Bilinski et al. 1981 a). Trichocyst matrix decondensation displays the requirement for Ca^{2+} both in vivo and in vitro. Mg and free phosphate are strong inhibitors of in vitro decondensation. Anti-calmodulin drugs have no effect in vitro. Acidic pH favours the condensed state, while basic pH favours expansion (Matt and Plattner 1983; Garofalo and Satir 1984). Bilinski et al. (1981 a) have suggested that the trichocyst matrix is kept in the condensed state by interaction with free phosphate ions (and possible Mg). Ca^{2+} influx at the moment of triggering would precipitate the phosphates thereby inducing decondensation. The whole decondensation process would not require ATP per se (Bilinski et al. 1981 a; Vilmart-Seuwen et al. 1986), but would simply correspond to a reorganization of the elaborate macromolecular matrix assemblage using conformational energy stored during biogenesis.

6.1.3 Membrane Retrieval and Trichocyst Renewal

In most exocytotic systems, the membrane of the secretory vesicle becomes incorporated into the plasma membrane after the fusion and discharge step (Palade 1975). This is compensated by active endocytosis of selected portions of the plasma membranes in order to maintain a stable cell surface area (see Heuser and Reese 1973; Farquhar 1983). The major characteristic of the *Paramecium* system, first clearly demonstrated by Hausmann and Allen (1976), is that the bulk of the trichocyst membrane does not become incorporated into the plasma membrane but is instead pinched off, fragmented within the cytoplasm and eventually recycled (see Fig. 9). This process, which lasts 5–10 min,

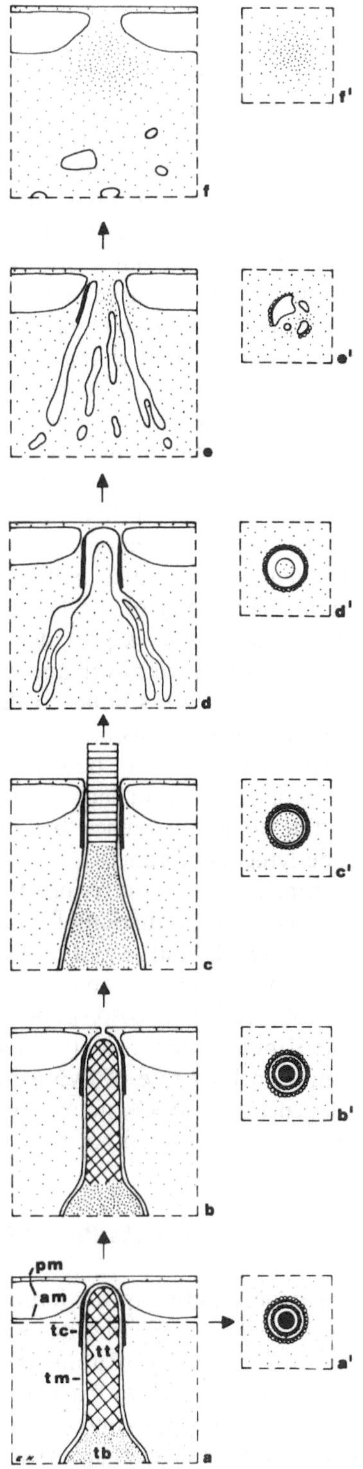

Fig. 9. Schematic presentation of the trichocyst expulsion process (**a** – **c** and **a'** – **c'**) and the destruction process of the trichocyst membranes (**d** – **f** and **d'** – **f'**) in longitudinal (*left column*) and in cross-sections (*right column*) in *Paramecium*. *am* Alveolar membrane; *pm* plasma membrane; *tb* trichocyst body; *tc* tubular collar; *tm* trichocyst membrane; *tt* trichocyst tip (Hausmann and Allen 1976)

was first studied by transmission electron microscopy of cells induced to discharge and, later, the fate of retrieved vesicles was followed by adding horseradish peroxidase (HRP) to the discharge medium (Allen and Fok 1980, 1984). The membrane resealing process itself is extremely rapid (less than a second; Plattner and Deitmer 1985c). The lack of intermixing of trichocyst and plasma membranes, which is virtually absolute (Plattner et al. 1985b), seems to be due to the conspicuous landmarks and structural barriers that separate the two systems, i.e. the collar which surrounds the trichocyst tip appears to come in contact with the outer ring of particles of the plasma membrane. The collar is resorbed after discharge, while the outer ring of particles remains in the plasma membrane as collapsed "parentheses" after resealing. Hence, the collar may not only provide a rigid channel through which the secretory material is squeezed, it may also serve as a membrane diffusion frontier (Pouphile et al. 1986; see Gumbiner and Louvard 1985, for a general discussion). Plattner et al. (1985b) have shown that membrane retrieval does not involve coated pits and that neither microtubules nor actin microfilaments appear to be involved.

Allen and Fok (1984) were able to follow the HRP marker incorporated after trichocyst discharge into several successive membrane compartments eventually ending in secondary lysosomes and digestive vacuoles after transiting in the vicinity of the Golgi apparatus where they may fuse with primary lysosomes. This work provides additional evidence for the occurrence of selective membrane retrieval and sorting mechanisms in *Paramecium* (see Allen 1984, for review), similar to those occurring in higher eukaryotes (Farquhar 1983, 1985). The possible complete recycling of "old" trichocyst membranes into new ones could not be established, however, but this may be due to breakdown of the HRP or lack of firm anchorage of the enzyme to the membranes leading to a divergence between the enzyme's destination and that of the underlying membrane.

Finally, trichocyst re-insertion has become amenable to quantitative analysis since the introduction of AED. Pape and Plattner (1985) have shown that after massive discharge, old trichocyst insertion sites are re-used and undergo the transitions described above with the possible identification of a short-lived "adaptation" stage corresponding to sites of docked trichocysts and circularized outer rings lacking a rosette. Redocking of trichocysts goes on for several hours with a docking rate of two to three organelles per minute (in cells displaying 9-h generation time) and a maximum of empty docking sites (membrane ghosts removed, plug present) at 1 h.

6.2 A Discussion of the Role of Calcium and the Mechanism of Firing

The understanding of the ionic requirements for exocytosis in *Paramecium* and the formulation of hypotheses on the mechanisms of "stimulus-secretion coupling" (Douglas and Rubin 1961) in this unicellular organism have relied heavily on concepts drawn from the analysis of neurotransmitter release in neurons and secretion in exo- and endocrine cells. The two key elements of the neuronal system with respect to our discussion are the demonstration that (1) release of neurotransmitter is activated by a transient increase in cytosolic Ca^{2+} above 10^{-6} M. (2) The Ca^{2+} ions responsible for this rise most probably originate from the extracellular space and penetrate through voltage-sensitive Ca^{2+} channels that are specifically located at the nerve terminals and are activated by the propagating nervous impulse (Katz 1969; Baker 1972). These notions were extended to typical endocrine cells several of which were also found to be excitable and critically dependent on extracellular calcium influx for exocytosis. However, it was later found that some secretory cells do not possess voltage-dependent Ca^{2+} channels, although their secretory products are released by Ca^{2+}-dependent exocytosis. Ca^{2+} release from internal stores or flux through channels not giving rise to action potentials were then advocated (see Ozawa and Sand 1986 for an extensive review). Further elaboration has resulted from the discovery of a variety of second messengers (cyclic nucleotides, phosphatidyl inositol metabolites, etc.) that act either synergistically or independently from Ca^{2+} (Gomperts 1986). In addition, a family of Ca^{2+}-binding proteins often mediating Ca^{2+} effects has been identified (Dedman 1986; Hutton 1986).

Careful critical examination of the experimental data is therefore necessary to establish the role of calcium in the activation of a cellular response (Campbell 1983). Åkerman and Nicholls (1983) have drawn a useful list of the arguments in favour of the role of calcium in transmitter release. We have added a column to their table to indicate the present state of the corresponding evidence in *Paramecium* (discussed below). It must be recalled that in *Paramecium*, much of the data is obtained by using non-natural triggering agents.

Transmitter release	Trichocyst exocytosis
1. Removal of Ca^{2+} inhibits transmission	Yes
2. Microinjection of Ca^{2+} into nerve terminals causes transmitter release in the absence of electrical impulses	No
3. An inward electrical current carried by Ca^{2+} ions occurs in activated nerve terminals	?
4. An increase in the cytosolic Ca^{2+} activity occurs upon activation of nerve terminals	Probably
5. Transmitter liberation from isolated nerve endings is activated by Ca^{2+} ionophores such as A23187	Yes

The evidence on the successive points in *Paramecium* is as follows:

1. Early work on Ca^{2+} removal by EDTA or EGTA chelation (Matt et al. 1978) followed later by using carefully buffered solutions clearly establishes that extracellular free Ca^{2+} is absolutely required for exocytosis (Plattner et al. 1985a; Cohen, pers. commun.) at least when using the usual triggering agents or AED. It should be noted that these experiments are delicate to perform since strong Ca^{2+} deprivation in the extracellular medium is rapidly lethal to *Paramecium*.

2. A striking observation is that microinjection of massive concentrations of $CaCl_2$ into living cells does not induce trichocyst discharge (Kersken et al. 1986; Perasso, pers. commun.). That the Ca^{2+} ion is not immediately sequestered (Baker et al. 1985) is evidenced by the fact that substantial deciliation of the cell occurs (a process known to be induced by Ca^{2+} shocks: Adoutte et al. 1980). This response is in marked contrast to that of most secretory cells and suggests that the trichocyst compartment may be "autonomous", that is, not able to "sense" the cytosolic Ca^{2+} concentration. It would only be receptive to the external Ca^{2+} ions or to Ca^{2+} released by a compartment that is in direct contact with the organelle.

3. The electrophysiology of secretion is still rudimentary in *Paramecium* contrary to the situation in *Didinium* in which the availability of a "natural" inducer (in the form of a *Paramecium!*) has allowed detailed electrophysiological dissection of the exocytotic steps (Hara et al. 1985). The work on *Didinium* leads to the conclusion that "the previous view that Ca^{2+} entry into the proboscis in association with the Ca-dependent depolarizing response cause the discharge of the organelles (Hara and Asai 1980) is no longer credible". A similar situation may hold for *Paramecium.* First, *Paramecium,* like *Didinium* does not respond to depolarizing or hyperpolarizing currents by trichocyst discharge unless they are extremely strong (Kung and Saimi 1982). In addition, a variety of agents inducing ciliary reversal (i.e. causing an increase in [Ca i] in the ciliary compartment) do not induce trichocyst release; thus "either there is a large difference in the threshold Ca^{2+} concentration for the two functions or the Ca^{2+} entering through the voltage-sensitive Ca-channel on the ciliary membrane stays near the ciliary lumen" (Kung and Saimi 1982). The second hypothesis, that of compartmentation of Ca^{2+} ions penetrating through the voltage-sensitive channel within the cilium and out of the somatic space is reinforced by several lines of experimental evidence recently summarized by Machemer (1986). Second, a variety of mutants partially or totally lacking the voltage-sensitive Ca channel responsible for ciliary reversal (Kung and Saimi 1982; Ramanathan et al., Chap. 15, this Vol.) display normal exocytosis (Adoutte et al. 1981; Plattner et al. 1985a). Deciliated cells also display normal exocytotic performance (Plattner et al. 1984). It is at least clear therefore that if Ca^{2+} enters the cells during exocytosis it does not do so through the voltage-sensitive Ca^{2+} channels that are responsible for *Paramecium*'s excitability. In fact, the sum of the electrophysiological evidence accumulated in *Paramecium* suggests that voltage-sensitive channels are not involved in trichocyst release. In *Didinium,* Hara et al. came to the conclusion that during discharge evoked by a prey the Ca^{2+} originates from intracellular stores.

On the whole then, one of the key elements of the transmitter model is still unproven and possibly absent in *Paramecium.*

4. The increase in cytosolic Ca^{2+} activity upon exocytotic triggering in *Paramecium* has been essentially deduced from the observation that trichocyst discharge, whatever the agent used, is paralleled by ciliary reversal, a process that is well known to depend on a rise in free Ca^{2+}. Unfortunately, most of the discharge-inducing agents also cause reversal in *nd* strains (Matt et al. 1980), indicating that ciliary reversal is independent of exocytosis per se and that the

agents probably affect the voltage-sensitive Ca^{2+} channels. Using the *d4-500* pawn mutant (totally lacking voltage-sensitive ciliary Ca channels), we could nevertheless observe in preliminary experiments that this mutant displays ciliary reversal upon trichocyst discharge (Adoutte, unpublished). The possibility that massive exocytosis causes some membrane damage and unspecific Ca^{2+} leakage into the cell has not been excluded, however. On the whole it thus seems likely but not rigorously proven that an increase in [Ca i] parallels trichocyst release.

5. Ca^{2+} ionophores have a powerful triggering activity in *Paramecium*, again strongly suggestive of a role for Ca^{2+} in exocytosis, whatever its "geographic" origin is (Plattner 1974, 1976; Satir and Oberg 1978; Matt et al. 1980).

This comparison with neurotransmitter release indicates that several critical points concerning the role of calcium in trichocyst exocytosis are not definitely settled. While it is likely that Ca^{2+} plays a crucial role in trichocyst matrix expansion both in vitro and in vivo, the origin of the Ca^{2+} ions in vivo and their routes of transit remain debatable. There are, however, four additional sets of data relevant to this discussion that can be considered at this point.

1. Several groups have carried out Ca-localization experiments at the ultrastructural level using various cytochemical methods. Plattner and Fuchs (1975) using the Oschmann and Wall (1972) method, documented intense Ca^{2+} and P deposits in trichocysts undergoing discharge (mostly in their tip region and specifically on the inner sheath), while "resting" organelles showed much less labelling. The data suggested massive Ca^{2+} influx into the organelle during expansion, possibly in combination with an ATPase system. The results were confirmed and extended by Fisher et al. (1976). The pyroantimonate labelling method (Wick and Hepler 1982) also yields intense labelling of trichocyst tips (Stelly et al. 1987). These data are in agreement with the notion of a Ca^{2+} influx within the trichocyst compartment at the time of decondensation.

2. Several Ca-ATPase activities have been described in relation to *Paramecium* surface structures.

First, a Ca^{2+}-stimulated ATPase firmly bound to the ciliary membrane has been recently characterized by Travis and Nelson (1986) and is possibly identical to that characterized in ciliary membranes by Doughty and Kaneshiro (1983, 1985). A second potent Ca^{2+}-ATPase, distinct from the tightly bound one, is recovered as a soluble enzyme in the "ciliary supernatant fraction" after deciliation induced by a Ca^{2+} shock (Riddle et al. 1982). After purification, two major bands at 68 and 53 kD are obtained on SDS gels. Their peptide maps are quite similar. The substrate and ionic optima of the enzyme have been characterized in detail and antibodies raised to the purified enzyme (Levin et al. 1987). This activity may be identical to the "deciliation supernatant ATPase" of Doughty and Kaneshiro (1983) and to the ATPase loosely bound to ciliary membranes of Andrivon et al. (1983). The function of these ATPases has not been definitely established yet. Ca^{2+}-ATPase activities more specifically associated with the "somatic membranes" (i.e. cortical, non-ciliary) have been described by Noguchi et al. (1979), Bilinski et al. (1981b) and Doughty and Kaneshiro (1983, 1985). The comprehensive work of the last authors clearly establishes the distinctness of the pellicle activities from the ciliary, soluble or intracellular organelles bound ones. In summary, a number of Ca^{2+}-ATPases (identified as bands on Triton gels) are specifically associated with a cortical fraction containing trichocysts.

Plattner et al. (1977, 1980) have provided an interesting clue to the possible function of at least some of these cortical ATPases. They demonstrated cytochemically that a Ca^{2+}-ATPase activity was associated with the exocytosis site and was strictly correlated with the presence of the rosette of intramembranous particles and/or connecting material. It is therefore suggested that an enzymatic activity is associated with the particles (and connecting material) with the possible function of maintaining a low Ca^{2+} level close to the trichocyst membrane in order to prevent uncontrolled firing. In this sense the rosette and connecting material would play the role of an "anti-fusion" device as discussed in detail by Hausmann (1978).

3. In close relation to the work just described, recent results from our laboratory indicate that highly purified cortical fractions of *Paramecium*, devoid of cilia and trichocysts but retaining intact alveoli, actively sequester Ca^{2+} in an ATP- and Mg^{2+}-dependent manner (Stelly et al. 1987). This fits well with the X-ray microanalysis data of Schmitz et al. (1985) showing substantial amounts of Ca^{2+} in the pellicle of *P. caudatum.* Ciliate alveoli may therefore correspond to closed surface membrane compartments playing an important role in the regulation of intracellular Ca^{2+} concentration. This finding has implications for exocytosis in that it suggests both a possible intracellular source for Ca^{2+} in exocytosis, located immediately adjacent to the trichocyst compartment and, in unstimulated cells, a possible sink for Ca^{2+}, favouring the maintenance of a low Ca^{2+} environment. The ultrastrucrural analogies between *Tetrahymena* alveoli and the Ca^{2+}-sequestering cisterns of nerve and muscle cells have been emphasized by Satir and Wissig (1982).

4. Anti-calmodulin drugs have been shown to prevent exocytosis by inhibition of matrix expansion in vivo (Garofalo et al. 1983), but not in vitro (Garofalo and Satir 1984). This and other results have led Garofalo and Satir (1984) to suggest that the Ca-calmodulin complex acts at the trichocyst membrane level to control Ca^{2+} penetration into the matrix. It is of interest that Momayezi et al. (1986) have recently identified calmodulin in trichocyst membranes by electron microscope immunocytochemistry.

Altogether, this set of data fits well with the suggested role of calcium in trichocyst exocytosis. An overall picture, admittedly still hypothetical in some aspects, starts to emerge: the trichocyst would correspond to a rather tightly insulated organelle in ionic terms. The lumen of the organelle would be kept low in Ca^{2+} through the joint operation of a Ca^{2+}-ATPase and regulation of Ca^{2+} diffusion by calmodulin-controlled proteins in the membrane. Fusion of the organelle's membrane with the plasma membrane would occur upon triggering by the natural inducer, using a specialized device, that of the rosette of intramembranous particles and underlying connecting material. Extracellular Ca^{2+} would rush in and promote explosive decondensation of the trichocyst matrix. The possibility of an early Ca^{2+} influx preceding membrane fusion, and originating from an intracellular or an extracellular source, remains open at the moment. Also, completely unknown for the moment is the possible role of secondary messengers (IP3, diacylglycerol, etc.). The speed of the whole discharge process in the millisecond range makes it somewhat unlikely that elaborate enzymatic processes are involved, although it is now known that IP3 release and effect can be extremely rapid (see Volpe et al. 1986, for a discussion of this point). The very rapid dephosphorylation of a 65-kD protein (Gilligan and Satir 1982) has been recently demonstrated to be associated with exocytosis performance under AED triggering (Zieseniss and Plattner 1985). The physiological role of this enzymatic step has not been fitted into the whole secretory process yet but it provides a promising line for further investigation. Finally, the possible effects of calcium on elements other than the trichocyst matrix are also open to speculation. By analogy with neurotransmitter release (Kelly et al. 1979) or secretion in exo- and endocrine cells (Hutton 1986), a variety of possible targets for Ca^{2+} can be suggested. In particular Ca^{2+} may play an important role in triggering the reorganization of the intramembranous cytoskeletal elements ("connecting material"). After firing of the elongated matrix through the rigid channel provided by the collar, rapid membrane resealing occurs and the trichocyst membrane is completely retrieved within minutes. During the following hour, local reorganizations render the empty site ready for the docking of a new trichocyst.

7 Unsolved Problems and Conclusions

The overall cycle of trichocysts as described in this chapter is summarized in Fig. 10. Two types of questions are raised by the data: a broad, "preliminary" question concerning trichocyst function and a set of specific questions concerning various steps of the trichocyst cycle. Concerning trichocyst function we have not progressed much beyond the conclusions reached by the great protozoologist Maupas in 1883: "Le rôle des trichocystes offensifs est si évident qu'il est inutile d'insister plus longuement. Pour les trichocystes défensifs, nous ne possédons aucune observation directe; mais leur rôle s'explique de soi-même par leur structure, par leur disposition périphérique et par l'impossibilité de leur en trouver un autre"[1].

We now know that under laboratory conditions, mutants with profound defects in trichocyst morphogenesis can survive, although they all have a secondary effect syndrome in the form of abnormal distribution of micro- and macronuclei at division (Ruiz et al. 1976; Cohen et al. 1980). A fairly high proportion (8/28) of a set of wild stocks kept for many years in the laboratory were found to be defective in trichocyst discharge (Nyberg 1978), although most appear to correspond to the "late" anomaly type (non-discharge) which have normal nuclear division (Cohen and Beisson 1980). Clearly then, the ability to discharge trichocysts is not vital for a *Paramecium* in a non-competitive environment. An essential role in membrane biogenesis or functioning is therefore

[1] "The role of offensive trichocysts (i.e. those of *Didinium*) is so obvious that it does not require further elaboration. As for defensive trichocysts (i.e. those of *Paramecium*), we lack direct observations; however, their role appears as self-evident by looking at their structure, their peripheral location and by the fact that it is impossible to attribute them any other function".

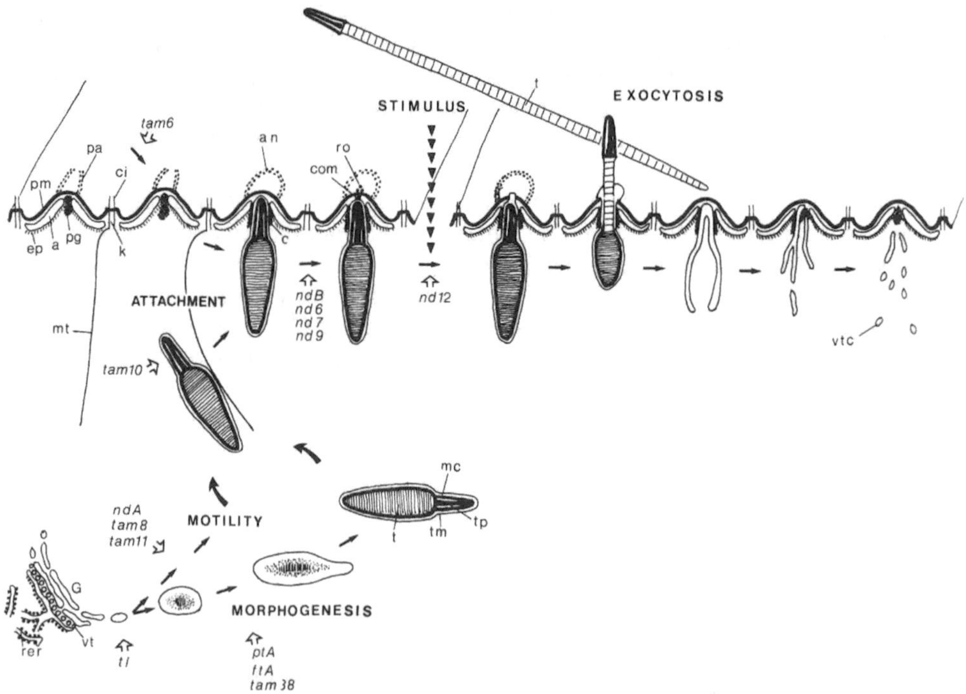

Fig. 10. A summary diagram of the trichocyst cycle in *Paramecium* including the point of impact of the major mutational lesions. The steps of trichocyst biogenesis, motility, docking firing and membrane retrieval are schematically displayed (not to scale). *a* Pellicular alveoli; *an* outer ring of intramembranous particles; *c* collar; *cil* cilium; *com* connecting material; *ep* epiplasm; *G* Golgi apparatus; *k* kinetosome; *mt* microtubule; *mc* outer sheath of trichocyst tip; *pa* parenthesis of intramembranous particles; *pg* plug; *pm* plasma membrane; *ro* rosette of intramembranous particles; *tm* trichocyst membrane; *tp* trichocyst tip; *rer* rough endoplasmic reticulum; *vt* transition vesicles; *vtc* membrane retrieval vesicles. See text for details (Cohen 1984)

quite unlikely and we are left with functions related to the interaction of the cell with its natural environment. By analogy with the clearly identified defensive or aggressive function of extrusomes in other ciliates, my preference, following Maupas, is for a similar function in *Paramecium*, although I am aware that trichocyst discharge is of little effect against some of its voracious predators (Wessenberg and Antipa 1970; Pollack 1974). In closing the discussion of this long-standing puzzle it may be mentioned that Dr. Miyake has suggested a role for trichocyst discharge in exchange of information (chemical signals?) between cells of a *Paramecium* community (pers. comm.) and that Lüthe et al. (1986) have found endogeneous lectins in trichocyst tips.

The specific questions concern unsolved aspects of the successive steps of the trichocyst cycle. From organization of the trichocyst genes to the final membrane fusion and expansion events there are now a series of precise molecular questions that can be approached, most of which have been raised throughout this review. They are presently centered around three main issues: (1) understanding the intimate molecular architecture of the trichocyst, in relation to the complexity of the trychinins and their genes, and its elaboration during trichocyst morphogenesis; (2) understanding the intracellular transit of trichocyst proteins and the mechanisms of their segregation into a distinct subcellular compartment; (3) defining a precise molecular function for the components of the sites of membrane fusion and definitely clarifying the role of calcium and possibly other messengers in fusion and discharge. It is hoped that the combination of the unique advantages of the organism and the system with the modern tools of cellular and molecular biology will continue to provide information of broad significance.

Acknowledgements. I gratefully acknowledge many ideas and comments provided by Janine Beisson, Jean Cohen and Linda Sperling. I thank all the authors who have granted me authorization to use their pictures and diagrams as well as Francine Iftode, Jean Cohen, Tadeus Gulik and Linda Sperling for unpublished micrographs and drawings. Helmut Plattner and Klaus Hausmann kindly provided careful critical reading of the manuscript. Preprints and unpublished data have been generously provided by Dave Nelson. I am grateful to Cécile Couanon for much help in the preparation of the manuscript and to Hans-Dieter Görtz for his patience. The work from our laboratory has been supported by the Université Paris-Sud, the C.N.R.S. and the Ministère de la Recherche et de l'Enseignement Supérieur (Grant No. 86 C 0591).

References

Adoutte A, Ramanathan R, Lewis RM, Dute RR, Ling KY, Kung C, Nelson DL (1980) Biochemical studies on the excitable membrane of *Paramecium tetraurelia.* III. Proteins of cilia and ciliary membranes. J Cell Biol 84:717–738

Adoutte A, Ling KY, Forte M, Ramanathan R, Nelson D, Kung C (1981) Ionic channels of *Paramecium:* from genetics and electrophysiology to biochemistry. J Physiol Paris 77:1145–1159

Adoutte A, Garreau de Loubresse N, Beisson J (1984) Proteolytic cleavage and maturation of the crystalline secretion products of *Paramecium.* J Mol Biol 180:1065–1081

Åkerman KEO, Nicholls DG (1983) Ca^{2+} transport and the regulation of transmitter release in isolated nerve endings. TIBS 8:63–64

Allen RD (1971) Fine structure of membranous and microfibrillar systems in the cortex of *Paramecium caudatum.* J Cell Biol 49:1–20

Allen RD (1984) *Paramecium* phagosome membrane: from oral region to cytoproct and back again. J Protozool 31:1–6

Allen RD, Fok AK (1980) Membrane recycling and endocytosis in *Paramecium* confirmed by horseradish peroxidase pulse-chase studies. J Cell Sci 45:131–145

Allen RD, Fok AK (1984) Membrane behavior of exocytic vesicles. III. Flow of horseradish peroxidase labeled trichocyst membrane remnants in *Paramecium.* Eur J Cell Biol 35:27–34

Allen RD, Hausmann K (1976) Membrane behavior of exocytic vesicles. I. The ultrastructure of *Paramecium* trichocysts in freeze-fracture preparations. J Ultrastruct Res 54:224–234

Anderer R, Hausmann K (1977) Properties and structure of isolated extrusive organelles. J Ultrastruct Res 60:21–26

Andrivon C, Brugerolle G, Delachambre D (1983) A specific Ca^{2+}-ATPase in the ciliary membrane of *Paramecium tetraurelia.* Biol Cell 47:351–364

Aufderheide KJ (1977) Saltatory motility of uninserted trichocysts and mitochondria in *Paramecium tetraurelia.* Science 198:299–300

Aufderheide KJ (1978a) Motility events of trichocyst insertion in *Paramecium tetraurelia.* J Protozool 25:362–365

Aufderheide KJ (1978b) Genetic aspects of intracellular motility: cortical localization and insertion of trichocysts in *Paramecium tetraurelia.* J Cell Sci 31:259–273

Aufderheide KJ (1978c) The effective sites of some mutations affecting exocytosis in *Paramecium tetraurelia.* Mol Gen Genet 165:199–205

Baker PF (1972) Transport and metabolism of calcium ions in nerve. Progr Biophys Mol Biol 24:177–223

Baker PF, Knight DE, Umbach JA (1985) Calcium clamp of the intracellular environment. Cell Calcium 6:5–14

Bannister LH (1972) The structure of trichocysts in *Paramecium caudatum.* J Cell Sci 11:899–929

Beisson J, Rossignol M (1975) Movements and positioning of organelles in *Paramecium aurelia.* In Puiseux-Dao S (ed) Molecular biology of nucleocytoplasmic relationships. Elsevier, Amsterdam, pp 291–294

Beisson J, Lefort-Tran M, Pouphile M, Rossignol M, Satir B (1976) Genetic analysis of membrane differentiation in *Paramecium.* Freeze-fracture study of the trichocyst cycle in wild-type and mutant strains. J Cell Biol 69:126–143

Beisson J, Cohen J, Lefort-Tran M, Pouphile M, Rossignol M (1980) Control of membrane fusion in exocytosis. Physiological studies on a *Paramecium* mutant blocked in the final step of the trichocyst extrusion process. J Cell Biol 85:213–227

Bilinski M, Plattner H, Matt H (1981a) Secretory protein decondensation as a distinct, Ca^{2+}-mediated event during the final steps of exocytosis in *Paramecium* cells. J Cell Biol 88:179–188

Bilinski M, Plattner H, Tiggemann R (1981b) Isolation of surface membranes from normal and exocytotic mutant strains of *Paramecium tetraurelia.* Ultrastructural and biochemical characterization. Eur J Cell Biol 24:108–115

Campbell AK (ed) (1983) Intracellular calcium. Its universal role as regulator. John Wiley & Sons, New York Chichester Brisbane Toronto Singapore, pp 1–556

Cohen J (1984) Cytosquelette et relations cortex-noyaux chez la paramécie. Thesis, Orsay Univ, pp 1–180

Cohen J, Beisson J (1980) Genetic analysis of the relationships between the cell surface and the nuclei in *Paramecium tetraurelia.* Genetics 95:797–818

Cohen J, Beisson J, Tucker JB (1980) Abnormal microtubule deployment during defective macronuclear division in a *Paramecium* mutant. J Cell Sci 44:153–167

Cohen J, Adoutte A, Grandchamp S, Houdebine LM, Beisson J (1982) Immunocytochemical study of microtubular structures throughout the cell cycle of *Paramecium.* Biol Cell 44:35–44

Dedman JR (1986) Mediation of intracellular calcium: variances on a common theme. Cell Calcium 7:297–307

De Lisle RC, Williams JA (1986) Regulation of membrane fusion in secretory exocytosis. Annu Rev Physiol 48:225–238

Doughty MJ, Kaneshiro ES (1983) Divalent cation-dependent ATPase activities associated with cilia and other subcellular fractions of *Paramecium:* an electrophoretic characterization on Triton-polyacrylamide gels. J Protozool 30:565–573

Doughty MJ, Kaneshiro ES (1985) Divalent cation-dependent ATPase activities in ciliary membranes and other surface structures in *Paramecium tetraurelia:* comparative in vitro studies. Arch Biochem Biophys 238:118–128

Douglas WW, Rubin WP (1961) The role of calcium in the secretory response of the adrenal medulla to acetylcholine. J Physiol London 159:40–57

Ehret CF, Haller G de (1963) Origin, development, and maturation of organelles and organelle systems of the cell surface in *Paramecium.* J Ultrastruct Res (Suppl) 6:1–42

Ehret CF, Powers EL (1959) The cell surface of *Paramecium.* Int Rev Cytol 8:97–133

Estève JC (1972) L'appareil de Golgi des Ciliés. Ultrastructure, particulièrement chez *Paramecium.* J Protozool 19:609–618

Estève JC (1974) Cytochimie ultrastructurale du trichocyste in situ chez *Paramecium caudatum.* Protistologica 10:371–378

Farquhar MG (1983) Intracellular membrane traffic: pathways, carriers, and sorting devises. Meth Enzymol 98:1–13

Farquhar MG (1985) Progress in unraveling pathways of Golgi traffic. Annu Rev Cell Biol 1:447–488
Finkelstein A, Zimmerberg J, Cohen FS (1986) Osmotic swelling of vesicles: its role in the fusion of vesicles with planar phospholipid bilayer membranes and its possible role in exocytosis. Annu Rev Physiol 48:163–174
Fisher G, Kaneshiro ES, Peters PD (1976) Divalent cation affinity sites in *Paramecium aurelia.* J Cell Biol 69:429–442
Garofalo RS, Satir BH (1984) *Paramecium* secretory granule content: quantitative studies on in vitro expansion and its regulation by calcium and pH. J Cell Biol 99:2193–2199
Garofalo RS, Knowles JKS, Satir BH (1978) Restoration of secretory capacity in a non-discharge *Paramecium* mutant by micro-injection of wild-type cytoplasmic factor(s). J Cell Biol 79:245a (Abstr)
Garofalo RS, Gilligan DM, Satir BH (1983) Calmodulin antagonists inhibit secretion in *Paramecium.* J Cell Biol 96:1072–1081
Garreau de Loubresse N, Adoutte A, Beisson J (1983) Maturation des produits de secrétion chez la paramécie: effets de la monensine et de mutations. Biol Cell 48:52a
Geraghty DE, Messing J, Rubenstein I (1982) Sequence analysis and comparison of cDNAs of the zein multigene family. EMBO J 1:1329–1335
Gilligan DM, Satir BH (1982) Protein phosphorylation/dephosphorylation and stimulus-secretion coupling in wild type and mutant *Paramecium.* J Biol Chem 257:13903–13906
Gilligan DM, Satir BH (1983) Stimulation and inhibition of secretion in *Paramecium:* role of divalent cations. J Cell Biol 97:224–234
Comperts BD (1986) Calcium shares the limelight in stimulus-secretion coupling. Trends Biochem Sci 11:290–292
Gumbiner B, Louvard D (1985) Localized barriers in the plasma membrane: a common way to form domains. Trends Biochem Sci 10:435–438
Hara R, Asai H (1980) Electrophysiological responses of *Didinium nasutum* to *Paramecium* capture and mechanical stimulation. Nature (London) 283:869–870
Hara R, Asai H, Naitoh Y (1985) Electrical responses of the carnivorous ciliate *Didinium nasutum* in relation to discharge of the extrusive organelles. J Exp Biol 119:211–224
Hausmann K (1978) Extrusive organelles in protists. In: Bourne GH, Danielli JF (eds) International review of cytology, vol 52. Academic Press, New York San Francisco London, pp 197–276
Hausmann K, Allen RD (1976) Membrane behavior of exocytic vesicles. II. Fate of the trichocyst membranes in *Paramecium* after induced trichocyst discharge. J Cell Biol 69:313–326
Hausmann W, Stockem W, Wohlfarth-Bottermann KE (1972a) Cytologische Studien an Trichocysten. I. Die Feinstruktur der gestreckten Spindeltrichocyste von *Paramecium caudatum.* Cytobiologie 5:208–227
Hausmann K, Stockem W, Wohlfarth-Bottermann KE (1972b) Cytologische Studien an Trichocysten. II. Die Feinstruktur ruhender und gehemmter Spindeltrichocysten von *Paramecium caudatum.* Cytobiologie 5:228–246
Heuser JE, Reese TS (1973) Evidence for recycling of synaptic vesicle membrane during transmitter release at the frog neuromuscular junction. J Cell Biol 57:315–344
Holz RW (1986) The role of osmotic forces in exocytosis from adrenal chromaffin cells. Annu Rev Physiol 48:175–189
Hovasse R, Mignot JP (1975) Trichocystes et organites analogues chez les protistes. Ann Biol 14:397–422
Hutton JC (1986) Calcium-binding proteins and secretion. Cell Calcium 7:339–352
Janisch R (1972) Pellicle of *Paramecium caudatum* as revealed by freeze-etching. J Protozool 19:470–473
Jurand A, Saxena DM (1974) Ultrastructural features of the trichless mutant of *Paramecium aurelia.* Acta Protozool 12:307–312
Jurand A, Selman GG (eds) (1969) The anatomy of *Paramecium aurelia.* MacMillan St. Martin, London, pp 1–218
Katz B (ed) (1969) The release of neural transmitter substances. Liverpool Univ Press, pp 1–60

Kelly RB, Deutsch JW, Carlson SS, Wagner JA (1979) Biochemistry of neurotransmitter release. Annu Rev Neurosci 2:399–446

Kersken H, Tiggemann R, Westphal C, Plattner H (1984) The secretory contents of *Paramecium tetraurelia* trichocysts: ultrastructural-cytochemical characterization. J Histochem Cytochem 32:179–192

Kersken H, Momayezi M, Braun C, Plattner H (1986) Filamentous actin in *Paramecium* cells: functional and structural changes correlated with phalloidin affinity labeling in vivo. J Histochem Cytochem 34:455–465

Kung C, Saimi Y (1982) The physiological basis of taxes in *Paramecium*. Annu Rev Physiol 44:519–534

Lefort-Tran M, Aufderheide K, Pouphile M, Rossignol M, Beisson J (1981) Control of exocytotic processes: cytological and physiological studies of trichocyst mutants in *Paramecium tetraurelia*. J Cell Biol 88:301–311

Levin AE, Travis SM, DeVito LD, Nelson DL (1987) Purification and characterization of a calcium-dependent ATPase from *Paramecium tetraurelia*. J Biol Chem (submitted)

Lüthe N, Plattner P, Haacke G, Walther P, Müller M (1986) Lectin binding sites in *Paramecium tetraurelia* cells. I. Labeling analysis predominantly of secretory components. Histochemistry 85:365–376

Lucy JA, Ahkong QF (1986) An osmotic model for the fusion of biological membranes. FEBS Lett 199:1–11

Machemer H (1986) Electromotor coupling in cilia. In: Lüttgau HC (ed) Membrane control of cellular activity, Fortschr Zool/Progr Zool 33:205–250

Matt H, Plattner H (1983) Decoupling of exocytotic membrane fusion from protein discharge in *Paramecium* cells. Cell Biol Int Rep 7:1025–1031

Matt H, Bilinski M, Plattner H (1978) Adenosine triphosphate, calcium and temperature requirements for the final steps of exocytosis in *Paramecium* cells. J Cell Sci 32:67–86

Matt H, Plattner H, Reichel K, Lefort-Tran M, Beisson J (1980) Genetic dissection of the final exocytosis steps in *Paramecium tetraurelia* cells: trigger analyses. J Cell Sci 46:41–60

Maupas E (1883) Contribution à l'étude morphologique et anatomique des infusoires ciliés. Arch Zool Exp Gen 1:427–664

Miller DM, Jahn TL, Fonseca JR (1968) Anodal contraction of *Paramecium* protoplasm. J Protozool 15:493–497

Momayezi M, Kersken H, Gras U, Vilmart-Seuwen J, Plattner H (1986) Calmodulin in *Paramecium tetraurelia:* localization from the in vivo to the ultrastructural level. J Histochem Cytochem 34:1621–1638

Noguchi M, Inoué H, Kubo K (1979) Calcium-activated adenosine triphosphatase activity of pellicles from *Paramecium caudatum*. J Biochem 85:367–373

Nyberg D (1978) Genetic analysis of trichocyst discharge of the wild stocks of *Paramecium tetraurelia*. J Protozool 25:107–112

Olbricht K, Plattner H, Matt H (1984) Synchronous exocytosis in *Paramecium* cells. II. Intramembranous changes analysed by freeze-fracturing. Exp Cell Res 151:14–20

Ornberg RL, Reese TS (1981) Beginning of exocytosis captured by rapid-freezing of *Limulus* amebocytes. J Cell Biol 90:40–54

Oschmann JL, Wall BJ (1972) Calcium binding to intestinal membranes. J Cell Biol 55:58–73

Ozawa S, Sand O (1986) Electrophysiology of excitable endocrine cells. Physiol Rev 66:887–952

Palade G (1975) Intracellular aspects of the process of protein synthesis. Science 189:347–358

Pape R, Plattner H (1985) Synchronous exocytosis in *Paramecium* cells. V. Ultrastructural adaptation phenomena during re-insertion of secretory organelles. Eur J Cell Biol 36:38–47

Peterson JB, Heuser JE, Nelson DL (1987) Dissociation and reassociation of trichocyst proteins: biochemical and ultrastructural studies. J Cell Sci 87:3–25

Pitelka DR (ed) (1963) Electron-microscopic structure of protozoa. Pergamon, New York Oxford London Paris, pp 1–269

Pitelka DR (1965) New observations on cortical ultrastructure in *Paramecium*. J Microsc 4:373–394

Plattner H (1974) Intramembranous changes on cationophore-triggered exocytosis in *Paramecium.* Nature (London) 252:722–724
Plattner H (1976) Membrane disruption, fusion and resealing in the course of exocytosis in *Paramecium* cells. Exp Cell Res 103:431–435
Plattner H (1981) Membrane behaviour during exocytosis. Cell Biol Int Rep 5:435–459
Plattner H, Fuchs S (1975) X-ray microanalysis of calcium binding sites in *Paramecium* with special reference to exocytosis. Histochemistry 45:23–47
Plattner H, Miller F, Bachmann L (1973) Membrane specializations in the form of regular membrane-to-membrane attachment sites in *Paramecium.* A correlated freeze-etching and ultrathin-sectioning analysis. J Cell Sci 13:687–719
Plattner H, Reichel K, Matt H (1977) Bivalent-cation-stimulated ATPase activity at preformed exocytosis sites in *Paramecium* coincides with membrane-intercalated particle aggregates. Nature (London) 267:702–704
Plattner H, Reichel K, Matt H, Beisson J, Lefort-Tran M, Pouphile M (1980) Genetic dissection of the final exocytosis steps in *Paramecium tetraurelia* cells: cytochemical determination of Ca^{2+}-ATPase activity over preformed exocytosis sites. J Cell Sci 46:17–40
Plattner H, Westphal C, Tiggemann R (1982) Cytoskeleton-secretory vesicle interactions during the docking of secretory vesicles at the cell membrane in *Paramecium tetraurelia* cells. J Cell Biol 92:368–377
Plattner H, Matt H, Kersken H, Haacke B, Stürzl R (1984) Synchronous exocytosis in *Paramecium* cells. I. A novel approach. Exp Cell Res 151:6–13
Plattner H, Stürzl R, Matt H (1985a) Synchronous exocytosis in *Paramecium* cells. IV. Polyamino compounds as potent trigger agents for repeatable trigger-redocking cycles. Eur J Cell Biol 36:32–37
Plattner H, Pape R, Haacke B, Olbricht K, Westphal C, Kersken H (1985b) Synchronous exocytosis in *Paramecium* cells. VI. Ultrastructural analysis of membrane resealing and retrieval. J Cell Sci 77:1–17
Plattner H, Deitmer JW (1985c) A synchronous exo-endocytosis system: the *Paramecium* cell as analyzed by structural and electrophysiological methods. Eur J Cell Biol 36:51 a
Pollack S (1974) Mutations affecting the trichocysts in *Paramecium aurelia.* I. Morphology and description of the mutants. J Protozool 21:352–362
Pollack S, Steers E Jr (1973) Thermal solubilization of the trichocysts from *Paramecium aurelia.* Exp Cell Res 78:186–190
Pouphile M, Lefort-Tran M, Plattner H, Rossignol M, Beisson J (1986) Genetic dissection of the morphogenesis of exocytosis sites in *Paramecium.* Biol Cell 56:151–162
Rand RP, Parsegian VA (1986) Mimicry and mechanism in phospholipid models of membrane fusion. Annu Rev Physiol 48:201–212
Rauh JJ, Nelson DL (1981) Calmodulin is a major component of extruded trichocysts from *Paramecium tetraurelia.* J Cell Biol 91:860–865
Riddle LM, Rauh JJ, Nelson DL (1982) A Ca^{2+}-activated ATPase specifically released by Ca^{2+} shock from *Paramecium tetraurelia.* Biochim Biophys Acta 688:525–540
Ruiz F, Adoutte A, Rossignol M, Beisson J (1976) Genetic analysis of morphogenetic processes in *Paramecium.* I. A mutation affecting trichocyst formation and nuclear division. Genet Res Cambr 27:109–122
Satir BH, Oberg SG (1978) *Paramecium* fusion rosettes: possible function as Ca^{2+} gates. Science 199:536–538
Satir BH, Schooley C, Kung C (1972) Internal membrane specializations in *P. aurelia.* J Cell Biol 55:227a
Satir B, Wissig S (1982) Alveolar sacs of *Tetrahymena:* ultrastructural characteristics and similarities to subsurface cisterns of muscle and herre. J Cell Sci 55:13–33
Schmitz N, Meyer R, Zierold K (1985) X-ray microanalysis in cryosections of natively frozen *Paramecium caudatum* with regard to ion distribution in ciliates. Scanning Electron Microscopy I:433–445
Selman GG, Jurand A (1970) Trichocyst development during the fission cycle of *Paramecium.* J Gen Microbiol 60:365–372
Sonneborn TM (1970) Gene action in development. Proc R Soc London Ser B 176:347–366

Sperling L, Tardieu A, Gulick-Krzywicki T (1987) Structural studies of *Paramecium* trichocysts: determination of the crystal lattice by X-ray diffraction and freeze-fracture electron microscopy. J Cell Biol 105:1649–1662

Steers E Jr, Beisson J, Marchesi VT (1969) A structural protein extracted from the trichocysts of *Paramecium aurelia*. Exp Cell Res 57:392–396

Stelly N, Mauger JP, Adoutte A, Claret M (1987) Le cortex de la Paramécie: un nouveau compartiment calcique. Biol Cell 60:94 a

Tindall SH (1986) Selection of chemical spacers to improve isoelectric focusing resolving power: implications for use in two-dimensional electrophoresis. Anal Biochem 159:287–294

Travis SM, Nelson DL (1986) Characterization of Ca^{2+}- or Mg^{2+}-ATPase of the excitable ciliary membrane from *Paramecium tetraurelia:* comparison with a soluble Ca^{2+}-dependent ATPase. Biochim Biophys Acta 862:39–48

Vilmart J, Plattner H (1983) Membrane-integrated proteins at preformed exocytosis sites. J Histochem Cytochem 31:626–632

Vilmart-Seuwen J, Kersken H, Stürzl R, Plattner H (1986) ATP keeps exocytosis sites in a primed state but is not required for membrane fusion: an analysis with *Paramecium* cells in vivo and in vitro. J Cell Biol 103:1279–1288

Volpe P, Di Virgilio F, Pozzan T, Salviati G (1986) Rôle of inositol 1,4,5-triphosphate in excitation-contraction coupling in skeletal muscle. FEBS Lett 197:1–4

Wessenberg H, Antipa G (1970) Capture and ingestion of *Paramecium* by *Didinium nasutum*. J Protozool 17:250–270

Westphal C, Plattner H (1981 a) Ultrastructural analysis of the cell membrane-secretory organelle interaction zone in *Paramecium tetraurelia* cells. I. In situ characterization by electron "staining" and enzymatic digestion. Biol Cell 42:125–140

Westphal C, Plattner H (1981 b) Ultrastructural analysis of the cell membrane-secretory organelle interaction zone in *Paramecium tetraurelia* cells. II. Changes involved in the attachment of secretory organelles. Biol Cell 42:141–146

White J, Kielian M, Helenius A (1983) Membrane fusion proteins of enveloped animal viruses. Q Rev Biophys 16:151–195

Wick SM, Hepler PK (1982) Selective localization of intracellular Ca^{2+} with potassium antimonate. J Histochem Cytochem 30:1190–1204

Wohlfarth-Bottermann KE (1953) Experimentelle und elektronenoptische Untersuchungen zur Funktion der Trichocysten von *Paramecium caudatum*. Arch Protistenkd 98:169–226

Yusa A (1963) An electron microscope study on regeneration of trichocysts in *Paramecium caudatum*. J Protozool 10:253–262

Zieseniss E, Plattner H (1985) Synchronous exocytosis in *Paramecium* cells involves very rapid ($\leqq 1$ s), reversible dephosphorylation of a 65-kD phosphoprotein in exocytosis-competent strains. J Cell Biol 101:2028–2035

Chapter 21 The Cytoskeleton

JEAN COHEN and JANINE BEISSON[1]

CONTENTS

1 Introduction

The cytoskeleton of ciliates, composed of a variety of fibrillar and tubular networks, has long been described by electron microscopy and its role in cell motility and maintenance of cell shape recognized (see Allen, Chap. 2, this Vol.). Interest in cell architecture was renewed a little over a decade ago, when the development of immunocytochemical methods (Lazarides and Weber 1974; Weber 1976) disclosed the presence of microtubules, actin microfilaments and intermediate filament networks spanning the cytoplasm of mammalian cells. With the description of the nucleoskeleton (Franke et al. 1981) and of fibrous networks underlying plasma membranes (Branton 1981), the concept of cy-

1 Centre de Génétique Moléculaire, C.N.R.S., F-91190 Gif-sur-Yvette, France

Paramecium. Ed. by H.-D. Görtz

Table 1. Cytoskeletal elements identified in *Paramecium*

Cytoskeletal element	Fine structure and localization	Main references [a]
Basal bodies, cilia	Microtubules and associated structures in the cortex and the gullet	5, 10, 14, 17, 19, 28
Ribbons (transverse and post ciliary)	4 to 6 microtubules associated with each basal body	7, 10, 14, 16, 19
Cytospindle	Bundles of 4 to 20 longitudinal microtubules in the cortex of dividing cells	7, 10, 14, 23 28
Kinetodesmal fibers	Striated fibers anchored at the base of basal bodies	1, 10, 12, 14, 18, 19, 28
Epiplasm	Granular layer apposed to the inner membrane of subpellicular alveoli	1, 10, 14, 28
Outer lattice	Network of granulofibrillar material running along the top of surface ridges	1, 9, 10, 18, 27
Striated bands	Microfilaments connecting epiplasm of adjacent units	1, 10, 28
Infraciliary lattice	Network of bundles of microfilaments at the ecto-/endoplasmic boundary	1, 10, 14, 18, 19, 27, 28
Contractile ring	Bundle of microfilaments in the fission furrow	8, 14
Cord	Bundle of microfilaments on the left side of the gullet	2, 14, 28
Filamentous reticulum	Network of microfilaments with nodes underlying the gullet	2, 10, 14, 19, 28
Cytopharyngeal ribbons	Microtubules radiating from the gullet towards the center of the cell	2, 19, 28
Postoral fibers	Bundles of microtubules running from the gullet towards the posterior pole of the cell	7, 10, 14, 15 19, 28

toskeleton evolved and can now be considered to encompass any fibrous, filamentous or tubular cellular structure insoluble in non-ionic detergents. As more and more organisms or cell types are examined, the diversity of cytoskeletons appears quite large, be it for the geometry of similar structures (for instance microtubule arrays, see Tucker 1979) or for the chemical composition of equivalent structures in different cells, for instance intermediate filaments (Anderton 1981; Lazarides 1982).

Among these diverse cytoskeletal networks, microtubules and actin microfilaments are ubiquitous in eukaryotes and have been the most studied. Their major component molecules have been purified and sequenced, tubulin and actin genes have been cloned and sequenced in a number of species and the dynamics of assembly-disassembly of the networks studied in vitro and in vivo. Altogether, these studies have led to several important notions: (1) The most ubiquitous cytoskeletal proteins, tubulins and actins, are highly conserved.

Table 1 (continued)

Cytoskeletal element	Fine structure and localization	Main references[a]
Internal meshwork composed of:	Isolated microtubules spanning the whole cytoplasm	this paper
cortical microtubules	Microtubules probably anchored at basal bodies perpendicular to the cortex	20, this paper
endoplasmic meshwork	Microtubules likely to be in continuity with cytopharyngeal ribbons	this paper
Backbone of contractile vacuoles (pore and roots)	Microtubules anchored in the cortex with a plug of actin microfilaments	7, 8, 10, 13, 14, 19, 28
Backbone of the cytoproct	Microtubules anchored in the cortex with a plug of actin microfilaments	3, 7, 8, 19, 28
Microfilaments associated with phagocytosis	Actin microfilaments around forming pouch and newly formed digestive vacuoles	4, 8
Macronuclear spindle	Longitudinal microtubules under the nuclear envelope	7, 22, 25
Mitotic spindle	Spindle of microtubules in the micronuclei	7, 10, 21, 26
Separation spindle	Bundle of macrotubules in the micronuclear envelope after mitosis	6, 7, 11, 24, 26

[a] References: 1. Allen 1971; 2. Allen 1974; 3. Allen and Wolf 1974; 4. Allen and Fok 1983a; 5. Chatton and Lwoff 1936; 6. Cohen et al. 1980; 7. Cohen et al. 1982; 8. Cohen et al. 1984; 9. Cohen et al. 1987; 10. Ehret and McArdle 1974; 11. Eichenlaub-Ritter and Tucker 1984; 12. Fernandez-Galiano 1966; 13. Hausman and Allen 1977; 14. Jurand and Selman 1969; 15. Lünd 1933; 16. Lynn 1981; 17. Machemer, Chap. 14, this Vol.; 18. Parducz, 1962; 19. Pitelka 1969; 20. Plattner et al. 1982; 21. Stevenson and Lloyd 1971a; 22. Stevenson and Lloyd 1971b; 23. Sundararaman and Hanson 1976; 24. Tucker 1979; 25. Tucker et al. 1980; 26. Tucker et al. 1985; 27. Gelei von 1937; 28. Wichterman 1986.

(2) The cytoskeletal networks undergo cell cycle-dependent reorganizations which involve a regulated equilibrium between monomers and supramolecular structures, while numerous associated proteins control assembly, disassembly and interactions with other organelles, as well as conversion of one type of network into another one (e.g. cytoplasmic network vs mitotic spindle in the case of microtubules). (3) Although microtubules and microfilaments can be self-assembled in vitro, the development and maintenance of their in vivo geometry obviously requires various (and generally still unknown) signals and molecular interactions. (4) These networks play an essential role in the control of cell shape, intracellular movements and cell motility.

In this context, the diverse cytoskeletal structures (listed in Table 1) present in *Paramecium*, as well as in other protozoa (Grain 1986), raise interesting questions concerning their nature, function, dynamics and evolutionary conservation.

Except for microtubules, whose distinctive architecture is unambiguously identified by electron microscopy, little or no information was available on the chemical composition or dynamics of the other cytologically defined networks until the problem was attacked a few years ago by cell fractionation and by immunocytochemical methods. Immunocytochemistry, whose irreplaceable power makes it the method of choice in the study of the cytoskeleton, has two main advantages. Firstly, a structure labelled by an antibody can be both visualized at the whole cell level by immunofluorescence and analyzed by electron microscopy to ascertain the site and specificity of the decoration. Secondly, many antibodies (but not all, unfortunately) react with their antigen whether in the native state or denatured on immunoblots after SDS-polyacrylamide gel electrophoresis of cell extracts. The structure recognized can therefore be correlated with the polypeptides involved, opening the way towards chemical characterization or purification of the antigen(s).

In the immunological study of *Paramecium* cytoskeleton, two strategies were followed. (1) In light of the assumed evolutionary conservation of the major cytoskeletal proteins, we screened on *Paramecium* a number of antibodies raised against cytoskeletal proteins of other organisms. This approach turned out to be often (although not always) deceptive and disclosed or confirmed the wide evolutionary divergence between protozoa and metazoa and even among protozoa (see Williams 1986a). (2) Another approach was to raise antibodies against cytoskeletal proteins of *Paramecium* separated on SDS gels from purified cortex (Keryer et al., in prep.) or axonemes (Cohen et al. 1982).

This chapter presents the data available on the chemical characterization of a few cytoskeletal proteins, and on the dynamics and possible functions of the major networks, as deduced from observations of wild-type and morphological mutants.

2 Elements of the Cytoskeleton of *Paramecium tetraurelia*

2.1 Microtubules and Tubulin

2.1.1 Organization and Dynamics of Microtubules

Among the cytoskeletal elements listed in Table 1, a number contain microtubules as a major component. Twelfe microtubular arrays can be distinguished cytologically: (1) basal bodies, (2) cilia, (3) transverse and postciliary ribbons associated with each basal body, (4) backbone of contractile vacuole pores and roots, (5) backbone of the cytoproct, (6) postoral fibers, (7) cytopharyngeal ribbons, (8) internal meshwork, (9) cytospindle (referred to as suprakinetodesmal microtubules by Sundararaman and Hanson 1976), (10) intranuclear microtubules in the dividing macronucleus (Tucker et al. 1980), (11) mitotic spindle within the micronucleus and (12) intranuclear macrotubules [30 nm diameter (Tucker 1979; Tucker et al. 1985), with more than 13 protofilaments (Eichenlaub-Ritter and Tucker 1984)] in the micronuclear separation spindle.

Table 2. Properties of the 12 different microtubular systems of *Paramecium*

	Cell cycle dependence			Sensitivity[a] to		Reactivity with different antibodies[b]					
	Permanent	Interphase-specific	Division-specific	Cold	Noco-dazole	(1)	(2)	(3)	(4)	(5)	(6)
Basal bodies	+	−	−	−	−	+	+	+	+	+	±
Cilia	+	−	−	−	−	+	+	±	−	−	−
Ribbons	+	−	−	−	−	+	+	+	+	+	±
Backbone of contractile vacuoles	+	−	−	−	−	+	+	+	+	+	+
Backbone of cytoproct	+	−	−	−	−	±	+	−	+		+
Postoral fibers	−	+	−	−	−	+	+	+	+	+	+
Cytopharyngeal ribbons	+	−	−	−	−	±	+	+	+	+	+
Internal meshwork	−	+	−	+	+	±	+	−			+
Cytospindle	−	−	+	−	±	+	+	+	+	+	+
Macronuclear spindle	−	−	+	−	±	±	+	−			+
Mitotic spindle	−	−	+	+	±	±	+	+			+
Separation spindle	−	−	+	−	±	+	+	+			+

[a] The data on sensitivity to cold or nocodazole were obtained in the laboratory (Keryer, Rossignol and Torres, pers. commun.).

[b] The different antibodies tested were:

(1) Anti-axonemal tubulin of *Paramecium* (Cohen et al. 1982) and anti-axonemal tubulin of *Tetrahymena* (Barahona and Rodrigues-Pousada 1983). Data with this latter antiserum were from A. Adoutte (pers. commun.).

(2) Anti-α and anti-β-tubulin available from Amersham (Blose et al. 1982).

(3) Monoclonal antibody 6-11B-1 (kindly provided by G. Piperno) against a specific epitope corresponding to acetylation of α-tubulin (Piperno and Fuller, 1985) on lysine residue (L'Hernault and Rosenbaum 1985).

(4) Rat monoclonal antibody raised against yeast tubulin (Kilmartin et al. 1982), commercially available from Sera-Lab, which was shown to react with the tyrosylated form of α-tubulin (Wehland et al. 1983). In addition to microtubules, this antibody labels the trichocysts of *Paramecium* and largely obscures the interior of the cell (indicated by a blank in the column).

(5) Polyclonal anti-tubulin from pig brain (Karsenti et al. 1978), a generous gift of D. Pantaloni, which reacts not only with microtubules but also with the trichocysts of *Paramecium*, and then masks the interior of the cell (indicated by a blank in the column).

(6) Monoclonal antibody, B1B2, obtained in the laboratory by immunizing mice with whole cell extracts (Deregnaucourt et al., in prep.), which is specific for microtubules as controlled by immunofluorescence and immunoelectron microscopy (Garreau de Loubresse, unpubl.), but whose protein specificity is still unknown, as it does not react on Western blots.

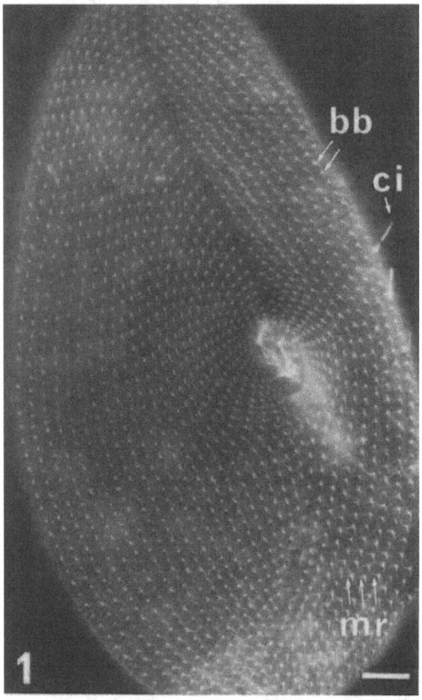
bb
ci
mr
1

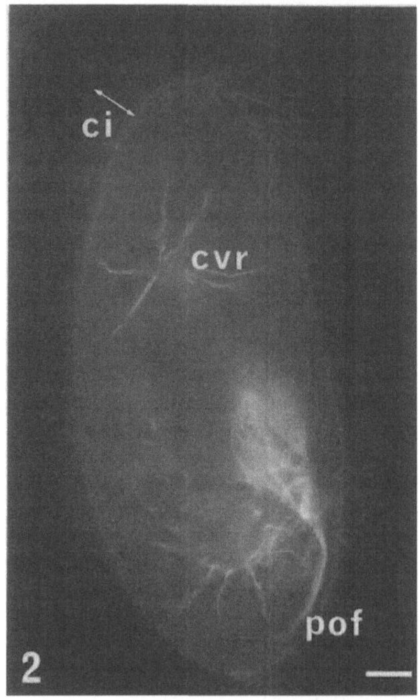
ci
cvr
pof
2

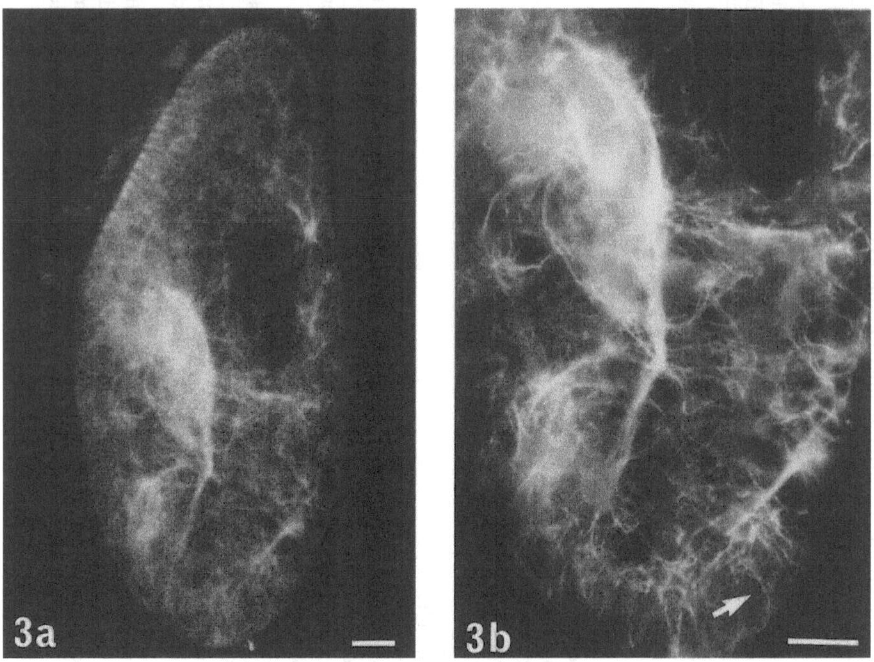
3a
3b

In order to visualize microtubules at the whole cell level, various anti-tubulin antibodies were used (Table 2) and all decorated at least some of the microtubule arrays. As microtubules contain proteins other than tubulins (microtubule associated proteins: MAPs, see Olmsted et al. 1984, for instance) corresponding antibodies could also serve as microtubule markers. However, no labelling was obtained in limited experiments using a few antibodies against well-characterized MAPs (MAP1, MAP2, Tau).

Examples of decoration are shown in Fig. 1 – 5 (see also Cohen et al. 1982; Cohen 1984). Aside from antibodies which label all the microtubular arrays of *Paramecium* (anti-α and anti-β-tubulin of Blose et al. 1982), other antibodies preferentially decorate some microtubule subsets and not others: anti-axonemal tubulin from both *Paramecium* and *Tetrahymena* preferentially label axonemes, cortical microtubules (Fig. 1) and thick bundles of intracellular microtubules such as postoral fibers, contractile vacuole roots and the micronuclear separation spindles. They decorate poorly, or not at all, the micronuclear spindle, the microtubules within the dividing macronucleus and the internal meshwork. Conversely, the anti-α-tyrosylated tubulin of yeast and the anti-pig brain tubulin decorate the axonemes only weakly or not at all but label most of the other arrays. The internal meshwork is also strongly decorated by the anti-*Paramecium* monoclonal antibody B1B2 (Fig. 3) but not by the monoclonal 6-11B-1 against *Chlamydomonas* acetylated α-tubulin which binds to most of the other microtubules arrays (see Fig. 2).

These differences in immunological reactivity (listed in Table 2, along with differences in sensitivity to depolymerizing agents) among the 12 microtubule arrays may reflect either differences in accessibility to antibodies or differences in the tubulin molecules composing the various microtubular systems as will be discussed below.

A new important observation reported here is the visualization of the internal meshwork. This meshwork seems to be anchored or nucleated at two sites, the gullet and the basal bodies (Fig. 3). In the ectoplasmic zone, individual microtubules probably originating from basal bodies (Plattner et al. 1982)

Figs. 1 – 3. Main microtubule systems in interphase cells

Fig. 1. Immunofluorescent decoration of the cortex of a deciliated cell given by the anti-*Paramecium* axonemal tubulin (Cohen et al. 1982). *Ci* cilium remaining despite the deciliation treatment; *bb* basal body; *mr* microtubular ribbons. Bar = 10 µm. (Courtesy of F. Ruiz)

Fig. 2. Decoration by the monoclonal 6-11B-1 against acetylated α-tubulin. The picture is focussed in the center of the cell, where postoral fibers (*pof*) and contractile vacuole roots (*cvr*) are visible. This labelling indicates that these structures likely contain acetylated α-tubulin molecules. Note that cilia (*ci*) are poorly labelled. Bar = 10 µm

Fig. 3 a,b. Immunofluorescent labelling given by the monoclonal antibody B1B2 specific for internal microtubules. **a** General view of the cell. **b** Enlargement of the posterior part of the same cell. Two domains can be distinguished according to their organization: an ectoplasmic one where microtubules arise and run perpendicularly from the cortex along approximately one trichocyst length (*arrow*) and an endoplasmic one where microtubules emanating from the gullet span the entire volume of the cytoplasm in a more sinuous way. Bar = 10 µm

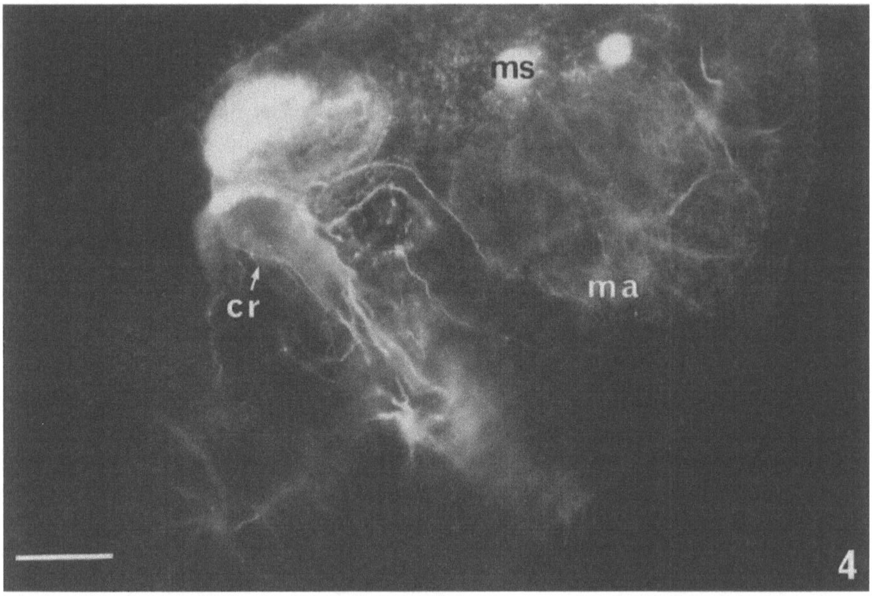

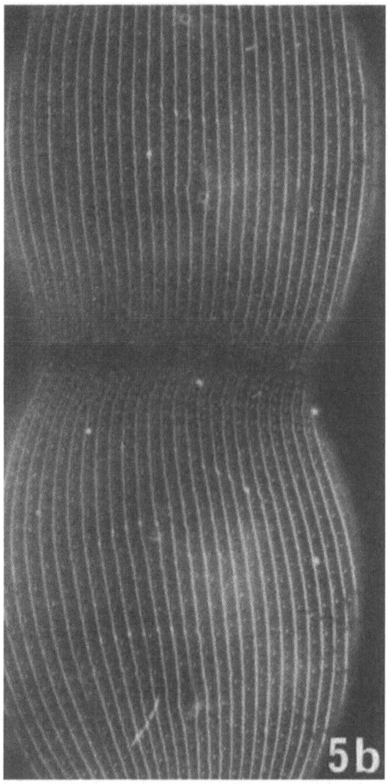

Figs. 4 and 5. Main microtubular systems in dividing cells

Fig. 4. Early dividing deciliated cell labelled by the monoclonal anti-α-tubulin (Blose et al. 1982). In order to mask cilia which are highly reactive with this antibody, the cells were first treated with the anti-axonemal tubulin antibody (rabbit serum) then incubated in the monoclonal antibody and in the fluorescent anti-mouse Ig second antibody. Microtubules of the internal meshwork, normally decorated by the antibody, have almost completely depolymerized. Cytopharyngeal ribbons (*cr*) are still present. At this stage microtubules spanning the rounded macronucleus (*ma*) are clearly visible (see also Tucker et al. 1980, for an electron microscope study) and mitotic spindles (*ms*) within the micronuclei (out of focus) are brightly stained. The fluorescence of contractile vacuole roots (out of focus) overlaps the labelling of intramacronuclear microtubules. Bar = 10 μm

run perpendicular to the cortex over a few micrometers (approximately one trichocyst length), while in the endoplasm they appear more sinuous and entangle with the network arising from several bundles (most likely the cytopharyngeal ribbons) which radiate from the gullet. It has not yet been possible to determine whether the microtubules emanating from the gullet terminate at the basal bodies (or vice versa) or whether they are distinct sets of microtubules.

The dynamics of the different microtubular systems can be followed with the appropriate antibodies throughout the cell cycle. During interphase, aside from the permanent arrays (cilia, basal bodies and their microtubular ribbons, contractile vacuole pores and their roots, cytopharyngeal ribbons), the postoral fibers and the internal meshwork are present. During division, the postoral fibers and most of the internal meshwork disappear while micro- and macronuclear spindles (Fig. 4) and the cytospindle (Fig.5) develop.

2.1.2 Tubulin

Although methods for purification of both axonemal and cytoplasmic tubulins have quite recently been developed (A.M. Hill, N. Levilliers and A. Adoutte, pers. commun.), no physico-chemical data on purified *Paramecium* tubulins are yet available and they have been so far only characterized immunologically and electrophoretically in 1D- and 2D-gels of whole cells or of different subfractions (axonemes, cortices, see Adoutte et al. 1980; Cohen et al. 1982).

The available data can be compared to that known for other ciliates. Among the two forms of tubulins, α and β (55-kD polypeptides), β-tubulin is the more conserved, as observed by peptide mapping and electrophoretic mobility of tubulins of distant species (Little et al. 1981). In some protists and ciliates, in particular (Suprenant et al. 1985; Grain 1986), the α-tubulin has a higher electrophoretic mobility than the β-tubulin. The same situation has been observed in *Paramecium* (Cohen et al. 1982; Adoutte et al. 1984).

Isoelectric focussing of tubulins of *Tetrahymena* revealed a microheterogeneity with five isoforms for α and two isoforms for β (Suprenant et al. 1985), while Callahan et al. (1984) and Barahona et al. (1985) showed respectively, that there were only one α-tubulin gene and two β-tubulin genes in this organism. Among the five isoforms of α-tubulin, three are specific to the axoneme and two are specific to the cytoplasmic tubulins (Suprenant et al. 1985). This suggests, at least for the α-tubulin, that the isoforms derive from post- translational modifications of a unique protein, in particular during axonemal assembly, as is the case for tubulins of the flagella of *Chlamydomonas* (Brunke et al. 1982).

Fig. 5a,b. Anti-axonemal tubulin decoration of deciliated dividing cells showing the cytospindle. Bar = 10 µm. **a** Ventral face at an early stage of division when the microtubular bundles of the cytospindle still grow as dicontinuous segments (*arrows*). **b** Dorsal face of a divider at a later stage where the bundles are continuous. (Courtesy of F. Ruiz)

One of the two *Stylonychia* α-tubulin genes has been sequenced (Helftenbein 1985) and shows a great homology in protein sequence (85%) with chicken α-tubulin.

The sequence conservation and likely diversity of tubulin isoforms in *Paramecium* can be deduced from the immunological data listed in Table 2 which support three conclusions:

1. Both α- and β-tubulins from *Paramecium* whole cell extracts or cortices cross-react on immunoblots with all the anti-tubulins tested, thus attesting to the overall sequence conservation of the molecules.

2. The antibodies raised against *Paramecium* and *Tetrahymena* axonemal tubulins do not cross-react with higher eukaryotic cytoplasmic tubulins. Adoutte et al. (1985) showed that the anti-axonemal tubulin of *Paramecium* recognized one (or very few) epitope(s) on the tubulin molecule and that this epitope was present in cilia and flagella in almost all species, from ciliates to mammals. These observations suggest a higher evolutionary conservation of axonemal tubulins than of cytoplasmic ones.

3. Complementary circumstantial evidence suggests that there are two types of tubulin in *Paramecium:* axonemal and cortical tubulin vs cytoplasmic tubulin since one set of antibodies preferentially decorates axonemal and cortical microtubules, while another set of antibodies preferentially decorates cytoplasmic ones (Table 2). The question is therefore raised whether these two types of tubulin differ in primary sequence or in post-translational modifications. This question will be answered by identification of *Paramecium* tubulin gene(s) and characterization of isoforms. In any case, it is worth noting that *Paramecium* tubulins can undergo the same type of post-translational modifications (at least acetylation and tyrosylation/detyrosylation) found in other organisms.

2.2 Microfilaments and Actin

2.2.1 Organization and Dynamics of Actin Microfilaments

The presence of actin microfilaments in *Paramecium* has been reported by the group of Plattner in Konstanz and by our group in Gif. Plattner and co-workers first described actin as mainly associated with cortical elements, cilia, basal bodies, infraciliary lattice, on the basis of cytological observations using an anti-actin of *Paramecium* that they prepared, heavy meromyosin (HMM, a proteolytic fragment of myosin which specifically binds to actin microfilaments) and DNase I (Tiggemann and Plattner 1981; Tiggemann et al. 1981). In contrast, using HMM, we did not detect either by fluorescence or by electron microscopy any microfilaments in the cortical structures. HMM binding was observed (1) associated with forming vacuoles and young food vacuoles (Fig. 6) as found in *Tetrahymena* by Méténier (1984), (2) around the contractile vacuole pores, (3) around the cytoproct and (4) as a sparse meshwork spanning the cytoplasm (Cohen et al. 1984). In recent publications (Kersken et al. 1986a, b), Plattner and co-workers also described actin associated with food vacuoles us-

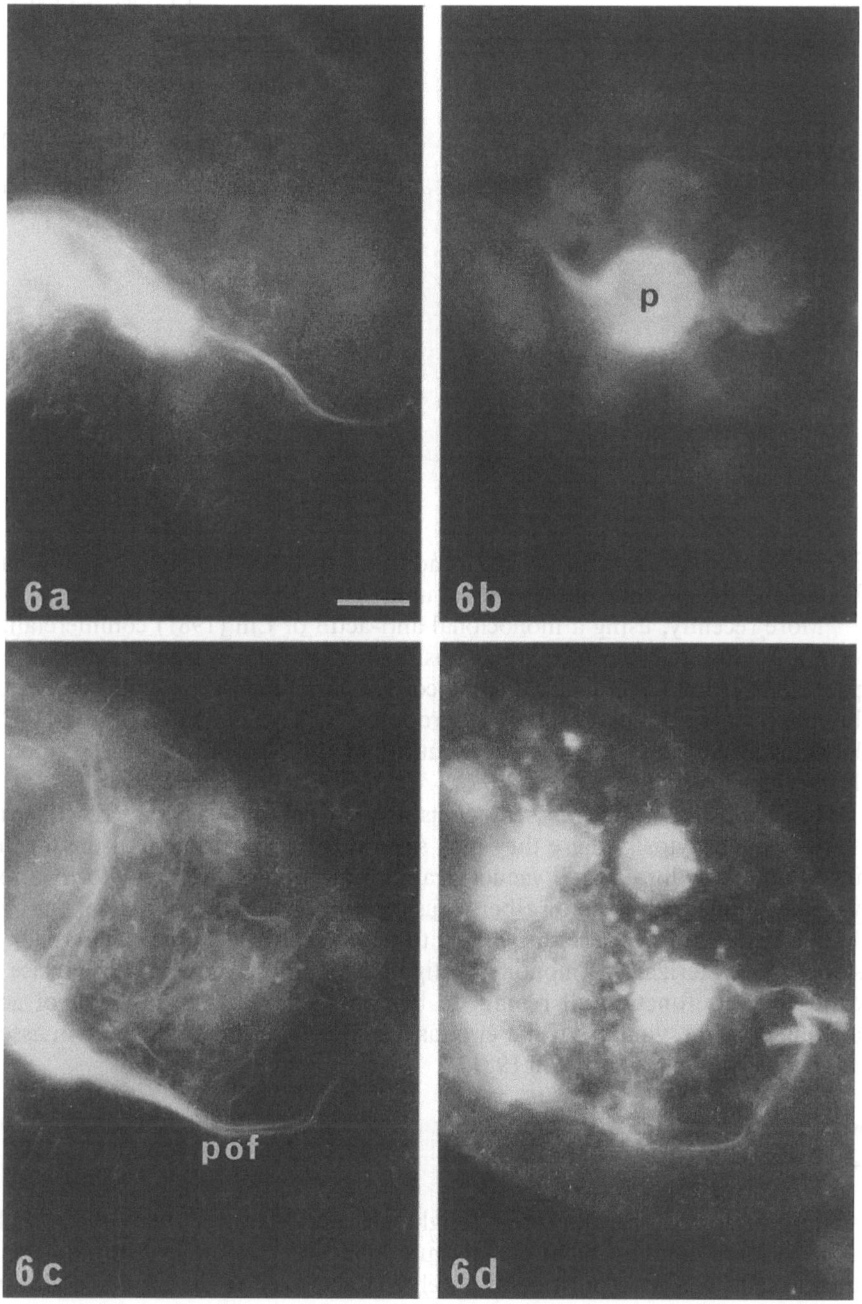

Fig. 6a – d. Actin microfilament dynamics during phagocytosis. Double labelling of cells at different stages of ingestion using our anti-axonemal tubulin labelled by rhodamine (**a, c**) and fluorescein-coupled HMM (**b, d**). Bar = 10 µm. **a, b** Early stage of ingestion. The forming pouch (*p*) is growing by addition of small vesicles which fuse with it. Note that the HMM labelling is highly concentrated around this pouch. **c, d** A few seconds after the pouch has constricted into a food vacuole, it slides along the postoral fibers (*pof*) down to the posterior part of the cell where it enters the cytoplasmic flow along a sinuous path. Actin microfilaments follow exactly the track of this newly formed food vacuole. (See Cohen et al. 1984 for details)

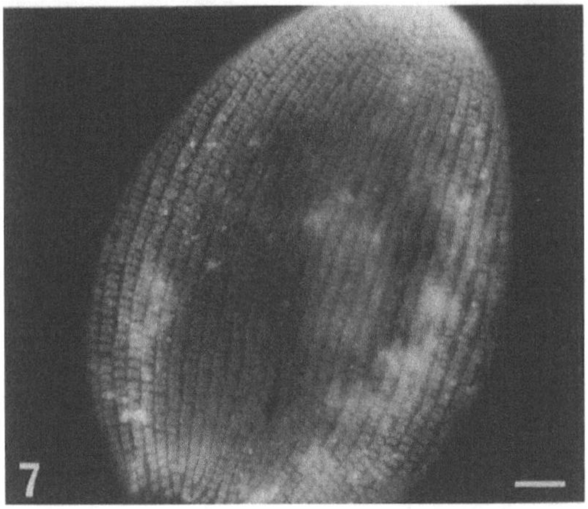

Fig. 7. Immunofluorescent decoration of the epiplasm "scales" of a just-divided cell given by the monoclonal anti-actin antibody of Lin (1981). Note the clear individualization of cortical units. Bar = 10μm. (Courtesy of F. Ruiz)

ing phalloidin as an actin probe, in addition to other labelling (mainly basal bodies and the "outline of the cortical units").

More recently, using a monoclonal anti-actin of Lin (1981) commercialized by Amersham and an antibody against a supposed *Paramecium* actin we prepared (see below), we did find some cortical labelling in the epiplasm (Fig. 7) but not in the infraciliary lattice as proposed by Tiggemann et al. (1981). This labelling could correspond to the "outline of the cortical units" described by Kersken et al. (1986a).

In conclusion, actin microfilaments are undoubtedly associated with early steps of phagocytosis where they play several roles: constriction of the forming ingestion pouch into a food vacuole, rapid migration of the newly formed food vacuole along the postoral fibers, fusion with acidosomes (Allen and Fok 1983a–c). In addition, actin is likely to be present in the cortex. However, its state of polymerization (monomer, oligomer or microfilament), its precise location and its function all remain to be characterized. The presence of actin oligomers in the vicinity of the epiplasm may evoke the actin/spectrin association in erythrocytes (Branton 1981).

2.2.2 Actin

Actin, a highly conserved protein (Hightower and Meagher 1986) of 42–43 kD and the main component of cytoplasmic microfilaments and thin filaments in muscle cells, has been purified and well characterized in metazoa and also in a few protozoa such as *Acanthamoeba*, where it is a major protein (see Grain 1986, for a review). In ciliates, actin which is not at all abundant, has yet to be characterized biochemically while the available data suggest that it has diverged significantly from that of other eukaryotes.

In *Tetrahymena*, a first attempt to isolate actin, according to usual methods, gave a protein of 38 kD able to form 14-nm filaments (Numata et al. 1980). However, in more recent publications, the authors retracted their statement and indicated that the 14-nm filaments derived from a 49-kD protein contaminating the 38-kD protein preparation (Numata and Watanabe 1982). The authors nevertheless maintain that the 38-kD protein might be *Tetrahymena* actin. Independently, Mitchell et al. (1985) and Mitchell and Zimmerman (1985) isolated an actinlike protein of 47.5-kD from *Tetrahymena*. However, cycles of polymerization-depolymerization gave filaments of 8 – 12 nm which cannot be decorated with heavy meromyosin, as should be the case for actin microfilaments. Recently, Cupples and Pearlman (1986) isolated and sequenced the unique actin gene from *Tetrahymena*. They found that it encodes a protein of 43 kD whose sequence has 75% homology with yeast actin. Although this protein is still unidentified biochemically, the sequence data would suggest that neither the 38-kD nor the 47.5-kD polypeptides are *Tetrahymena* actin.

In another ciliate, *Oxytricha fallax*, an actin gene was sequenced (Kaine and Spear 1982) and the deduced protein sequence shown to display only 62% homology with yeast actin. We compared the sequences of *Oxytricha* and *Tetrahymena* actin genes, which appear to share only 62% homology. This suggests huge evolutionary distances within the ciliates, as well as between ciliates and higher eukaryotes.

In *Paramecium*, the first report of purification of actin was published by Tiggemann and Plattner (1981) using undetailed but classical protocols. However, no biochemical characterization of the protein was provided. In our laboratory (Lazard and Cohen, unpubl. results; Cohen 1984), we attempted to isolate actin by affinity on DNase I Sepharose, according to Lindberg and Eriksson (1971), Lazarides and Lindberg (1974), Dhermy et al. (1978) and Zechel (1980). This method yielded a single 43-kD protein which therefore appeared to be a good candidate for *Paramecium* actin. However, peptide mapping showed that only three peptides comigrated in partial digests of the 43-kD *Paramecium* polypeptide and rabbit muscle actin. Furthermore, no cross-reaction of our 43-kD protein was obtained with an anti-frog muscle actin (Gounon and Karsenti 1981) on immunoblots nor of our anti-43-kD protein with rabbit muscle actin on immunoblots or with rat fibroblast stress fibers in situ. We therefore concluded that the 43-kD polypeptide was not actin. However, we recently found the monoclonal anti-actin antibody of Lin (1981) to specifically decorate the epiplasm or its vicinity (Fig. 7). Interestingly, using stringent incubation conditions (see Sect. 5), our anti-43-kD protein significantly labelled the same structure despite a high fluorescence background. The question is therefore reopened: our 43-kD polypeptide might after all be *Paramecium* actin. Evidently, additional biochemical data are needed to draw a definitive conclusion.

2.3 Other Cytoskeletal Networks

Aside from microtubules and actin microfilaments, higher eukaryotes contain intermediate filaments. Limited attempts to characterize such filaments in *Paramecium* with heterologous immunological probes gave negative or uncertain results. Among the numerous networks of fibrillar or granulofibrillar

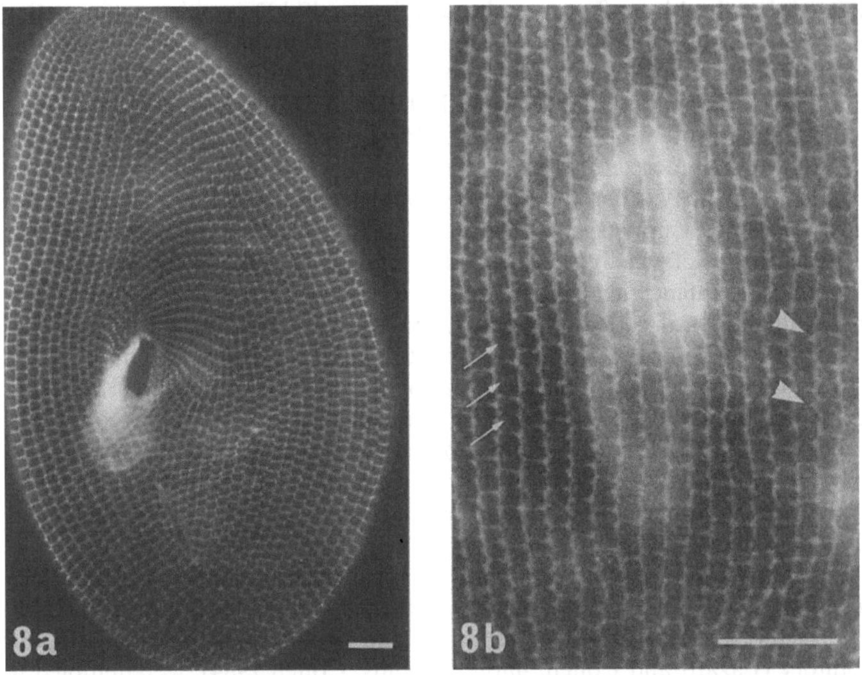
8a
8b

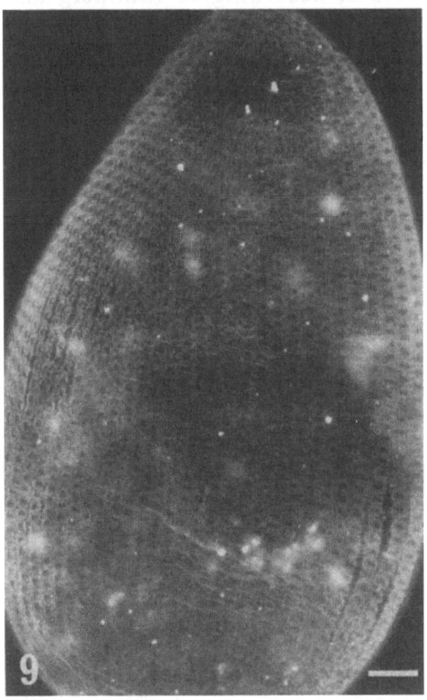
9

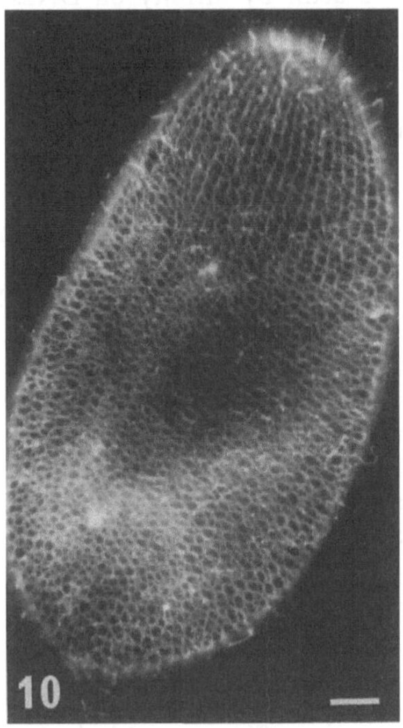
10

material, four have been studied immunologically or cytologically, their dynamics is known and they now seem amenable to biochemical characterization: the outer lattice, the epiplasm, the infraciliary lattice and the kinetodesmal fibers. In addition, electrophoretic and immunological data on the filamentous reticulum are available. In contrast, nothing is known of the striated bands, cord and contractile ring.

2.3.1 *Outer Lattice*

The outer lattice is a submembranar fibrous network running all over the cell, at the top of the surface ridges. It happens to be specifically decorated, as judged by immunofluorescence (Fig. 8) and immunogold labelling in electron microscopy (Cohen et al. 1987), by a monoclonal antibody, CC212, raised against ciliated cortices of quail oviduct cells and characterized as an anti-myosin of smooth muscle or non-muscle cells (Klotz et al. 1986). The overall pattern of the outer lattice, and its evolution throughout the cell cycle was thus amenable to analysis (Cohen et al. 1987). Its meshes delineate the periphery of each cortical unit. Its dynamics present original features: during division, outer lattice growth is achieved without disruption of the pre-existing meshes by longitudinal elongation (Fig. 12a). New transverse partitions between newly formed cortical units appear only later (Fig. 8b), after the cell has completed division. The coincidence of the outer lattice with the surface relief and its persistence throughout the cell cycle suggest that its function could be to stabilize the surface relief.

Western blot experiments on whole cell extracts using CC212 as a probe revealed two main bands at 130 and 50 kD, suggesting that these two polypeptides are components of the outer lattice. No evidence that either of the polypeptides was related to myosin was obtained (Cohen et al. 1987).

Despite the ultrastructural similarities between *P. caudatum* (Allen 1971) and *P. tetraurelia*, a notably weaker reaction of CC212 on *P. caudatum* outer lattice was observed (A. Fleury, pers. commun.).

◄

Fig. 8a,b. Immunofluorescent labelling of the outer lattice given by the monoclonal CC212. Bar = 10 μm. **a** Ventral side of an interphase cell. Each mesh of the lattice delimits the contour of a cortical unit. **b** Enlarged portion of a cell 5 min after division. Note the elongated cortical units (*arrows*) where buds of the future transverse partitions (*arrowheads*) begin to appear. (**b** Courtesy of F. Ruiz)

Fig. 9. Visualization of the epiplasm of an interphase cell using the anti-"band 4" antibody. The partition of the cortex into "scales" is less obvious than with the anti-actin of Lin (1981) probably because the decoration is not of the epiplasm itself, but of fibrils emanating from it. Bar = 10 μm. (Courtesy of F. Ruiz)

Fig. 10. Visualization of the infraciliary lattice of an interphase cell using the monoclonal antibody 3D12 under non-stringent conditions. The infraciliary lattice is much less regular than the outer lattice (Fig. 8). Bar = 10 μm. (Courtesy of N. Garreau de Loubresse)

2.3.2 Epiplasm

Two different immunological probes were found to bind to the epiplasm or to material associated with the epiplasm.

1. Among the antibodies raised against the major polypeptides of purified cortices (Keryer et al., in prep.), the antiserum obtained by immunizing a rabbit with "band 4" (a 120-kD polypeptide) specifically decorates the epiplasm, as judged by immunofluorescence microscopy (Fig. 9). At the electron microscope level, the labelling was localized on thin fibrils emanating from the epiplasm rather than on the epiplasm per se (Keryer et al., in prep.).

2. More recently, the monoclonal antibody 2G11 (one of the numerous antibodies raised against purified mammalian centrosomes in M. Bornens' laboratory) was shown to specifically label the epiplasm per se in fluorescence and electron microscopy (Keryer et al., in prep.). This antibody gives a pattern similar to the one obtained with the anti-actin of Lin (1981) shown in Fig. 7.

The two antibodies, and more strikingly 2G11, demonstrate that, as already noted by Allen (1971), the epiplasm is not a continuous layer all over the cell surface but is composed of individual scales corresponding to the cortical units. Immunofluorescence observations on dividing cells show that multiplication of cortical units occurs by transverse fragmentation of pre-existing scales. Experiments to characterize the polypeptide(s) recognized by 2G11 are underway.

An epiplasmic layer is present and has been purified in several protozoa: the ciliate *Pseudomicrothorax dubius* (Peck 1977), the dinoflagellate *Noctiluca scintillans* (Metivier 1986) and the euglenoid *Euglena acus* (Bricheux 1987). Despite the diversity in the number and relative electrophoretic mobilities of the polypeptides so obtained, strong immunological cross-reactivity between the three species was observed (Bricheux 1987) indicating a certain three-dimensional homology and/or sequence conservation of the antigen. In *Tetrahymena pyriformis,* although the epiplasm has not been purified, Williams et al. (1979) could assign two proteins of 122 and 145 kD to this structure. These proteins were shown to undergo significant variations in their electrophoretic mobilities in different species of *Tetrahymena* (Williams et al. 1984).

2.3.3 Infraciliary Lattice

The infraciliary lattice, a network of bundles of filaments running under the cortex, was first visualized in immunofluorescence and immunoelectron microscopy by 3D12 (Fig. 10), a monoclonal antibody from the same fusion as CC212, under non-stringent conditions (see Sect. 5). Using 3D12, a cycle of disassembly-reassembly of the infraciliary lattice was observed during division (Garreau de Loubresse et al., in preparation). Reassembly seems to proceed from numerous independent foci. No reaction on Western blot could be obtained with this antibody.

More recently, an antibody raised against the 22–23 kD polypeptides of the ecto-/endoplasmic boundary (EEB, a massive filamentous array) of the ciliate *Isotrica prostoma* (Vigués and Grolière 1985) was shown to cross-react with the

Paramecium infraciliary lattice in immunoelectron microscopy (Viguès, pers. commun.). The use of this anti-EEB in immunofluorescence confirmed all the data on the organization and dynamics of the infraciliary lattice previously obtained with 3D12 (Garreau de Loubresse et al., in prep.).

These authors observed that in lysates of cells treated by Triton X100 at low ionic strength, whole shrunken infraciliary lattices were present. Purification of the structure and characterization of its components are underway. The capacity of the infraciliary lattice to contract in the presence of Ca^{2+} (noted by Kersken et al. 1986 b) along with the above observations and the immunological cross-reaction of the structure with the anti EEB antibodies suggest that some component(s) of the infraciliary lattice might display Ca^{2+}-dependent contractile properties analogous to those of the 22–23 kD polypeptides of *Isotricha* (Viguès and Grolière 1985) and those present in the spasmonemes of Vorticellids (Amos 1975). Ultrastructural homologies of the *Paramecium* infraciliary lattice with filamentous bundles in the contractile heterotrich *Spirostomum* have also been noted by Pitelka (1969).

2.3.4 Kinetodesmal Fibers

No immunological probes specific for these structures are yet available. However, kinetodesmal fibers can be observed at the whole cell level, either in vivo (Aufderheide 1986) or after silver staining (Fernandez-Galiano 1966, 1976). By the latter method, regression of old fibers and growth of new ones from old and new basal bodies was observed during cell division (Fernandez-Galiano 1978; Iftode and Adoutte, pers. commun.). Along with the microtubular arrays which are labile during certain phases of the cell cycle and with the infraciliary lattice, the kinetodesmal fibers are the only cytoskeletal structures which undergo a cell cycle-dependent regression and regrowth.

Preliminary experiments of purification of *Paramecium* kinetodesmal fibers indicated that they are composed of five main polypeptides in the range of 30–38 kD (Sperling and Rossignol, pers. commun.).

For comparison, the kinetodesmal fibers of *Tetrahymena* were found to contain a 250-kD polypeptide (Williams et al. 1979) which appears to be constant in all the species studied (Williams et al. 1984), while Hyams and King (1985), using an antiserum raised against *Tetrahymena* kinetodesmal fibers, identified two polypeptides of 76 and 47 kD as components of the structure. In *Naegleria* flagellates, a major protein of the rootlets has an apparent molecular weight of 170 kD (Larson and Dingle 1981). Despite the similarity in fine structure of kinetodesmal fibers of different species, the major proteins of the structure appear to be different, at least in molecular weight.

2.3.5 Filamentous Reticulum

The filamentous reticulum was specifically labelled by immunofluorescence and immunoelectron microscopy using the monoclonal CC212 (Cohen et al.

1987; Garreau de Loubresse, unpubl.) which also decorates the outer lattice. However, it has not been possible to show continuity or homology between these two networks. The antibody recognizes two bands of 130 and 50 kD on immunoblots. It is not presently known whether one of these bands corresponds to the outer lattice and the other to the filamentous reticulum or whether the two bands are components of both structures.

Independently, in experiments of cell fractionation, purified cortices and purified gullets could be obtained (Keryer et al., in preparation). SDS-gel analysis of the two fractions showed that two bands in the range of 60–65 kD (bands 10 and 11) appeared enriched in the gullet fraction. Antibodies raised against these two polypeptides indeed decorate the gullet by immunofluorescence, and more precisely the filamentous reticulum as revealed by immunoelectron microscopy (F. Iftode, R. Jeanmaire, N. Stelly and A. Adoutte, pers. commun.).

In *Tetrahymena,* antibodies raised against oral filaments [which seem to be homologous to the filamentous reticulum of *Paramecium* (Williams 1986c)] have been shown to recognize two polypeptides of 87 and 80 kD (Williams et al. 1986). The anti-87 kD antiserum also decorates the gullet of *Paramecium,* which suggests that the filamentous reticulum of both species might share common determinants.

3 General Properties of the Cytoskeleton

3.1 Organization of the Cytoskeleton

The cytoskeleton of *Paramecium* (and of ciliates in general) differs from that of other cell types, in particular mammalian cells, by the fact that it is essentially cortical. Most of the networks run entirely in the cortex: epiplasm, cortical actin, outer lattice, striated bands, infraciliary lattice, kinetodesmal fibers, cilia, basal bodies and their microtubular ribbons, as well as the cytospindle. Other networks, although internal, are nucleated in the cortex (or in the gullet, a differentiated part of the cortex): the contractile vacuole roots, the postoral fibers, the cytopharyngeal ribbons and the internal microtubular meshwork. Even for actin microfilaments surrounding digestive vacuoles, the same notion can be extended since each vacuole has its origin in a pouch attached to the gullet where actin microfilaments are nucleated.

Concerning the true cortical cytoskeletal networks, they fall into two classes according to their relations to basal bodies and to their overall organization. The first class is composed of discontinuous elements which are repeated in each cortical unit and are directly associated with basal bodies: cilia, microtubular ribbons, kinetodesmal fibers and possibly striated bands are nucleated on the basal body itself or from the "peri-basal body material", while epiplasm scales are tightly bound to basal bodies. In contrast, the second class includes the outer and infraciliary lattices which form continuous networks all over the cell surface and show no obvious physical connection with basal bodies.

The originality of *Paramecium* and probably of all ciliates is that the organizing centers of the cytoskeleton are neither unique (they are present in thousands of exemplars) nor central in the cell (they are located around basal bodies, in the cortex). This particularity could, however, be shared by ciliated epithelial cells of metazoa, where the organization of the cytoskeleton seems to depend upon the cortex rather than the center of the cell (Sandoz et al. 1983, for example).

3.2 Functions of the Cytoskeleton

Aside from the rather obvious function of certain structures (cilia in locomotion and feeding, postoral fibers in guiding newly formed food vacuoles, macro- and micronuclear spindles in distributing genetic material between the daughter cells and actin microfilaments in food vacuole formation and transit), a function for other cytoskeletal elements can be or has been proposed on the basis of cytological observations. The kinetodesmal fibers and the cytospindle might stabilize the kinety pattern; the outer lattice might help to maintain the surface relief (see Cohen et al. 1987); the epiplasm might stabilize or guide cortical unit organization; the infraciliary lattice could maintain cell shape (see also Sibley and Hanson 1974); the internal meshwork (microtubules, actin) might be involved in cyclosis, intracellular movement of vesicles (trichocysts, digestive vacuoles) and nuclei. These functions can be ascertained by experiments introducing either physiological or genetic perturbations of cytoskeletal elements.

3.2.1 Action of Drugs Affecting Cytoskeletal Structures

Treatment of cells by drugs which specifically bind to tubulin or actin has been widely used in studies of the cytoskeleton: colchicine and nocodazole depolymerize microtubules while taxol stabilizes them and vinblastin affects them by crystallizing tubulin monomers; cytochalasins depolymerize actin microfilaments, while phalloidin stabilizes them.

Prudence is required in the interpretation of in vivo experiments since side effects are often observed. Phalloidin can provoke "gelation" of the cytoplasm thus interfering non-specifically with many processes such as ciliary beating (Kersken et al. 1986b). However, the following apparently significant effects of several of these drugs on *Paramecium* were observed.

As indicated in Table 2, not all microtubular arrays are sensitive to depolymerizing drugs. Depolymerization by nocodazole of the internal microtubular meshwork affects macronuclear shape and movements (A. Torres, pers. commun.). Nocodazole also partly inhibits growth of the cytospindle yielding an exaggerated curvature of kineties in the daughter cells (Cohen et al. 1982). It also blocks the elongation of intramacronuclear microtubules and of the macrotubules of the micronuclear separation spindles, yielding unequal repartition of macro- and micronuclei between the daughter cells (as also observed with vinblastin, Ruiz et al. 1976).

As for actin microfilaments, the only effect of cytochalasins was to immediately block the formation of the ingestion pouch (Cohen et al. 1984) and the fusion of young food vacuoles with acidosomes (Allen and Fok 1983c). Trichocyst saltatory motion, a phenomenon first described by Aufderheide (1977), is also blocked by cytochalasin (Aufderheide, personal communication). No effect on contractile vacuole pulsations or cytoproct function was observed despite the presence of HMM-decorated material in these structures (Cohen et al. 1984).

3.2.2 Study of Mutants

Among the available morphological mutants, none has yet been shown to be specifically defective in a cytoskeletal protein. However, some display typical alterations in the organization of their cytoskeleton which provide information on the role of particular networks.

1. Several mutations designated as *screwy* (Sonneborn 1974) have a twisted cell shape with corresponding curved kineties. Phenocopies of *screwy* are produced by application of nocodazole to wild-type cells during division. Both mutant cells and nocodazole-treated wild-type cells display defective assembly of their cytospindle near the cleavage furrow (Cohen et al. 1982; Rossignol and Beisson, unpubl.). These observations support the conclusion that the cytospindle limits the torsion normally occurring between the future daughter cells during division, or relieves the torsion after division.

2. The various mutants characterized by defective or normal but *unattached* trichocysts all display similar abnormalities in nuclear positioning and movements (Ruiz et al. 1976; Cohen and Beisson 1980; Cohen 1984). It was found, using either the monoclonal B1B2 or the monoclonal anti-α-tubulin (Blose et al. 1982), that the internal microtubule meshwork of these mutants displayed a common anomaly in its ectoplasmic organization. Figure 11 shows that in these mutants all ectoplasmic microtubules run parallel to the surface instead of reaching it perpendicularly. The strict correlation between this defect in the organization of the internal microtubule meshwork and the defective movements of nuclei in the mutants, regardless of the primary mutational defect (Cohen and Beisson 1980; Cohen 1984), confirms the likely role of this network in controlling nuclear movements, as is the case in *Aspergillus* (Oakley and Morris 1980), sea urchin eggs (Schatten et al. 1982) and other cell types (reviewed by Cohen 1984).

3. The thermosensitive mutant *sm19* (Ruiz et al. 1987) yields smaller and smaller cells with fewer and fewer basal bodies when shifted to the non-permissive temperature (35 °C). It was demonstrated that the primary defect of *sm19* is in basal body duplication. It was also shown that secondarily, i.e. after two to three divisions at 35° C, this mutant displays strong correlated alterations in the outer lattice (Fig. 12) and in the relief of the ridges (Cohen et al. 1987). These observations confirm the role of the outer lattice in sustaining the surface relief.

In conclusion, both physiological and genetic approaches have allowed us to uncover the role of several structures and conversely to find a physical support

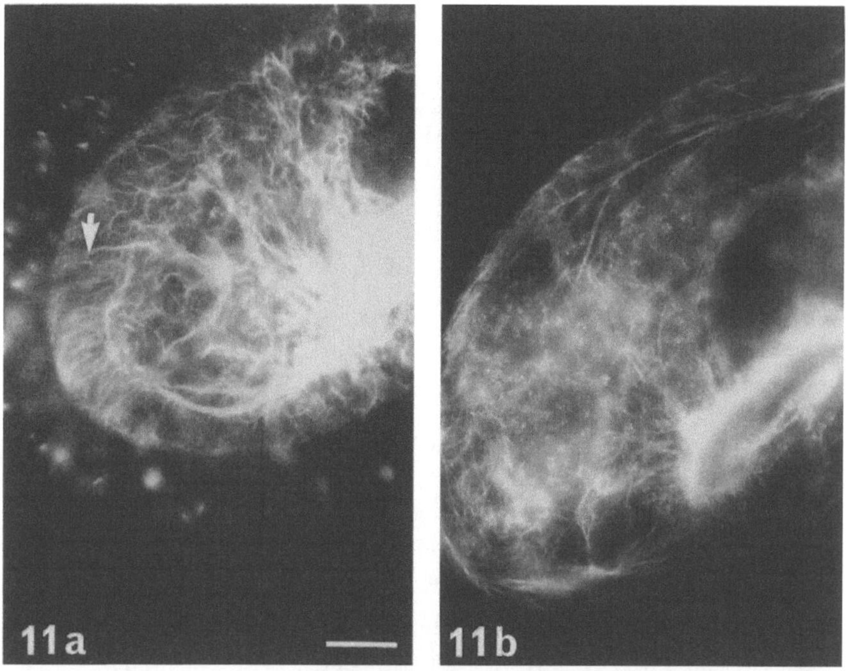

Fig. 11 a,b. Comparison of the organization of the internal meshwork of (**a**) wild-type and (**b**) *tam 38* mutant cells as viewed with the monoclonal antibody B1B2. Note, as frequently observed with this antibody, that inconsistent reactions on ciliary tips are obtained as observed in **a** and not in **b**. Bar = 10 µm. **a** In wild-type cells, the ectoplasmic zone described in Fig. 3 is clearly visible (*arrow*). **b** In the mutant *tam38,* lacking trichocysts in its cortex, microtubules run parallel to the surface instead of plunging perpendicularly into the endoplasm

for some functions. However, concerning intracellular movements, such as cyclosis or contractile vacuole pulsation, the comprehension of the mechanism is much less advanced.

3.3 Dynamics of the Cytoskeleton

During division, the multiplication or reorganization of the different cytoskeletal elements proceed by a variety of mechanisms. While "old" basal bodies, their microtubular ribbons and contractile vacuole pores and roots are totally conserved, kinetodesmal fibers are all formed de novo at each division. The epiplasm "grows" and is reorganized around pre-existing material, as is also the case for the outer lattice which remains continuous throughout the cell cycle. In contrast, the infraciliary lattice is almost completely disassembled and reassembled as is most of the internal microtubule meshwork and the kinetodesmal fibers. It would thus appear that the outermost structures (basal bodies, ribbons, epiplasm, outer lattice) are more permanent than the innermost ones (kinetodesmal fibers, infraciliary lattice, cytoplasmic microtubule meshwork).

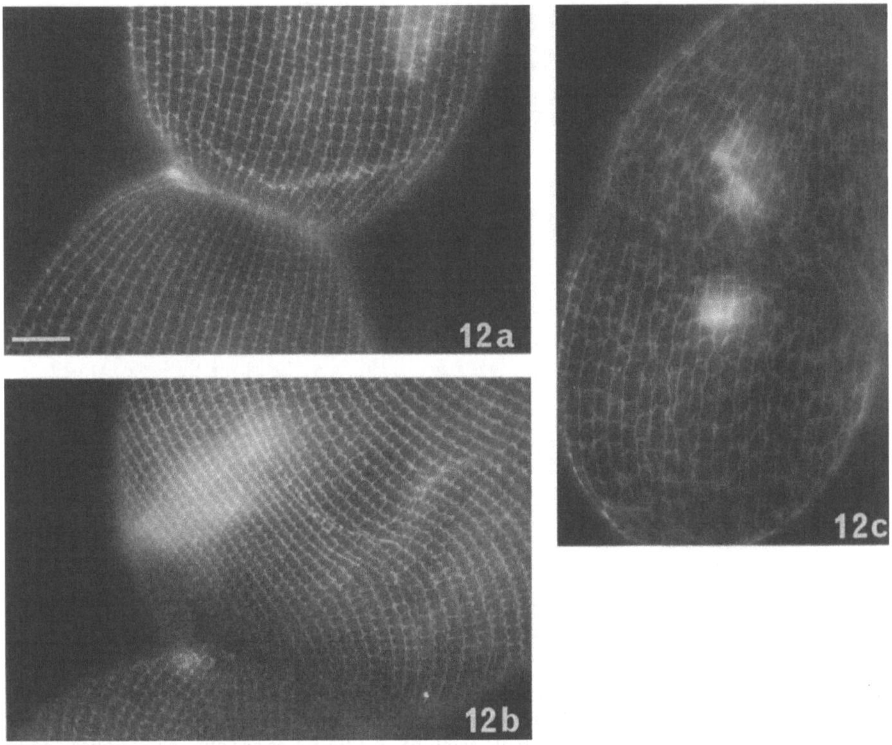

Fig. 12a–c. Immunodecoration of the outer lattice of a dividing cell of three different genotypes by the monoclonal CC212. **a** Wild-type; **b** *screwy;* **c** *sm19* (after 4–5 generations at the non-permissive temperature) mutant cells. Note in the *screwy* cell the general torsion of the pattern although the process of elongation and partition of the meshes appears normal. In contrast, in the *sm19* cell, no such torsion is observed but the local pattern of the meshes is quite disorganized. Bar = 10 μm. (**a, c** Courtesy of F. Ruiz; **b** courtesy of M. Rossignol)

Despite the different fates of individual structures or networks, the existence of a number of spatio-temporal relationships in the reorganization of different structures (see Table 3) reveal precise interactions between structures and demonstrate complex integrative mechanisms of the cytoskeleton as a whole.

Inverse correlations between assembly of division-specific structures and disassembly of other structures are observed. The postoral fibers and the internal meshwork disappear when the nuclear spindles and the cytospindle polymerize. This suggests that the two groups of microtubular systems are interconverted through a common pool of free tubulin, as is the case in mammalian cells (see for example Weber 1976). Another inverse correlation, between the rigid microtubule bundles of the cytospindle and the kinetodesmal fibers, suggests that the first one can functionally replace the others during division since the respective locations of these networks in the ridges of the cortical relief is nearly the same.

In contrast, growth and reorganization of the epiplasm and the outer lattice proceed strictly in parallel: both structures first elongate as basal bodies dupli-

Table 3. Stability of cytoskeletal elements during the cell cycle [a]

Networks	Permanent	Disorganized during division	Appear during division
Internal microtubular	Backbone of contractile vacuoles Backbone of cytoproct	Postoral fibers Internal meshwork	Macronuclear spindle Mitotic spindle Separation spindle
Cortical microtubular	Basal bodies Cilia Ribbons Gullet		Cytospindle
Cortical non-microtubular	Outer lattice Epiplasm	Kinetodesmal fibers Infraciliary lattice Actin microfilaments involved in phagocytosis	Contractile ring

[a] No data on the striated bands, the filamentous reticulum or the cord throughout the cell cycle are available.

cate, buds of the outer lattice appear when epiplasm scales begin to fragment and new transverse partitions of the outer lattice are completed when new epiplasm scales become individualized. This indicates the interdependence of the two structures.

The study of morphological mutants confirms these relationships and can provide new data on the interdependence of the cytoskeletal networks.

1. In the case of inverted kineties, both in *Paramecium* (Beisson and Sonneborn 1965) and *Tetrahymena* (Ng and Frankel 1977), kinetodesmal fibers and transverse and postciliary ribbons have been shown to develop with reversed polarity, indicating that the development, relative position and angle of these appendages are determined solely by the intrinsic polarity of the basal body. This is confirmed by observations on the mutant *kin241* (Sonneborn 1974; Beisson et al. 1976; Grandchamp and Beisson 1981) in which erratic basal bodies are found to lie outside of rows and in any orientation: anti-tubulin decoration always shows that the transverse and postciliary ribbons retain their respective position (our unpubl. observations).

2. Assembly of the cytospindle, on the right side of basal body rows, above the kinetodesmal fibers, retains this polarity along the inverted rows which occur occasionally in wild-type cells or more frequently in mutants such as *kin241* (Cohen et al. 1982; our unpubl. observations). The same observation was reported in the case of inverted kineties in *Tetrahymena,* where the longitudinal microtubular bands (a structure homologous to the cytospindle of *Paramecium* but which is permanent throughout the cell cycle) are localized according to the polarity of the neighbouring basal bodies (Ng and Frankel 1977).

3. The length of transverse and most strikingly postciliary ribbons depends on the size of the cortical unit, as observed in the mutant *sm19* where larger units, displaying longer ribbons, are formed (Fig. 13; Ruiz et al. 1987).

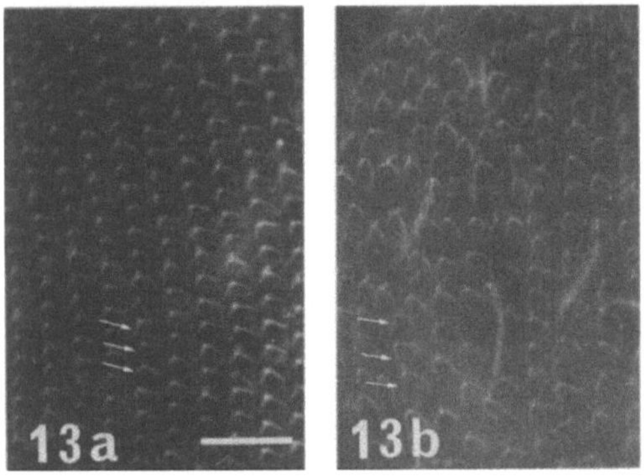

Fig. 13 a,b. Comparison of the size of the microtubular ribbons in the cortex of (**a**) wild-type and (**b**) *sm19* mutant cells grown at the non-permissive temperature for 4–5 generations as viewed with the anti-axonemal tubulin antibody. Note that *sm19* displays larger cortical units with longer (especially postciliary) ribbons (*arrows*). Bar = 10 µm. (**b** Courtesy of F. Ruiz)

4. The shapes of epiplasm scales and outer lattice meshes are always strictly complementary, as observed in the mutants *sm19* (Ruiz, unpubl. data) or *kin241* where striking irregularities in the two networks can be observed.

5. Although outer and infraciliary lattices do not seem directly connected to basal bodies and undergo quite different modes of reorganization at division, their pattern depends on the arrangement and spacing of basal bodies. In the *screwy* mutants with curved kineties, a corresponding torsion of the outer lattice (Rossignol, unpubl. observations; see Fig. 12b) and the infraciliary lattice is observed, without any other defect in their pattern. In the mutant *sm19*, disorganization of the outer lattice (Fig. 12c) appears as a secondary effect of basal body underduplication and increased spacing (Cohen et al. 1987). A similar dependence of oral filaments upon basal body arrangement in the gullet has been reported for the mutant *NP1* of *Tetrahymena thermophila* (Williams 1986b).

These observations provide new evidence for the importance of nearest neighbour interactions (Sonneborn 1975; Aufderheide et al. 1980) in the assembly of cortical structures, and point out the determining role of basal body arrangement.

4 Conclusion

The application of immunocytochemical methods to *Paramecium* and the search for antibodies specifically binding to defined components of its cytoskeleton have provided new insights into overall cellular organization and dynamics. The major new observations concern the visualization of an internal microtubule meshwork comparable to that of higher eukaryotic cells (spanning the whole cytoplasmic volume, disassembled during division, depolymerized by nocodazole or cold), the identification of actin in the filamentous arrays surrounding forming vacuoles and young food vacuoles and as a sparse internal meshwork, the visualization of the outer lattice, the infraciliary lattice and the epiplasm and of their various modes of reorganization during division, and the demonstration that basal bodies play a central role in the organization of all cytoskeletal networks including those (outer and infraciliary lattices) which do not seem to be physically connected to them. Further studies on the cytoskeleton of wild-type cells and morphological mutants should lead to some understanding of the interlocking events involved in duplication of cortical units and cell morphogenesis.

The biochemical and molecular genetics study of *Paramecium* cytoskeleton is developing. Characterization of the tubulin and actin genes and of the diversity of tubulin isoforms should be soon worked out. The antibodies specifically binding to the epiplasm, infraciliary lattice and outer lattice should allow us to characterize some of the components of these structures and examine their possible evolutionary relationships with structures found in other protozoa or metazoa. Purification of cortex and gullet fractions has been achieved and antibodies against a few cortical polypeptides have been obtained (Keryer et al., in prep.). The specific recognition of other polypeptides by various probes such as antibodies against phosphorylated epitopes (Keryer et al. 1987) has been demonstrated. Monoclonal antibodies raised against purified cortices have been obtained (J.-C. Cléraut, R. Jeanmaire, R. Perasso and A. Adoutte, pers. commun.). It can be therefore expected that the next *Paramecium* book will provide a wealth of data on the biochemistry and molecular genetics of *Paramecium* cytoskeleton.

Finally, it can be pointed out that *Paramecium*, presently lingering behind with respect to biochemical and molecular genetic aspects, may be a most suitable system to study the control of cytoskeleton geometry, a still practically unexplored domain. To our knowledge, the mutational defects observed in the organization of cytoplasmic microtubule networks, epiplasm, infraciliary and outer lattices are the first examples of altered geometry of cytoskeletal structures, most probably independent of defects in their major component molecules.

5 Technical Note

The method worked out in our laboratory for immunofluorescence studies (Cohen et al. 1982, 1984, 1987) consisted in cell permeabilization then direct incubation in antibodies (generally without fixation) in buffers which stabilize the cytoskeleton such as PIPES 65 mM, HEPES 25 mM, EGTA 10 mM, $MgCl_2$ 2 mM, pH 6.9 (PHEM, Schliwa and Van Blerkom 1981) or non-saline phosphate buffer (Na phosphate 6 mM, EGTA 10 mM, $MgCl_2$ 2 mM, pH 7.0, first used for HMM decoration, Cohen et al. 1984). This method provided a quite satisfactory conservation of cellular organization and an apparently good specificity of labelling as controlled by electron microscopy. However, it is generally admitted that standard conditions for stringent antigen-antibody affinity require 150 mM salt in the incubation buffer. We therefore modified our initial protocols as follows: after permeabilization in the appropriate buffer (PHEM or phosphate), cells are fixed for 1 h in 2% paraformaldehyde in the same buffer, then transferred into phosphate-buffered saline (PBS) for antibody treatment. This modification did not affect in any way the labelling given by most of the antibodies. However, two of them, 3D12 (Sect. 2.3.3) and the anti-43 kD

(Sect. 2.2.2), displayed different decorations under the two conditions: 3D12 decorates the infraciliary lattice only under non-stringent conditions (low ionic strength), perhaps via non-specific interactions (although the antibody represents an unambiguous "stain" for the structure). In contrast, the anti-43 kD, a crude antiserum giving a high background level in both types of conditions, specifically binds to the epiplasm only under stringent conditions (PBS).

Acknowledgements. We thank Dr. G. Piperno for giving us a sample of his monoclonal antibody 6-11B-1. Dr. D. Pantaloni for affinity-purified anti-pig brain tubulin antibodies, Drs. M. Bornens, C. Klotz and D. Sandoz for permitting us to screen a collection of monoclonal antibodies on *Paramecium,* among which are CC212 and 3D12, Dr. C. Deregnaucourt for her monoclonal antibody B1B2, and Drs. P. Gounon and E. Karsenti for their anti-frog muscle actin. Drs. A. Adoutte and L. Sperling are gratefully acknowledged for critical reading of the manuscript.

References

Adoutte A, Ramanathan R, Lewis RM, Dute RR, Ling KY, Kung C, Nelson DL (1980) Biochemical studies of the excitable membrane of *Paramecium tetraurelia.* III. Proteins of cilia and ciliary membranes. J Cell Biol 84:717–738

Adoutte A, Claisse M, Cance J (1984) Tubulin evolution: an electrophoretic and immunological analysis. Orig Life 13:177–182

Adoutte A, Claisse M, Maunoury R, Beisson J (1985) Tubulin evolution: conservation of protozoan specific epitopes in the ciliary tubulins of vertebrates. J Mol Evol 22:220–229

Allen RD (1971) Fine structure of membranous and microfibrillar systems in the cortex of *Paramecium caudatum.* J Cell Biol 49:1–20

Allen RD (1974) Food vacuole membrane with microtubule-associated membrane transfer in *Paramecium.* J Cell Biol 63:904–922

Allen RD, Fok AK (1983a) Phagosome fusion vesicles of *Paramecium.* I. Thin section morphology. Eur J Cell Biol 29:150–158

Allen RD, Fok AK (1983b) Phagosome fusion vesicles of *Paramecium.* II. Freeze-fracture evidence for membrane replacement. Eur J Cell Biol 29:159–165

Allen RD, Fok AK (1983c) Non-lysosomal vesicles (acidosomes) are involved in phagosome acidification in *Paramecium.* J Cell Biol 97:566–570

Allen RD, Wolf RW (1974) The cytoproct of *Paramecium caudatum.* Structure and function, microtubules, and fate of food vacuole membranes. J Cell Sci 14:611–631

Amos WB (1975) Contraction and calcium binding in the vorticellid ciliates. In: Inoué S, Stephen RE (eds) Molecules and cell movement. Raven, New York, pp 411–436

Anderton BH (1981) Intermediate filaments: a family of homologous structures. J Muscle Res Cell Motil 2:141–166

Aufderheide KJ (1977) Saltatory motility of uninserted trichocysts and mitochondria in *Paramecium tetraurelia.* Science 198:299–300

Aufderheide KJ (1986) Identification of the basal bodies and kinetodesmal fibers in living cell of *Paramecium tetraurelia* Sonneborn, 1975 and *Paramecium sonneborni* Aufderheide, Daggett and Nerad, 1983. J Protozool 33:77–80

Aufderheide KJ, Frankel J, Williams NE (1980) Formation and positioning of surface-related structures in protozoa. Microbiol Rev 44:252–302

Barahona I, Rodrigues-Pousada C (1983) Site of tubulin synthesis in *Tetrahymena pyriformis.* FEBS Lett 158:271–275

Barahoma I, Galego L, Jacquet M, Rodrigues-Pousada C (1985) Cloning of two β-tubulin related genes in *Tetrahymena pyriformis.* FEBS Lett 188:227–232

Beisson J, Sonneborn TM (1965) Cytoplasmic inheritance of the organization of the cell cortex in *Paramecium aurelia.* PNAS 53:275–282

Beisson J, Rossignol M, Ruiz F, Adoutte A, Grandchamp S (1976) Genetic analysis of morphogenetic processes in *Paramecium:* a mutation affecting cortical pattern and nuclear reorganization (Abstr.). J Protozool 23:3A
Blose SH, Meltzer DI, Feramisco JR (1982) 10 nm filaments induced to collapse in cells microinjected with antibodies against tubulin (Abstr.). J Cell Biol 95:229a
Branton D (1981) Membrane cytoskeletal interactions in the human erythrocyte. Cold Spring Harbor Symp Quant Biol 46:1–5
Bricheux G (1987) Étude biochimique et immunologique des protéines du complexe membrane-cytosquelette des euglènes. Thesis, Univ Clermont II, Fr
Brunke KJ, Collis PS, Weeks DP (1982) Post-translational modification of tubulin is dependent on organelle assembly. Nature (London) 297:516–518
Callahan RC, Shalke G, Gorovsky MA (1984) Developmental rearrangements associated with a single type of expressed α-tubulin gene in *Tetrahymena.* Cell 36:441–445
Chatton E, Lwoff A (1936) Techniques pour l'étude des Protozoaires spécialement de leurs structures superficielles (cinétosome et argyrome). Bull Soc Fr Microsc 5:25–39
Cohen J (1984) Cytosquelette et relations cortex-noyaux chez la paramécie. Thesis, Univ Paris XI Centre d'Orsay, Fr
Cohen J, Beisson J (1980) Genetic analysis of the relationships between the cell surface and the nuclei in *Paramecium tetraurelia.* Genetics 95:799–818
Cohen J, Beisson J, Tucker JB (1980) Abnormal microtubule deployment during defective macronuclear division in a *Paramecium* mutant. J Cell Sci 44:153–167
Cohen J, Adoutte A, Grandchamp S, Houdebine L-M, Beisson J (1982) Immunocytochemical study of microtubular structures throughout the cell cycle of *Paramecium.* Biol Cell 44:35–44
Cohen J, Garreau de Loubresse N, Beisson J (1984) Actin microfilaments in *Paramecium:* localization and role in intracellular movements. Cell Motil 4:443–468
Cohen J, Garreau de Loubresse N, Klotz C, Ruiz F, Bordes N, Sandoz D, Bornens M, Beisson J (1987) Organization and dynamics of a cortical fibrous network of *Paramecium:* the outer lattice. Cell Motil 7:315–324
Cupples CG, Pearlman EE (1986) Isolation and characterization of the actin gene from *Tetrahymena thermophila.* PNAS 83:5160–5164
Dhermy D, Bournier O, Boivin P (1978) Partially polymerized erythrocyte actin obtained by affinity chromatography on DNase Sepharose. Comparison with rabbit skeletal muscle actin and role of calcium. BBRC 85:906–915
Ehret CF, McArdle EW (1974) The structure of *Paramecium* as viewed from its constituent levels of organization. In: Wagtendonk WJ van (ed) *Paramecium,* a current survey. Elsevier, Amsterdam, pp 263–337
Eichenlaub-Ritter U, Tucker JB (1984) Microtubules with more than 13 protofilaments in the dividing nuclei of ciliates. Nature (London) 307:60–62
Fernandez-Galiano D (1966) Une nouvelle méthode pour la mise en évidence de l'infraciliature des ciliés. Protistologica 2:35–37
Fernandez-Galiano D (1976) Silver impregnation of ciliated protozoa: procedure yielding good results with the pyridinated silver carbonate method. Trans Am Microsc Soc 95:557–560
Fernandez-Galiano D (1978) Le comportement des cinétodesmes pendant la division de *Paramecium caudatum.* Protistologica 14:2961–2964
Franke WW, Scheer U, Krohne G, Jarasch E-D (1981) The nuclear envelope and the architecture of the nuclear periphery. J Cell Biol 91:39s–50s
Gelei G von (1937) Ein neues Fibrillensystem im Ectoplasma von *Paramecium;* zugleich ein Vergleich zwischen dem neuen und dem alten Gittersystem. Arch Protistenk 89:133–162
Gounon P, Karsenti E (1981) Involvement of contractile proteins in changes in consistency of oocyte nucleoplasm in the newt *Pleurodeles waltii.* J Cell Biol 88:410–421
Grain J (1986) The cytoskeleton in protists: nature, structure and functions. Int Rev Cytol 104:153–249
Grandchamp S, Beisson J (1981) Positional control of nuclear differentiation in *Paramecium.* Dev Biol 81:336–341

Hausmann K, Allen RD (1977) Membranes and microtubules of the excretory apparatus of *Paramecium caudatum*. Cytobiology 15:303–320

Helftenbein E (1985) Nucleotide sequence of a macronuclear DNA molecule coding for α-tubulin from the ciliate *Stylonychia lemnae*. Special codon usage: TAA is not a translation termination codon. Nucl Acid Res 13:415–433

Hightower RC, Meagher RB (1986) The molecular evolution of actin. Genetics 114:315–332

Hyams G, King AC (1985) Identification of proteins of the striated rootlet of *Tetrahymena* by immunofluorescence microscopy and immunoblotting with an anti-rootlet serum. Eur J Cell Biol 38:102–105

Jurand A, Selman GG (1969) The anatomy of *Paramecium aurelia*. McMillan, London, 218 pp

Kaine BP, Spear BB (1982) Nucleotide sequence of a macronuclear gene for actin in *Oxytricha fallax*. Nature (London) 295:430–432

Karsenti E, Guilbet B, Bornens M, Avrameas S, Whalen R, Pantaloni D (1978) Detection of tubulin and actin in various cell lines by an immunoperoxidase technique. J Histochem Cytochem 26:934–947

Kersken H, Vilmart-Seuwen J, Momayesi M, Plattner H (1986a) Filamentous actin in *Paramecium* cells: mapping by phalloidin affinity labelling in vivo and in vitro. J Histochem Cytochem 34:443–454

Kersken H, Momayesi M, Braun C, Plattner H (1986b) Filamentous actin in *Paramecium* cells: functional and structural changes correlated with phalloidin affinity labelling in vivo. J Histochem Cytochem 34:455–465

Keryer G, Davis FM, Rao PN, Beisson J (1987) Protein phosphorylation and dynamics of cytoskeletal structures associated with basal bodies in *Paramecium*. Cell Motil 8:44–54

Kilmartin JV, Wright B, Milstein C (1982) Rat monoclonal anti-tubulin antibodies derived by using a new non-secreting rat cell line. J Cell Biol 93:576–582

Klotz C, Bordes N, Lainé MC, Sandoz D, Bornens M (1986) Myosin at the apical pole of ciliated epithelial cells as revealed by a monoclonal antibody. J Cell Biol 103:613–619

Larson DL, Dingle AD (1981) Isolation, ultrastructure and protein composition of the flagellar rootlet of *Naegleria gruberi*. J Cell Biol 89:424–432

Lazarides E (1982) Intermediate filaments: a chemical heterogeneous developmentally regulated class of proteins. Annu Rev Biochem 51:219–250

Lazarides E, Lindberg U (1974) Actin is the naturally occurring inhibitor of desoxyribonuclease I. PNAS 71:4742–4746

Lazarides E, Weber K (1974) Actin antibody: the specific visualization of actin filaments in non-muscle cells. PNAS 71:2268–2272

L'Hernault SW, Rosenbaum JL (1985) *Chlamydomonas* α-tubulin is posttranslationally modified by acetylation on the ε-amino group of a lysin. Biochemistry 24:473–478

Lin JJC (1981) Monoclonal antibodies against myofibrillar components of rat skeletal muscle decorate the intermediate filaments of cultured cells. PNAS 78:2335–2339

Lindberg U, Eriksson S (1971) Purification from crude extract by affinity chromatography of the inhibitor of deoxyribonuclease. Eur J Biochem 18:474–479

Little M, Luduena RF, Langford GM, Asnes CF, Farrel K (1981) Comparison of proteolytic cleavage pattern of α-tubulins and β-tubulins from taxonomically distant species. J Mol Biol 149:95–107

Lünd EE (1933) A correlation of the silverline and neuromotor systems of *Paramecium*. Univ Cal Publ Zool 39:35–75

Lynn DH (1981) The organization and evolution of microtubular organelles in ciliated protozoa. Biol Rev 56:243–292

Méténier G (1984) Actin in *Tetrahymena paravorax:* ultrastructural localization of HMM-binding filaments in glycerinated cells. J Protozool 31:205–215

Metivier C (1986) La motilité chez le dinoflagellé évolué *Noctiluca scintillans*. Organisation structurale. Régulation ionique. Caractérisation biochimique et immunologique des protéines corticales. Thesis, Univ Paris VI, Fr

Mitchell EJ, Zimmerman AM (1985) Biochemical evidence for the presence of an actin protein in *Tetrahymena pyriformis*. J Cell Sci 73:279–297

Mitchell EJ, Zimmerman S, Zimmerman AM (1985) Actin and tubulin in *Tetrahymena*. Can J Biochem Cell Biol 63:389–396

Ng SF, Frankel J (1977) 180° rotation of ciliary rows and its morphogenetic implications in *Tetrahymena pyriformis*. PNAS 74:1115–1119

Numata O, Watanabe Y (1982) In vitro assembly and disassembly of 14-nm filament from *Tetrahymena pyriformis*. The protein component of 14-nm filament is a 49000-dalton protein. J Biochem 91:1563–1573

Numata O, Yasuda T, Hirabayashi T, Watanabe Y (1980) A new fiber forming protein from *Tetrahymena pyriformis*. Exp Cell Res 129:223–230

Oakley BR, Morris NR (1980) Nuclear movement is β-tubulin-dependent in *Aspergillus nidulans*. Cell 19:255–262

Olmsted JB, Cox JV, Asnes C, Parysek LM, Lyon HD (1984) Cellular regulation of microtubule organization. J Cell Biol 99:28s–32s

Parducz B (1962) On a new concept of cortical organization in *Paramecium*. Acta Biol Hung 13:299–322

Peck RK (1977) The ultrastructure of the somatic cortex of *Pseudomicrothorax dubius:* structure and function of the epiplasm in ciliated protozoa. J Cell Sci 25:367–385

Piperno G, Fuller MT (1985) Monoclonal antibodies specific for an acetylated form of α-tubulin recognize the antigen in cilia and flagella from a variety of organisms. J Cell Biol 101:2085–2094

Pitelka DR (1969) Fibrillar systems in protozoa. In: Tze-Tuan Chen (ed) Research in protozoology, vol 3. Pergamon, New York, pp 279–388

Plattner H, Westphal C, Tiggemann R (1982) Cytoskeleton-secretory vesicle interaction during the docking of secretory vesicles at the cell membrane in *Paramecium tetraurelia* cells. J Cell Biol 92:368–377

Ruiz F, Adoutte A, Rossignol M, Beisson J (1976) Genetic analysis of morphogenetic processes in *Paramecium*. I. A mutation affecting trichocyst formation and nuclear division. Genet Res Cambr 27:109–122

Ruiz F, Garreau de Loubresse N, Beisson J (1987) A mutation affecting basal body duplication and cell shape in *Paramecium*. J Cell Biol 104:417–431

Sandoz D, Gounon P, Karsenti E, Boisvieux-Ulrich E, Lainé MC, Paulin D (1983) Organization of intermediate filaments in ciliated cells from quail oviduct. J Submicrosc Cytol 15:323–326

Schatten G, Schatten H, Bestor TH, Balczon R (1982) Taxol inhibits the nuclear movements during fertilization and induce asters in unfertilized sea urchin eggs. J Cell Biol 94:455–465

Schliwa M, Van Blerkom J (1981) Structural interaction of cytoskeletal components. J Cell Biol 90:222–235

Sibley JT, Hanson ED (1974) Identification and function of a subcortical cytoskeleton in *Paramecium*. Arch Protistenk 116:221–235

Sonneborn TM (1974) *Paramecium aurelia*. In: King RC (ed) Handbook of genetics, vol 2. Plenum, New York, pp 469–594

Sonneborn TM (1975) Positional information and nearest neighbor interactions in relation to spatial patterns in ciliates. Ann Biol 14:565–584

Stevenson I, Lloyd FP (1971 a) Ultrastructure of nuclear division in *Paramecium aurelia*. I. Mitosis in the micronucleus. Aust J Biol Sci 24:963–975

Stevenson I, Lloyd FP (1971 b) Ultrastructure of nuclear division in *Paramecium aurelia*. II. Amitosis of the macronucleus. Aust J Biol Sci 24:977–987

Sundararaman V, Hanson ED (1976) Longitudinal microtubules and their functions during asexual reproduction in *Paramecium tetraurelia*. Genet Res Camb 27:205–211

Suprenant KA, Hays E, Le Cluyse E, Dentler WL (1985) Multiple forms of tubulin in the cilia and cytoplasm of *Tetrahymena thermophila*. PNAS 82:6908–6912

Tiggemann R, Plattner H (1981) Localization of actin in the cortex of *Paramecium tetraurelia* by immuno and affinity fluorescence microscopy. Eur J Cell Biol 24:184–190

Tiggemann R, Plattner H, Rashed I, Baeuerle P, Wachter E (1981) Quantitative data on peroxidatic markers for electron microscopy with a note on actin identification in *Paramecium* cells. J Histochem Cytochem 29:1387–1396

Tucker JB (1979) Spatial organization of microtubules. In: Roberts KR, Hyams JS (eds) Microtubules. Academic Press, London New York, pp 315–357

Tucker JB, Beisson J, Roche DLJ, Cohen J (1980) Microtubules and control of macronuclear "amitosis" in *Paramecium*. J Cell Sci 44:135–151

Tucker JB, Mattews SA, Hendry KAK, Mackie JB, Roche DLJ (1985) Spindle microtubule differentiation on deployment during micronuclear mitosis in *Paramecium*. J Cell Biol 101:1966–1976

Viguès B, Grolière C-A (1985) Evidence for a Ca^{2+}-binding protein associated to non-actin microfilamentous systems in two ciliated protozoans. Exp Cell Res 159:366–376

Weber K (1976) Visualization of tubulin-containing structures by immunofluorescence microscopy. Cytoplasmic microtubules, mitotic figures and vinblastine-induced paracrystals. In: Goldman R, Pollard T, Rosenbaum J (eds) Cell motility. Cold Spring Harbor Lab, New York, pp 403–417

Wehland J, Willingham MC, Sandoval IV (1983) A rat monoclonal antibody reacting specifically with the tyrosylated form of α-tubulin. I. Biochemical characterization, effects on microtubule polymerization in vitro, and microtubule polymerization and organization in vivo. J Cell Biol 97:1467–1475

Wichterman R (1986) The biology of *Paramecium*, 2nd edn. Plenum, New York London, 599 pp

Williams NE (1986a) Evolutionary change in cytoskeletal proteins and cell architecture in lower eukaryotes. Progr Protistol 1:309–324

Williams NE (1986b) The assembly of ciliates. J Protozool 33:313–316

Williams NE (1986c) The nature and organization of filaments in the oral apparatus of *Tetrahymena*. J Protozool 33:352–358

Williams NE, Vaudaux PE, Skriver L (1979) Cytoskeletal proteins of the cell surface in *Tetrahymena*. I. Identification and localization of major proteins. Exp Cell Res 123: 311–320

Williams NE, Buhse HE Jr, Smith MG (1984) Protein similarities in the genus *Tetrahymena* and a description of *Tetrahymena leucophrys* n.sp. J Protozool 31:313–321

Williams NE, Honts JE, Graeff RW (1986) Oral filament proteins and their regulation in *Tetrahymena pyriformis*. Exp Cell Res 164:295–310

Zechel K (1980) Isolation of polymerization-competent cytoplasmic actin by affinity chromatography on immobilized DNase I using formamide as eluant. Eur J Biochem 110: 343–348

Chapter 22 Endocytobiosis

H.-D. GÖRTZ[1]

CONTENTS

1 Introduction

Paramecium may be colonized by a variety of pro- or eukaryotic microorganisms. The nature of these endocytobioses ranges from mutualism to parasitism, and the endocytobionts may colonize different compartments of the host cytoplasm or even the micro- or macronuclei (for details of the cytology of *Paramecium* see Allen, Chap. 2, this Vol.). However, the basic questions of host–endocytobiont interactions are similar in the different associations, and the purpose of this chapter is to describe the characteristics of endocytobiosis in *Paramecium*, to discuss its significance, and to review recent advances rather than to give a catalog of the various endocytobionts.

The term endocytobiosis is used here because it specifies the localization of the guest organisms more precisely than the term endosymbiosis. Also, the term endocytobiosis is more neutral with regard to possible mutualistic or parasitic effects. The organisms colonizing the host cells are called endocytobionts, although the term endosymbiont is still frequently used for endocytobionts in *Paramecium* (e.g., it is the historical term used for kappa particles).

1 Zoologisches Institut der Universität, D-4400 Münster, FRG

Paramecium. Ed. by H.-D. Görtz

2 Categories of Endocytobiosis

Endocytobioses may be classified on the basis of various features. One possibility is to use the nature of the association, but in many cases it is difficult to decide whether a symbiosis is mutualistic, parasitic, or in between these extremes (see also Sect. 5). In the following, the degree of adaptation, maintenance and infectivity of endocytobionts is used as a classification with the categories "accidental invasions", "permanent endocytobioses", and "endocytobioses with highly infectious microorganisms".

2.1 Accidental Invasions

Some endocytobioses in *Paramecium* which have been found only once may belong to this category, where the invading organisms appear to be poorly adapted and possibly pathogenic. Examples are a few cases of infections by fungi (see the reviews by Ball 1969; Wichterman 1986). More recent observations are infections of *P. bursaria,* which had lost the symbiotic algae, with yeast or bacteria (Görtz 1982) and a bacterial infection of *P. caudatum* (Dieckmann 1981). Eventually, such temporary infections of a host population could develop into stable endocytobioses, similar to the amazing development of the endocytobiosis of *Amoeba proteus* with X-bacteria (Jeon 1983).

2.2 Permanent Endocytobioses

Various organisms live permanently in *Paramecium* cells. These endocytobionts are well adapted, and not obviously harmful under natural conditions. They need not be highly infectious. Most of the known endocytobionts in paramecia belong to this category; they have been listed and described in a number of reviews (Soldo 1974; Görtz 1983; Preer and Preer 1984; Wichterman 1986). In the following, only some of the well-known examples will be mentioned to show the extent and variability of this phenomenon.

Caedibacter taeniospiralis, one of the best-known bacterial endocytobionts, lives in the cytoplasm of *P. tetraurelia,* not surrounded by a host membrane. The most striking feature of the *Caedibacter* species (formerly called kappa particles, see Preer et al. 1974; Preer and Preer 1984) and some other endocytobionts is their killing activity. When released from their host into the medium and ingested by a paramecium which does not harbor bacterial symbionts of this type the paramecium is killed. Endocytobionts (also called killer particles) of killer paramecia are extensively treated by Quackenbush (Chap. 23, this Vol.).

Lyticum flagellatum (lambda particles) and *L. sinuosum,* both with flagella, are killer particles which are enclosed in vacuoles, whereas *Tectibacter* (delta particles) which does not show a killing activity is covered by electron-dense material on the outer membrane (Beale et al. 1969) and is not enclosed in a host vacuole. The epsilon particles in *P. multimicronucleatum* are found in bulbous

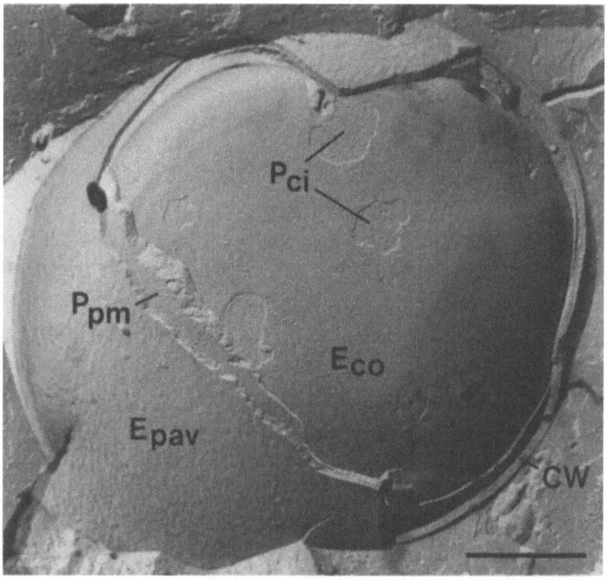

Fig. 1. Freeze-fractured perialgal vacuole and enclosed endosymbiotic *Chlorella* cell. The fracture plane has exposed the successive convex membrane faces: the smooth E-face of the perialgal vacuole membrane (E_{pav}), which is poor in particles; the P-face of the algal plasma membrane (P_{pm}); and the E-face of the outer chloroplast envelope membrane (E_{co}) through "windows" in which the P-face of the inner envelope membrane (P_{ci}) is visible. Following the curve of the cross-fractured perialgal vacuole membrane, the very small distance between the algal cell wall (*CW*) and the vacuole membrane can be seen. ×17000; bar = 1 µm. (Micrograph kindly provided by R. Meier, Heidelberg)

distensions of the outer membrane of the nuclei (Jenkins 1970), and also other bacterial endocytobionts are often situated closely to the nuclei. However, most endocytobionts do not penetrate the nuclear envelope. An exception is a bacterium in the cytoplasm of a *P. sexaurelia* strain, which eventually invades the macronucleus (Görtz, 1983, 1986). Permanently inhabiting the macronucleus of *P. caudatum* is a kappa-like bacterium (Estève 1978; Schmidt and Görtz 1987; see also Quackenbush, Chap. 23, this Vol.). Endonucleobionts are not surrounded by host membranes inside the nuclei. The significance of being enclosed by a vacuole in the host cytoplasm or not is not clear, although it has been argued (Smith 1979) that encirclement by a host membrane would only disappear, if at all, in a later stage of the evolution of an endocytobiosis.

A stable and mutualistic endocytobiosis is that of *P. bursaria* with a *Chlorella* species. The algae, which are often called zoochlorellae, are situated singly in a vacuole (Karakashian et al. 1968), where they are tightly surrounded by the host membrane (Fig. 1). *P. bursaria* does not depend on its algae, and the symbionts can be killed by various means, e.g., by prolonged growth in the dark, treatment of the cells with an inhibitor of photosynthesis, or with cycloheximide (Karakashian 1975; Reisser 1976; Tsukii and Hiwatashi, cited by Iwatsuki and Naitoh 1981; Weis, personal communication). The resulting aposymbiotic paramecia can be reinfected by exsymbiotic algae or even by other microorganisms (see Sect. 3). The taxonomic state of symbiotic *Chlorella* strains is not clear (Weis and Berry 1974; Reisser et al. 1985).

2.3 Endocytobioses with Highly Infectious Microorganisms

Experimental infections of paramecia have been achieved by microinjection or by the addition of isolated endocytobionts to the medium (Koizumi and Preer 1966; Gibson 1973; Preer et al. 1974; Soldo 1974). Natural infections, however, appear to be rare with most of the known endocytobionts and will not be necessary for those which are permanently maintained by their host; rather, the high infectivity seems to be a parasitic attitude.

Interestingly, highly infectious endocytobionts (bacteria or flagellates) of paramecium all invade the nuclei. They are nucleus-specific, specialized to infect either the micro- or macronucleus. All infectious bacterial endonucleobionts belong to the genus *Holospora* (Preer and Preer 1984; Görtz 1986). They show a developmental cycle with an infectious form (up to 20-μm-long) and a morphologically different reproductive form (Fig. 2) (Görtz and Dieckmann 1980; Görtz 1986).

Another infectious endonucleobiont is the flagellate *Leptomonas karyophilus,* which has been found in the macronucleus of *P. trichium* (Gillies and Hanson 1963). Similar flagellates are also found in the macronuclei of other ciliates (Lee et al. 1985; Görtz and Dieckmann 1987). The precise invasion of either the micro- or macronucleus is certainly one of the most interesting features of endocytobiosis.

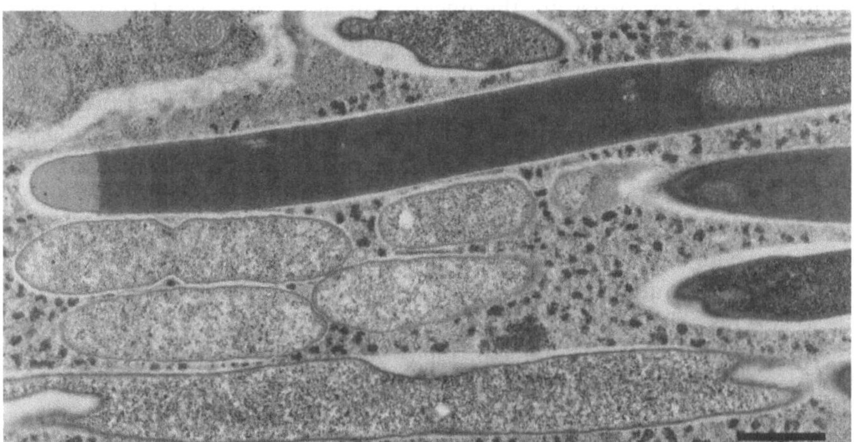

Fig. 2. Endonculeobiotic bacteria, *Holospora obtusa,* in the macronucleus of *Paramecium caudatum.* ×12000; bar = 1 μm. (Micrograph kindly provided by M. Wiemann, Münster)

3 Establishment of Endocytobionts: Infection and Integration

Generally, the first step of the establishment of a new endocytobiont appears to be the ingestion of the prospective endocytobiont by the prospective host. In the next step the endocytobiont must leave the food vacuole (phagosome) and find the final site in its host cell. This appears to be easy when the invading organism is motile like *L. karyophilus* (Gillies and Hanson 1963; Freedman and Hanson 1970). Once in the food vacuole these flagellates move vigorously, leave the food vacuole, and actively swim in the cytoplasm. It is, however, unknown whether the food vacuole membrane is perforated or not. In the cytoplasm the parasites appear to probe against the macronuclear membrane and enter the nucleus by an unknown mechanism.

Holospora species and *Chlorella* apparently do not disrupt the membrane of the food vacuole. Instead, *Holospora* causes an invagination of the vacuole-membrane and is sluiced out into a developing stockinglike distension. After having left the food vacuole this bacterium, which is nonmotile itself, is transported into the micro- or macronucleus (Ossipov and Podlipaev 1977; Podlipaev and Ossipov 1979; Görtz 1986). The transport mechanism and mode of entry into the nucleus is still under investigation.

For *Chlorella,* it is not exactly known how the alga leaves the food vacuole. Either the whole food vacuole could develop into a perialgal vacuole, or the alga could get out by causing a budding of the food vacuole, perhaps similar to *Holospora* (for a discussion of these hypotheses, see Meier et al. 1984; Reisser et al. 1985).

Infections with *Holospora* are effective and rapid (Ossipov 1981; Görtz 1983). An infecting *H. obtusa* can invade the macronucleus within 10 min after addition to the culture medium (Fujishima and Görtz 1983), and there can be no doubt that *Holospora* has specific signals for the recognition and infection of the macronucleus (e.g., by *H. obtusa*) or the micronucleus (e.g., by *H. elegans*). The infectious form of *Holospora* has a peculiar cytology. The cytoplasmic part only fills about half of the bacterium, while the other half is filled with electron-dense material of unknown nature and a less electron-dense tip at one pole of the cell (Görtz and Dieckmann 1980; Görtz 1981; Schmidt et al. 1987). The bacterium always leaves the food vacuole and enters the nucleus with this pole first (Görtz 1981, 1986; Fujishima and Fujita 1985). Fine fibrillar connections between the outer bacterial membrane and the host membrane which are seen during the infection process could be induced by components localized in the polar tip of *Holospora.*

The infection process with endocytobionts other than *Holospora, Chlorella,* or *L. karyophilus* has not been investigated in *Paramecium,* and, actually, most other endocytobionts are not or only weakly infectious. In these cases we do not know when and how the original invasion occurred. For some bacteria inhabiting the nuclei it has been speculated that they could have invaded their host nuclei by a coinfection together with one of the *Holospora* species, which were subsequently excluded by competition (Görtz and Freiburg 1984; Görtz 1986).

Those endocytobionts inhabiting the cytoplasm and not encircled by a host membrane (e.g., *Caedibacter* sp. and *Tectibacter* sp.) must have left the food vacuole by perforating the vacuole membrane or must have lost an initially present host membrane in a later step (Smith 1979).

Once ingested, a prospective endocytobiont must leave the food vacuole within a few minutes or has to prevent a fusion of the food vacuole (phagosome) with primary lysosomes unless it can resist the lysosomal enzymes. For *P. caudatum* it has been found that primary lysosomes fuse with the newly formed food vacuoles within less than 10 min (see Fok and Allen, Chap. 19, this Vol.). *Holospora* can get into a host nucleus within 10 min after addition to the culture medium, suggesting that this bacterium can leave the food vacuole long before a fusion with primary lysosomes takes place. It is not known whether a fusion with acidosomes takes place before *Holospora* leaves the food vacuole, but it has been argued that acidification might even be needed to trigger the development of the infectious form of this bacterium (Gromov et al. 1974; Görtz 1983). The beginning of this development, which ends in the nucleus with the differentiation into the reproductive form, can be observed in the food vacuoles.

For the infection of aposymbiotic *P. bursaria* with *Chlorella* it is not known whether the algae eventually prevent a fusion with lysosomes or escape the food vacuole before this fusion takes place (Meier et al. 1984). Most of the ingested algae are, however, digested. According to Weis (1980), the process to take up symbiotic algae is an inductive process. Fifteen minutes after a first ingestion of *Chlorella,* which are digested, further ingested algae are enclosed in perialgal vacuoles and maintained as endocytobionts. Weis (1980) suggested the significance of maltose, secreted in large quantities by symbiotic *Chlorella* (see Sect. 5), and carbohydrate components on the algal surface as signals for the induction. However, treatment of *Chlorella* with lectins, specific antibodies, or carbohydrate- or protein-degrading enzymes, which do not impair the maltose excretion, inhibits the infection (Reisser 1983). So, maltose apparently is not the only signal.

Aposymbiotic *P. bursaria* are also infected by various other microorganisms (Görtz 1982; Reisser 1987). This implies that the signals inducing the uptake of a prospective endocytobiont by this paramecium cannot be very specific. However, although a great variety of endocytobionts may be established in *Paramecium,* they are maintained to different degrees.

4 Maintenance of Endocytobionts and Adaptations

Host and endocytobionts must meet some basic requirements for a stable association. The host must not kill or exclude an invaded microorganism and must supply its guest with sufficient quantities of appropriate metabolites. The endocytobiont must not seriously harm or even kill its host. Rather, it must grow with a rate similar to the host. During a division of the host cell a regular distribution of endocytobionts must be guaranteed. Some of these and further requirements are, in principle, prerequisites for the start of an endocytobiosis. Later, some of these requirements or qualifications may improve and result in a better adaptation, evolved over many generations. For many endocytobioses in paramecium the existence of such adaptations can only be assumed, because the associations are stable. In some cases adaptations and features necessary for maintaining and stabilizing the host-endocytobiont unit have been found and investigated.

Genetic crosses of killer paramecia with nonkiller paramecia revealed that the persistence of the endocytobionts (killer particles) depends on the nuclear genes of the host cell. It could even be shown in the heterozygous condition of certain alleles, about one-half the number of endocytobionts in homozygous cells were present (Chao 1953; Balsley 1967). A dependence on host genes was also observed for *Holospora* species (Preer 1969; Gibson 1973; Fujishima, pers. commun.).

The nature of these genes is not known in any of the cases, and it is not even clear whether the gene products support or inhibit the growth (Preer et al. 1974). It is of interest that most bacterial endocytobionts cannot be cultured extracellularly, indicating that they depend on certain metabolites supplied by the host paramecium. The reported successful cultivation of *Lyticum flagellatum* (lambda particles) and *Pseudocaedibacter conjugatus* (mu particles) required a complex medium and achieved very low maximum population densities in culture (Van Wagtendonk et al. 1963; Williams 1971).

In agreement with these observations is the small genome size of some endocytobiotic bacteria in *Paramecium.* The genomes are considerably smaller than those of free-living bacteria which suggests a loss of a number of their original genes and indicates that they have reduced metabolic capabilities (Soldo 1974; Soldo et al. 1983). Although the genome of *Caedibacter* is not so small, it was somewhat unexpected that the translational systems of host paramecia and endocytobiotic *Caedibacter varicaedens* are not shared, and that all major proteins found in the endocytobiont are synthesized in this endocytobiont (Schmidt 1984). Also, a transfer of genes from endocytobionts to the host genome has not been shown in *Paramecium.*

As a further consequence of prolonged intracellular existence several endocytobionts of ciliates have multicopy genomes (5 to 20 copies per cell) Soldo and Godoy 1974; Schmidt and Heckmann 1980; Soldo 1983). This is a feature also of mitochondria and chloroplasts, whereas free-living bacteria generally contain only a single copy of the genome. Provided that the segregation of endocytobionts during the host cell division is balanced, the divisions of endocytobionts need not be strictly coupled with their growth and DNA replication. An advantage of multicopy genomes could be that endocytobionts are then less sensitive to mutations (Soldo 1983, 1985). The small genome size indicates that these endocytobionts have kept mainly those genes whose products cannot be replaced by gene products of the host cell; a defect caused by mutation in these few genes would be fatal.

Free-living ciliates constantly multiply more or less rapidly depending on the environmental conditions. Therefore, during the host cell divisions endocytobionts have to be distributed to the daughter cells, even if they are infectious. Otherwise the endocytobionts will be diluted out of the host population sooner or later. The mechanisms for distribution of endocytobionts into the daugther cells can be best observed for endonucleobionts. During the division of their host nuclei *Holospora* bacteria of the reproductive stage are transported into the host nuclei, while the bacteria of the infectious stage are collected in the connecting piece of the dividing host nucleus (Ossipov 1981; Görtz 1986). These infectious forms are later excreted and can infect other paramecia.

Little is known about the interaction of *Holospora* with the host metabolism, and it is not even clear whether the endonucleobionts digest host chromatin or use free monomers from the nuclei. Apparently, the paramecium reacts to the presence of the endonucleobionts. Freiburg (1985) has shown that the RNA synthesis in *Holospora*-infected macronuclei is about five times higher than that of uninfected macronuclei, and he could prove that this was the activity of host enzyme. This was not clear for a higher proteinase activity also observed in infected nuclei.

Just as the presence of high activities of lytic enzymes inside functional nuclei is somewhat puzzling, it is strange that bacterial endonucleobionts can be lysed in the host nucleus and thereby can be completely lost without destroying the nucleus or affecting the vitality of the host cell. A lysis of *Holospora obtusa* due to unknown reasons is eventually observed in the macronuclei of *P. caudatum* (Görtz 1986) or can be induced experimentally by antibiotics.

Symbiotic *Chlorellae* in *P. bursaria* are not digested (Karakashian and Rudzinska 1981; Reisser 1983) even when the host cells starve. Perialgal vacuoles apparently do not fuse with primary lysosomes, because the membranes of these vacuoles are different from food vacuole membranes and probably cannot be recognized by the primary lysosomes (Meier et al. 1984). On the other hand, the membrane of the perialgal vacuole appears to permit the exchange of various metabolites, e.g., certain sugars (see Sect. 5). Results obtained by Albers et al. (1983) indicate the possible existence of a glutamine–glutamate shuttle between *Chlorella* and its host cell. Exsymbiotic *Chlorella* can grow with nitrate as sole nitrogen source when kept in pure culture. On the contrary, nitrate is not assimilated by *P. bursaria* with symbiotic algae (Reisser 1976; Albers et al. 1983).

5 Adaptive Value of Endocytobionts

Paramecia can live without their endocytobionts. Selective advantages of bacterial endocytobioses for the host cell are not obvious. However, it is difficult to judge the adaptive value of endocytobionts from laboratory cultures. In natural environments a certain endocytobiont, which may be harmful and even fatal for its host under laboratory conditions, can be beneficial, e.g., by killing competitors feeding on identical prey organisms (see Landis, Chap. 24, this Vol.). At present, the killing mechanism is not understood, and it is also not known, what makes a "killer" paramecium resistant to its own toxin (see Quackenbush, Chap. 23, this Vol.).

It could be assumed that *Paramecium* is supplied with vitamins or essential metabolites by its endocytobionts. Actually, Soldo and Godoy (1973) reported that *P. octaurelia* harboring *Lyticum flagellatum* (lambda particles) does not require folic acid, whereas endocytobiont-free cells do. Although *Paramecium* will get sufficient folic acid with its natural food, the provision of a host cell with a vitamin may be an early step to make an endocytobiosis essential.

This is the only known case where bacterial endocytobionts supply their host with metabolites. *Chlorella,* an algal endocytobiont, excretes large amounts of sugars, more than 90% of it maltose, which is used by the host cell (Muscatine et al. 1967; Brown and Nielsen 1974). Free-living *Chlorella* species not excreting maltose are not induced to do so in experimentally achieved endocytobiosis (Reisser 1987).

Algal photosynthesis can fix respiratory CO_2 of the *Paramecium* and produces oxygen which is used by host respiration (Reisser 1980). At sufficient light intensities the respiratory oxygen demand of the host cell is covered by the photosynthesis of *Chlorella* and oxygen is actually released into the medium. The production of oxygen may be one of the reasons why endocytobiotic *Chlorella* even induces new behavioral traits in *P. bursaria:* green *P. bursaria* show a step-down photophobic reaction; in contrast to aposymbiotic *P. bursaria,* they can gather in a light spot (photoaccumulation) (Iwatsuki and Naitoh 1981; Reisser and Häder 1984; Cronkite 1986).

As another consequence, green *P. bursaria* is significantly less subject to attack by *Didinium nasutum,* a ciliate predator of paramecia, than colorless cells (Berger 1980). Also, aposymbiotic *P. bursaria* are more easily infected by other microorganisms. The presence of *Chlorella* in the cell can prevent such infections. Moreover, when an aposymbiotic cell is colonized by yeasts or bacteria and provided with exsymbiotic *Chlorella* in the medium, the algae are ingested and taken as endocytobionts. In the following days yeasts or bacteria are displaced (Görtz 1982), whereas the algae are maintained. Infections of green paramecia with bacteria or yeasts are not successful.

It is obvious that the endocytobiosis with *Chlorella* brings about a selective advantage for *P. bursaria.* The host-alga unit is almost autotrophic, and, provided there is sufficient light, it is possible to forget about a culture for several months without feeding it. Also, the algal endocytobiont enables the host organism to establish ecological strategies quite different from other *Paramecium* species (see Landis, Chap. 24; Nyberg, Chap. 3, this Vol.). On the other hand, *Chlorella* is nourished by the host cell when deprived of light and is only lost after prolonged culture in the dark. Thus, in conclusion, the endocytobiosis of *P. bursaria* and *Chlorella* can be regarded as truely mutualistic. However, both partners have not lost the capability to grow independently of each other.

The resistance of a host cell against additional infections has also been observed in *P. caudatum* harboring a *Holospora* species (Heckmann 1977; Görtz 1983). This may be a general phenomenon of endocytobiosis, but in the case of *Holospora* this resistance is possibly directed specifically against another *Holospora* species, while infections with different bacteria may be successful. Resident *Holospora* can even be replaced by invading bacteria (Görtz and Freiburg 1984; Görtz 1986, 1987), although paramecia which are colonized by more than one endocytobiont species are commonly observed in samples of natural populations (see Sect. 2).

The resistance to pathogenic microorganisms can, of course, be of adaptive value for a host harboring endocytobionts in cytoplasm or nuclei. However, whereas the location in the nuclei may be of special advantage for the endonucleobiont, it has been found that the presence of microorganisms in the nuclei is often harmful to the host under laboratory conditions (Esteve 1978; Wichterman 1986). Surprisingly, an infection of micronuclei may be more harmful than one of the macronuclei. *P. caudatum* infected with *H. elegans* in the micronucleus can still conjugate, and the micronucleus can even undergo meiosis, but the exconjugants are not viable (Görtz and Fujishima 1983). This indicates that infected micronuclei can no longer give rise to new functional macronuclei, which is a parallel to the parasitic castration of metazoa. Nevertheless, however, paramecia with infected micronuclei have a higher fission rate than amicronucleate cells, indicating that infected micronuclei may still have a function for the vegetative life (for the function of micronuclei in vegetative life, see Mikami, Chap. 8, this Vol.).

6 Conclusions and Perspectives

Paramecium appears to be an ideal host for numerous microorganisms which have developed various adaptations for the intracellular way of life. Some of the endocytobionts are highly infectious, whereas others show only weak or no infectivity but are permanently maintained. At present, we are only beginning to understand intercellular recognition and communication in a few endocytobioses in animal cells. Therefore, with the various methods available now the opportunity must be taken to investigate the recognition and regulation mechanisms between *Paramecium* and its symbionts. Some examples of questions that can be approached are: Why are algal endocytobionts never digested by *P. bursaria?* Why are killer paramecia resistant to their own toxin and how do endonucleobionts recognize their host nucleus? Experimental infections of *Paramecium* with bacteria or other organisms are easily achieved. This provides the opportunity for detailed studies of the infection mechanisms and for the search of the signals and receptors involved in the process. The results may also help to understand infections of metazoan cells with pathogenic germs.

Better knowledge of endocytobioses is also necessary to understand the ecological significance, e.g., of the killer phenomenon, the *P. bursaria – Chlorella* symbiosis or the endonucleobiosis with *Holospora,* and the study of some of these associations can also contribute to the discussion of the serial endosymbiosis theory of the evolution of the eukaryotic cell.

Acknowledgments. I thank Drs. R.L. Quackenbush and H.J. Schmidt for critically reading the manuscript. I am indebted to the Deutsche Forschungsgemeinschaft for support of my work reported in this chapter (SFB 310, C 2).

References

Albers D, Wiessner W, Reisser W (1983) Nitrogen metabolism of the green *Paramecium bursaria* endosymbiosis. In: Schenk HEA, Schwemmler W (eds) Endocytobiosis, vol 2. de Gruyter, Berlin New York, p 553

Ball GH (1969) Organisms living on and in protozoa. In: Chen T-T (ed) Research in protozoology, vol 3. Pergamon, London New York, pp 565–718

Balsley M (1967) Dependence of the kappa particles of stock 7 of *Paramecium aurelia* on a single gene. Genetics 54:320

Beale GH, Jurand A, Preer JR (1969) The classes of endosymbionts of *Paramecium aurelia.* J Cell Sci 5:65–91

Berger J (1980) Feeding behavior of *Didinium nasutum* on *Paramecium bursaria* with normal or apochlorotic zoochlorellae. J Gen Microbiol 118:397–404

Brown JA, Nielsen PJ (1974) Transfer of photosynthetically produced carbohydrate from endosymbiotic chlorellae to *Paramecium bursaria.* J Protozool 21:569–570

Chao PK (1953) Kappa concentration per cell in relation to the life cycle, genotype and mating type in *Paramecium aurelia,* variety 4. Proc Natl Acad Sci USA 39:103–113

Cronkite DL (1986) Relations between photobehavior and ionically stimulated swimming behavior in *Paramecium bursaria.* J Protozool 33:52–56

Dieckmann J (1981) An infectious bacterium in the cytoplasm of *Paramecium caudatum.* Abstr Int Congr Protozool 6th Warsaw, p 77

Estève J-D (1978) Une population de type (killer) chez *Paramecium caudatum* (Ehrenberg). Protistologica 14:201–207

Freedman LJ, Hanson ED (1970) Further observations on the life cycle of *Leptomonas karyophilus.* J Protozool 17 (Suppl):18

Freiburg M (1985) Isolation and characterization of macronuclei of *Paramecium caudatum* infected with the macronucleus-specific bacterium *Holospora obtusa.* J Cell Sci 73:389–398

Fujishima M, Fujita M (1985) Infection and maintenance of *Holospora obtusa,* a macronucleus-specific bacterium of the ciliate *Paramecium caudatum.* J Cell Sci 76:179–187

Fujishima M, Görtz H-D (1983) Infection of macronuclear anlagen of *Paramecium caudatum* with the macronucleus-specific symbiont *Holospora obtusa.* J Cell Sci 64:137–146

Gibson I (1973) Transplantation of killer endosymbionts of *Paramecium.* Nature (London) 241:127–129

Gillies C, Hanson ED (1963) A new species of *Leptomonas* parasitizing the macronucleus of *Paramecium trichium.* J Protozool 10:467–473

Görtz H-D (1981) Interaction of *Paramecium caudatum* with the infectious form of the micronucleus-specific *Holospora elegans.* Abstr Int Congr Protozool 6th Warsaw, p 122

Görtz H-D (1982) Infection of *Paramecium bursaria* with bacteria and yeasts. J Cell Sci 58:445–453

Görtz H-D (1983) Endonuclear symbionts in ciliates. Int Rev Cytol (Suppl) 14:145–176

Görtz H-D (1986) Endonucleobiosis in ciliates. Int Rev Cytol 102:169–213

Görtz H-D (1987) Different endocytobionts simultaneously colonizing ciliate cells. Ann NY Acad Sci 503:261–268

Görtz H-D, Dieckmann J (1980) Life cycle and infectivity of *Holospora elegans,* a micronucleus-specific symbiont of *Paramecium caudatum* Ehrenberg. Protistologica 16:591–603

Görtz H-D, Dieckmann J (1987) *Leptomonas ciliatorum* n. sp. (Kinetoplastida, Trypanosomatidae) in the macronucleus of a hypotrichous ciliate. J Protozool 34:259–263

Görtz H-D, Freiburg M (1984) Bacterial symbionts in the micronucleus of *Paramecium bursaria.* Endocytol C Res 1:37–46

Görtz H-D, Fujishima M (1983) Conjugation and meiosis of *Paramecium caudatum* infected with the micronucleus-specific bacterium *Holospora elegans.* Eur J Cell Biol 32:86–91

Gromov BV, Ossipov DV, Mamkaeva KA (1974) Characteristics of the reaction to extreme pH value of elongated cell spores of symbiotic bacteria in the macronucleus of *Paramecium caudatum.* Dokl Akad Nauk SSSR 217:712–714

Heckmann K (1977) Endosymbionten von Protozoen. Proc Int Congr Protozool 5th, New York, 160–164

Iwatsuki K, Naitoh Y (1981) The role of symbiotic *Chlorella* in photoresponses of *Paramecium bursaria.* Proc Jpn Acad 57 (Ser B):318–323
Jenkins RA (1970) The fine structure of a nuclear envelope associated endosymbiont of *Paramecium.* J Gen Microbiol 61:355–359
Jeon KW (1983) Integration of bacterial endosymbionts in amoebae. Int Rev Cytol (Suppl) 14:29–47
Karakashian MW (1975) Symbiosis in *Paramecium bursaria.* Symp Soc Exp Biol 29:145–173
Karakashian SJ (1963) Growth of *Paramecium bursaria* as influenced by the presence of algal symbionts. Physiol Zool 36:52–68
Karakashian SJ, Karakashian MW (1965) Evolution and symbiosis in the genus *Chlorella* and related algae. Evolution 19:368–377
Karakashian SJ, Rudzinska MA (1981) Inhibition of lysosomal fusion with symbiont-containing vacuoles in *Paramecium bursaria.* Exp Cell Res 131:387–393
Karakashian SJ, Karakashian MW, Rudzinska MA (1968) Electron microscopic observations on the symbiosis of *Paramecium bursaria* and its intracellular algae. J Protozool 15:113–128
Koizumi S, Preer JR, Jr (1966) Transfer of cytoplasm by microinjection in *Paramecium aurelia.* J Protozool 13 (Suppl):11–12
Lee JJ, Soldo AT, Reisser W, Lee MJ, Jeon KW, Görtz H-D (1985) The extent of algal and bacterial endosymbioses in protozoa. J Protozool 32:391–403
Meier R, Lefort-Tran M, Pouphile M, Reisser W, Wiessner W (1984) Comparative freeze-fracture study of perialgal and digestive vacuoles in *Paramecium bursaria.* J Cell Sci 73:121–140
Muscatine L, Karakashian SJ, Karakashian MW (1967) Soluble extracellular products of algae symbiotic with a ciliate, a sponge and a mutant *Hydra.* Comp Biochem Physiol 20:1–12
Ossipov DV (1981) Problems of nuclear heteromorphism of the unicellular organisms. Nauka, Leningrad
Ossipov DV, Podlipaev SA (1977) Electron microscope examinations of early stages of infection of *Paramecium caudatum* by bacterial symbionts of the macronucleus (iota-bacteria). Acta Protozool 16:289–308
Podlipaev SA, Ossipov DV (1979) Early stages of infection of *Paramecium caudatum* micronuclei by symbiotic bacteria – omega-particles (electron microscope examination). Acta Protozool 18:465–480
Preer LB (1969) Alpha, an infectious macronuclear symbiont of *Paramecium aurelia.* J Protozool 16:570–578
Preer JR, Jr, Preer LB (1984) Endosymbionts of protozoa. In: Krieg NR (ed) Bergey's manual of systematic bacteriology, vol 1. Williams & Wilkins, Baltimore, p 795
Preer JR, Jr, Preer LB, Jurand A (1974) Kappa and other endosymbionts in *Paramecium aurelia.* Bacteriol Rev 38:113–163
Reisser W (1976) Die stoffwechselphysiologischen Beziehungen zwischen *Paramecium bursaria.* Ehrbg. und *Chlorella* spec. in der *Paramecium bursaria*-Symbiose. I. Der Stickstoff- und der Kohlenstoff-Stoffwechsel. Arch Mikrobiol 107:357–360
Reisser W (1980) The regulation of the algal population size in *Paramecium bursaria.* In: Schenk HEA, Schwemmler W (eds) Endocytobiology, vol 2. de Gruyter, Berlin New York, pp 97–104
Reisser W (1983) Cell-to-cell recognition in green *Paramecium bursaria:* the role of algal surface structures. In: Schenk HEA, Schwemmler W (eds) Endocytobiology, vol 2. de Gruyter, Berlin New York, pp 554
Reisser W (1987) Naturally occurring and artificially established associations of ciliates and algae: models for different steps in the evolution of stable endosymbioses. Ann NY Acad Sci 503:316–329
Reisser W, Häder DP (1984) Role of endosymbiotic algae in photokinesis and photophobic responses of ciliates. Photochem Photobiol 39:673–678
Reisser W, Meier R, Görtz H-D, Jeon KW (1985) Establishment, maintenance, and integration mechanisms of endosymbionts in protozoa. J Protozool 32:383–390
Schmidt HJ (1984) Studies on protein synthesis in kappa particles. J Gen Microbiol 130:1517–1523

Schmidt HJ, Görtz H-D (1987) Killer particles in the macronucleus of *Paramecium caudatum.* Ann NY Acad Sci 503:580–583

Schmidt HJ, Heckmann K (1980) DNA of omikron. In: Schenk HEA, Schwemmler W (eds) Endocytobiology, vol 2. de Gruyter, Berlin New York, pp 401–407

Schmidt HJ, Freiburg M, Görtz H-D (1987) Comparison of the infectious forms of two bacterial endonucleobionts, *Holospora elegans* and *H. obtusa,* from the ciliate *Paramecium caudatum.* Microbios 49:189–197

Smith DC (1979) From extracellular to intracellular: the establishment of a symbiosis. Proc Soc London Ser B 204:115–130

Soldo AT (1974) Intracellular particles in *Paramecium.* In: van Wagtendonk WJ (ed) *Paramecium* – a current survey. Elsevier, Amsterdam London New York, pp 377–442

Soldo AT (1983) The biology of the xenosome, an intracellular symbiont. Int Rev Cytol (Suppl) 14:79–109

Soldo AT (1985) The ciliate protozoan and its guest. Bio Essays 4:86–90

Soldo AT, Godoy GA (1973) Observations on the production of folic acid by symbiont lambda particles of *Paramecium aurelia* stock 299. J Protozool 20:502

Soldo AT, Godoy GA (1974) The molecular complexity of mu and pi symbiont DNA of *Paramecium aurelia.* Nucleic Acids Res 1:387–396

Soldo AT, Brickson SA, Lavin F (1983) The size and structure of the DNA genome of symbiont xenosome particles in the ciliate *Parauronema acutum.* J Gen Microbiol 129:1317–1325

Van Wagtendonk WJ, Clark JA, Godoy GA (1963) The biological status of lambda and related particles in *Paramecium aurelia.* Proc Natl Acad Sci USA 50:835–838

Weis D (1980) Hypothesis: free maltose and algal cell surface sugars are signals in the infection of *Paramecium bursaria* by algae. In: Schwemmler W, Schenk HEA (eds) Endocytobiology, vol 1. de Gruyter, Berlin New York, pp 105–112

Weis D (1983) Infection of *Paramecium bursaria* as an inductive process. In: Schenk HEA, Schwemmler W (eds) Endocytobiology, vol 2. de Gruyter, Berlin New York, pp 523–532

Weis D, Berry JL (1974) Correlation of host mating behavior with species of endosymbiotic algae in *Paramecium bursaria.* J Protozool 21:416

Wichterman R (1986) The biology of *Paramecium,* 2nd edn. Plenum, New York

Williams J (1971) The growth in vitro of killer particles of *Paramecium aurelia* and the axenic culture of this protozoan. J Gen Microbiol 68:253–262

Chapter 23 Endosymbionts of Killer Paramecia

Robert L. Quackenbush[1]

Contents

1 Introduction

Killer paramecia were first reported by T.M. Sonneborn (1938). Subsequently, he determined that the genetic determinants for the killer trait were located in the cytoplasm and concluded that they were cytoplasmic genes which he referred to as "plasmagenes" (Sonneborn 1943). However, studies conducted by Preer and others demonstrated that *Paramecium* killer traits were actually determined by bacterial endosymbionts residing in the cytoplasms of killer paramecia (Preer et al. 1974). Several excellent reviews on the bacterial endosymbionts of paramecia have been published (Gibson 1974; Preer et al. 1974; Soldo 1974). The purpose of this chapter is to provide an update to those reviews with respect to the endosymbionts of killer paramecia.

2 Taxonomy of Bacterial Endosymbionts of Killer Paramecia

Many different kinds of endosymbiotic bacteria are known to confer killer traits upon host paramecia (with one exception all known strains of killer paramecia belong to the *P. aurelia* species complex). Until recently, these endosymbionts were known only by their Greek letter names. In the 1974 review by Preer et al.,

1 Department of Microbiology, University of South Dakota, Vermillion, SD 57069, USA

Paramecium. Ed. by H.-D. Görtz

binomial names were proposed to replace the Greek letter names and to acknowledge the bacterial nature of these organisms. This nomenclature was subsequently revised by Quackenbush (1978) on the basis of DNA-DNA hybridization data that demonstrated clear phylogenetic divisions within some of the taxons proposed by Preer et al. (1974). The symbionts mu, nu, pi, gamma, and kappa were all originally assigned to a single genus, *Caedobacter*. However, hybridization data demonstrated little or no sequence homology among the groups assigned to this genus (Quackenbush 1977a, 1978). The pi particles, which were originally characterized as mutants of kappa, were shown not to possess significant homology with their supposed parent strain of kappa. However, DNA from pi particles did hybridize to a significant extent with DNA from mu and nu particles (which also failed to demonstrate homology with DNA from kappa particles). Thus, it was proposed that mu, nu, pi, and gamma be included in a new genus, *Pseudocaedobacter*. It is clear now that pi particles are not mutants of kappa; instead, they were originally coresidents of the same host and became apparent only when the coresident kappa strain died out.

DNA-DNA hybridization data further revealed that there is considerable diversity even within kappa particles as a group. In fact DNA from class 7 kappa is less than 10% homologous with DNA from class 51 kappa. Within the classes of kappa, as proposed by Preer et al. (1972), a significant amount of sequence homology exists. However, between classes significant homology has been observed only between classes 7, 562, and 1038, and classes 1038 and 51. Although it would have been justifiable on the basis of hybridization data to split kappa into four different genera, it was proposed that all kappa be included in a single genus, *Caedobacter*, because it is not possible to distinguish these organisms from one another using simple laboratory procedures and since they share several unique characteristics. They are all obligate endosymbionts that confer killer traits upon their host paramecia and they all produce R bodies. The genus is presently split into four species (Table 1) on the bases of DNA sequence homologies, R-body morphology (Table 2), host species, and type of killer trait conferred upon the host organism.

The correct names of these two genera are spelled slightly different than originally proposed to conform with standards set by the International Code of Nomenclature of Bacteria (Lapage et al. 1975). The officially recognized spellings of the genus names are: *Caedibacter* and *Pseudocaedibacter* (Preer and Preer 1982, 1984).

It is anticipated that a fifth species of *Caedibacter*, *C. caryophila*, will be added (Schmidt, Görtz, and Quackenbush 1987). This bacterium, which was first described by Estève (1978), is unique for several reasons. It is a macronuclear endosymbiont of *P. caudatum*. All of the other species of *Caedibacter* live in the cytoplasm of *P. biaurelia* or *P. tetraurelia*. In addition, it appears that the R body produced by *C. caryophila* is significantly different from those produced by the other species with respect to size and morphology (see below).

The symbionts of paramecia that confer killer traits and their distinguishing characteristics are listed in Table 1. A more complete description of these organisms is available in several earlier publications (Preer et al. 1974; Quackenbush 1978; Preer and Preer 1984).

Table 1. Key to the taxa of the endosymbionts of killer paramecia

I. Capable of R-body production

Genus I. *Caedibacter*

A. Cytoplasmic endosymbionts of members of the *P. aurelia* species complex
  1. Produce type 7 R bodies
    a) Symbionts of *P. biaurelia*
    b) Mediate mate killing
      C. paraconjugatus

    b*) Cell-cell contact between killer host and sensitive cell not required for killing to occur
      C. varicaedens

    c) R bodies associated with icosahedral bacteriophage capsids
    d) Bacteriophage capsids adhere to R body
      C. varicaedens subsp. 7
    d*) Bacteriophage capsids do not adhere to R body
      C. varicaedens subsp. 562
    c*) Icosahedral bacteriophage capsids not produced
      C. varicaedens subsp. 1038
    a*) Symbionts of *P. tetraurelia*
      C. pseudomutans
  2. Produce type 51 R bodies
    C. taeniospiralis

B. Macronuclear endosymbionts of *P. caudatum*
    C. caryophila

II. R bodies not produced

A. Possess flagella

Genus II. *Lyticum*

  1. Straight rods
    L. flagellatum

  2. Curved or spiral rods
    L. sinuosum

B. Small Gram-negative rods that do not possess flagella

Genus III. *Pseudocaedibacter*

  1. Mediate mate killing.
    Psc. conjugatus

  2. Physical contact between killer host and sensitive cell not required for killing to occur.
    Psc. minutus

Table 2. Key for identification of R-body types

I. Edges running along the length of the R-body ribbon are parallel. As a result, in the natural rolled-up form, the R body appears as nearly perfect cylinder

A. R body unrolls from the inside in a telescoping fashion

1. R bodies visible when viewed by light microscopy (> 0.3 μm in diameter and length in rolled-up form)

a) R body not associated with icosahedral bacteriophage particles and possesses ribbon width < 0.6 μm. Genetic determinants are plasmid-borne. Observed only in *C. taeniospiralis*

Type 51

b) R body associated with icosahedral bacteriophage particles and possesses ribbon width of approx. 0.8 μm. Observed only in *C. caryophila*

Type Cc

2. R bodies cannot be detected by light microscopy (< 0.3 μm in diameter and length in rolled-up form). Observed only in *Ps. taeniospiralis*

Type Pt

B. R body unrolls from the outside. Usually associated with icosahedral bacteriophages. Size similar to type 51. Outer terminus is blunt. Occurs in cytoplasmic endosymbionts of *P. biaurelia* and *P. tetraurelia*

Type 7

II. Edges running along the length of the R-body ribbon are not parallel. The unrolled ribbon is widest at the center of its length and tapers progressively to the termini forming very acute (approx. 15°) angles. These R bodies are quite large, approx. 0.8 μm-long in the rolled-up form, and they unroll from the outside. Occur in *Ps. avenae*

Type Pa

3 R Bodies, Extrachromosomal Elements, and Killer Traits

Killer traits and the bacteria that determine them have been studied for almost 50 years. Unfortunately, only limited progress has been made toward understanding the nature of the components required for toxigenicity. The mechanisms involved in killing of sensitive paramecia, as well as those involved in conferring resistance upon host paramecia to killing, mediated by its own symbionts, essentially remain unknown. However, considerable progress has been made toward understanding some traits that are required for expression of killer traits mediated by members of the genus *Caedibacter*.

Members of the genus *Caedibacter* have three characteristics in common: they are all obligate endosymbionts of paramecia; the all mediate expression of a killer trait in their host organisms; and they all produce R bodies. Expression of the killer trait by paramecia infected by kappa (*Caedibacter*) is dependent upon exposure of uninfected or sensitive paramecia to toxin-bearing forms of kappa. Generally, this is achieved by the release of kappa from the killer host into the environment making the toxin-bearing endosymbiont available for ingestion by sensitive paramecia. If toxin-bearing kappa are ingested by a sensitive paramecium the killing reaction is initiated when the kappa particle is digested in the food vacuole and the toxin is released. The prelethal effects that ensue vary according to which species or, in the case of *C. varicaedens*, which strain is involved. *C. paraconjugatus* is unique among kappa because physical contact, with subsequent cytoplasmic exchange, between the killer host and the sensitive paramecium must take place for killing to occur. Killer paramecia are generally resistant to the toxic activities of their endosymbionts, but not to those of kappa belonging to other species.

R-body production is the one trait that distinguishes *Caedibacter* from the other types of symbionts of paramecia and thus is the hallmark of the genus. R bodies are long (up to 20 µm), proteinaceous ribbons that are rolled up inside the bacterial cell to form a hollow cylindrical structure that, in all but one species, has a diameter of about 0.4 µm and a length of about 0.4 µm (Fig. 1). Three types of T bodies occur in *Caedibacter* (Table 2): type 51 R bodies (Fig. 2) are produced only by *C. taeniospiralis;* type 7 R bodies produced by *C. varicaedens, C. pseudomutans*, and *C. paraconjugatus;* and a third type of R body, which will be referred to as a type Cc R body, is produced by *C. caryophila.* Type 51 and type Cc R bodies differ from type 7 R bodies in that they will unroll in response to changes in pH (to less than 6.5). Type 51 and type Cm R bodies both unroll in a telescoping fashion from the inside, whereas type 7 R bodies unroll from the outside. Both the inner and outer termini of type 51 R-body ribbons form acute angles. The outer termini of type Cc and type 7 R bodies are blunt. Type Cc R bodies are significantly larger than the other two R-body types (approx. 0.8 µm in width and diameter versus a maximum of 0.5 µm for type 7 or 51).

Although the nature and mode(s) of action of endosymbiont-derived toxin(s) involved in killing of sensitive paramecia are unknown, R bodies have been suggested to play a key role in killing mediated by *Caedibacter.* Mueller (1963) and Smith-Sonneborn and van Wagtendonk (1964) demonstrated that only *Caedibacter* containing R bodies were toxic to sensitive paramecia. In addition, R bodies purified from certain strains of *C. varicaedens* have been shown to be toxic to sensitive paramecia (Preer and Preer 1964; Preer et al. 1972). The appearance of R bodies in *Caedibacter* is closely associated with the appearance of bacteriophage capsids (Fig. 1) or structures that appear to be products of capsid subunit assembly (Preer and Jurand 1968; Grimes and Preer 1971; Preer et al. 1974). Thus, for many years it has been believed that R-body synthesis is probably essential to expression of killer traits by *Caedibacter* and that the genetic determinants for these traits are extrachromosomal.

Dilts (1976) demonstrated the presence of a 42-kb plasmid in *C. taeniospiralis* 51. She further demonstrated that this plasmid was highly amplified in R body-containing cells. In a normal population of *C. taeniospiralis*, of which only about 5% of the individuals contain R bodies, approximately 5% of the total DNA was plasmid. However, populations enriched in R body-containing individuals (to about 60%) also were found to be highly enriched with respect to plasmid content (accounting for approximately 50% of the total DNA). These

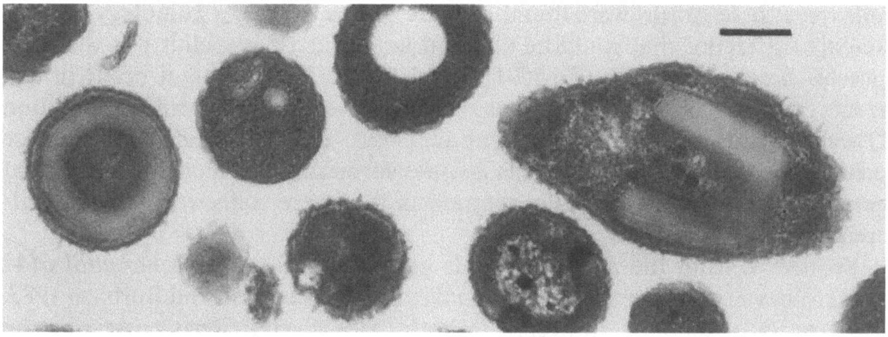

Fig. 1. Electron micrograph of sectioned *C. varicaedens* 310. Cells with R bodies are on the *left* (cross-section) and *right* (logitudinal section) portions of the figure. Several icosahedral bacteriophages are associated with the R body on the right. Bar = 0.2 μm

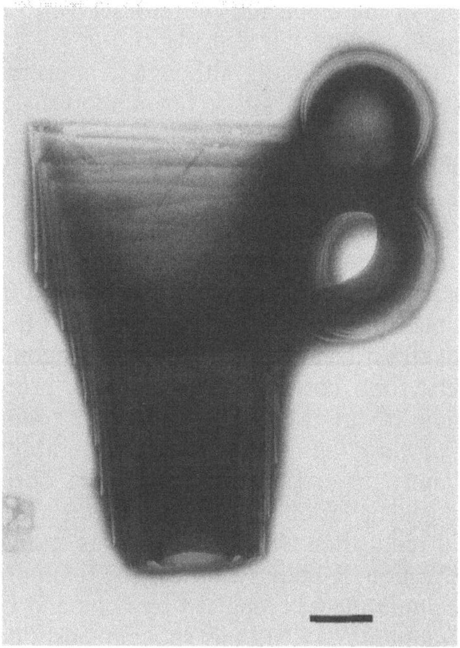

Fig. 2. Electron micrograph of three negatively stained type 51 R bodies; all are partially unrolled, two projecting toward the viewer and the other lying on its side (Quackenbush and Burbach 1983). Bar = 0.2 μm

results further support the hypothesis that the genetic determinants for R-body production may be extrachromosomal.

Some strains of *C. taeniospiralis* carry at least two types of plasmids. However, in every case one plasmid species (major plasmid species) was present in much higher copy number than the coresident plasmid(s) (minor plasmid species) (Quackenbush et al. 1982; personal observations). The major plasmid species from eight strains of *C. taeniospiralis* have been mapped by restriction endonuclease analysis (Quackenbush et al. 1982; Quackenbush 1983). The plasmids from these strains were found to share a high degree of homology. In fact, the only difference that could be detected among these plasmids was the presence of three transposons (Tn 4501, Tn 4502, and Tn 4503) that occur in different combinations and positions according to the plasmid in question (Quackenbush 1983; Quackenbush et al. 1986). The high degree of sequence conservation among these plasmids is quite surprising when one considers that they come from isolates of killer paramecia that were collected from four different continents.

Sequences from the major plasmids of two strains of *C. taeniospiralis* (47 and 116) have been cloned in *Escherichia coli* (Quackenbush and Burbach 1983; Pond and Quackenbush, in prep.). In both cases *E. coli* clones possessing the ability to produce R bodies were recovered. The sequences coding for R-body production from both sources were determined to be identical with respect to their restriction maps and their respective locations on the parent plasmids. None of the R body-producing *E. coli* clones exhibited toxic activity against sensitive paramecia. Furthermore, none of the sequences cloned from other regions of these plasmids were observed to express toxigenicity when cloned into *E. coli.*

The above results, although demonstrating that R-body production is an extrachromosomal trait, do not support the hypothesis that R bodies are required for expression of the killer trait and that the killer trait is itself extrachromosomal. However, Widmayer (1965) reported a mutant of *C. taeniospiralis* 51, generated by exposure to X-rays, that could not confer a killer trait upon its host but did confer resistance. In addition, this mutant, *C. taeniospiralis* 51m43, possessed a severely limited ability to produce R bodies. Dilts (1977) observed that this mutant strain differed from its parental strain with respect to its plasmid content. She observed that the R body-coding plasmid of the mutant strain commonly occurred in oligomeric forms rather than in the monomeric form observed in all plasmid preparations obtained from the parental strain. It was also observed that the proportion of plasmid DNA occurring in *C. taeniospiralis* 51m43 was considerably less than that found in the parental strain (Dilts and Quackenbush, pers. observations). These results indicate that the genetic lesion in *C. taeniospiralis* 51m43 probably caused a defect in plasmid replication which perhaps was essential for the induction of plasmid amplification, R-body synthesis, and toxin production. A second tox$^-$ mutant of *C. taeniospiralis* was reported by Dilts and Quackenbush (1986). This mutant, *C. taeniospiralis* 169-1, was spontaneously derived from *C. taeniospiralis* 169. *C. taeniospiralis* 169-1 was similar to 51m43 in that it could no longer produce R bodies and yet it conferred resistance to killing upon its ciliate host. However, the two mutants differed with respect to genetic lesion. In the case of 169-1, the mutation was due to the movement of a transposon (Tn 4503) into the actual R body-coding sequence of the R body-coding plasmid pKAP169, to generate pKAP169-1. The above evidence provides proof that R-body production is required for the expression of toxigenicity and that toxigenicity is, at least in part, an extrachromosomal trait. At present, the possibility that one or more components required for expression of toxigenicity are determined on the endosymbiont chromosome cannot be disregarded.

4 Kappaphages and R Body-Coding Plasmids

As presented above, the evidence strongly indicates that two of the three traits required for a bacterium to be admitted to the genus *Caedibacter*, R body, and toxin production, are probably extrachromosomally determined. In the case of *C. taeniospiralis*, it has been demonstrated that plasmids carry the genetic determinants for type 51 R-body synthesis and that they are required for expression of toxigenicity.

Plasmids have not been detected in any strains of the other species of *Caedibacter* nor do they produce type 51 R bodies (Preer et al. 1974; Quackenbush 1978). All of the other species produce type 7 R bodies, except *C. caryophila* which produces type Cc R bodies. The synthesis of these R bodies (with the single exception of those produced by *C. varicaedens* 1038 which will be discussed below) are clearly associated with the appearance of icosahedral viral capsids (Preer and Jurand 1968; Grimes and Preer 1971; Preer et al. 1972). In most cases these capsids contain nucleic acid. In the case of *C. varicaedens* 562, it has been demonstrated that the capsids are filled with double-stranded DNA (Preer et al. 1971) having a kinetic complexity slightly greater than that of bacteriophage lambda (Quackenbush 1977b). It is clear that these virus-like particles (kappaphages) are indeed bacteriophages. However, it also seems likely, as suggested by Preer et al. (1974), that these kappaphages may be defective since they appear to have lost the ability to lyse their host cells upon maturation of the virions. The fact that kappaphages appear to be induced at the same time R-body synthesis is induced indicates; (1) that kappaphages are temperate bacteriophages and (2) that genetic determinants for R-body synthesis (and possibly toxin synthesis) are contained in the kappaphage genomes.

A system of nomenclature for kappaphages has been proposed (Quackenbush et al. 1987) in which the name of each kappaphage strain would consist of a prefix which would be the initials of the host species, followed by the strain number of the host species. Thus, the kappaphage of *C. varicaedens* 310 is referred to as kappaphage Cv310 and that of *C. pseudomutans* 51m1 is kappaphage Cps51m1.

Relationships among the kappaphage genomes and between R body-coding plasmids and kappaphage genomes have been studied (Quackenbush 1978; Quackenbush et al. 1987). The results of these studies, which involved restriction endonuclease analyses and DNA-DNA hybridizations, demonstrate not only that R body-coding plasmids share little or no homology with kappaphages (with two possible exceptions, kappaphages Cv576 and Cv1038, to be discussed below), but that there is also considerable diversity among the kappaphages.

In all cases, close general relationships between R body-coding plasmids and kappaphages could not be detected. However, two instances of localized homology were detected: Southern blots containing DNA from *C. varicaedens* 576 hybridized strongly with Tn4503 probe DNA derived from the R body-coding plasmid of *C. taeniospiralis*, pKAP47; and Southern blots containing DNA from *C. varicaedens* 1038 hybridized strongly with a DNA probe consisting of the R body-coding sequence derived from pKAP47 (Quackenbush et al.

1987). At present, it is not clear whether the probe DNAs were hybridizing to kappaphage or chromosomal sequences. It is of particular interest to determine the source of sequence from *C. varicaedens* 1038 that shares homology with the type 51 R-body-coding sequence. *C. varicaedens* 1038 differs from all other strains of that species in that its R bodies are not associated with icosahedral phage particles. Instead, R bodies of strain 1038 are associated with helical structures that are similar to those associated with type 51 R bodies (Preer et al. 1972). Thus, R bodies of strain 1038, which otherwise clearly exhibit characteristics of type 7 R bodies, may represent an evolutionary intermediate between type 7 and type 51 R bodies. Since the synthesis of type 51 R bodies is an extrachromosomal trait and R bodies of the other species of *Caedibacter* are generally associated with the induction of kappaphages, it seems likely that the R-body-coding sequence of strain 1038 (presumably the sequence possessing homology with the type 51 R-body-coding sequence) would be located in the kappaphage Cv1038 genome. Isolation and cloning of this sequence would probably open the door to identification of the other type 7 R-body-coding sequences.

As mentioned above, the data from hybridization and restriction enzyme analyses demonstrate that there is considerable diversity among the kappaphages. However, these data do reveal close relationships among certain groups of kappaphages. In fact, when kappaphages are grouped into separate classes on the basis of relatedness to one another, the pattern is exactly the same as the classification scheme proposed by Preer et al. (1972) for classifying their host bacteria.

5 Genes Required for Type 51 R-Body Synthesis and Their Expression

The sequences required for type 51 R-body synthesis from *C. taeniospiralis* strains 47 (Quackenbush and Burbach 1983) and 116 (Pond and Quackenbush, in prep.) have been cloned and expressed in *E. coli.* Physical maps, generated by restriction endonuclease analysis, of the R-body-coding sequences from these two strains were found to be identical. The amount of sequence required for synthesis and assembly of fully functional type 51 R bodies in *E. coli* is less than 2000 base pairs (Kanabrocki et al. 1986a; Pond and Quackenbush, in prep.).

Identification of the polypeptides required for R-body synthesis and assembly was accomplished by minicell analysis of the expression of pBQ65 (a recombinant plasmid containing the type 51 R-body-coding sequence from *C. taeniospiralis* 47) and its deletion derivatives (Kanabrocki et al. 1986a). At least three polypeptides are required. These polypeptides migrate in SDS-polyacrylamide gels as if they had molecular weights of about 18, 13, and 10 kD. However, projection of their molecular weights from DNA sequence data indicates that their actual molecular weights are only about 60% of the observed values (Heruth and Quackenbush, in prep.). Pulse-chase analysis of the expression of pBQ65 in minicells indicates that the 13-kD polypeptide is probably the major structural subunit of type 51 R bodies and that the 18- and 10-kD polypeptides are required for assembly, but not structure (Kanabrocki et al. 1986a). The assembly process apparently involves posttranslational processing of the 13-kD polypeptide into at least six separate polypeptide species (on the basis of pI) where each pI species actually is a pair of polypeptides that differ slightly with respect to molecular mass. After posttranslational processing, the 13-kD polypeptides are assembled into progressively larger and larger oligomers with a mature R body presumably being the final product. The precise roles played by the 10- and 18-kD polypeptides in this process are not well understood. However, it appears likely that the 10-kD polypeptide is involved in creating highly stable (probably covalent) cross-links between the structural subunits (13-kD polypeptides) during the polymerization process.

The precise locations of the genes encoding the 10-, 13-, and 18-kD polypeptides have been determined by restriction mapping and minicell analysis of deletion derivatives of pBQ65 (Kanabrocki et al. 1986a) and have been confirmed by DNA sequence data (Heruth and Quackenbush, in prep.). The three genes are probably independent transcriptional units as there are several hundred base pairs between each gene and each is preceded by sequences that resemble the consensus sequence for *E. coli* promoters. They are transcribed in the same direction. Their order, in the direction of transcription, begins with the gene encoding the 18-kD polypeptide followed by 13-kD and then 10-kD. Deletion of any one of these genes results in the loss of the ability to produce R bodies. It is possible that a fourth essential gene, encoding a very small polypeptide, located between the genes encoding the 13- and 10-kD polypeptides could be present, but the evidence for its existence is incomplete at this time (Heruth, Pond, and Quackenbush, unpubl. observations).

6 Relationships Among R Bodies

There are at least three distinct types of R bodies that occur in *Caedibacter* (Table 2). Even though R bodies differ with respect to morphology and response to chemical stimuli, the unique nature of these structures and the organisms in which they occur would lead one to suspect that their basic structures would be quite similar. This is not the case. As indicated above, the genetic determinants of type 51 R bodies share no close homology with those of type 7 R bodies (with the single exception of those from *C. varicaedens* 1038). However, there does appear to be some relationship between the polypeptides associated with the different R body types 7 and 51. Significant antigenic cross-reactivity in reactions involving antisera against R bodies from *C. varicaedens* strains 7 and 562 with Western blots containing proteins from type 51 R bodies have been observed (FR Pond, unpubl. observations). No significant antigenic cross-reactivity between type 51 and type Cm R bodies were observed.

R bodies, in addition to being produced by all species of *Caedibacter*, are also produced by two free-living bacteria, *Pseudomonas taeniospiralis* (Lalucat et al. 1979) and *Ps. avenae* (Wells and Horne 1983). These R bodies have also been found to be distinct from the other R-body types antigenically and with respect to genetic determinants (Kanabrocki et al. 1986b; Lalucat et al. 1986;

Pond and Quackenbush, unpubl. observations). Thus, there are at least five distinct types of R bodies, three produced by *Caedibacter* and two that occur in two different species of *Pseudomonas* (Table 2). I propose that these two additional R-body types be designated type Pa and type Pt.

It is difficult to believe that these structures did not have a common evolutionary origin. The key to understanding the relationships among the R-body types lies in characterization of their respective genetic determinants.

7 Summary

Considerable diversity exists among endosymbiotic bacteria that confer killer traits upon their host paramecia. As suggested by Preer et al. (1974) several species, or genera, of endosymbionts can occur within a single host *Paramecium.* In fact, it is precisely this situation that caused early workers in this field to incorrectly describe coresident endosymbionts (*Psc. falsus* 51 and *C. pseudomutans*) as mutants of *C. taeniospiralis* 51.

Superficially, endosymbionts commonly known as kappa particles appear to be closely related. However, DNA-DNA hybridization data demonstrated that considerable phylogenetic diversity exists within the group. As a result, this group, comprising the genus *Caedibacter,* was split into several different species. At present, there are four official species of *Caedibacter* and a fifth that is being submitted for induction into the genus.

The single trait that distinguishes *Caedibacter* from the other killer endosymbionts is the ability to produce R bodies. In the case of *C. taeniospiralis* it has been demonstrated that R-body synthesis, although not being the sole requirement, is required for expression of the killer trait. In addition, the genetic determinants of R bodies in *Caedibacter* are probably extrachromosomal. In *C. taeniospiralis* R bodies are known to be plasmid-encoded traits and there is strong indirect evidence indicating that R bodies of the other species of kappa are encoded on bacteriophage genomes. The R bodies themselves, which also are superficially very similar, exhibit considerable diversity with respect to their genetic determinants and antigenic structure.

Considerable progress has been made in describing the biology of these endosymbionts. However, as pointed out earlier, we have made limited progress toward understanding the major questions that come to mind when one thinks of these organisms. That is: what is the nature and mode of action of the toxin(s); and how do the host paramecia resist the toxic activities of their own endosymbionts? Some of the research described above perhaps has opened some doors that may lead to answers to these question with respect to the kappa particles.

Acknowledgments. I would like thank Finn Pond for his comments on this manuscript, Drs. Hans-Dieter Görtz and Helmut Schmidt for communicating results on *C. caryophila,* and Drs. Lawrence Fine and Jerome Howard for stimulating discussions. This work was supported by Public Health Service grant GM 36293 from the National Institutes of Health and National Science Foundation grant Int 8420591.

References

Dilts JA (1976) Covalently closed circular DNA in kappa endosymbionts of *Paramecium aurelia*. Genet Res 27:161–170

Dilts JA (1977) Chromosomal and extrachromosomal DNA from 4 bacterial endosymbionts of *Paramecium aurelia*. J Bacteriol 129:888–894

Dilts JA, Quackenbush RL (1986) A mutation in the R body-coding sequence destroys expression of the killer trait in *Paramecium tetraurelia*. Science 232:641–643

Estève JC (1978) Une population de type «killer» chez *Paramecium caudatum* (Ehrenberg). Protistologica 10:371–378

Gibson I (1974) The endosymbionts of *Paramecium*. Crit Rev Microbiol 3:243–273

Grimes GW, Preer JR, Jr (1971) Further observations on the correlation between kappa and phage-like particles on *Paramecium*. Genet Res 18:115–116

Kanabrocki JA, Quackenbush RL, Pond FR (1986a) Organization and expression of genetic determinants for synthesis and assembly of type 51 R bodies. J Bacteriol 168:40–48

Kanabrocki JA, Lalucat J, Cox BJ, Quackenbush RL (1986b) A comparative study of R bodies and their genetic determinants: relationship of type 51 R bodies to R bodies produced by *Pseudomonas taeniospiralis*. J Bacteriol 168 (in press)

Lalucat J, Meyer O, Mayer F, Pares R, Schlegel HG (1979) R-bodies in newly isolated free-living hydrogen-oxidizing bacteria. Arch Microbiol 121:9–15

Lalucat J, Wells B, Gibson I (1986) Relationships between R bodies of certain bacteria. Micron Microsc Acta 17:243–245

Lapage SP, Sneath PHA, Lessel EF, Skerman VBD, Seeliger HPR, Clark WA (1975) International code of nomenclature of bacteria – bacteriological code. Am Soc Microbiol, Washington, DC

Mueller JA (1963) Separation of kappa particles with infective activity and identification of infective particles in *Paramecium aurelia*. Exp Cell Res 30:492–508

Preer JR, Jr, Jurand A (1968) The relation between virus-like particles and R bodies of *Paramecium aurelia*. Genet Res 12:331–340

Preer JR, Jr, Preer LB (1982) Revival of names of protozoon endosymbionts and proposal of *Holospora caryophila* nom. nov. Int J Syst Bacteriol 32:140–141

Preer JR, Jr, Preer LB (1984) Endosymbionts of protozoa. In: Krieg NR (ed) Bergey's manual of systematic bacteriology, vol 1. Williams & Wilkins, Baltimore, pp 795–811

Preer JR, Jr, Preer LB, Rudman B, Jurand A (1971) Isolation and composition of bacteriophage-like particles from kappa of killer paramecia. Mol Gen Genet 111:202–208

Preer JR, Jr, Preer LB, Jurand A (1974) Kappa and other endosymbionts of *Paramecium aurelia*. Bacteriol Rev 38:113–163

Preer LB, Preer JR, Jr (1964) Killing activity from lysed particles of *Paramecium*. Genet Res 5:230–239

Preer LB, Jurand A, Preer JR, Jr, Rudman BM (1972) The classes of kappa in *Paramecium aurelia*. J Cell Sci 11:581–600

Quackenbush RL (1977a) Phylogenetic relationships of bacterial endosymbionts of *Paramecium aurelia:* polynucleotide sequence relationships of 51 kappa and its mutants. J Bacteriol 129:895–900

Quackenbush RL (1977b) Genetic relationships among endosymbiotic bacteria of *Paramecium aurelia*. PhD Diss, Ind Univ, Bloomington

Quackenbush RL (1978) Genetic relationships among bacterial endosymbionts of *Paramecium aurelia:* deoxynucleotide sequence relationships among members of *Caedobacter*. J Gen Microbiol 108:181–187

Quackenbush RL (1983) Plasmids from bacterial endosymbionts of hump-killer paramecia. Plasmid 9:298–306

Quackenbush RL, Burbach JA (1983) Cloning and expression of DNA sequences associated with the killer trait of *Paramecium tetraurelia* stock 47. Proc Natl Acad Sci USA 80:250–254

Quackenbush RL, Dilts JA, Maser RL (1982) Physical map of a plasmid from *Caedibacter taeniospiralis* 51. J Bacteriol 152:939–942

Quackenbush RL, Dilts JA, Cox BJ (1986) Transposonlike elements in *Caedibacter taeniospiralis.* J Bacteriol 166:349–352

Quackenbush RL, Cox BJ, Kanabrocki JA (1987) Extrachromosomal elements of extrachromosomal elements of *Paramecium* and their extrachromosomal elements. In: Wickner RB, Hinnebusch A, Lambowitz A, Gunsalus IC, Hollaender A (eds) Extrachromosomal elements of lower eukaryotes. Plenum, New York, pp 265–278

Smith-Sonneborn JE, Van Wagtendonk WJ (1964) Purification and chemical characterization of *kappa* of stock 51, *Paramecium aurelia.* Exp Cell Res 33:50–59

Soldo AT (1974) Intracellular particles in *Paramecium.* In: Van Wagtendonk WJ (ed) *Paramecium:* a current survey. Elsevier, Amsterdam New York, pp 377–442

Sonneborn TM (1938) Mating types in *Paramecium aurelia:* diverse conditions for mating in different stocks; occurrence, number and interrelations of the types. Proc Am Philos Soc 79:411–434

Sonneborn TM (1943) Gene and cytoplasm. I. The determination and inheritance of the killer character in variety 4 of *P. aurelia.* II. The bearing of determination and inheritance in *P. aurelia* on problems in cytoplasmic inheritance, pneumococcus transformations, mutations and development. Proc Natl Acad Sci USA 29:329–343

Wells B, Horne RW (1983) The ultrastructure of *Pseudomonas avenae.* II. Intracellular refractile (R-body) structure. Micron Microsc Acta 14:329–344

Widmayer D (1965) A non-killer resistant kappa and its bearing on the interpretation of kappa in *Paramecium aurelia.* Genetics 51:613–623

Chapter 24 Ecology

Wayne G. Landis[1]

Contents

1 Introduction

1.1 Importance of Protozoan Ecology

Only in the last few years has the importance of protozoa in aquatic ecosystems been recognized. Protozoa act as a major ecosystem unit and are important in both assimulative and degradative pathways as a harvester of prokaryotic and eukaryotic microorganisms (Bamforth 1981). In fact, protozoan community structure reflects the general health of many aquatic ecosystems (Cairns et al. 1980; Henebry and Cairns 1984; Pratt and Cairns 1985a). Protozoa are also known to contain enzymes which can hydrolyze a variety of organofluorophosphates (Landis et al. 1985, 1986, 1987). Indeed, protozoa may play an important role in communities engineered to degrade such chemical

1 Environmental Toxicology Branch, Toxicology Division, Chemical Research, Development and Engineering Center, Aberdeen Proving Grounds, MD 21010-5423, USA

Paramecium. Ed. by H.-D. Görtz

pollutants as PCBs, organophosphates, and herbicides either in situ or as part of a waste treatment process. Before an efficient process can be designed, it is necessary to acquire detailed knowledge of the interactions among protozoa, their role in the community, and their responses to physiochemical changes in the environment.

1.2 Ecology of Paramecium

Members of the genus *Paramecium* are among the best studied of the free-living protozoa. As this volume demonstrates, almost every facet of their physiology, cytology, morphology, genetics, breeding systems, movement, and symbionts have been extensively studied. However, the natural history and ecology of this group of diverse organisms have been relatively ignored. The majority of ecological studies dealing with paramecia have used the organism as a laboratory model.

Since the 1930s *Paramecium* have been used as tools to investigate competition, predator-prey interactions, and the relationship between ecosystem complexity and stability. The earliest experiments were the classic studies of Gause (1934) investigating the competitive interactions between *P. aurelia, P. caudatum,* and *P. bursaria.* Predators such as *Didinium* were also introduced. Experiments of this genre are still being carried out (see Sects. 3.2 and 3.3). Recently, systems including *Paramecium* have also been used to examine the hypothesis that ecosystem complexity is related to ecosystem stability. Unfortunately, an interest in the investigation of the ecology of these organisms in their natural environment has not been as apparent.

Until Hairston's (1958, 1967) initial studies concerning the distribution of the various species of the *P. aurelia* complex native to the Ann Arbor area, no true ecological study of the genus *Paramecium* had been undertaken. A major fault was the belated realization that the varieties (later called syngens) of the various species were in fact biological species. Sonneborn's (1957) seminal thesis on the breeding systems and species problem in *Paramecium* exacerbated the problem. Sonneborn realized the biological uniqueness of the varieties, but declined to provide binomial names because identification was (and remains) a problem. In the late 1960s and 1970s a number of field studies examined the presence of competitive interactions, population regulation, and the evolution and role of the killer trait. Widespread recognition of syngens as bona fide biological species (Sonneborn 1975) has pointed out the variety of material available for the investigation of speciation. As interest in the ecology of protozoa increases, the population biology of *Paramecium* can serve as a model for other ciliate protozoa.

1.3 Nomenclature

The purpose of this chapter is to review our present knowledge of the ecology of the genus *Paramecium.* In each discussion it must be realized that the genus is actually a group of species complexes (Hairston 1958; Sonneborn 1975; Corliss and Daggett 1983). Although only the species of the *P. aurelia* complexes have been given binomial names, the so-called syngens of *P. bursaria, P. caudatum,* and *P. multimicronucleatum,* are also biological species (see also Chap. 3 by Nyberg and Chap. 4 by Tsukii, this Vol.) and are so treated in the following discussion. Unfortunately, in the older literature a species was either not identified or was referred to as a syngen or variety. In this chapter current nomenclature is used when the species can be identified from the literature or when the organism is referred to as a member of a particular species complex.

The remaining sections cover the distribution of the species complexes of the genus *Paramecium,* their ecology, and natural history. Finally, I present several areas of research which deserve further effort. Since the preceding chapters have reviewed virtually every aspect of the biology of *Paramecium,* it is fitting that this chapter is the last. It is often easy to forget that the aspects discussed in previous sections have been shaped by evolutionary forces to create ecological strategies to ensure survival. In this chapter I will endeavor to summarize these strategies.

2 Distribution

2.1 Habitats

Paramecium are ubiquitous freshwater organisms which can survive a variety of environmental conditions with collections possible over temperatures of 28° – 3° C (Landis, unpubl. data; Hairston 1967). Habitats include swamps, seeps, bogs, rivers, streams, ponds, and lakes (Wichterman 1953; Sonneborn 1957, 1975; Hairston 1967; Landis 1981).

2.2 Geographic Distribution

Known geographic distributions of the sibling species of the various species complexes assuredly is, in part, a reflection of the distribution of researchers working on that particular type of *Paramecium.* As of the compilation of this chapter, no systematic sampling has been undertaken to adequately describe the range and distribution of species complexes of *Paramecium.* In addition, it has long been known that many recognized species of the *P. aurelia* complex (e.g., *P. tetraurelia*) are composed of populations which, when mated, exhibit an F_2 mortality of up to 95% (Dippell 1954; Sonneborn 1957). Sibling species composed of strong inbreeders may themselves be comprised of populations of nascent species. Conversely, populations of the various species of the *P. bursaria* complex are able to interbreed and produce viable F_2s (Sonneborn 1957). The listing below is a summary of the distributions of the recognized sibling species.

Paramecium aurelia complex (Sonneborn 1975):

P. primaurelia: Has the widest observed distribution of the *P. aurelia* complex. This species has been collected from North America, Central America, South America, Europe, Australia, Hawaii, and Japan.

P. biaurelia: The species is cosmopolitan. Stocks have been obtained from North America, South America, Europe, Middle East, Japan, and New Zealand.

P.triaurelia: Previously reported as limited to North America (Alaska to Florida), this species has recently been isolated from Europe (Spain and Rumania) (Przybós 1980).

P. tetraurelia: A wide-ranging group known for high F_2 mortality in crosses from geographically distant populations. Species found in North America, Central America, South America, Europe, Australia, and Japan.

P. pentaurelia: Collections made in North America (from north-central to south-east) and Western Australia.

P. sexaurelia: This species has been sampled from North America, the Caribbean, Africa, and the Indian subcontinent.

P. septaurelia: Species found in southeastern North America (Alabama and Florida).

P. octaurelia: Known from southeastern North America, Central America, and Africa.

P. novaurelia: Limited to Europe including the British Isles.

P. decaurelia: Only one stock collected from North America (Florida).

P. undecaurelia: Collected from southern North America (Texas and Mississippi).

P. dodecaurelia: Sampled from the Gulf of Mexico area of North America.

P. tredecaurelia: Widely distributed although only three stocks are known. The stocks originate from Europe (France), Madagascar, and North America (Mexico).

P. quadecaurelia: This species is known from one Australian collection.

P. sonneborni: Recently discovered species from North America (Texas) (Aufderheide et al. 1983).

P. bursaria complex (Sonneborn 1957):

P. bursaria species-1: Collected from Asia (China).

P. bursaria species 2 and species 3: Known from collections in North America (United States).

P. bursaria species 4, species 5, and species 6: Isolated from Europe, species 5 is known from only one collection.

P. caudatum complex (Sonneborn 1957):

P. caudatum species 1 and species 3: Isolated from Asia (Japan and North America (United States).

P. caudatum species 2: Collected from North and South America.
P. caudatum species 4, species 5, species 6, species 7, species 9, species 10, species 11, species 14: All known from North America. Some have very restricted distributions.
P. caudatum species 8: Found in North America and Europe (Scotland).
P. caudatum species 15 and species 16: Restricted to Europe.

P. multimicronucleatum complex (Sonneborn 1957; Daggett and Nevad 1985):

P. multimicronucleatum species 1: Isolated from Europe (Pavlovsk, USSR), North America.
P. multimicronucleatum species 2: Isolated from Central America (Panama) and North America.
P. multimicronucleatum species 3: Stock 12 isolated from North America (Mexico).
P. multimicronucleatum species 4: Known from North America (Texas).

2.3 Migration and Transport

As demonstrated above, paramecia are known throughout the world, and such species as *P. primaurelia, P. biaurelia,* and *P. tetraurelia* are apparently ubiquitous. Many other species are known from widely separate sites; for example, *P. pentaurelia* has been isolated from both North America and Australia. The very close association between the killer endosymbionts and the *P. aurelia* complex is also widespread geographically. Hump-type killer *P. tetraurelia,* like stock 51, have been isolated from North America, Central America, Asia, and Australia.

Spinner-killer *P. biaurelia,* like stock 7, are known from North America and Europe. Yet paramecia are not known to encyst and are quickly desiccated. Although paramecia may be transported by aquatic insects, waterfowl, or migratory mammals, such transport has not been conclusively demonstrated. Such a mechanism also seems highly unlikely over oceanic distances to inland sites. Preer (1977) has suggested that the present distributions of paramecia and the killer trait may be relics of their distribution before the breakup of Pangea. Modifications to this distribution could have occurred during the formation of land bridges (such as the one across the Bering Sea during the last glaciation). However, it must be considered that such a widespread distribution of several species of *Paramecium* points to a longevity on the geological time scale. This hypothesis is not unique among protozoa as a similar hypothesis has been proposed to describe the worldwide distribution of the *Tetrahymena pyriformis* complex (Elliott 1973) and marine interstitial ciliates (Corliss and Hartwig 1977).

3 Experimental Ecology

Among the earliest ecological studies using paramecia were those investigating competition and predator–prey interactions in the laboratory. These studies have probably had a greater impact on ecological theory and science than any of the field research described in the next section. Gause's work in the 1930s set the stage for the establishment of the competition theory as a basic pillar of ecology. Recently, the importance of competition in ecoystems has been the topic of a great deal of debate (den Boer 1986). Indeed, recent field work questions the importance of competitive interactions in regulating the community structure of *Paramecium* except in the case of killers (Landis 1981, 1982, 1986). Interestingly, studies conducted with paramecia helped to question the almost axiomatically held belief that an increase in community diversity led to an increase in community stability. The next few sections review those studies relevant to wild populations of paramecia.

3.1 Competition Studies

The earliest studies on the competitive interactions of paramecia were those by Gause (1934). In these studies unknown species of the *P. aurelia* and *P. caudatum* complexes were grown in mixed culture. An inoculating loop was used to inoculate the Osterhaut's medium with bacteria. *P. caudatum* in these studies gradually was replaced in the culture by the *P. aurelia.* These findings were contradicted by the work of Vandermeer (1969). In experiments using *P. triaurelia* and an isolate of *P. caudatum* collected near Ann Arbor, the outcome was coexistence of the two species. Vandermeer used the bacterized Cerophyll medium commonly used in *Paramecium* research as modified by Gill (1972a). Experiments conducted by Gause (1935) with mixed cultures of *P. bursaria* and *P. aurelia* demonstrated that coexistence was possible because the *P. bursaria* had a competitive advantage in the bottom of the culture where they fed on the yeast. In the upper layers of the culture *P. aurelia* had the advantage since it reportedly fed on the bacterial populations. Only a slight depression of each population was noticed since the species were separated physically and used different resources. In Vandermeer's (1969) experiments, *P. bursaria* were always driven to near extinction by *P. triaurelia* and the *P. caudatum* species.

Conflicting results such as these are not restricted to experiments between species complexes. Hairston and Kellerman (1965) conducted competition experiments between *P. biaurelia* and *P. triaurelia* using dilute media at a variety of temperatures, food concentrations, and initial numbers of competing species. Some of the stocks of paramecia were originally collected near Ann Arbor; others were derived from Sonneborn's laboratory. In all cases *P. biaurelia* outcompeted *P. triaurelia,* usually driving *P. triaurelia* to extinction. A similar set of experiments was conducted by Gill (1972a) using *P. pentaurelia* as the competitor for *P. biaurelia.* Competitive interactions among the stocks derived from the Ann Arbor area were studied at 15°, 20°, and 25° C. In all cases using field stocks, *P. pentaurelia* became extinct. Experiments using laboratory stocks, conducted at 15° and 20°C, however, demonstrated coexistence, although the intrinsic rates of increase of these stocks were the same as the field stocks. Gill hypothesized that a killer endosymbiont such as kappa, which are known to inhabit *P. biaurelia,* may have been responsible for the competitive advantage of the wild stock of *P. biaurelia.* Unfortunately, neither an examination of the cytoplasm nor a simple test for killing activity was conducted. More recently, killer paramecia were demonstrated to have a competitive advantage over sensitive paramecia at concentrations similar to those found in nature (Landis 1981). This was shown with both killer and sensitive stocks of *P. tetraurelia* recently derived from nature and laboratory stocks (51k and 51s) identical except for the presence of killer endosymbionts. Gill (1972a) hypothesized that killer endosymbionts are the major means of inter- and intraspecific competition among paramecia. As discussed in Section 4, the importance of the killer trait may be very situation-specific.

3.2 Predator-Prey Interactions

Interestingly, Berger (1980) has reported that apoclorotic *P. bursaria* were ingested by *Didinium nasutum* much more frequently than the normal chlorella-containing *P. bursaria.* Apparently, the symbionts confer protection against predation by at least one protozoan predator *Didinium.* Many other studies (Gause 1934; Hairston et al. 1968; Luckenbill 1973) have confirmed the fact that coexistence among paramecia and ciliate predators such as *Didinium* and Woodruffia is not possible unless the paramecia can find refuge in space or time.

Paramecia are predators of bacterial, algal, and yeast populations (Wichterman 1953). Certain chemicals attract paramecia (Van Houten et al. 1982) and presumably, in nature, indicate an increased concentration of resources. The feeding rates for *P. caudatum* and *P. trichium* have also been determined by Fenchel (1980). Ciliates require a high food concentration, but are able to ingest large quantities of food. Fenchel hypothesizes that bactivorous ciliates are opportunists, fugitives exploiting spatially and temporally patchy resources. As detailed in the next section, this hypothesis is upheld to a large degree by field research.

4 Field Ecology

The study of the population ecology of *Paramecium* primarily has been limited to species of the *P. aurelia* complex. However, members of the *P. bursaria, P. multimicronucleatum,* and *P. caudatum* complexes have also been collected during these studies. No dedicated study of these species complexes has been undertaken but the incidental collections made during the studies of *P. aurelia* do provide interesting data.

4.1 Seasonal Distribution

P. primaurelia, P. biaurelia, P. triaurelia, P. tetraurelia, and *P. pentaurelia* could be collected near Bloomington, Indiana from May until freezing of the sampling sites. Unidentified species of the *P. bursaria, P. multimicronucleatum,* and *P. caudatum* complexes were also obtained in these series of collections, but much less often. Although 1-liter samples were taken through the ice and after the spring melt, no paramecia were obtained before May (Landis 1981, 1982, unpubl. data). Extensive sampling conducted in the Ann Arbor, Michigan vicinity showed a similar pattern. In both river and seep environments paramecia are rare during the winter months with population increases observed in the April–May time frame (Hairston and Kellerman 1964; Hairston 1958, 1967). Members of the *P. aurelia* complex have been collected in Florida during early January (Landis, unpubl. data), but no other information has been obtained from tropical climates.

4.2 Spatial and Temporal Distribution

The observed spatial distribution of the species of the *P. aurelia* complexes is clumped (contagious). Samples taken in a river and seep environment by Hairston (1967; Hairston and Kellerman 1964) demonstrated that the numbers of *P. biaurelia* differed greatly in samples taken only a few meters apart. This variability was also expressed in sampling tubes permanently placed in a woodland seep (Gill and Hairston 1972).

In a series of field studies Landis (1981, 1982) investigated the spatial and temporal distribution of *Paramecium* in two small ponds near Bloomington, Indiana. A grid of 0.5 m^2 area with 16 collecting sites was used to ensure repeated samplings of the same sites. Small 5-ml samples were taken at the mud-water interface and at the surface. Individual species of the *P. aurelia* complex as well as the complex as a group exhibited a statistically significant contagious (clumped) distribution. In comparison, the distribution of the *P. bursaria* complex was much more dispersed.

The *P. bursaria* and *P. aurelia* complexes are distributed quite differently in the water column. Hairston (1967; Hairston and Kellerman 1964) and Landis (1981) reported that species belonging to the *P. aurelia* complex are found at the mud-water interface. In a much more limited series of samples this has also proven to be the case in the *P. caudatum* complex (Landis, unpubl. results). Conversely, *P. bursaria* are found in samples taken at the mud-water interface and the surface of the littoral zones of small ponds (Landis 1982). Members of the *P. aurelia, P. caudatum,* and *P. bursaria* complexes exhibit fluctuating populations over a period of 1 week or less (Hairston and Kellerman 1964; Hairston 1967; Gill and Hairston 1972; Landis 1981, 1982). *P. aurelia* and *P. caudatum* have been found to co-occur with high densities of each (Hairston and Kellerman 1964; Landis, unpublished data). In contrast, almost no temporal overlap occurs between the *P. aurelia* and the *P. bursaria* complexes. Only twice in 85 collections of 16 samples each did *P. aurelia* and *P. bursaria* appear in the same series of samples. Taylor (1979), collecting ciliates within a small Ontario pond to observed community-level relationships, found that the potentially autotrophic *P. bursaria* is part of cluster of protozoan species separate from that of *P. caudatum*. Collections made in Michigan and Georgia by Pratt and Cairns (1985b) to determine functional relationships in protozoan assemblages also placed *P. bursaria* in a functional group separate from the other heterotrophic members of the genus *Paramecium*.

4.3 Population Regulation and Interspecific Interactions

Populations of the sensitive members of the *P. aurelia* complex seem to be regulated by density-independent factors. Hairston (1967), during a series of field experiments to simulate rainfall runoff into the seep study area, concluded that the *P. biaurelia* population was food limited. As counts of bacteria in the seep were not performed, it is possible that a limiting essential nutrient was increased following runoff. However, it seems likely that the populations were resource limited.

The different species of *Paramecium* can often be grown in the same medium. An underlying assumption of much of the early work on the ecology of paramecia is that competition is important in regulating populations in nature. If competition exclusion occurs in nature, a negative association among species would be expected. Landis (1981) constructed 2 × 2 tables for chi-square analysis of the collections made of the *P. aurelia* complex and an unknown species of *P. multimicronucleatum*. V-scores were used to measure how often two species occur together relative to a purely chance association (Pielou 1977). Comparisons were made only if the species had been collected from five or more samples. The only statistically significant associations had V-scores, indicating that the species occurred together more often than would be expected by chance. Another indication that competitive exclusion is not occurring is the lack of correlation between population density and species diversity for the sensitive species of *P. aurelia* (Landis 1981).

Paramecium appears to be an example of Hutchinson's famous "paradox of the plankton" (Hutchinson 1961). Planktonic algae coexist although they use essentially the same resources; light, carbon dioxide, and inorganic nutrients. A hypothesis put forth by Hutchinson (1961) and refined by Richerson et al. (1970) resolves this paradox. Essentially, the hypothesis is that spatial and temporal heterogeneity allow very similar species to coexist. The relative numbers of species present in the community is dependent on the number and occurrence of resource patches in time and space and on the ability of species to locate and colonize the patches.

Tilman (1980, 1982; Titman 1976) has codified this competition for resources in a spatially and temporally heterogeneous environment using simple graphical methods. Growth rates, resource utilization rates, resource supply and types are incorporated to divide the graph into regions of competitive equilibria and exclusion. The addition of temporal and spatial heterogeneity is incorporated by using 99% confidence ellipses to describe variability. This work

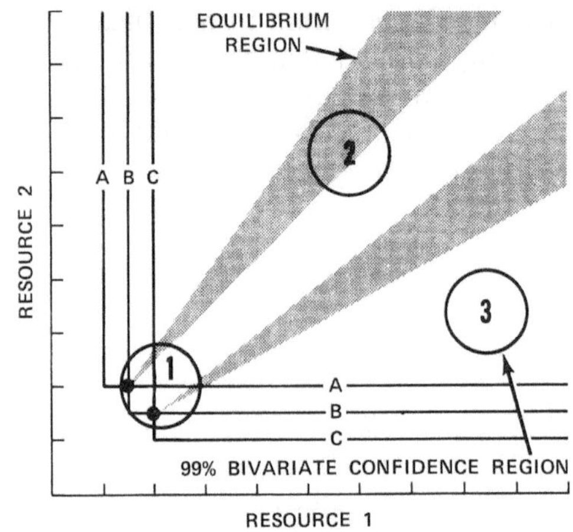

Fig. 1. Resource competition modeling graphically examines competitive relationships. The *circles* are 95% confidence intervals of the resource supplies. The *shaded areas* are equilibrium regions where competition will result in the coexistence of both species. In *region 1* all species can coexist. In *region 2,* only two of the three can coexist even with the environmental heterogeneity. *Region 3* does not overlap any of the equilibrium regions. Extinction of all but one species occurs

demonstrates how coexistence among many species is possible given a limited amount of resources or sufficient temporal or spatial heterogeneity. An example using essential resource types is presented in Fig. 1. Three species can coexist even if only two resources are specified. In confidence region 1 the equilibrium regions for both species combinations are included. Due to environmental variability, all three species will coexist. In confidence region 2 which includes only one equilibrium region, two species will survive, but the third is driven to extinction. In confidence region 3 no equilibrium region is included and one species can drive the other two to extinction. This model describes a mechanism by which several species of sensitive *Paramecium* can coexist.

Paramecia exhibit behavior in the laboratory which suggests they are adapted to exploiting a rapidly changing environment. Well-fed paramecia rise to the top of a test tube. In nature this would aid in the dispersion of the organisms to new resource patches. As paramecia are starved they settle to the bottom which, in nature, is host to high bacterial populations.

P. bursaria, because of their symbiotic relationship with zoochlorellae, are unlikely to be as dependent on a specific resource patch. Indeed, the spatial distribution of *P. bursaria* is much more dispersed than that of the *P. aurelia* complex.

Killer *Paramecium* (Quakenbush, Chap. 23, this Vol.) may be an exception to the general rule just presented. *P. tetraurelia* containing 51 type kappa were found in the highest concentration (116 paramecia ml^{-1}) of any species during the 3-year study of Landis (1981). No sensitive species were found in samples containing killers.

4.4 Ecology of Endosymbiosis

As demonstrated in previous chapters, paramecia are hosts to both prokaryotic and eukaryotic symbionts. Natural selection for the survival of the host is not a necessary condition for these relationships. As L'Herêtier (1970) demonstrated for the cytoplasmically inherited carbon dioxide sensitivity in *Drosophila,* so long as the trait is occasionally transmitted in the sperm, it can be maintained in the population in the face of considerable selection. In a recent study, Landis (1987) analyzed the factors which control the existence of endosymbionts in paramecia. Transmission by cytoplasmic exchange and infection can maintain a symbiont even in the face of selective pressure. The highly infectious nuclear symbionts of the *P. aurelia* and *P. caudatum* complexes may be such an example. However, the killer endosymbionts of the *P. aurelia* complex and the endosymbiotic chlorella of *P. bursaria* species do not seem to fit this category. In fact, the hosts have evolved mechanisms to ensure the stability of the host–symbiont relationship and their life-history strategies reflect its importance.

The killer trait in the *P. aurelia* complex is among the classical examples of cytoplasmic inheritance (Sonneborn 1943, 1959; Preer et al. 1974; Preer and Preer 1977). Landis (1981, 1986, 1987) has demonstrated that, at least in the case of kappa, *Caedibacter taeniospiralis* (51 type kappa), and *C. varicaedens* (7 type kappa), natural selection is the mechanism which maintains the trait. Perhaps the most extreme case is the host–symbiont relationship among *P. tetraurelia* and *C. taeniospiralis.* In many respects the 51 type kappa is evolutionarily degenerative compared to the 7 type (see also Quackenbush, Chap. 23, this Vol.). Phage particles are no longer found in conjunction with the R-bodies in 51 type kappa as compared to 7 type kappa. Subtle differences in R-body structure also exist, yet the killing mechanism appears to be conserved. Paramecia containing kappa also contain genes which control the maintenance of the symbiont. Organisms that are *Kk* contain half the number of bacteria as *KK* cells. A heterozygous cell which undergoes autogamy has a 50:50 probability of becoming *kk* and losing the killer trait. Conjugation is therefore a mechanism that potentially leads to an increased rate of loss of the killer trait from a population by the formation of heterozygous organisms. Obligate self-fertilization may be adaptive in populations that take advantage of the killer trait in competition for resource patches. In fact, *P. tetraurelia* exhibit a skewed distribution of mating types. The even mating type of *P. tetraurelia* is rare. The mating type of 25 clones of *P. tetraurelia* isolated from two ponds near Bloomington, Indiana were determined to all be of the odd mating type (M. Schneller and W. Landis, unpubl. data). Conjugation must be rare. Restriction to the odd mating type may be in part an adaptation to the maintenance of the killer trait.

The *Chlorella* sp. and *P. bursaria* interaction is another example of a symbiotic relationship that plays a crucial role in the life-history strategy of the organism. Although the *Paramecium* can be cured of their algal symbionts in the laboratory they have never been found in nature to be symbiont-free (Reisser 1981; see also Görtz, Chap. 22, this vol.). Endosymbotic *Chlorella* species are biochemically different from their free-living relatives (Reisser and Benseler 1981). The algae produce maltose which acts as an energy source suitable for the reproduction of the *P. bursaria* host even in a illuminated minimal salt solution (Reisser 1981). Protection against some predators has also been demonstrated (Berger 1980). Additionally, the photosynthetic pigments of the algae cause the *P. bursaria* to photoaccumulate, ensuring their movement toward the light (Niess et al. 1981; Cronkite 1986). As a good host the *P. bursaria* cell protects the algal cell from digestion, recognizing the symbiotic *Chlorella* sp. as different from other species of algae encountered and ingested by the organism (Reisser 1981). The ecology of *P. bursaria* reflects the impact of the symbiotic association. As demonstrated above, *P. bursaria* are distributed near randomly, can be found throughout the water column, are of a community cluster very different from that of the heterotrophic paramecia, and are outbreeders. Many of these characteristics are likely due to the autotrophic potential of the *P. bursaria* complex.

4.5 Life-History Strategies

Several attempts have been made to fit the genetics, breeding systems, and ecology of *Paramecium* into an explanation of the evolution of their life-history strategies. Sonneborn (1955, 1957) noted the contrasting breeding systems in the genus, especially between the *P. aurelia* and *P. bursaria* complexes. *P. aurelia* are inbreeders, in general, having short periods of immaturity, low mortality among inbreeding pairs, high F_2 mortality in crosses between distant populations, and autogamy. In contrast, *P. bursaria* have long periods of immaturity, high mortality when inbred, low mortality in crosses between distant populations, and do not undergo autogamy (Jennings 1939, 1944, b, c). Gill and Hairston (1972) described a model of the population biology of the *P. aurelia* complex based on their extensive sampling and experimental manipulation of *P. biaurelia* in a woodland seep. They described an organism capable of colonizing resource patches in which the carrying capacity varies over space and time. Predation or competition are not seen as important factors in controlling population size. Unfortunately, in this author's eyes, they stated that their study may be an exception and that the members of the *P. aurelia* complex, in general, are climax species living in stable environments such as rivers, sloughs, and ponds. An implicit assumption is that these aquatic systems are more homogeneous, as perceived by a microorganism, than the seep. Furthermore, they stated that competitive interactions mediate relative abundances. The studies of Landis (1981) refute this for at least one assumed "stable" environment, two ponds of approximately 20 years of age in central Indiana. No evidence of competitive exclusion was found except in the case of killer paramecia. Indeed, the three species found most commonly in the samples varied in abundance much in the same way as did the *P. biaurelia* populations in the seep studied by Hairston and Gill.

A recently published and more extensively developed description of the life-history strategies of *Paramecium* (Landis 1986) incorporates the genetic diversity, breeding systems, physiology, and ecology of these diverse organisms. The *P. bursaria* and *P. aurelia* complexes are used for comparative purposes as examples of the ends of a broad spectrum (Fig. 2). Table 1 contrasts these species complexes. *P. bursaria* is an outbreeder with multiple mating types, a long period of immaturity, and high genetic diversity. The spatial distribution of *P. bursaria* is very close to random and the organisms can be found both at

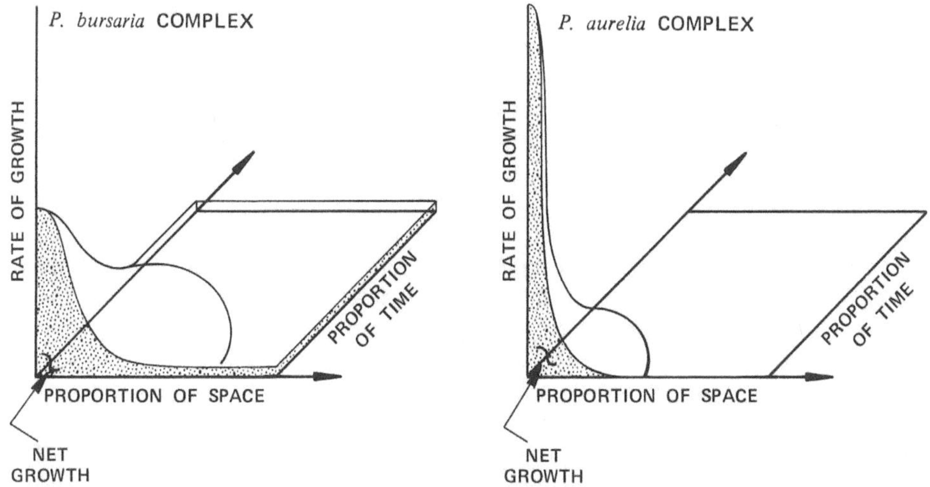

Fig. 2. Evolutionary strategies. The *P. aurelia* complex is hypothesized to colonize a specialized resource patch in which it can reproduce rapidly. In contrast, the maximum reproductive rate of *P. bursaria* may not be as rapid as *P. aurelia*, but *P. bursaria* can reproduce over a broad range of environmental conditions due, in part, to its symbiotic chlorella (Landis 1986)

Table 1. Comparison of the characteristics of the *P. aurelia* and *P. bursaria* complexes

Characteristic	Complex	
	P. bursaria	*P. aurelia*
Breeding system	*Outbreeder:* Multiple mating types. Long periods of immaturity, does not undergo autogamy	*Inbreeder:* Two mating types. Short periods of immaturity, undergoes autogamy
Genetic diversity	*High:* Recessive lethals known to occur	*Low:* Based on electrophosphate data
Spatial distribution	*Near random:* Indices of contagion indicate dispersed population. Occurs throughout water column	*Contagious:* Indices of contagion indicate very clumped distribution. Collected at mud-water interface
Rate of Speciation	*Low:* Only five to six species identified	*High:* Fifteen species identified. *P. primaurelia* and *P. tetraurelia* composed of populations that when mated exhibit F_2 mortality of 90%
Endosymbionts	*Photosynthetic eukaryochlorella* provide maltose to the host and provide protection from protozoan predators	*Bacterial symbionts* confer the killer trait in some instances. Also host to several infectious macronuclear and other nonkiller symbionts

the mud-water interface and at the surface. The endosymbiotic chlorellae of *P. bursaria* make the complex less dependent on specific resource patches. Additionally, *P. bursaria* have retained the ability to ingest and assimilate a variety of prey organisms. Apparently, the resource patch as perceived by *P. bursaria* is broad and widely distributed. The rate of speciation in the *P. bursaria* complex is also assumed to be low since only five or six species are currently known. In summary, the *P. bursaria* complex can be described as a generalist group, especially when compared to the *P. aurelia* complex.

Members of the *P. aurelia* complex are inbreeders, although considerable variety exists in regard to this trait. Genetic diversity within the species of the complex is low as measured by electrophoretic markers (Tait 1970; Allen et al. 1971, 1983; Adams and Allen 1975; Allen and Gibson 1975). Variability in surface antigen markers in wild populations of *P. novaurelis* has been noted (Pringle 1956; Pringle and Beale 1960). Chromosome number does vary greatly among populations of *P. tetraurelia,* but the F_2 mortality rate is so high that these populations are likely to represent nascent species (Dippell 1954). The spatial distribution of the *P. aurelia* complex is very clumped and they are found almost exclusively at the mud-water interface.

A pattern emerges from a comparison of the *P. bursaria* and *P. aurelia* complexes. The *P. bursaria* complex can be described as a generalist, able to survive under a variety of conditions. Genetic diversity within such a species would enhance its ability to exist in a wide range of environments. An outbreeding system would act to maintain the genetic diversity of a population. Unfortunately, lack of specialization would generally mean a loss of efficiency in the utilization of a particular resource. The spatial distribution of a generalist is also likely to be more random than that of a specialist since the environment is perceived as more uniform. Presence of photosynthetic chlorella also broadens the range of environments suitable for the *P. bursaria* species (Fig. 2). On the other hand, the *P. aurelia* complex can be seen as composed of relative specialists. Specialization increases efficency in utilizing a particular resource, but the cost is a decrease in genetic variety. The general strategy of the *P. aurelia* is hypothesized to be a high rate of reproduction under a comparatively narrow range of conditions.

Almost no data are available to reconcile the breeding systems of the other species complexes of *Paramecium.* Limited sampling indicates that in many ways, the breeding system of the *P. caudatum* resembles that of the *P. aurelia* complex (Sonneborn 1957) and that these complexes have a similar life-style (see also Chap. 3 by Nyberg and Chap. 4 by Tsukii, this Vol.). *P. multimicronucleatum* is more of an outbreeder, although to a lesser extent than the *P. bursaria* complex. This fact leads to speculation that the complex exhibits a different life-history strategy than the other nonphotosynthetic species of the *Paramecium,* but the necessary sampling data do not exist.

5 Research Needs

I would like to conclude this chapter with a list of some of the research needs for the study of the ecology of *Paramecium*. These topics are also important to protozoan population biology, genetics, ecology, and evolution.

Resolution of the nomenclatural confusion in the genus *Paramecium* is a must. The *P. aurelia* complex has been recognized as a group of reproductively isolated populations for over 50 years. The *P. caudatum* and *P. bursaria* complexes have almost as long a history, yet the naming of these species has not been accomplished. Indeed, it is now difficult to trace the literature on a particular species because of the various names attributed to each. Even if mating is the only reliable method of identifying the species, these reproductively isolated populations must be recognized as biological species. This task is enormous: a reliable means of identification must be set up, names recognized, and type cultures set aside in a repository such as the American Type Culture Collection. Nonetheless, this must be done if the nomenclature of protozoa is to be ordered so that it reflects biological reality.

As workable methods of identification are created and the nomenclature set, a sampling program to establish the ranges of the various species of paramecia would answer many basic questions about the distribution of these ubiquitous protists. The biogeography and population genetics of these organisms may set the stage for the study of the evolution of microbial populations.

Extensive seasonal and microspatial sampling of the *P. caudatum*, *P. multimicronucleatum*, and *P. bursaria* complexes is necessary for comparison to the *P. aurelia* complex. Perhaps the relationship between breeding system, genetic diversity, and ecology proposed by Landis (1986) for the *P. aurelia* and *P. bursaria* complexes will be held as a general theory or dashed upon the stones. Considering the paucity of knowledge, any information from a properly designed study will be important.

We must begin to understand the limnological processes exhibited at the scale of perception of a 100-μm organism. How homogeneous is the littoral zone and how rapidly do conditions change? Are there important changes in hardness, pH, or temperature on centimeter scale? These questions are essential to understanding the prokaryotic and eukaryotic microorganisms which constitute a majority of the biomass in aquatic ecosystems.

Acknowledgments. I would like to thank Margaret Durst for her patience in typing and editing this manuscript. Nancy A. Chester and Renee M. Tauber extracted material from the library following the guidance of the computer searches of Lester Miller.

References

Adams J, Allen SL (1975) Genetic polymorphism and differentiation. In: Markert CL (ed) Isozymes, genetics, and evolution, vol 4. Academic Press, London New York, pp 867–882

Allen SL, Gibson I (1975) Syngenic variation for enzymes of *Paramecium aurelia*. In: Markert CL (ed) Isozymes, genetics, and evolution, vol 4. Academic Press, London New York, pp 882–897

Allen SL, Byrne B, Cronkite DL (1971) Intersyngenic variations in the esterase of bacterized *Paramecium aurelia.* Biochem Gen 5:135–150
Allen SL, Nevad TA, Rushford CL (1983) Intraspecies variability: the esterases and acid phosphatase of four species of the *Paramecium aurelia* complex. J Protozool 30:131–143
Aufderheide KJ, Daggett P-M, Nevad TA (1983) *Paramecium sonneborni* n.sp., a new member of the *Paramecium aurelia* species-complex. J Protozool 30:128–130
Bamforth SS (1981) Protist biogeography. J Protozool 28:2–9
Berger J (1980) Feeding behavior of *Didinium nasutum* on *Paramecium bursaria* with normal or apochlorotic zoochlorellae. J Gen Microb 118:397–404
Cairns J, Hart K, Henebry MS (1980) The effects of sublethal dose of copper sulfate on the colonization rates of freshwater protozoan communities. Am Midl Nat 104:94–101
Corliss JC, Dagget P (1983) *"Paramecium aurelia"* and *"Tetrahymena pyriformis"*: current status of the taxonomy and nomenclature of these popularly known and widely used ciliates. Protistologica 19:307–322
Corliss JO, Hartwig E (1977) The "primitive" interstitial ciliates: their ecology, nuclear uniqueness and postulated place in the evolution and systematics of the phylum Ciliophora. Mikrofauna Meeresboden 61:59–83
Cronkite D (1986) Relations between photobehavior and ionically stimulated swimming behavior in *Paramecium bursaria.* J Protozool 33:52–55
Daggett PM, Nevad TA (1985) American Type Culture Collection Catalogue of Protists–Algae and Protozoa Sixteenth Edition, 1985. American Type Culture Collection, Rockville, pp 22–23
den Boer PJ (1986) The present status of the competitive exclusion principal. Trends Ecol Evol 1:25–28
Dippell RV (1954) A preliminary report on the chromosomal constitution of certain variety 4 races of *Paramecium aurelia.* Caryologia 6 (Suppl):1109–1111
Elliot AM (1973) Lifecycle and distribution of *Tetrahymena.* In: Elliott AM (ed) Biology of *Tetrahymena.* Dowden, Hutchinson, & Ross, Stroudsburg, PA, pp 259–286
Fenchel T (1980) Suspension feeding in ciliated protozoa: feeding rates and their ecological significance. Microb Ecol 6:13–25
Gause GF (1934) The struggle for existence. William & Wilkins, Baltimore
Gause GF (1935) Vérifications expérimentales de la théorie mathématique de la lutte pour la vie. Hermann, Paris
Gill DE (1972a) Intrinsic rates of increase, saturation densities, and competitive ability. I. An experiment with *Paramecium.* Am Nat 106:461–471
Gill DE (1972b) Density dependence and population regulation in laboratory cultures of *Paramecium.* Ecology 53:701–708
Gill DE, Hairston NG (1972) The dynamics of a natural population of *Paramecium* and the role of interspecific competition in community structure. J Anim Ecol 41:134–139
Hairston NG (1958) Observations on the ecology of *Paramecium,* with comments on the species problem. Evolution 12:440–450
Hairston NG (1967) Studies on the limitations of a natural population of *Paramecium aurelia.* Ecology 48:904–910
Hairston NG, Kellerman SL (1964) *Paramecium* ecology: electromigration for field samples and observations on density. Ecology 45:373–376
Hairston NG, Kellerman SL (1965) Competition between varities 2 and 3 of *Paramecium aurelia:* the influence of temperature in a food-limited system. Ecology 46:134–139
Hairston NG, Allan JD, Colwell RK, Futuyma DD, Howell J, Lubin MD, Mathias J, Vandermeer JH (1968) The relationship between species diversity and stability: an experimental approach with protozoa and bacteria. Ecology 49:1091–1101
Henebry MS, Cairns J, Jr (1984) Protozoan colonization rates and trophic status of some freshwater wetland lakes. J Protozool 31:456–467
Hutchinson GE (1961) The paradox of the plankton. Am Nat 95:137–143
Jennings HS (1939) Genetics of *Paramecium bursaria.* I. Mating types and groups, their interrelations and distribution; mating behavior and self sterility. Genetics 24:202–233
Jennings HS (1944a) *Paramecium bursaria:* life history. II. Age and death of clones in relation to the results of conjugation. J Exp Zool 96:17–52

Jennings HS (1944 b) *Paramecium bursaria:* life history. III. Repeated conjugations in the same stock and different ages, with and without inbreeding, in relation to mortality at conjugation. J Exp Zool 96:243–273
Jennings HS (1944 c) *Paramecium bursaria:* life history. IV. Relation of inbreeding to mortality of exconjugant clones. J Exptl Zool 97:165–197
Landis WG (1981) The ecology, role of the killer trait, and interactions of five species of the *Paramecium aurelia* complex inhabiting the littoral zone. Can J Zool 59:1734–1743
Landis WG (1982) The spatial and temporal distribution of *Paramecium bursaria* in the littoral zone. J Protozool 29 (2):159–161
Landis WG (1986) The interplay among ecology, breeding system, and genetics in the *Paramecium aurelia* and *Paramecium bursaria* complexes. Progr Protistol 1:225–245
Landis WG (1987) Factors determining the frequency of the killer trait within populations of the *Paramecium aurelia* complex. Genetics 115:197–205
Landis WG, Savage RE, Jr, Hoskin FCG (1985) An organofluorophosphate hydrolyzing activity in *Tetrahymena thermophila.* J Protozool 32:517–519
Landis WG, Haley MV, Johnson DW (1986) Kinetics of the DFPase activity in *Tetrahymena thermophila.* J Protozool 33 (2):216–218
Landis WG, Durst HD, Savage RE, Jr, Haley DM, Haley MV, Johnson DW (1987) Discovery of multiple organofluorophosphate hydrolyzing activities in the protozoan *Tetrahymena thermophila.* J Appl Tox 7:35–41
L'Herétier P (1970) *Drosophila* virus and their roles as evolutionary factors. Evol Biol 4:185–209
Luckenbill LS (1973) Coexistence in laboratory populations of *Paramecium aurelia* and its predator *Didinium nasutum.* Ecology 54:1320–1327
Niess D, Reisser W, Wiessner W (1981) The role of endosymbiotic algae in photoaccumulation of green *Paramecium bursaria.* Planta 152:268–271
Pielou EC (1977) Mathematical ecology. John Wiley & Sons, New York, pp 203–211
Pratt JR, Cairns J, Jr (1985 a) Long-term patterns of protozoan colonization in Douglas Lake, Michigan. J Protozool 32:95–99
Pratt JR, Cairns J, Jr (1985 b) Functional groups in the protozoa: roles in differing ecosystems. J Protozool 32:415–423
Preer JR, Jr (1977) The killer system in *Paramecium* – kappa and its viruses. Microbiology. Am Soc Microbiol, pp 576–578
Preer JR, Jr, Preer LB, Jurand A (1974) Kappa and other endosymbionts in *Paramecium aurelia.* Bacteriol Rev 38:113–163
Preer LB, Preer JR, Jr (1977) Inheritance of Infectious Elements. Cell Biology: A comprehensive treatise 1:319–373
Pringle CR (1956) Antigenic variation in *Paramecium aurelia,* variety 9. Z Abst-Vererbungsl 87:421–430
Pringle CR, Beale GH (1960) Antigenic polymorphism in a wild population of *Paramecium aurelia.* Genet Res (Camber) 1:62–68
Przybós (1980) Species of the *Paramecium aurelia* complex in Spain. J Protozool 27:89 a
Reisser W (1981) Host–symbiont interaction in *Paramecium bursaria:* physiological and morphological features and their evolutionary significance. Ber Dtsch Bot Ges 94:557–563
Reisser W, Benseler W (1981) Comparative studies on photosynthetic enzymes of the symbiotic *Chlorella* from *Paramecium bursaria* and other symbiotic and non-symbiotic *Chlorella* strains. Arch Microbiol 129:178–180
Richerson P, Armstrong R, Goldman CR (1970) Contemporaneous disequilibrium, a new hypothesis to explain the "paradox of the plankton." Proc Natl Acad Sci USA 67:1710–1714
Sonneborn TM (1955) Heredity, development and evolution in *Paramecium.* Nature (Lóndon) 175:1100–1102
Sonneborn TM (1957) Breeding systems, reproductive methods, and species problems in protozoa. In: Mayr E (ed) The species problem. Am Assoc Adv Sci, pp 155–324
Sonneborn TM (1959) Kappa and related particles in *Paramecium.* Adv Virus Res 6:229–356
Sonneborn TM (1975) The *Paramecium aurelia* complex of fourteen sibling species. Trans Am Microsc Soc 94 (2):155–178

Tait A (1970) Enzyme variation between syngens in *Paramecium aurelia.* Biochem Gen 4:461–470
Taylor WD (1978) Maximum growth rate, size and commonness in a community of bactivorous ciliates. Oecologia (Berlin) 36:263–272
Taylor WD (1979) Sampling data on the bactivorous ciliates of a small pond compared to neutral models of community structure. Ecology 60:876–883
Tilman D (1980) Resources: a graphical-mechanistic approach to competition and predation. Am Nat 116:362–393
Tilman D (1982) Resource competition and community structure. Univ Press, Princeton
Titman D (1976) Ecological competition between algae: experimental confirmation of resource-based competition theory. Science 192:463–465
Van Houten J, Martel E, Kasch T (1982) Kinetic analysis of chemokinesis of *Paramecium.* J Protozool 29:226–230
Vandermeer JH (1969) The competitive structure of communities: an experimental approach with protozoa. Ecology 50:362–371
Wichterman R (1953) The biology of *Paramecium.* Blakiston, New York, p 527
Yonge WH, Jr, Cairns J, Jr (1976) Comparison between numbers and kinds of freshwater protozoans colonizing autoclaved and unautoclaved polyurethane foam substrates. Appl Environ Microbiol 31:750–753

Subject Index

MIX
Papier aus verantwortungsvollen Quellen
Paper from responsible sources
FSC® C105338

If you have any concerns about our products,
you can contact us on
ProductSafety@springernature.com

In case Publisher is established outside the EU,
the EU authorized representative is:
Springer Nature Customer Service Center GmbH
Europaplatz 3, 69115 Heidelberg, Germany

Printed by Libri Plureos GmbH
in Hamburg, Germany